SPSS® Statistics 17.0 Statistical Procedures Companion

Marija J. Norušis

Prentice Hall
A division of Pearson Education
1 Lake Street
Upper Saddle River, NJ 07458

For more information about SPSS® software products, please visit our Web site at *http://www.spss.com* or contact

SPSS Inc.
233 South Wacker Drive, 11th Floor
Chicago, IL 60606-6412
Tel: (312) 651-3000
Fax: (312) 651-3668

SPSS Statistics 17.0 Statistical Procedures Companion
Published by Prentice Hall Inc., Copyright © 2008
Upper Saddle River, NJ 07458

ISBN-13: 978-0-321-62141-2
ISBN-10: 0-321-62141-7

Preface

You have all of the essential components: software with enough options to make a whirling dervish take pause, a dataset of questionable virtue cast in a starring role, and a plot waiting to unfold. Directing all of this to a satisfying conclusion isn't easy. The goal of this book is to point your production in the right direction.

Whether you're new to statistical analysis and SPSS Statistics or are a well-seasoned user who remembers how awesome the powers of SPSS-X were in 1983, this book contains tips, warnings, and examples to help you better analyze your data. Although the underlying statistical concepts are reviewed, the emphasis is on practical matters: understanding the analyses that you can perform using SPSS Statistics. The book is a companion, not a taskmaster. You won't find step-by-step instructions for solving your problems. That's because every statistical problem is unique. Its solution depends on knowing the shortcomings and strengths of the available data and on understanding enough about the subject matter to pursue intelligent, sensible analyses. That's not always an easy task, even for a statistician. (I still recall the investigator who asked for help in analyzing a dataset that contained measurements taken from 20 dogs. When I asked why several dogs weighed so much more than the others, he sheepishly admitted that some of the dogs were actually pigs. That explained the outliers!)

Many readers of the *SPSS Statistics 17.0 Guide to Data Analysis*, an introductory book covering statistical concepts and the fundamentals of data analysis, asked for books providing additional statistical procedures. This book and the *SPSS Statistics 17.0 Advanced Statistical Procedures Companion* cover all of the statistical procedures in the Base system and the Regression Models and Advanced Models modules. The target audience is the data analyst, either an advanced student or a professional. The *Companions* offer practical suggestions and emphasize topics that arise when you are analyzing real data for presentations, reports, and dissertations, instead of the pristine data of homework assignments.

For detailed information about all three books, including sample chapters, please visit my Web site at *http://www.norusis.com.*

Contents of the Statistical Procedures Companion

The first part of this book focuses on the most frequently used procedures in the SPSS Statistics Base system, plus the widely used Logistic Regression procedure from the Regression Models module. There are also introductory chapters on using the software, creating and cleaning data files, and creating new variables. Part 1 is introductory in nature and assumes little. Examples are detailed, and the pace is leisurely. The operational material should be sufficient to enable those without prior experience to use SPSS Statistics.

The second part of the book is devoted to more advanced topics: Ordinal Regression, the General Linear Models procedure, including univariate and multivariate analysis of variance and repeated measures, and the General Loglinear procedure. Part 2 is necessarily more complex and assumes familiarity with the underlying concepts of the general linear model. The material in this second part assumes that you have a certain amount of experience in analyzing data.

The accompanying CD-ROM contains most of the real-life datasets used in the book. The instructions at the end of each chapter for obtaining the output yield the same numerical results shown here. You can use the Pivot Table Editor to rearrange tables and label the contents in any way that suits you. A few of the tables in the book contain numbers produced by SPSS Statistics but in a tabular form that is not available; these are clearly identified in the instructions.

Contents of the Advanced Statistical Procedures Companion

The *SPSS Statistics 17.0 Advanced Statistical Procedures Companion* covers the following topics:

- Nonlinear regression to model population growth
- Logit loglinear models to study voting behavior
- Multinomial regression to predict the degree of support for spending money on space exploration
- Loglinear model selection to identify models that describe the relationships between happiness, marital status, and income
- Life table analysis to estimate the longevity of employees at a company
- Kaplan-Meier analysis to look at the distribution of survival times after diagnosis of a disease
- Cox proportional hazards models to evaluate the impact of age, histology, and stage of a disease on survival
- Variance components analysis to estimate components of variability

- Linear mixed models to study achievement scores when observations are correlated
- Weighted least-squares analysis to accommodate observations with unequal variability in a regression model
- Two-stage least-squares analysis to allow correlations between predictor variables and errors
- Probit regression models to estimate the relationship between dosage of a drug and the response rate
- Multidimensional scaling (ALSCAL) to construct objective scales of subjective attributes
- Generalized linear models that extend the General Linear Model to situations in which the dependent variable is not normally distributed and/or the relationship between the dependent variable and the independent variables is linear by way of a link function.
- Generalized estimating equations that extend the General Linear Model to accommodate correlated data.

Acknowledgments

I thank SPSS Statistics users who have read my books for many years. Their suggestions and encouragement have been invaluable. Please continue to send corrections and suggestions about my books (not the software!) to *marija@norusis.com.*

Good datasets are essential for any book that describes statistical analysis. I wish to thank George D'Elia of the State University of New York at Buffalo for data from his study on the impact of the Internet on public library use, Richard Shekelle for the Western Electric data, Roger Johnson for the body fat data, the Inter-University Consortium for Political and Social Research for the school truancy data, Judith Singer and John Willet for use of the datasets in their excellent book, Donald Hedeker for the smoking data, John Hartigan for the Olympic skating data, and the General Social Survey for a wealth of data on topics from anomie to astrology.

I am grateful to David Nichols and other members of the SPSS staff who participated in various ways in reviewing and producing the book. Finally, I thank Bruce Stephenson who devotedly read the book and insistently encouraged me to just finish it.

Marija J. Norušis

Warning: Drowsiness has been reported by fewer than 1 in 1,000,000 readers. Do not operate heavy machinery while reading this book.

Contents

Part 1

3 Introducing Data 37

4 Preparing Your Data 53

5 Transforming Your Data 67

6 Describing Your Data 83

7 Testing Hypotheses 111

8 T Tests 127

11 Correlation 197

12 Bivariate Linear Regression 217

13 Multiple Linear Regression 237

14 Discriminant Analysis 275

17 Factor Analysis 389

18 Reliability Analysis 427

Part 2

Chapter

1

Introduction

It's possible that instead of feeling grateful and elated for having access to the vast capabilities of SPSS Statistics, you feel overwhelmed by the deluge of menus and dialog boxes that spring up to greet you. Instead of exclaiming "Wow," you may be shouting "Woe!" Relax. Although SPSS Statistics has many capabilities, including data management and report generation, you don't have to become an expert in them to analyze data. You can restrict your attention to the statistical features of the software. This chapter is a brief overview of the statistical procedures covered in this book. Its aim is to help you identify the parts of SPSS Statistics that you'll need to learn to solve your problems effectively. Chapter 2 shows you how to use the software and the Help system.

Statistical Procedures Described in the Book

Most statistical features in SPSS Statistics are accessible from the Analyze menu that appears when you start the system. This menu is the big, but brief, picture of the types of analyses available in SPSS Statistics. (SPSS Statistics has several modules, and if you don't have all of them installed, the menu may be shorter.) If you point to a selection on the Analyze menu, another menu drops down from its side. Choosing one of those selections opens a dialog box in which you can specify the details of the analysis that you want. Because the Analyze menu is your entry point for statistical procedures in SPSS Statistics, you'll start with a quick tour through the menu.

Descriptive Statistics

Figure 1-1
Descriptive Statistics procedures

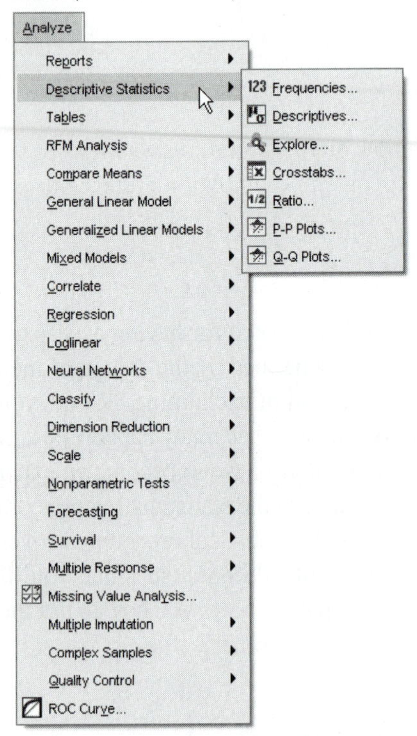

The first step in data analysis is an examination of the data. You want to make sure that your data are sound enough for you to proceed. You must check the values and summarize the features of your data prior to subjecting them to more dangerous interventions.

The procedures commonly used for this purpose appear under Descriptive Statistics on the Analyze menu. (Don't be confused by a common division between *descriptive* and *inferential* statistics; these procedures offer some inferential tests along with the descriptive statistics.)

Frequencies. Use this procedure to count the number of times each value of a variable occurs in your data. For example, how many people in your survey believe in life after death? You can also make pie charts and bar charts. For data that have a natural order, you can display histograms and basic descriptive statistics.

Descriptives. Use this procedure to calculate basic summary statistics, such as means and standard deviations. For example, what is the average income of employees at your company? What is the smallest salary? What is the largest?

Explore. Use this procedure to look at the distribution of a variable for several groups. Boxplots, stem-and-leaf plots, normal plots, tests of whether data come from a normal distribution, and tests of equality of variances are found here.

Crosstabs. Use this procedure to count how often various *combinations* of values occur for two or more variables. How many people in each region prefer your product to your competitor's product? Are men and women equally likely to claim that they would continue working if they became rich? You can also test whether two variables are independent, and you can compute coefficients that measure how strongly two variables are related.

Ratio. Use this procedure to calculate means and medians of ratios and their confidence intervals, as well as coefficients that measure the spread of ratios. If you're a real estate appraiser, these are some of the tools you need to prepare and present your reports. You can calculate the ratios of sale and appraised value and calculate appropriate summaries for a set of such ratios.

P-P and Q-Q Plots. Use these to examine whether your data are a sample from a specified distribution.

Compare Means

Figure 1-2
Compare Means procedures

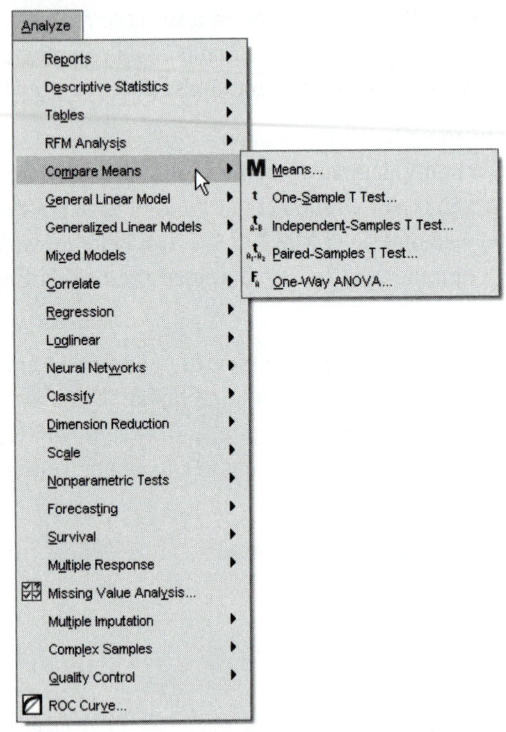

The mean is probably the most frequently reported statistic, and testing hypotheses about equality of population means is the most common statistical activity.

Means. Use this procedure to calculate means, as well as other summary statistics, for groups of cases classified by grouping variables. For example, you can calculate average salaries based on highest degree, gender, major, and region of the country.

One-Sample T Test. Use this procedure to test the hypothesis that your data are a sample from a population with a particular mean. For example, you test whether the average work week is really 40 hours or whether one-pound bags of carrots really weigh one pound.

Independent-Samples T Test. Use this procedure to test the hypothesis that two independent population means are equal. You test whether people with graduate degrees work the same number of hours as people with less formal education. You determine whether the average number of hours spent watching television differs for men and for women.

Paired-Samples T Test. Use this procedure to test that two related samples come from populations with the same mean. You test whether husbands and wives work the same average number of hours per week or earn the same incomes. You study the heights of fathers and sons.

One-Way ANOVA. Use this procedure to test the hypothesis that two or more groups come from populations with the same mean. You test that average standardized reading scores are equal for five different types of schools. You test that average income is the same for four regions of the country.

General Linear Model

Figure 1-3
General Linear Model procedures

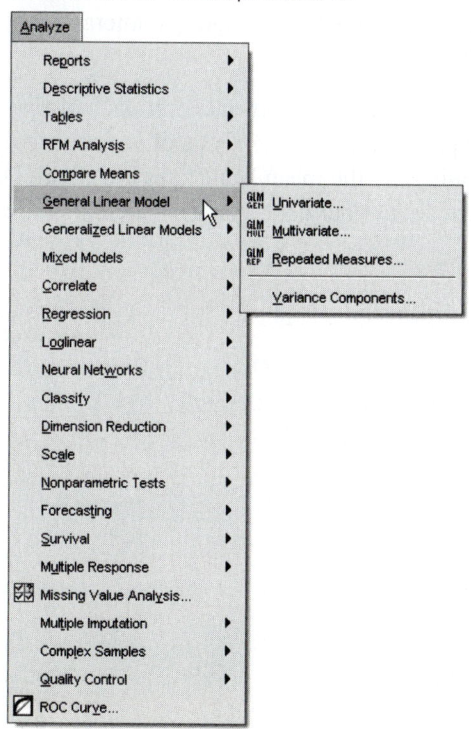

Analysis of variance and regression are both special cases of the general linear model, which is used to analyze data in which you have one or more quantitative dependent variables and one or more predictor variables that can be either quantitative or categorical. If you have one dependent variable, the analysis is called **univariate**. If you have two or more dependent variables, the analysis is **multivariate**. If one or more variables are recorded more than once, the design is a **repeated measures** analysis of variance.

Univariate. With this general linear model, you can study the relationship between students' test scores and variables such as type of school, qualification of the teachers, and socioeconomic status of the parents. You can also study changes between several time periods for the same students and can adjust for variables that might affect the values for a student. Using special procedures, you can compare means for different groups of cases.

Multivariate. With this general linear model, you can test the hypothesis that the average scores on four measures of balance are equal for men and women under three experimental conditions. You can adjust the comparison for different values of a covariate, such as age. You can also test hypotheses about the interactions of the variables that are used to form the groups.

Repeated Measures. If you measure a person's heart rate after each of five types of physical exertion, you can use repeated measures analysis of variance to test the hypothesis that the population values for the mean heart rates are equal. Depending on the assumptions that you make about your data, you can calculate univariate or multivariate tests.

General Loglinear Analysis

Figure 1-4
Loglinear procedures

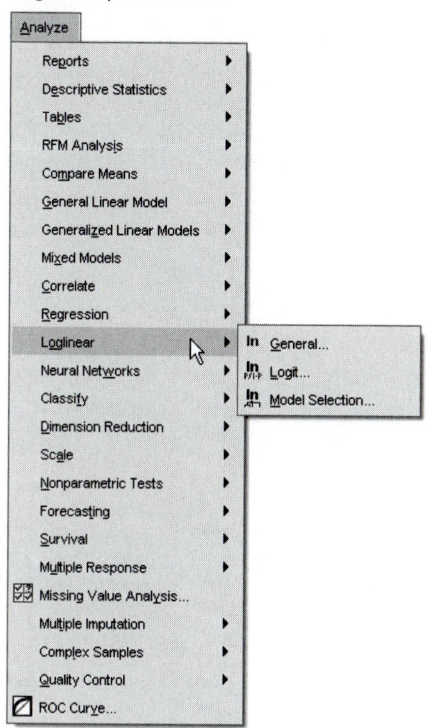

With the loglinear procedures, you can model the relationships among categorical variables.

General. Use this procedure to analyze the relationships between education, happiness, and marital status. The dependent variable is the count of the number of cases in each cell of a multiway crosstabulation.

Correlate

Figure 1-5
Correlate procedures

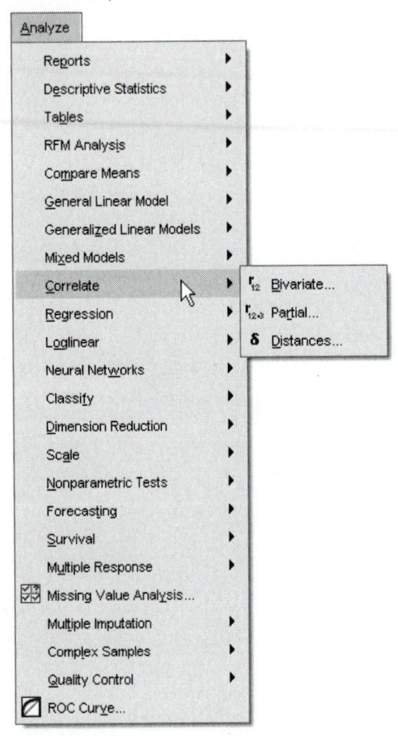

Although the term *correlation* is used commonly, it has a very precise statistical meaning.

Bivariate. Use this procedure to calculate a correlation coefficient that measures how strongly two variables are linearly related. This is your chance to examine whether there is a linear relationship between hours studied and score on a test or whether hours of TV viewing per night is linearly related to income.

Partial. When you calculate a partial correlation coefficient, you determine the strength of the linear relationship between two variables, while "controlling for" the effects of other variables. What's the relationship between hours studied and score on a test when the number of hours of sleep the night before is taken into account? Is there still a relationship between hours of TV viewing and income when age and years of education are held constant?

Regression

Figure 1-6
Regression procedures

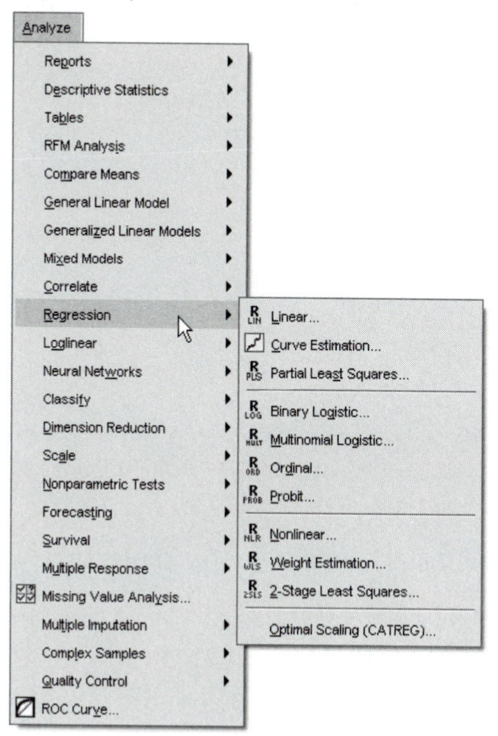

You use all of the procedures under Regression to build a model that predicts the values of a dependent variable from one or more predictor variables.

Linear. Use this procedure to build a statistical model that predicts a person's salary based on the individual's education, prior work experience, and length of time on the job.

Binary Logistic. If the dependent variable has only two values, indicating whether an event has or has not occurred, use this regression procedure to predict the probability that the event will occur based on the values for a set of predictor variables. Can you tell who will graduate from college based on high school scores and grades? What variables are related to whether a patient survives surgery?

Ordinal Regression. If the dependent variable is a categorical variable whose values can be ordered, use ordinal regression to predict the probability of each category based on values of a set of predictor variables. Can you predict the stage of a disease based on the severity of symptoms and various objective measures? Is there a relationship between satisfaction with a product, measured on a three-point scale, and various characteristics of the consumer, such as age, gender, and education?

Classify

Figure 1-7
Classify procedures

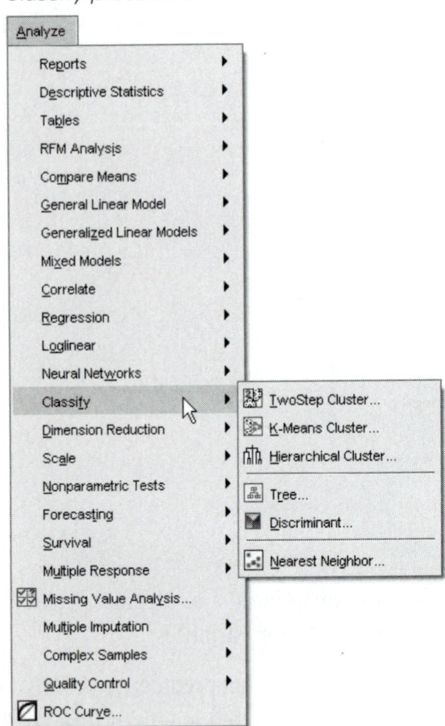

Some objects and people are more alike than others. This is useful information for matchmaking, targeting marketing campaigns, or diagnosing patients. Once you have selected the characteristics you'll use for identifying similar groups of cases, you can select from three SPSS Statistics procedures to identify clusters of similar cases.

TwoStep Cluster. If you have a very large number of cases, use this procedure to define clusters efficiently first by identifying very similar subgroups and then by forming clusters of these subgroups to arrive at a final solution.

K-Means Cluster. Use this procedure to iteratively form a specified number *k* of clusters by sequentially adding cases to one of *k* clusters based on how similar the case is to other cases in the cluster.

Hierarchical Cluster. In this procedure, every case starts out as a cluster of its own. At each step, the two most similar clusters are merged. This continues until all cases are in one large cluster. You examine the results at each step to see which number of clusters is best for you.

When you have two or more groups of cases and know which group the cases belong to, you can study how the groups differ and whether this information is useful for classifying new cases into one of the existing groups.

Discriminant. Use this procedure to estimate equations that separate the groups and tell you how well they work for classifying cases. For example, if you have a sample of people who claim they are republicans, democrats, and unaffiliated, you can determine whether these groups have different average incomes, levels of education, and ages. If so, you might be able to predict a person's political preference based on the demographic information.

Dimension Reduction

Figure 1-8
Dimension Reduction procedures

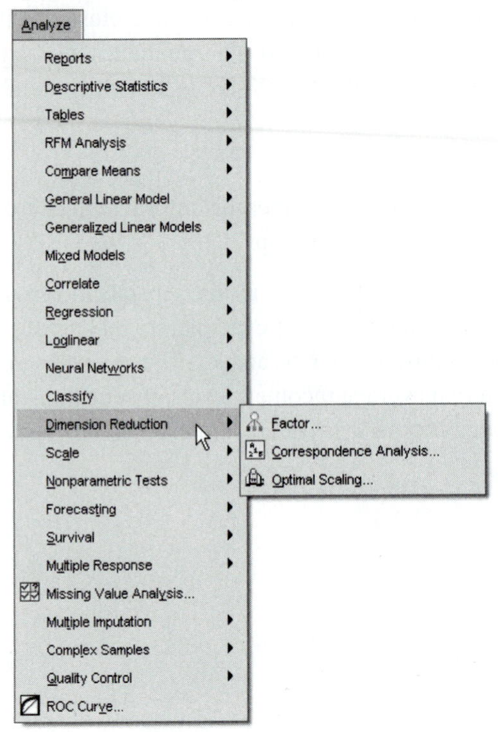

You can measure weight and blood pressure, but you can't weigh altruism and love. You can't see them either. You infer their existence from actions (and from words).

Factor. This procedure is a statistical method that models the observed correlations between variables with underlying "constructs," or "factors." For example, the correlations between variables such as volunteer work, charitable contributions, and community involvement may be explained by postulating the existence of an altruism factor. Factor analysis can help you to summarize many interrelated variables with a smaller number of unobservable factors.

Scale

Figure 1-9
Scale procedures

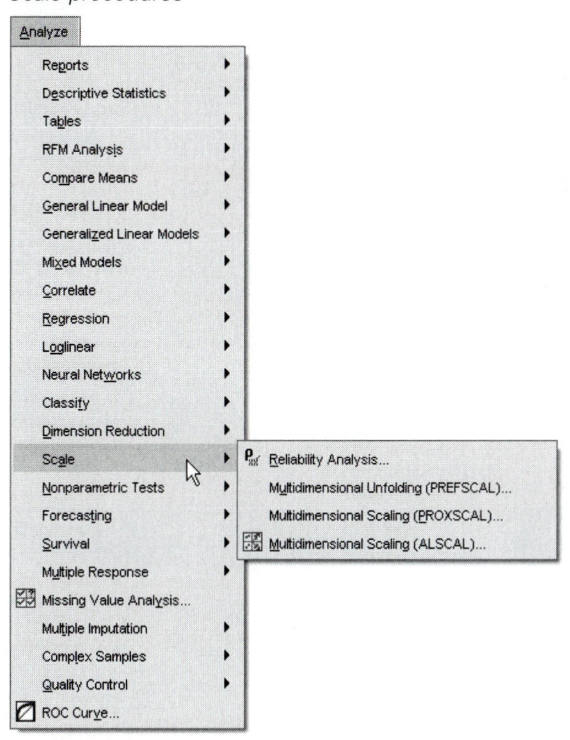

If you construct a test or questionnaire to measure a particular skill or attitude, you want all of the items to be related to the final score. If they are not, the scale may not be reliable.

Reliability Analysis. Use this procedure to compute statistics that measure the relationship of each item to the overall scale, along with coefficients that measure the reliability of the scale. For example, you can determine whether all subsets of a motor dexterity test are similarly related to the final score. You may decide to drop some items or scales that are not related to the rest.

Nonparametric Tests

Figure 1-10
Nonparametric Tests procedures

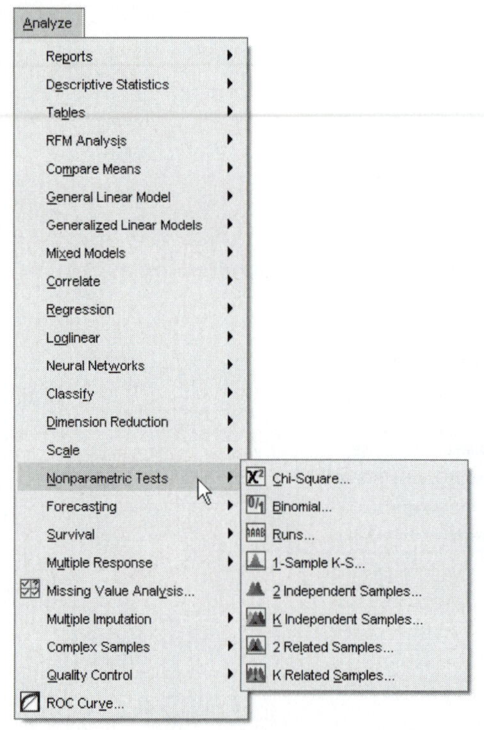

Many statistical procedures depend on the data coming from specific distributions. In particular, you'll often encounter the assumption of **normality** in statistical tests. If your data fail to meet the necessary assumptions, you may need to consider alternatives.

Some statistical procedures are **robust** to particular assumptions, meaning that violation of the assumption doesn't influence the results much. Sometimes, you can transform your data to better satisfy the assumptions.

Another strategy is to use a nonparametric test that requires less restrictive assumptions about the data. SPSS Statistics makes available quite a few of the most commonly used nonparametric tests, as shown in Figure 1-10.

Chapter

2

Getting to Know SPSS Statistics

With very little effort, you've trained your coffeemaker to brew your coffee while you sleep, your iPod to interface with your alarm clock and wake you, and your rental car to turn on those windshield wipers when you need them. Now you're ready to train SPSS Statistics to do statistical computations and charts exactly the way you want them done.

The amount of training required depends on your current needs. If you have a data file that is already in SPSS Statistics format and you don't care about the style and beauty of your results, a perfunctory acquaintance with some of the basic operations of SPSS Statistics will serve you. On the other hand, if you have to create your own data file and present your results in camera-ready copy, you'll have to delve deeper. Even with a lot of training, though, SPSS Statistics won't analyze your data for you, any more than your coffeemaker will sip the coffee it has brewed.

In this chapter, you'll find an overview of the SPSS Statistics system as a whole (what you see on your screen and how to get the software moving). This includes how to edit the results that SPSS Statistics presents in pivot tables and charts and how to use syntax so that you can repeat the same analyses again. You could figure most of this out for yourself. It's much easier than switching to daylight savings time on your new $10 digital wristwatch/cell phone/personal organizer; however, a quick orientation to using the software should help.

Your best guide for learning about the details of the SPSS Statistics software is the Help system. The Help system is always there, so you don't have to worry about storing details about the software in your personal memory. (That's for keeping track of the current name of your bank.) Throughout this book, you'll find pointers to the Help system that will give you more detailed information about managing your files, editing your output, and performing other tasks. This book won't repeat information contained in the Help system. It will show you how to liberate the wealth of information that waits patiently in the Help system.

Tip: To run SPSS Statistics, you have to know how to use the graphical user interface (GUI) that's part of most modern software. If you're afraid of mice, think of buttons as being sewn, and associate menus with restaurants, you need to gain a new perspective on the world. Ask a 10-year-old, read a book, or venture boldly into the Help system.

In a Nutshell

The basic steps for using SPSS Statistics are:

- Get data into SPSS Statistics (Chapter 3).
- Make sure the data are correct (Chapter 4).

  Warning: Skip this step at your own risk!
- Rearrange or transform the data (Chapter 5).
- Run a statistical or graphical procedure (Chapter 6–Chapter 24).
- Scrutinize the results.
- Edit the output to showcase (not change!) your results.

Tutorials

The best introduction to the operations of SPSS Statistics is found in the tutorials. Choosing Tutorial from the Help menu (see Figure 2-1) transports you to the world of the SPSS Statistics tutorials, a riveting and action-packed introduction to the SPSS Statistics experience. Things happen, boxes pop up, and you see actual menus, pivot tables, charts, annotations, and lots of explanations.

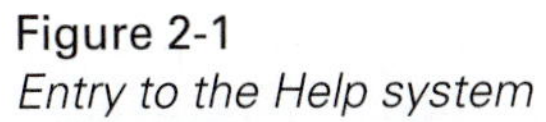

Figure 2-1
Entry to the Help system

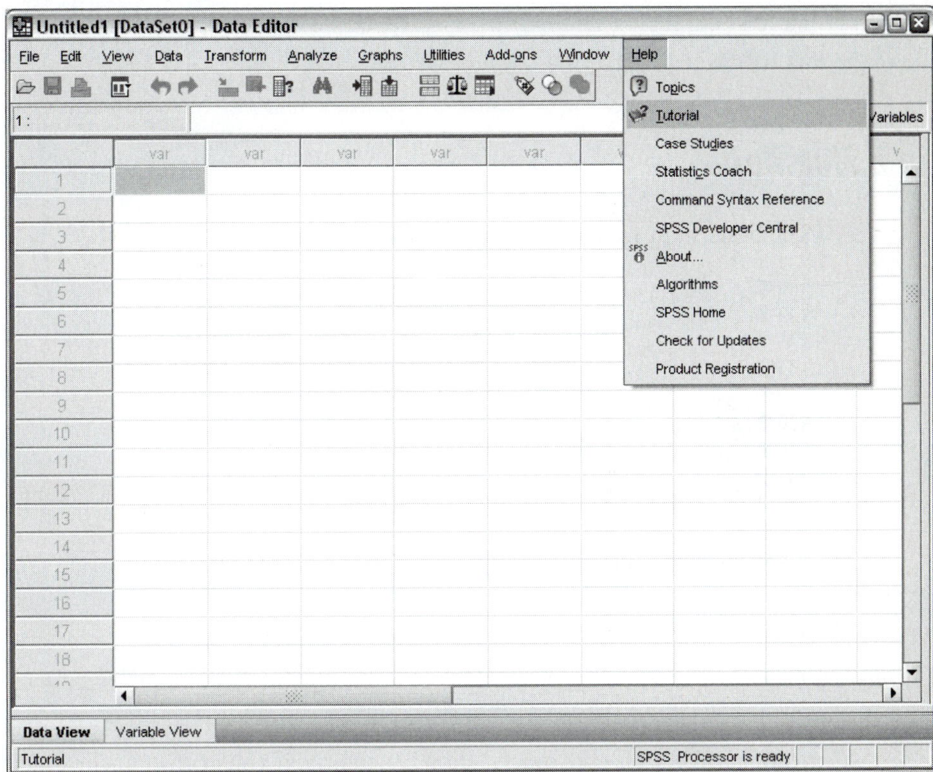

When you choose Tutorial, a table of contents appears. Click one of the book icons to display additional book icons and topics. To start the tutorial, click on the underlined text for the topic that you want. In the lower right corner, you see four buttons. Position your cursor over the buttons to display their ToolTips: Index, Table of Contents, Previous, and Next. Figure 2-2 shows part of the table of contents with the "Using the Data Editor" book icon open.

Figure 2-2
Tutorial table of contents

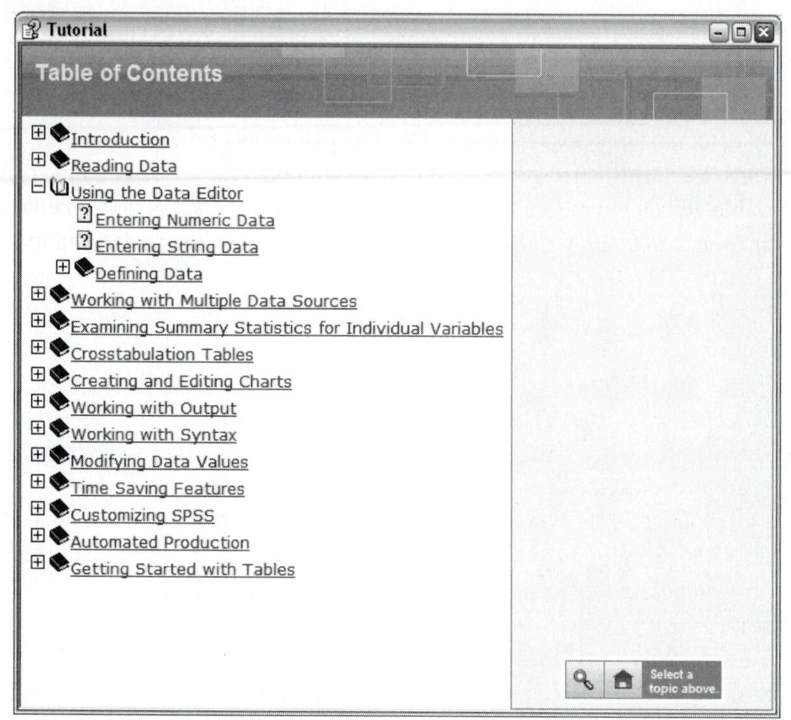

Warning: You won't see much action in a tutorial if you neglect to click the Next button (the right arrow) when you're ready to move to the next screen.

Windows of SPSS Statistics

When you first open SPSS Statistics, you are in the Data Editor window. This is where your data live once you tell SPSS Statistics where to find them. You can open an existing data file using File > Open, or you can enter data directly into the Data Editor, just as you would enter data into a spreadsheet. The data in the Data Editor are arranged in rows and columns. Each row corresponds to a case in your data file. Each column corresponds to a variable, which is a measurement or attribute of a case. Each of the variables has a name that is used to identify it to SPSS Statistics. (See Chapter 3 for a detailed introduction to data files.)

When you run an analysis, your results appear in an output Viewer window. The Viewer window contains pivot tables and graphs. A **pivot table** is an arrangement of numbers and labels in rows, columns, and additional layers. You can use the Pivot Table Editor to rearrange pivot tables in every conceivable way. With the Chart Editor, you can direct your creative energies to selecting colors, styles, line thicknesses, and other properties of a chart.

If you want to enter your own commands for running SPSS Statistics, instead of using the GUI dialog boxes, you can enter the instructions into a Syntax Editor window. You can also use the Paste button in GUI dialog boxes to have SPSS Statistics generate commands and paste them into a syntax file.

The Data Editor, Viewer, and Syntax Editor are the main types of windows used in SPSS Statistics. You can jump from one to another with the Window menu, as shown in Figure 2-3. The check mark indicates which window is active, or currently selected. You can have multiple data files, Viewer files, and syntax files open at the same time. The green cross at the upper left of the window designates the currently active file. That is, you are analyzing data from the data file that has a green cross. Output is generated to the Viewer file with the green cross. If you see an asterisk (*) before the filename, it means that there are unsaved changes in that file.

Tip: Most windows in SPSS Statistics have a toolbar across the top, below the menus. The toolbar buttons show cryptic pictures that identify the functions. Move the pointer to one of the buttons and pause for a second; a ToolTip will appear to tell you what the button actually does. The toolbar buttons usually duplicate functions that are available elsewhere. The handiest button on the toolbar is the Dialog Recall button. Click it, and a menu listing all of the dialog boxes that you've used in the current SPSS Statistics session drops down. Choose the dialog box that you want to revisit, and you can go directly to it without wading through the menu system.

Figure 2-3
Window menu showing Data Editor as the active window

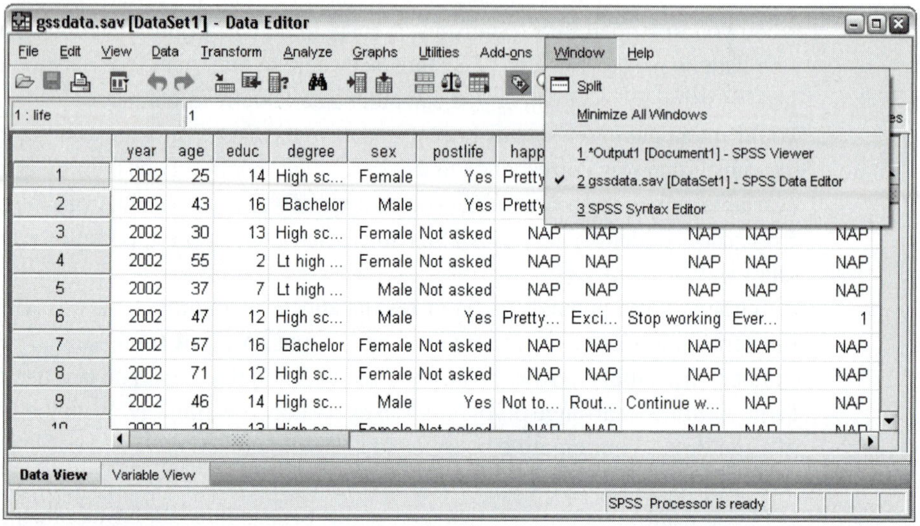

Warning: Sometimes what you can do in SPSS Statistics depends on where you are. For example, data editing capabilities are available only if you're in the Data Editor, and output editing is available only if you're in the output Viewer. You may have to switch your location to find the menu that you want.

Inside a Dialog Box

The menus across the top in Figure 2-3 organize your (SPSS Statistics-related) activities. Click any of them and you get a menu of tasks. For example, if you click Analyze > Descriptive Statistics, you see the menu shown in Figure 2-4. If you have data in the Data Editor, choosing an item that ends with an ellipsis (...) opens a dialog box.

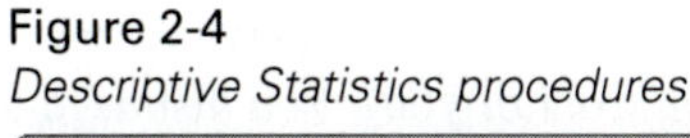

Figure 2-4
Descriptive Statistics procedures

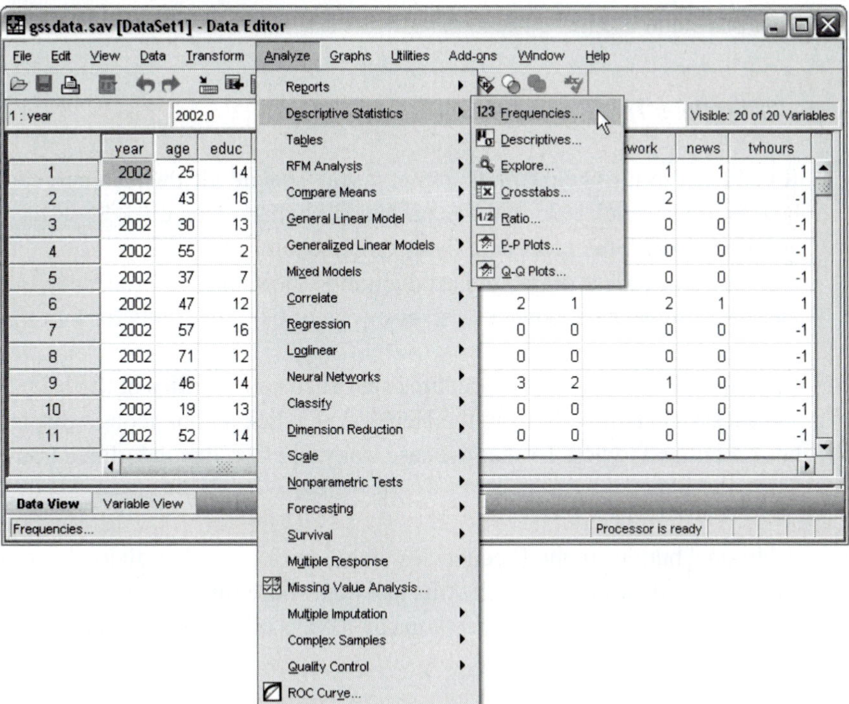

Choose Frequencies (shown in Figure 2-4), and the dialog box shown in Figure 2-5 opens. If you've opened this dialog box before, it will be partially filled in, and you may be able to specify your analysis with a few well-placed selections.

Figure 2-5
Frequencies dialog box

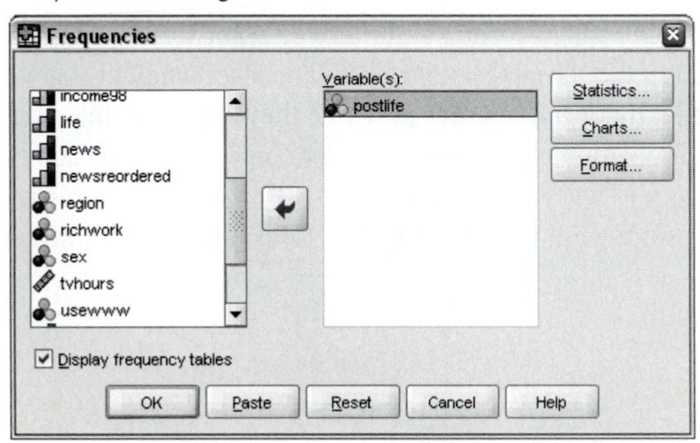

If this is your first visit to the Frequencies dialog box, things may take a bit longer, but the Help system is ever at your side (well, in front of you). You just need to swallow your pride and use it.

Help—What Does This Procedure Do?

If you need serious help (for example, if you have no idea what a procedure does or how to use it), click Help in the main dialog box, and then click Show Me. You'll be escorted through the procedure by a complete tutorial.

If you want a less dramatic introduction to the statistical procedure, click Help and then click To Obtain Frequency Tables or one of the other links for more information.

Tip: Tutorials for statistics procedures are called **case studies** in the Help system. You can access them directly from the Help menu. Click a book icon to display additional book icons and topics. To start the case study, click on the underlined text for the topic that you want.

Additional buttons in the Frequencies dialog box open other dialog boxes that control the output and the analyses. Clicking a button whose function you don't know does not endanger you or your data, since you can always click Cancel to make a hasty retreat.

Help—What Is This Variable?

Once you've determined that you want to run a procedure, you have to select the sacrificial variables. On the left side of the dialog box, you see the list of variables in your data file. You click a variable name to select it, and then click the right-arrow button to move the variable into the Variable(s) list.

Tip: You can use Edit > Options to specify whether you want variable names or variable labels to be displayed in the Variable(s) list. You can select whether you want them to be arranged alphabetically or in the order in which they appear in the data file. An advantage of displaying variables alphabetically by name is that you can type the first letter of the name that you want in the variable display and SPSS Statistics jumps to the first variable that starts with that letter, which saves a lot of scrolling in huge data files.

Even though you start off selecting really great names for variables (so you can easily remember what they are), you may later find yourself with some doubts: Is *age* the child's age? The mother's age? The cheese's age? Position the pointer over a variable name and click the right mouse button. A pop-up menu offers the Variable Information selection, as shown in Figure 2-6. Click it (either button works!) to display the variable information, as shown in Figure 2-7.

Figure 2-6
Frequencies dialog box with variable name pop-up

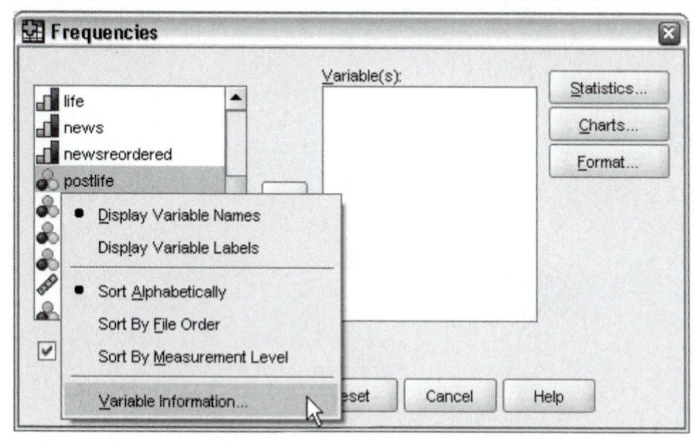

Figure 2-7
Frequencies dialog box with Variable Information pop-up on postlife, showing labels

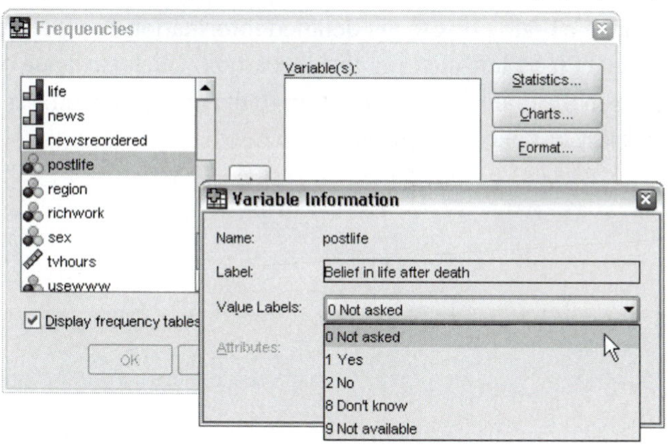

Help—What Is This Statistic?

Clicking the Statistics button in the Frequencies dialog box opens the Frequencies Statistics dialog box, as shown in Figure 2-8.

Figure 2-8
Frequencies Statistics dialog box

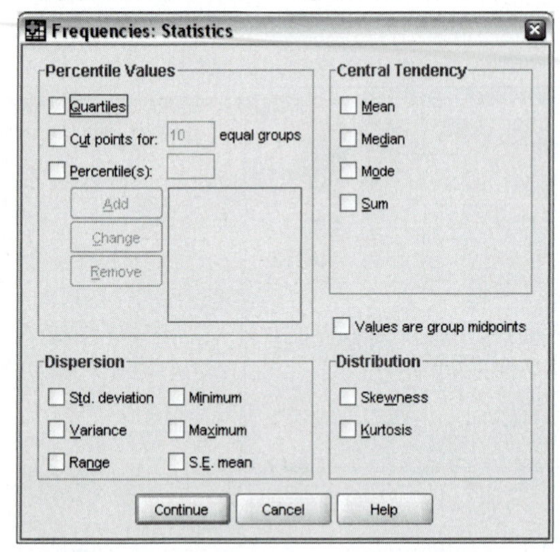

Click the Help button in a dialog box to get detailed information about its features. If a statistic suits you, select it by left-clicking the check box. Click Continue once you've made your selections in the dialog box. You return to the main Frequencies dialog box, where you can click OK or open another dialog box.

SPSS Statistics Viewer

Once you've clicked the OK button on a statistical procedure, your results appear in the SPSS Statistics Viewer window, as shown in Figure 2-9. The Viewer is divided into two panes: the outline pane (on the left) is an index to all of the pivot tables and charts in the Viewer, and the contents pane (on the right) displays the pivot tables and charts.

Figure 2-9
Viewer window

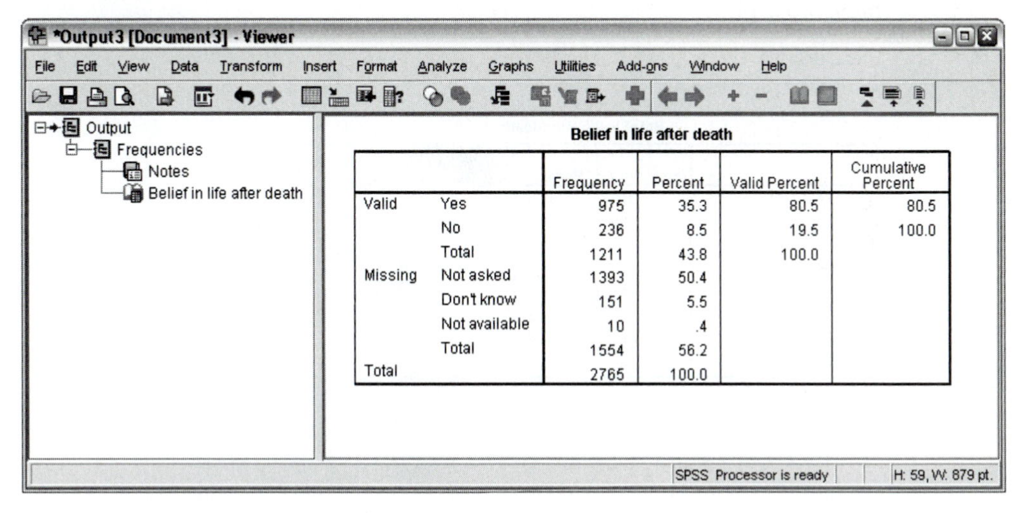

Belief in life after death

		Frequency	Percent	Valid Percent	Cumulative Percent
Valid	Yes	975	35.3	80.5	80.5
	No	236	8.5	19.5	100.0
	Total	1211	43.8	100.0	
Missing	Not asked	1393	50.4		
	Don't know	151	5.5		
	Not available	10	.4		
	Total	1554	56.2		
Total		2765	100.0		

Tip: If either of the panes is too narrow to read, don't despair. Just point the mouse at the dividing line, press and hold down the left button, and drag the line sideways.

Help—What Is in This Output?

The Help system doesn't abandon you once you've obtained the tabulations and statistics you requested. To obtain help in the SPSS Statistics Viewer, you must double-click a pivot table to activate it.

The border around an activated pivot table looks like the border shown in Figure 2-10. Right-click a row or column header for a pop-up menu, such as that shown in Figure 2-10 for the column labeled *Valid Percent*. Choose What's This? to see the pop-up text shown in Figure 2-11.

Figure 2-10
Activated pivot table

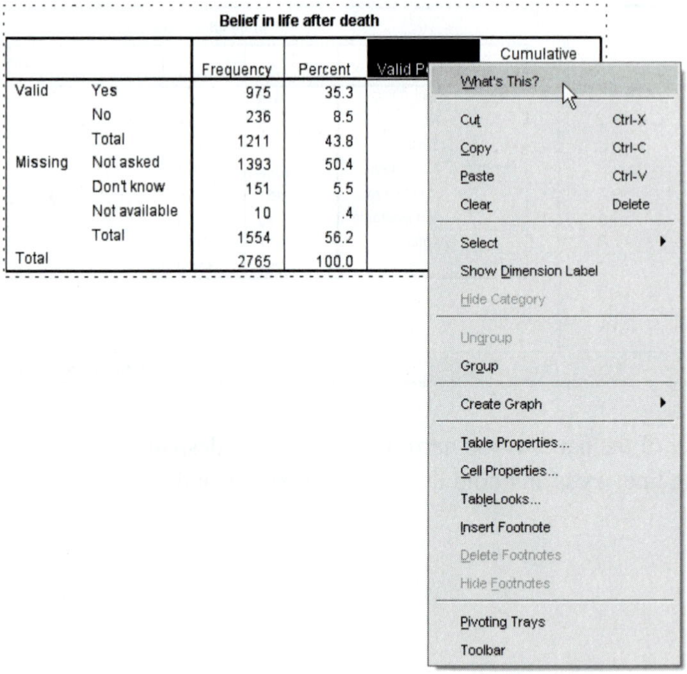

Belief in life after death

		Frequency	Percent	Valid P	Cumulative
Valid	Yes	975	35.3		
	No	236	8.5		
	Total	1211	43.8		
Missing	Not asked	1393	50.4		
	Don't know	151	5.5		
	Not available	10	.4		
	Total	1554	56.2		
Total		2765	100.0		

What's This?
Cut Ctrl-X
Copy Ctrl-C
Paste Ctrl-V
Clear Delete
Select
Show Dimension Label
Hide Category
Ungroup
Group
Create Graph
Table Properties...
Cell Properties...
TableLooks...
Insert Footnote
Delete Footnotes
Hide Footnotes
Pivoting Trays
Toolbar

Figure 2-11

Pop-up definition for Valid Percent label

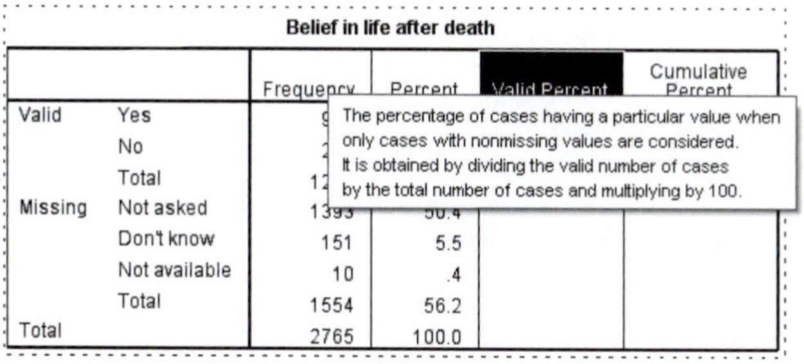

Belief in life after death

		Frequency	Percent	Valid Percent	Cumulative Percent
Valid	Yes	[illegible]	[illegible]	[illegible]	[illegible]
	No	[illegible]	[illegible]	[illegible]	[illegible]
	Total	[illegible]	[illegible]	[illegible]	[illegible]
Missing	Not asked	1393	50.4		
	Don't know	151	5.5		
	Not available	10	.4		
	Total	1554	56.2		
Total		2765	100.0		

The percentage of cases having a particular value when only cases with nonmissing values are considered. It is obtained by dividing the valid number of cases by the total number of cases and multiplying by 100.

Help—Where Did the Numbers Come from?

If you want to know how a particular statistic is computed by SPSS Statistics, you can examine the statistical algorithms used. From the Help menu, choose Algorithms. You will see a list of links to algorithms. Double-click the one that corresponds to the procedure you are interested in. You can also click the Help button in a main dialog box and then under Related Topics, click the procedure's algorithm.

Warning: A visit to the algorithms document is not for the timid. It's unlikely that you will find the formulas you remember so vividly from your college statistics class. There are many ways to compute the same statistic, and the algorithms used in the software are those that give the most accurate results for large datasets.

Getting More Help

If you choose Topics from the Help menu, you have access to the entire Help system. The pane on the left side (see Figure 2-12) has tabs at the top, like the tabs on file folders or complicated dialog boxes. Clicking the Contents tab opens the table of contents for the entire Help system. Clicking the Index tab brings up an index of Help topics, arranged alphabetically by name. Enter the keyword you're looking for and scroll through the entries until you find what you want. For example, you can scroll to find all entries in the Help system indexed under the word Frequencies. Or, if you want to know how variables are named in SPSS Statistics, type variable names and then click rules and then the Display button. If the term you selected appears in several different software models, you may have to make a further selection. You'll find the information shown on the right side in Figure 2-12. The **Search** tab is useful for finding specific pieces of information in the vast Help system.

Figure 2-12
Index of Help topics

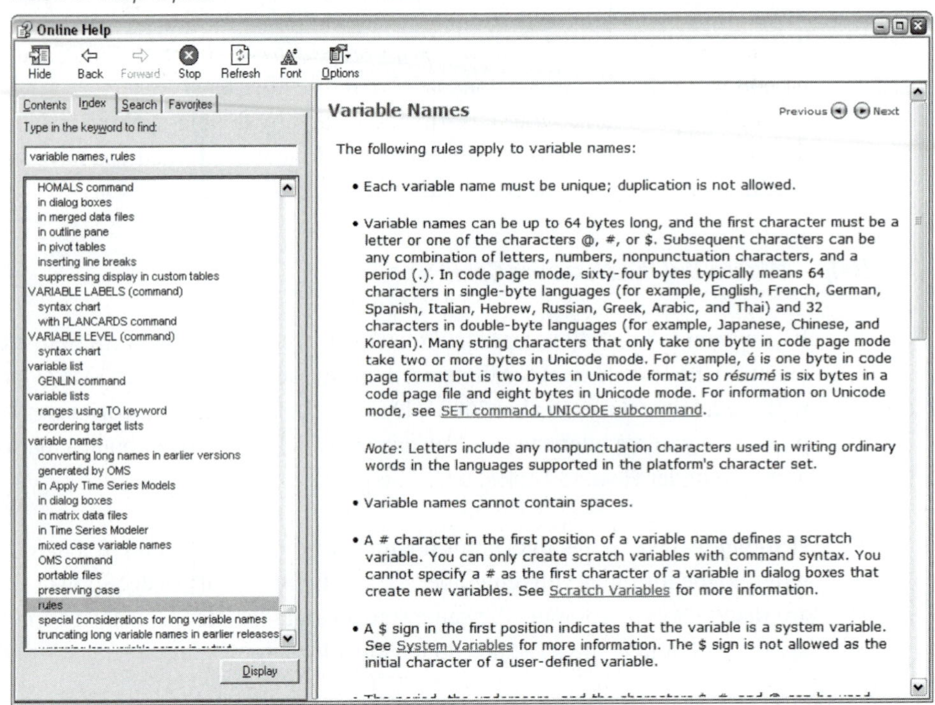

Tip: There is a newsgroup (*comp.soft-sys.stat.spss*), as well as a mailing list (*SPSSX-L*), in which you can ask (or answer) questions. Neither of these is sponsored by SPSS Inc.

Dressing Up Your Output

The best feature of pivot tables is that you have incredible control over the way they look. You can change labels, add titles, and rearrange their structure to suit your needs. The Pivot Table Editor makes it easy to customize your tables. You can then drag them into presentations, dissertations, and documents of all sorts.

Tip: Run the "Working with Output" tutorial for an overview of using the SPSS Statistics Viewer and editing pivot tables.

As a simple example of what you can do with the Pivot Table Editor, consider the table shown in Figure 2-13. It shows the number and percentage of men and of women who say they believe in life after death. You see that 75.5% of the 563 men and 84.9% of the 648 women surveyed say they believe in life after death.

To edit a pivot table, you must first activate it by double-clicking it. Note that the pivot table border will look like that shown in Figure 2-13 when you have successfully activated it.

Figure 2-13
Activated pivot table

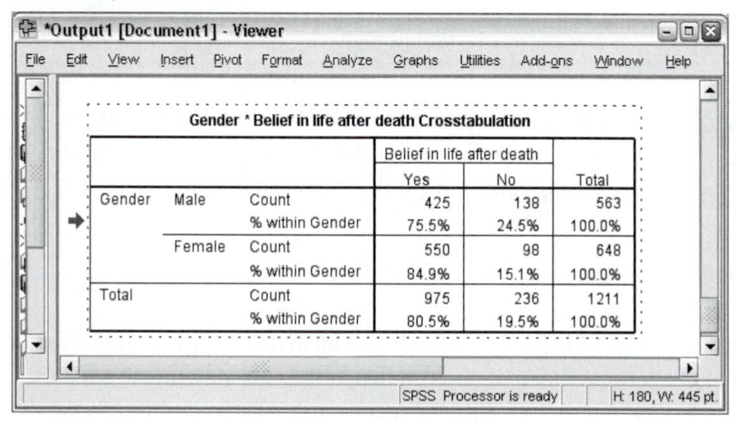

Gender * Belief in life after death Crosstabulation

			Belief in life after death		
			Yes	No	Total
Gender	Male	Count	425	138	563
		% within Gender	75.5%	24.5%	100.0%
	Female	Count	550	98	648
		% within Gender	84.9%	15.1%	100.0%
Total		Count	975	236	1211
		% within Gender	80.5%	19.5%	100.0%

Changing Text

Double-click any piece of text in an activated pivot table. The box surrounding it will enlarge, and the text will change colors. Enter your new text, and then click anywhere else in the table. To make an ugly title, such as the one shown in Figure 2-13, go away, select the title and press the Delete key on your keyboard. Figure 2-14 is Figure 2-13 with the title deleted and some of the labels changed.

Figure 2-14
New title and labels

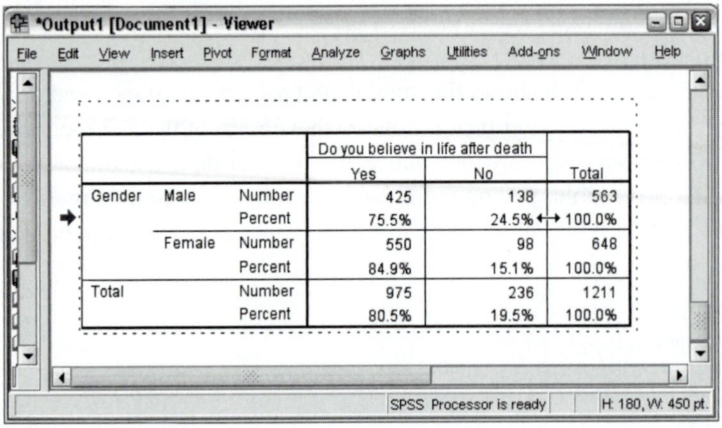

*Output1 [Document1] - Viewer

File Edit View Insert Pivot Format Analyze Graphs Utilities Add-ons Window Help

			Do you believe in life after death		
			Yes	No	Total
Gender	Male	Number	425	138	563
		Percent	75.5%	24.5%	100.0%
	Female	Number	550	98	648
		Percent	84.9%	15.1%	100.0%
Total		Number	975	236	1211
		Percent	80.5%	19.5%	100.0%

SPSS Processor is ready H: 180, W: 450 pt.

Warning: When you have an activated pivot table in the Viewer, the menus and their contents change. A Pivot menu appears, and the Help menu is simplified. Deactivate the pivot table by clicking outside of it to return to the regular SPSS Statistics menus.

Changing the Width of Columns

You can remove or add space to columns in an activated pivot table by positioning the cursor on any vertical line and clicking when the two-way arrow appears, as shown in Figure 2-14. Drag the line in either direction to make the columns wider or narrower.

Tip: For instructions on how to accomplish a particular pivot table maneuver, choose Help > Topics and then search for pivot table. You'll find a list of tasks. Select the one you want for details on what to do.

Changing the Number of Decimal Places

Use your mouse to select the cell entries with too many (or too few) decimal places. From the Format menu, choose Cell Properties and then Format Value. Select the number of decimal places that you want and click OK.

Showing or Hiding Cells

If you want to hide particular rows or columns in your table (such as counts or percentages), activate the table and then select the doomed row or column using Ctrl-Alt-click in the column heading or row label. From the View menu, choose Hide. To resurrect the row or column at a later time, activate the table and then from the View menu, choose Show All. If you are sure you'll never want to see the offending pieces, you can select them and press the Delete key, and they are gone forever.

Figure 2-15 is the table without the column totals and without the rows that show the number of cases.

Figure 2-15
Hidden rows and columns

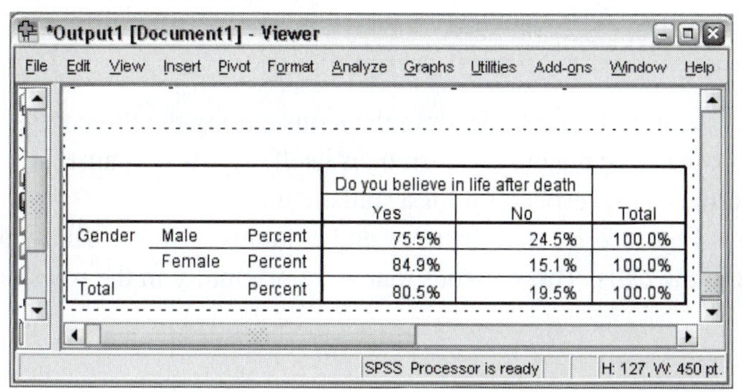

			Do you believe in life after death		Total
			Yes	No	
Gender	Male	Percent	75.5%	24.5%	100.0%
	Female	Percent	84.9%	15.1%	100.0%
Total		Percent	80.5%	19.5%	100.0%

Warning: SPSS Statistics does not recompute statistics when you reorganize a pivot table. Hiding or cutting a row or column leaves the total statistics unchanged, which may be misleading.

Rearranging the Rows, Columns, and Layers

One of the handiest tricks you can do with pivot tables is to switch the arrangement of the rows, columns, and statistics in a table. Activate (double-click) the pivot table, and from the Pivot menu, choose Pivoting Trays. (Alas! No refreshments.) A schematic representation of a pivot table (see Figure 2-16) appears, with three areas (trays) labeled Layer, Row, and Column. Colored icons in these trays represent the contents of the table, one for each variable and one for statistics. You use these icons to control the layout of the different statistics in the table.

Figure 2-16
Pivoting trays

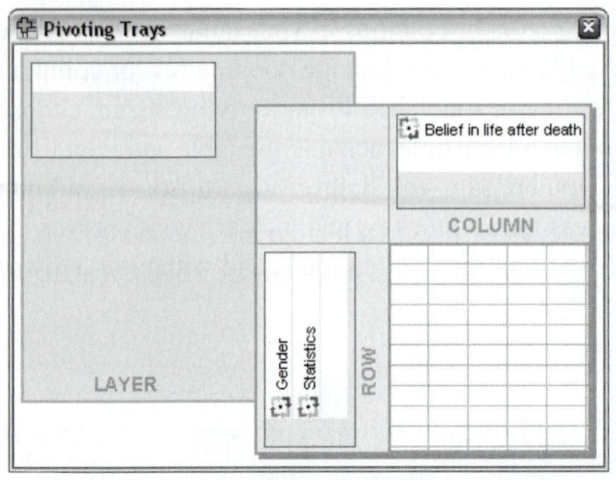

To change the structure of your table, drag the icons among the three trays. Each time you drag an icon, the table magically rearranges itself. Try it—it's amazing, but it's too abstract to describe easily, especially for a statistician.

Figure 2-17 is Figure 2-14 with the rows and columns switched and without any decimal points in the cells. The statistics that were originally in the rows now appear in columns.

Figure 2-17
Pivoted table

*Output1 [Document1] - Viewer

File Edit View Insert Pivot Format Analyze Graphs Utilities Add-ons Window Help

		Count			Percent		
		Gender			Gender		
		Male	Female	Total	Male	Female	Total
Do you believe in life after death	Yes	425	550	975	75%	85%	81%
	No	138	98	236	25%	15%	19%
Total		563	648	1211	100%	100%	100%

SPSS Processor is ready | H: 122, W: 616 pt.

Tip: You can save your output file by choosing File > Save when you are in the Viewer. The file type is **.spv*. You can open a saved output file by choosing File > Open > Output and then selecting the name of the file.

Editing Your Charts

Just as you can edit your pivot tables, you can edit charts as well. Double-click a chart in the Viewer to open it in a new Chart Editor window, as shown in Figure 2-18. The toolbar simplifies access to many functions.

Warning: To access some chart editing capabilities, such as identifying points on a scatterplot or changing the width of bars in a histogram, you must click an element of the chart to select it. For example, you must click any point in a scatterplot or any bar in a histogram or bar chart.

Figure 2-18
Chart Editor window

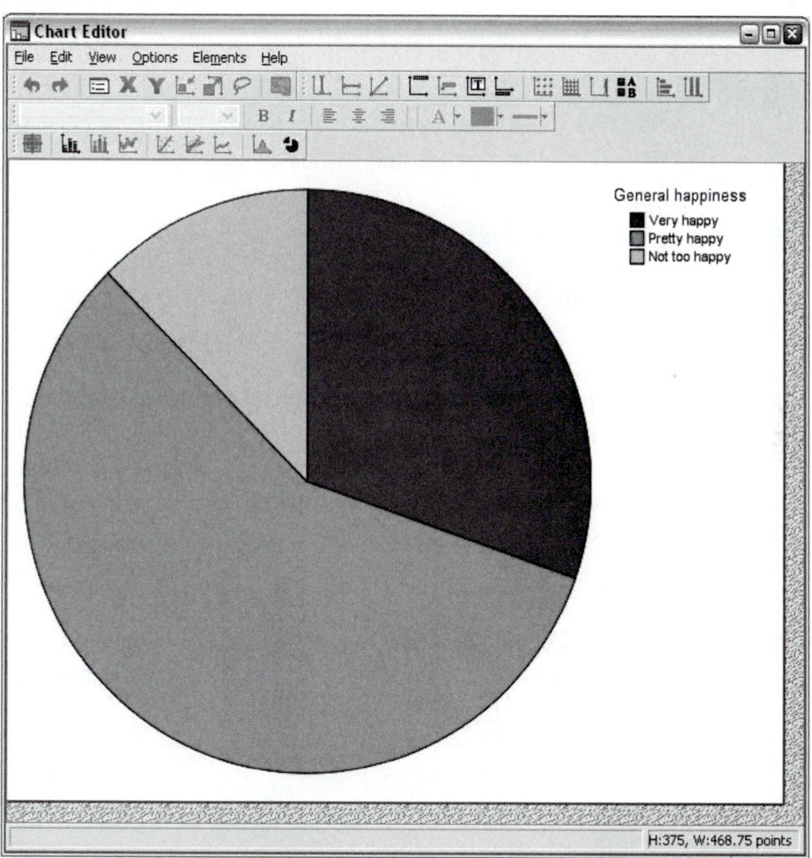

Some things that you may want to do that are common to all chart types are:

- Change labels (just double-click any text and substitute your own).
- Create reference lines.
- Change colors, line types, and sizes.

When you close the Chart Editor window, the original chart in the Viewer updates to show any changes that you made.

Tip: For detailed instructions, run the tutorial "Creating and Editing Charts."

Using Syntax

The simplest way to use SPSS Statistics is through the dialog box interface. However, there are situations in which you may want to use SPSS Statistics syntax to control your analyses. **SPSS Statistics syntax** consists of keywords strung together according to arcane rules that seemed reasonable and worked pretty well a couple of decades ago.

Tip: Syntax keywords and rules are documented in the *SPSS Statistics Command Syntax Reference*, which you can access through the Help menu. You can also run the tutorial "Working with Syntax."

The most common reasons for using command syntax are to repeat analyses quickly and easily or to keep a record of what you've done. For these purposes, click the Paste button in the procedure dialog box that specifies the variables and the analyses that you want. SPSS Statistics pastes the corresponding syntax into a Syntax Editor window. If you are unsure of whether your selection will result in the desired analysis, run your analysis using the dialog box interface, and once you have it correct, recall the dialog box and click the Paste button.

Warning: Clicking Paste only puts the commands in the Syntax Editor window. Make a selection from the Run menu in the Syntax Editor window to actually run the commands.

You can also have SPSS Statistics automatically save the syntax that corresponds to your dialog box selections in a **journal file**. From the Edit menu, choose Options. Then click the File Locations tab. Under Session Journal, select Record syntax in journal and click the Append button. Click Browse to change the name or location of the journal file from the default values.

In Figure 2-19, you see the results of clicking Paste in a Frequencies dialog box that generates a table and a pie chart and computes the mode for the variable *postlife*. You can run commands that are in a syntax file using the Run menu. Some procedures in SPSS Statistics have additional features that are available only through syntax.

Figure 2-19
Syntax Editor window

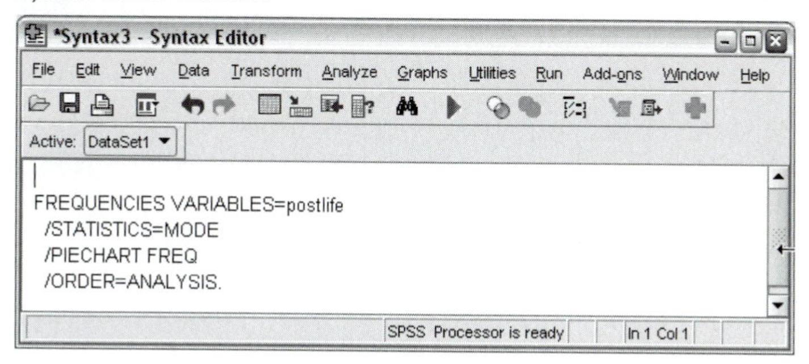

Tip: If you've created a syntax file, make sure that you save it when you exit SPSS Statistics.

Chapter

3

Introducing Data

Everything you do in SPSS Statistics is based on your data. That's why it's so important to set up your data file carefully. If you run the wrong analysis or drag the wrong variable into a dialog box, you can easily fix the problem. If your data file is deficient in some way, it won't matter that you spent hours analyzing the data. Everything you do may be just plain wrong. Given the starring role of the data file in your analysis, this chapter and Chapter 4 guide you in setting up your data file and getting it ready for analysis. These two chapters also provide an overview of some other features found on the Data menu.

Tip: In the Help system, run the tutorials "Reading Data" and "Using the Data Editor."

Vocabulary

You ask 120 people to rate, on a scale of 1 to 10, their satisfaction with your product. For each person, you record the following information:

Name	Joe Smith
Date of product purchase	12/25/2001
Rating	8
Age	32
Gender	Male
Income	None of your business

- Each person or object for whom you have information is called a **case**.
- Each piece of information that you record is called a **variable**. Name, date of purchase, rating, age, gender, and income are all variables.
- Each case has a **value** for each variable. For Joe Smith, the value for *rating* is 8. The value for *age* is 32.

The variables that you record may have values that are numbers or that contain alphabetical characters, or they may be **dates** in some month, day, and year format. (Age is a **numeric** variable. Name is called a **string**, or alphanumeric, variable. Dates are best stored in numeric variables with one of the "date formats" applied.)

Some variables, such as age, weight, and blood pressure, are inherently numeric. Other variables, such as state of residence or college attended, are inherently text. However, you get to decide what kinds of variables they are in your data file. For example, if you code state names from 1 to 50, *state* becomes a numeric variable. If you code age into ranges, such as less than 50 and greater than or equal to 50, you can use the letters *Y* and *O* to represent the categories. *Age* is then a string variable.

You may not have values for all of the variables for each case. You may have failed to record a response or you can't decipher what you wrote; you may not have asked a question or the respondent may have refused to answer. All of these are examples of **missing values**. In your data file, you should enter a code that indicates that a response is missing. For example, you might enter the code –1 if a respondent refuses to divulge his or her income. All SPSS Statistics procedures give special treatment to cases with missing values (for example, by not including them in statistical calculations).

Tip: If there are several reasons why information is missing, you should select a different code for each type of missing value so that you can analyze them individually. For example, you can use a code of –1 if a respondent refuses to give his or her income, a code of –2 if you didn't ask the income question, and a code of –3 if you can't read the answer to the income question.

Planning the Data File

Before you embark on creating a data file, you have to think about what information you're going to enter and in what form. You should assign each case a unique ID number so that you can match it with other information, such as the data forms. You'll find this essential when you have to check values or identify particular cases in your output.

Warning: Be sure to protect the identity of your subjects, especially if others will have access to the data.

Enter information in as much detail as possible and let the computer do all of the necessary computations. For example, enter dates and let the computer calculate time intervals between the dates. If you have heights and weights, let the computer calculate obesity measures. Enter actual ages and incomes and use the computer to calculate categories. This way, you can evaluate different groupings. Don't be afraid of redundant information. For example, if you asked for age and birth date, enter them both. They'll serve as a check on one another.

Getting Data into SPSS Statistics

The ease with which you can get data into the SPSS Statistics Data Editor depends on the format of your data. Regardless of the source, each variable must have a name and a type. If you don't specify these, SPSS Statistics names the variables starting with *VAR00001* and assigns to them the type *Numeric* (with two decimal places displayed).

Using SPSS Statistics Data Files

It's easiest if your data are already in an SPSS Statistics data file. (SPSS Statistics data files have a file extension of **.sav*). In this case, all you have to do is choose File > Open and select the filename; the Data Editor appears fully labeled and laden with data, as shown in Figure 3-1. This is not wishful thinking; many data files, such as those from the U.S. Census Bureau, the General Social Survey, and other large-scale surveys, are distributed in SPSS Statistics format. In these cases, you simply insert the CD or open the downloaded file and you're ready to start.

Figure 3-1
Data Editor in Data View

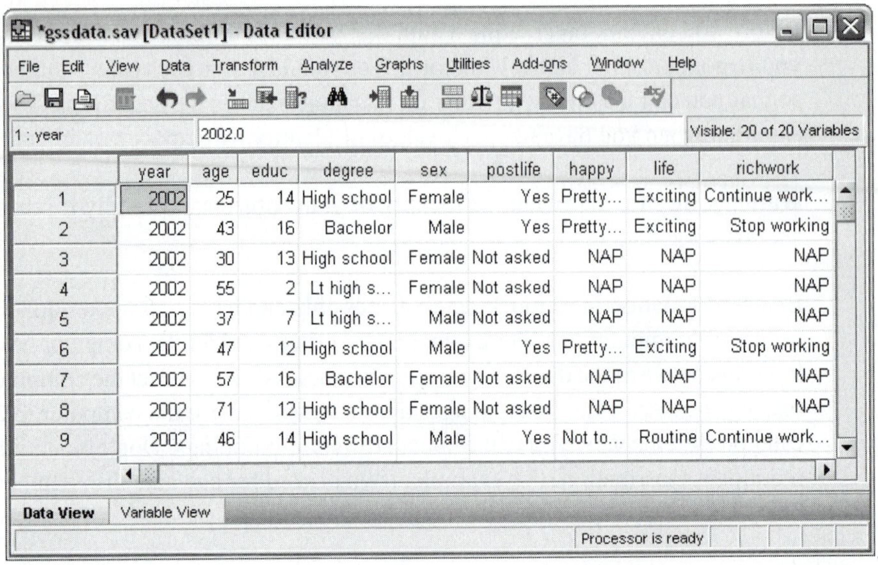

*gssdata.sav [DataSet1] - Data Editor

File Edit View Data Transform Analyze Graphs Utilities Add-ons Window Help

1 : year 2002.0 Visible: 20 of 20 Variables

	year	age	educ	degree	sex	postlife	happy	life	richwork
1	2002	25	14	High school	Female	Yes	Pretty...	Exciting	Continue work...
2	2002	43	16	Bachelor	Male	Yes	Pretty...	Exciting	Stop working
3	2002	30	13	High school	Female	Not asked	NAP	NAP	NAP
4	2002	55	2	Lt high s...	Female	Not asked	NAP	NAP	NAP
5	2002	37	7	Lt high s...	Male	Not asked	NAP	NAP	NAP
6	2002	47	12	High school	Male	Yes	Pretty...	Exciting	Stop working
7	2002	57	16	Bachelor	Female	Not asked	NAP	NAP	NAP
8	2002	71	12	High school	Female	Not asked	NAP	NAP	NAP
9	2002	46	14	High school	Male	Yes	Not to...	Routine	Continue work...

Data View Variable View

Processor is ready

Warning: Even if you get a ready-to-use SPSS Statistics data file, you can't assume that it has been carefully and correctly prepared. Treat it with skepticism and perform the data cleaning described in Chapter 4.

The contents of the Data Editor can be viewed in two different ways. The tabs shown at the bottom left of Figure 3-1 control whether you see information about the variables (Variable View) or whether you see the actual data values (Data View).

- When the Data Editor is showing Data View, each row contains values of the variables for one case. Each column contains the values of the cases for one variable.
- When the Data Editor is showing Variable View, each row describes the various properties of one of the variables in the data. (See Figure 3-2.) Each column shows a particular property (numeric or string, missing values defined, number of decimal places displayed in the Data Editor, and so on) of each of the variables in the file. The actual data values are not shown in Variable View.

In Data View, you can select whether you want to display the actual data values or the descriptive **value labels** that you have assigned to the values. For example, if you assign a label of *Male* to the code of 1, instead of seeing the value 1 in the Data Editor, you can see the label *Male*. To see labels instead of values, select Value Labels on the View menu; to go back to the actual data, deselect Value Labels on the View menu.

Figure 3-2 is an example of the Data Editor in Variable View. Instead of seeing data values, you see a row describing each variable in the data file. The columns contain information about the properties of the variable, whether it is numeric or string, what the missing value codes are, and so on.

Figure 3-2
Data Editor in Variable View

*gssdata.sav [DataSet1] - Data Editor

File Edit View Data Transform Analyze Graphs Utilities Add-ons Window Help

	Name	Type	Width	Decimals	Label	Values	Missing
1	year	Numeric	4	0	GSS year	None	None
2	age	Numeric	2	0	Age of respondent	{98, DK}...	0, 99, 98
3	educ	Numeric	2	0	Highest year of s...	{97, NAP}...	97, 99, 98
4	degree	Numeric	1	0	Highest degree	{0, Lt high s...	7, 9, 8
5	sex	Numeric	1	0	Gender	{1, Male}...	None
6	postlife	Numeric	1	0	Belief in life after ...	{0, Not aske...	0, 8, 9
7	happy	Numeric	1	0	General happiness	{0, NAP}...	0, 8, 9
8	life	Numeric	1	0	Is life exciting, ro...	{0, NAP}...	0, 8, 9
9	richwork	Numeric	1	0	If rich, continue or...	{0, NAP}...	0, 8, 9
10	news	Numeric	1	0	How often do you...	{0, NAP}...	0, 8, 9
11	tvhours	Numeric	2	0	Hours per day wa...	{-1, NAP}...	-1, 99, 98

Data View | **Variable View**

Processor is ready

Tip: Always keep a copy of your *original* data file. You never know when you may need it to untangle some problem that occurs.

Using Spreadsheet and Database Files

The next-best situation is if your data are already in machine-readable format but just not in an SPSS Statistics file. In this case, SPSS Statistics can probably open the file directly. In the Open File dialog box (choose File > Open > Data), select the file type of your data from the Files of Type drop-down list, and proceed to locate the file.

If you have trouble, run the tutorial "Reading Data" for more information about bringing data from spreadsheets and databases into SPSS Statistics.

Tip: If you bring data into SPSS Statistics from another source, make sure that you go through the data file, variable by variable, to insert labels, improve on variable names, and declare missing values. Make sure that all of the cases made it into the file correctly before you start analyzing the data.

Feeding Text Files to the Text Wizard

If your data are just text (numbers, letters, and characters saved in a text file, typically with a *.*txt* file extension), you should choose File > Read Text Data. SPSS Statistics prompts you to locate the file, and then it opens the Text Wizard, which will guide you in converting text data to an SPSS Statistics data file. The Text Wizard needs either a delimiter (such as a blank space, comma, or tab character) between data values, or it needs the data values to be in the same position for each case. If you have a special symbol delimiting the values, you just have to indicate what it is, and the wizard takes over. If you have the values laid out in the same position for each case, you will use a handy ruler carried by the wizard to designate where one variable ends and the next begins.

Tip: If your data are in a word-processing file format, you must save the file as an unformatted text file. If you can choose between entering data into a word processor or into the SPSS Statistics Data Editor, you're less likely to create problems for yourself if you use the Data Editor.

Typing Your Own Data

If your data aren't in a computer-readable format, you can enter them directly into the SPSS Statistics Data Editor. From the menus, choose File > New > Data. This opens the Data Editor in Data View. If you type a number into the first cell, SPSS Statistics will label that column with the variable name *VAR00001* (see Figure 3-3).

Surely you can think of a better name than that. To christen your variables with names of your own and to supply other variable characteristics, click the tab labeled Variable View, as shown at the bottom of Figure 3-3.

Figure 3-3
Entering data

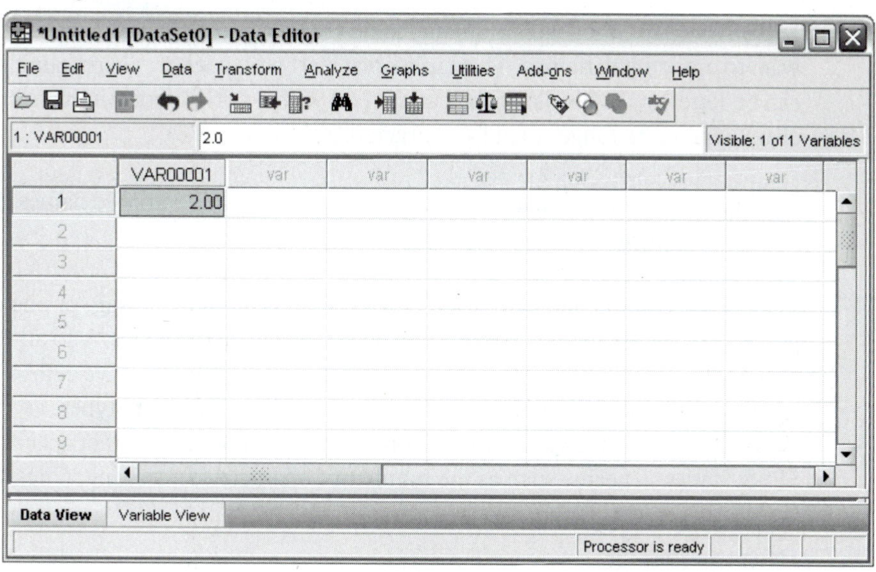

Figure 3-4 shows the Data Editor in Variable View with the variables defined for the survey described at the beginning of this chapter.

Figure 3-4
Data Editor in Variable View

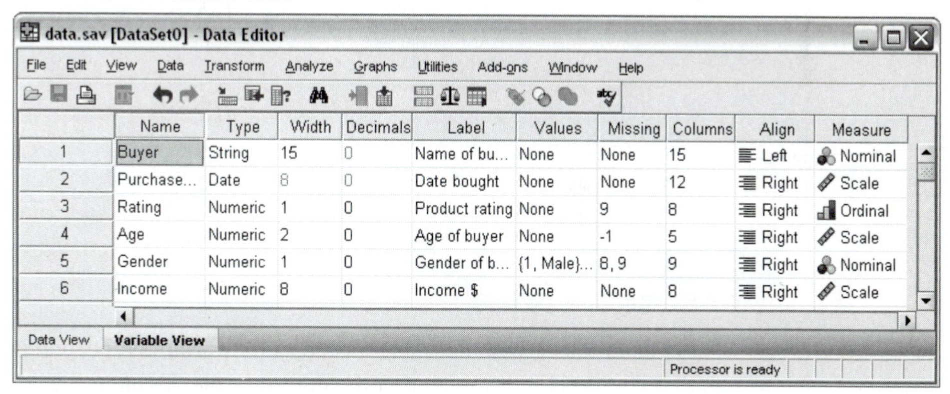

	Name	Type	Width	Decimals	Label	Values	Missing	Columns	Align	Measure
1	Buyer	String	15	0	Name of bu...	None	None	15	Left	Nominal
2	Purchase...	Date	8	0	Date bought	None	None	12	Right	Scale
3	Rating	Numeric	1	0	Product rating	None	9	8	Right	Ordinal
4	Age	Numeric	2	0	Age of buyer	None	-1	5	Right	Scale
5	Gender	Numeric	1	0	Gender of b...	{1, Male}...	8, 9	9	Right	Nominal
6	Income	Numeric	8	0	Income $	None	None	8	Right	Scale

Assigning Variable Names and Properties

In the *Name* column, enter a unique name for each variable in the order in which you want to enter the variables. The name must start with a letter; the remaining characters can be letters or digits. (A name cannot end with a period, contain blanks or special characters, or be longer than 64 characters.)

Tip: Try to use variable names that are descriptive. Avoid variable names that can be confused with each other, such as *jobsat* and *satjob*.

In the *Type* column, indicate whether a variable is numeric, string, or one of the other possible types. By default, SPSS Statistics assigns the type *Numeric* to all variables. To change the variable *Buyer* to a string variable, you click in the *Type* cell. A gray button appears in the cell. Click the button to display a list of variable types, as shown in Figure 3-5. Select the variable type you want and supply any further specifications that SPSS Statistics needs, such as the number of characters in a string. Click OK.

Figure 3-5
String declaration

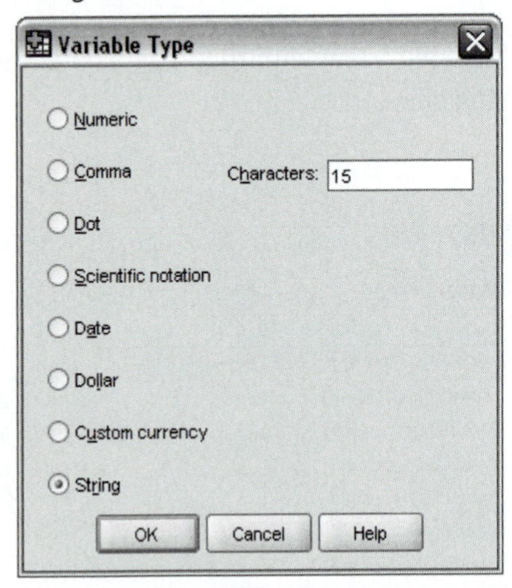

If you want to enter dates, you must select Date as the variable type, and then indicate the format that you will be using to enter the date variable. To enter the date 05/31/86, with the slashes in the field, you would click the format shown in Figure 3-6.

Figure 3-6
Date declaration

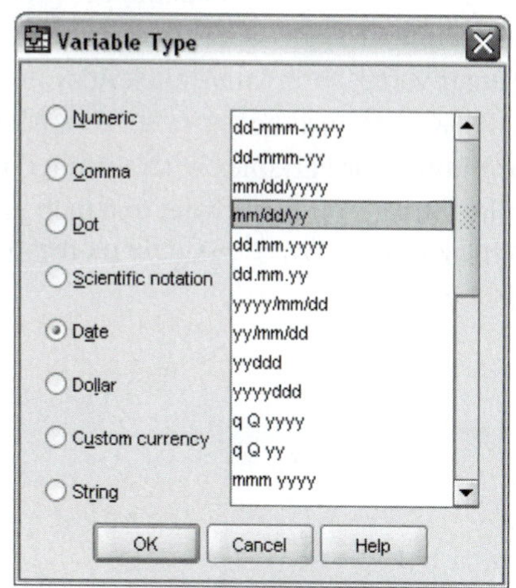

In the *Width* column, type the maximum number of characters or numbers you'll use, or bump the switch to increase or decrease the number shown. For variables that display numeric values (all but *String* and *Date*), specify the number of decimal places in the *Decimals* field. By default, SPSS Statistics assigns 8 as the width of all variables and 2 as the number of decimals.

Tip: Be sure to save your data file, even while you are still entering your variable information. You don't want to lose your work if someone trips over the power cord or if rogue software crashes Windows.

Assigning Descriptive Labels

Variable labels. In the *Label* column, assign descriptive text to the variable by clicking the cell and then entering the label. For example, if you named the customer satisfaction rating as *Satisfaction*, you can assign a more complete label, such as *Customer satisfaction with product*. This will be used to label output.

To increase the width of a column in Variable View, point at a divider between column headings, press and hold the mouse button, and drag the divider to the right.

Value labels. To label individual values, click the button in the *Value* column. This opens the dialog box shown in Figure 3-7. For example, you could assign the label *Male* to a code of 1, the label *Female* to a code of 2, the label *Not asked* to a code of 9.

The sequence of operations here is: enter value, enter label, click Add, and repeat for each value. When you're finished, click OK to record the entire set of labels.

Labels for individual values are useful only for variables with a limited number of categories whose codes aren't self-explanatory. You don't want to attach value labels to individual ages; however, you should label the missing value codes for all variables if you use more than one code.

Figure 3-7
Value labels for gender

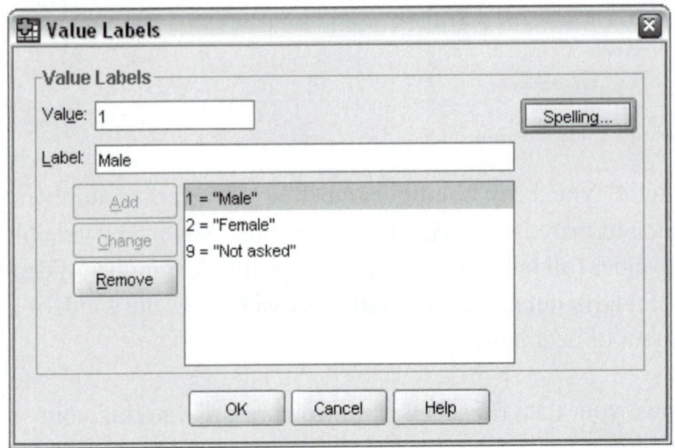

Tip: Take the time to assign value labels. You'll soon forget whether code 1 represents males or females.

Assigning Missing Values

To indicate which codes were used for each variable when information is not available, click in the *Missing* column and, as usual, click the gray button when it appears. Figure 3-8 assigns the codes 8 and 9 for missing values. Cases with these codes will be treated differently during statistical analysis. If you don't assign codes for missing values, even nonsensical values are accepted. A value of –1 for *age* would be considered a real age.

The missing-value codes that you assign to a variable are called **user-missing values**. **System-missing values** are assigned by SPSS Statistics to any blank numeric cell in the Data Editor or to any calculated value that is not defined. A system-missing value is indicated with a period (.).

You can't assign missing values to a string variable that is more than eight characters in width. For string variables, uppercase and lowercase letters are treated as distinct characters. This means that if you use the code *NA* (not available) as a missing-value code, entries coded as *na* will not be treated as missing. Also, if a string variable is three characters wide and the missing value code is only two characters wide, the placement of the two characters in the field of three affects what's considered missing. Blanks at the end of the field (trailing blanks) are ignored in missing-value specifications.

Figure 3-8
Missing values specification

Warning: Don't use a blank space as a missing value. Use a specific number or character to signify *I looked for this value and I don't know what it is*. Don't use missing-value codes that are between the smallest and largest valid values, even if these particular codes don't occur in the data.

Assigning Column Width

The number in the *Columns* column controls the variable's width in Data View. You can drag the column wider or narrower as necessary.

Assigning Levels of Measurement

Click a cell in the *Measure* column to assign a level of measurement to each variable. You have three choices (nominal, ordinal, and scale).

- If the values of a variable cannot be ordered in a meaningful way (for example, region, country of birth, favorite statistician), the variable is measured on a **nominal scale**.
- If the values of a variable can be ordered from smallest to largest (for example, level of satisfaction, degree of agreement with a statement) *but* the numeric distance between the values is not meaningful, the variable is measured on an **ordinal scale**.
- If the actual distance between values is interpretable (for example, age, weight, number of cars owned), the level of measurement is **scale**.

If you don't specify the scale, SPSS Statistics attempts to divine it based on characteristics of the data, but its judgment in this matter is fallible. For example, string variables are always designated as nominal. SPSS Statistics uses different icons for the three types of variables.

The scale on which a variable is measured does not necessarily dictate the appropriate statistical analysis for a variable. For example, an ID number assigned to subjects in an experiment is usually classified as a nominal variable. If the numbers are assigned sequentially, however, they can be plotted on a scale to see if subject responses change with time. Velleman and Wilkinson (1993) discuss the problems associated with stereotyping variables.

Tip: SPSS Statistics attempts to use the level of measurement of each variable to guide you. For example, SPSS Statistics will not let you calculate means for nominal variables. Certain statistical procedures don't allow string variables in particular fields in the dialog boxes. If you don't see the variable you're looking for in the variable list for a particular procedure, make sure that it has the correct level of measurement in the data definition. If you mistakenly identify a variable as nominal, it will be excluded from lists for some statistical procedures.

Entering the Data

Once you've set up the description of the variables, you're ready to enter data into the Data Editor. Click the Data View tab, and you'll see the window shown in Figure 3-9, which already shows some cases entered. Enter the values of the variables for each case. The Data Editor is in value-label view, meaning that as you enter a number or code, if there is a value label for it, the value label is displayed.

Figure 3-9
Data Editor in Data View

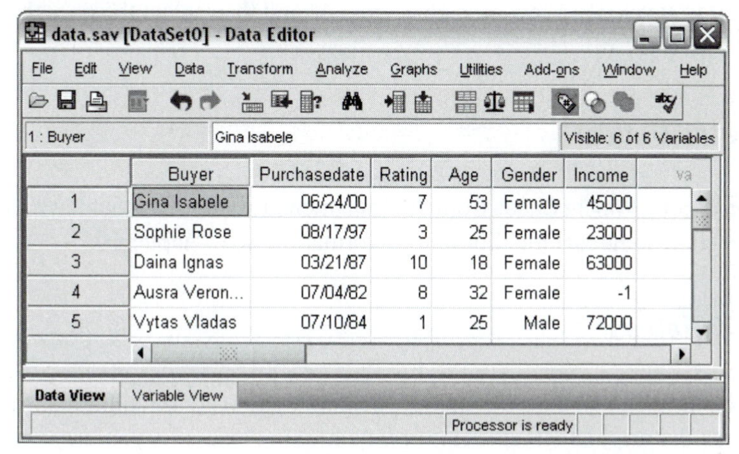

- To correct a mistake, click in the cell and enter the correct value.
- To insert a new case, click the case below which you want to insert it. From the Edit menu, choose Insert Cases. (There's a button for this on the toolbar, which you can identify by the ToolTips.)
- To add a new variable in Data View, click the variable name of the variable before which you want to insert the new variable. From the Edit menu, choose Insert Variable (or find the toolbar button).

Tip: Run the tutorial "Using the Data Editor" for more helpful hints about simplifying the variable-definition and data-entry processes.

Saving the Data File

At any point in the data-definition process, you can save your definitions and data by choosing Save from the File menu. You can also include text information in an SPSS Statistics data file by choosing Utilities > Data File Comments. Anyone using the file can read the text associated with it.

Tip: The Save As option on the File menu lets you specify which variables you want to save by clicking the Variables button. If there are a lot of variables in the data file and you want to save only a few, first click Drop All and then select the ones that you want to save.

Entering Tabulated Data

You can choose Weight Cases from the Data menu to enter data that have already been aggregated in some fashion. For more information, see Chapter 10.

Creating New Variables

Once you have data in the Data Editor, you may want to create additional variables that depend on the original variables in the data file. For example, you may want to create a variable that is the difference between the years of education for husband-and-wife pairs or a variable that groups ages into two categories (less than or equal to 50 and greater than 50). Chapter 5 provides an overview of how to do this in SPSS Statistics.

Saving Time When Dealing with Your Data

If you don't learn how to manage your data file with SPSS Statistics, you'll waste a lot of time and energy trying to obtain analyses for particular subsets of cases.

Selecting Cases for Analyses

If you want to perform analyses on a subset of your cases (for example, patients treated at a particular institution), you can choose Select Cases from the Data menu. All analyses are restricted to the cases that meet the criteria you specified. If you look at the Data Editor when Select Cases is in effect, you'll see lines through the cases that did not meet the selection criteria. Those cases are still there (unless you tell SPSS Statistics to delete them), but they aren't used in statistical or graphical procedures.

Warning: If you use the Select Cases dialog box to perform analyses for cases that meet a particular criterion, make sure that Filter Out Unselected Cases is selected in the Select Cases dialog box. If you select Delete Unselected Cases, all of the cases that don't meet the criterion will be permanently gone when you save the data file. (Aren't you glad you saved that original data file?!)

Repeating the Analysis for Different Groups of Cases

If you want to perform the same analysis for several groups of cases, choose Split File from the Data menu. A separate analysis is done for each combination of values of the variables specified in the Split File dialog box. You can also select how you want the output displayed—all output for each subgroup together or the same output for each subgroup together.

Chapter

4

Preparing Your Data

You know that someone has taken great care in creating the data file that you're going to analyze. You've been meticulous and very careful. Now it's time to abandon all of your preconceived notions about the qualities of your data and to look for problems that may later cause you grief (for example, when you're presenting a report or defending your dissertation). In this chapter, you'll find suggestions for looking for errors in your data file.

Do not be tempted by the lure of clicking a statistical procedure name that you recognize and by the immediate gratification of moving variable names into boxes and obtaining important-looking results. The best initial use of your time is to examine your data carefully. The morning of the big presentation or report is not when you want to discover that no one over the age of 21 is in your customer database or that the average individual purchased 540 tubes of toothpaste in a year.

Tip: The SPSS Statistics Data Preparation module is an extra-cost option especially designated for data cleaning.

Checking Variable Definitions

In the Data Editor, you supplied various pieces of information about each of the variables in your data file. Before proceeding any further, you want to make sure that you didn't make any mistakes. Although you can try to proof the information in the Data Editor itself, there are more convenient options. Using selections on the Utilities menu, you can display the properties and labels for each of the variables in the data file, although you have to return to the Data Editor to make corrections. If you want to check the data definitions, see the values that actually occur in the data file, and change labels and settings in one convenient location, choose Define Variable Properties from the Data menu.

Using the Utilities Menu

Choose Utilities > Variables to get data-definition information for each variable in the data file. Click a variable name on the left to view the information shown on the right (see Figure 4-1). You see the name and the variable label and value labels that you have assigned, as well as the missing-value codes. Numeric variables are identified as type *F*. The number before the period is the width of the field specified in the Data Editor, and the number after the period is the number of decimal places, if the number is greater than 0.

Make sure that all of your missing-value codes are correctly identified. For the variable *postlife* (*Belief in life after death*), the values 0, 8, and 9 are missing values. The nonmissing values are labeled as *Yes* for the code of 1 and *No* for the code of 2. The level of measurement is given as nominal, although a variable with exactly two valid values, such as *yes/no* or *agree/disagree*, satisfies the requirements for the scale level of measurement because there is only one interval.

Figure 4-1
Variable information for postlife

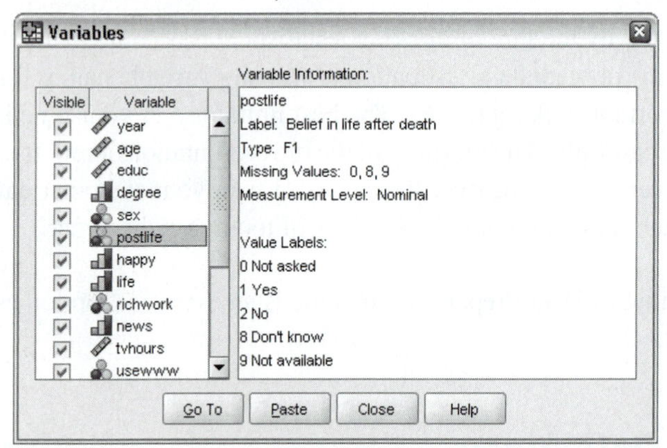

Tip: If you click Go To, you find yourself in the column of the Data Editor for the selected variable if the Data Editor is in Data View. To edit the variable information from the Data Editor in Data View, double-click the variable name at the top of that column. This takes you to the Variable View for that variable.

To get a listing of the information for all of the variables without having to select the variables individually, choose File > Display Data File Information > Working File. This lists variable information for the whole data file. The disadvantage is that you can't quickly go back to the Data Editor to fix mistakes. An advantage is that codes that are defined as missing are identified, so it's easier to check the labels.

Using Define Variable Properties

The quickest way to see what's in the data file and to fix variable definitions is to choose Data > Define Variable Properties. A dialog box similar to Figure 4-2 opens. Click the variable names or drag the variables that you want to be scanned into the Variables to Scan list. If you have scale variables with lots of values and drag them into the list, you'll see a table with lots of values, unless you've limited the number of cases to scan with the option at the bottom of Figure 4-2.

Figure 4-2
Define Variable Properties dialog box

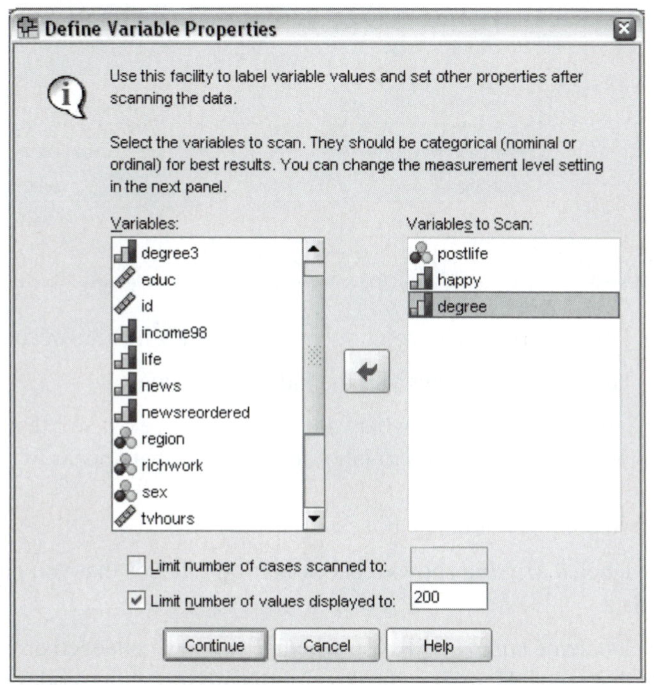

Once you've selected the variables for review, click Continue. You'll then see the "real" Define Variable Properties dialog box, as shown in Figure 4-3. Select any variable in the list to see the variable definitions in effect and the results of scanning the data. You can easily view or change the properties of the variable. You can switch among the different types of numeric data here, but you can't change variable types from numeric to string or vice versa. You must be in Variable View in the Data Editor to do something that radical.

Figure 4-3
Define Variable Properties results

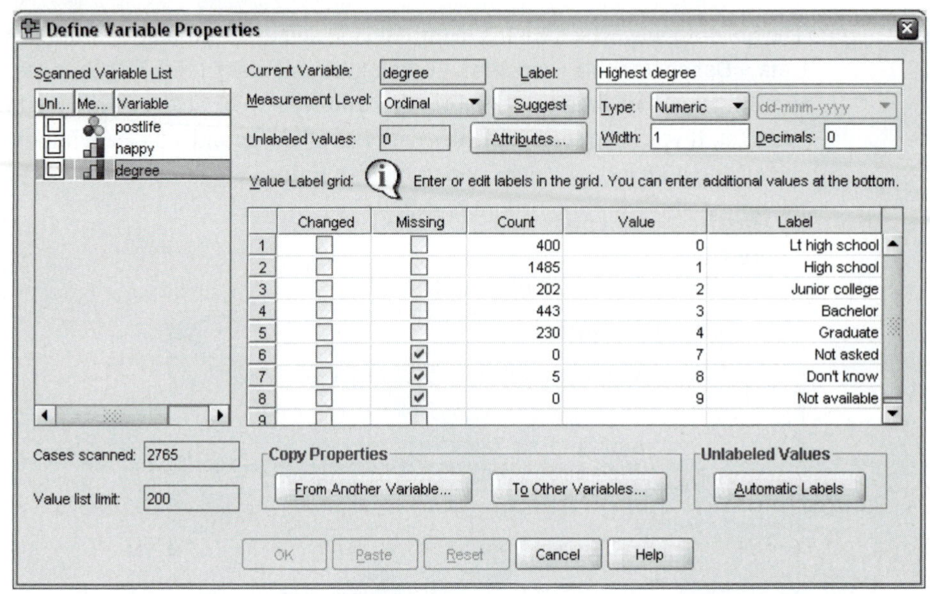

- The column labeled *Count* tells you how many times each value occurs.
- The column labeled *Value* shows you the individual value.
- The column labeled *Label* shows what value label, if any, exists for that value. You can see whether you've forgotten to label some values that appear in the data.

On the left are two columns of check boxes:

- The column labeled *Missing* shows a check mark for values that you have defined as missing data.
- The column labeled *Changed* tells you whether you have changed any of the attributes of that value. Here, it's checked for the missing-value codes because the value labels have been improved.

Warning: Don't click OK until you're finished examining all of the variables.

Checking Your Case Count

In acquiring your data and setting up the data file, you have no doubt spent considerable time and experienced aggravation of some type. Before you proceed any further, you want to make sure that the number of cases in your data file is neither too many nor too few but just right.

Eliminating Duplicate Cases

It's easy to enter the same case two times (or even more). To oust the clones, choose Data > Identify Duplicate Cases. The resulting dialog box (see Figure 4-4) offers all kinds of options, but, most of the time, you'll do one of two things.

Figure 4-4
Identify Duplicate Cases dialog box

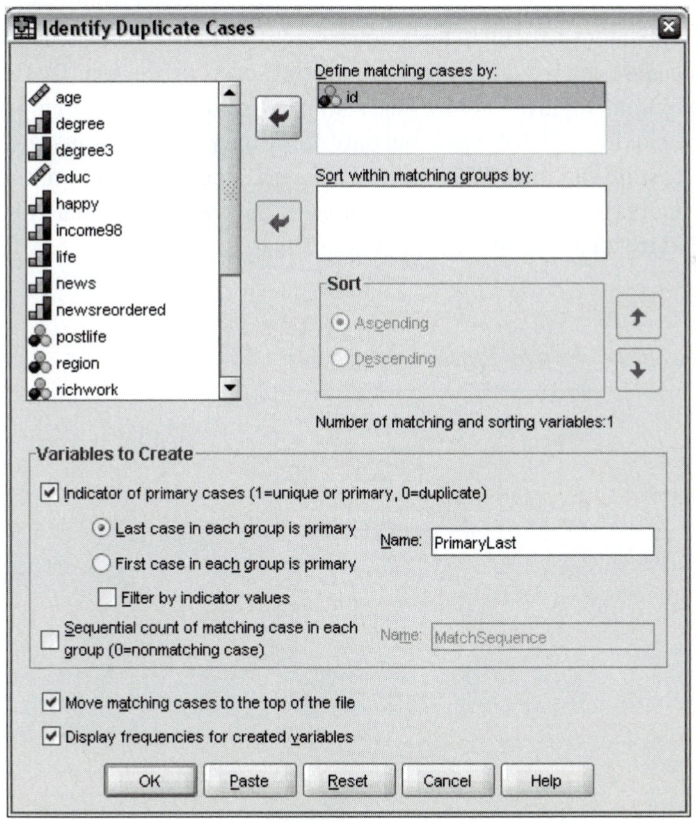

If you entered a (supposedly) unique ID variable for each case, move the name of that ID variable into the Define matching cases by list. If it takes more than one variable to guarantee uniqueness (for example, college and student ID), move all of these variables into the list. When you click OK, SPSS Statistics checks the file for cases that have duplicate values of the ID variable(s).

You can also check for cases that are completely identical. Use Ctrl-A (hold down the Ctrl key while you press A) to select all of the variables. Then move them into the Define matching cases by list and click OK.

Tip: Don't automatically discard cases with the same ID number unless all of the other values also match. It's possible that the problem is merely that a wrong ID number was entered.

Adding Missing Cases

It's considerably more difficult to track down cases that didn't make it into the data file. Run any procedure and look at the count of the total cases processed. That's always the first piece of output. Figure 4-5 shows the summary from the Crosstabs procedure. You see that the data file has 2,765 cases, but only 1,211 have valid (nonmissing) values for the *sex* and *postlife* variables. That's not of concern, since the design of the survey is such that not all people are asked all questions. If the count isn't what you think it should be and if you assigned sequential numbers to cases, you can look for missing ID numbers.

Figure 4-5
Case processing summary for Crosstabs procedure

	Cases					
	Valid		Missing		Total	
	N	%	N	%	N	%
Respondent's sex*Belief in life after death	1211	43.8%	1554	56.2%	2765	100%

Checking Data Values

Checking and correcting the variable definitions is the easy part of checking the data file because you know what they should be, or at least you can find out. It's much harder to check the actual data values. If you didn't enter the data values yourself, try to determine how they were added to the file. What quality control measures were taken? If possible, take a random sample of original records and compare them to the values in the file. If you find mistakes, you may have to check all of the values in the file before proceeding.

Warning: Data cleaning is not an excuse to get rid of data values that you don't like. You are looking for values that are obviously in error and need to be corrected or replaced with missing values. This is not the time to deal with unusual but correct data points. You'll deal with those during the actual data analysis phase.

Listing the Values

If you have a data file with a small number of cases and variables, one of the first steps should be to scan the data values for each case. You can easily see whether you've added an extra 0 to an age or skipped a value. If the data file is small enough to fit on a single screen, you can just look at the values in the Data Editor.

You can list values for selected variables using Analyze > Reports > Case Summaries. Click the names of the variables whose values you want listed to get a report such as the one shown in Figure 4-6. Correct any values that are obviously wrong and check suspicious values.

Figure 4-6
Listing from Case Summaries

		Case ID	Age	Hours worked last week
1		232	25	40
2		479	43	72
3		763	30	40
4		324	55	60
5		333	37	40
6		654	47	42
7		707	57	NAP
8		832	71	NAP
9		948	46	40
10		101	19	14
Total	N	10	10	8

Making Frequency Tables

For variables with a limited number of distinct values, make frequency tables that show you the number of times each value occurs in the data file. (For more information, see Chapter 6.) Ask yourself the following questions:

- *Are the codes that you used for missing values labeled as missing values in the frequency table?* Make sure that you're not forgetting some codes or including codes that you shouldn't. If the codes are not labeled, go back to the Data Editor and specify them as missing values. Rerun the table to make sure that you've gotten it right.
- *Do the value labels correctly match the codes?* For example, if you see that 90% of your customers are very dissatisfied with your product, before you hire expensive consultants to remedy the problem, make sure that you haven't made a mistake in assigning the labels. If something looks suspicious, check it out. Just because it's in a computer file, it doesn't mean that it's right.
- *Are all of the values in the table possible?* For example, if you asked the number of times a person has been married and you see values of –2, you know that's an error. If you asked people to rate something on a scale of 1 to 5 and you see values of 6, you've got a problem. If someone claims to watch TV for 25 hours a day, you've got a problem. Go back to the source and see if you can figure out what the correct values are. If you can't, replace them with codes for missing values.
- *Are there values that are possible but highly unlikely?* For example, if you see a case who claims to own 11 toasters, you want to check whether the value is correct. If the value is correct, you'll have to take that into account when analyzing the data.
- *Are there unexpectedly large or small counts for any of the values?* If you're studying the relationship of highest educational degree to subscription to Web services offered by your company and you see that no one in your sample has a college degree, suspect problems.

Tip: To search for a particular data value for a variable, access the Data Editor in Data View, select the variable that you want to search by clicking its name in the column heading, and then from the menus, choose Edit > Find. Enter the value you want.

Looking At the Distribution of Values

For a scale variable with too many values for a frequency table (for example, weight or income in dollars), you need different tools for checking the data values. Counting how often different values occur isn't useful anymore.

Are the Smallest and Largest Values Sensible?

One of the first tasks should be to check the smallest and largest values to make sure that they are plausible. You don't want to look at just the single largest and single smallest values; instead, you want to look at a certain percentage or number of cases with the largest and smallest values. That's much more efficient. There are several ways you can do this. The simplest, but most limited, way is to choose Analyze > Descriptive Statistics and then either choose Descriptives or Explore. Click Statistics in the Explore dialog box and select Outliers in the Explore Statistics dialog box. This produces the table shown in Figure 4-7—a list of cases with the five smallest and the five largest values for a particular variable. Values that are defined as *Missing* are not included in the table, so if you see missing values in the list, there's something wrong. Check the other values if they appear to be unusual.

In Figure 4-7, you see that one of the respondents claims to watch television 24 hours a day. You know that's not correct. It's possible that he or she understood the question to mean how many hours is the TV set on. When analyzing the TV variable, you'll have to decide what to do with people who have reported impossible values. In Figure 4-7, you see that there are only four cases with values of 16 hours or greater, and then there is a gap until 12 hours. You might want to set values greater than 12 hours to 12 hours when analyzing the data.

Figure 4-7
Extreme values from Explore

			Case Number	Value
HOURS PER DAY WATCHING TV	Highest	1	1149	24
		2	2003	22
		3	947	20
		4	1622	16
		5	510	12[1]
	Lowest	1	2727	0
		2	2652	0
		3	2568	0
		4	2472	0
		5	2380	0[2]

1. Only a partial list of cases with the value 12 are shown in the table of upper extremes.

2. Only a partial list of cases with the value 0 are shown in the table of lower extremes.

Is There Anything Strange about the Distribution of Values?

The next task is to examine the distribution of the values using histograms or stem-and-leaf plots. Make a stem-and-leaf plot (for small datasets) or a histogram of the data using either Graphs > Histogram or the Explore Plots dialog box. (For more information about histograms and stem-and-leaf plots, see Chapter 6.) You want to look for unusual patterns in your data. For example, look at the histogram of ages in Figure 4-8. Where have all of the 30-year-olds gone? Why are there no people above the age of 90? Were there really no people younger than 18 in the survey?

Figure 4-8
Histogram with missing ages

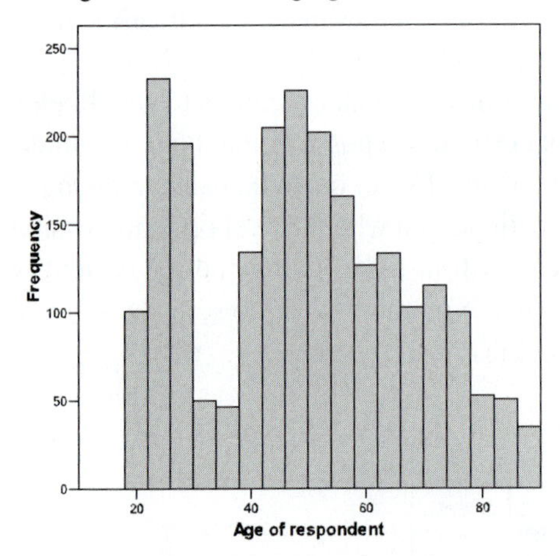

Looking At Combinations of Variables

Examining the variables individually is just the first step in checking your data. The next step is to look for errors that can be spotted only when you look at combinations of variables.

Are There Logical Impossibilities?

Think about whether there are sets of variables that cannot have particular combinations of values. For example, if you have a data file of hospital admissions, you can make a frequency table to count the reason for admission and the number of male and female admissions. Looking at these tables, you may not notice anything strange.

However, if you look at these two variables together in a Crosstabs table, you may uncover unusual events. (For a detailed discussion of Crosstabs, see Chapter 6.) You may find males giving birth to babies and women undergoing prostate surgery. You know these patients need special attention.

Sometimes, pairs of variables have values that must be ordered in a particular way. For example, if you ask a woman her current age, her age at first marriage, and the duration of her first marriage, you know that the current age must be greater than or equal to the age at first marriage. You also know that the age at first marriage plus the duration of first marriage cannot exceed the current age. Start by looking at the simplest relationship: Is the age at first marriage less than the current age?

You can plot the two variables on a scatterplot and look for cases that have unacceptable values. For example, look at Figure 4-9, a plot of age at first marriage and current age. You know that all of the points must fall on or above the identity line. You notice two unusual points. By choosing Data Label Mode from the Elements menu in the Chart Editor, you can identify them as individual cases 30 and 58. There's something wrong with the value of age at first marriage for case ID 58, since it shows the unlikely, but hopeful, value of around 100 years. This turns out to be a problem with values of 98 not being properly declared as missing. (That's what I found in my copy of the General Social Survey!) The second point, case ID 30, is more troublesome: The age at first marriage is greater than the current age. You can't tell which value is correct without going back to the underlying records.

Figure 4-9
Scatterplot of current age and age at first marriage

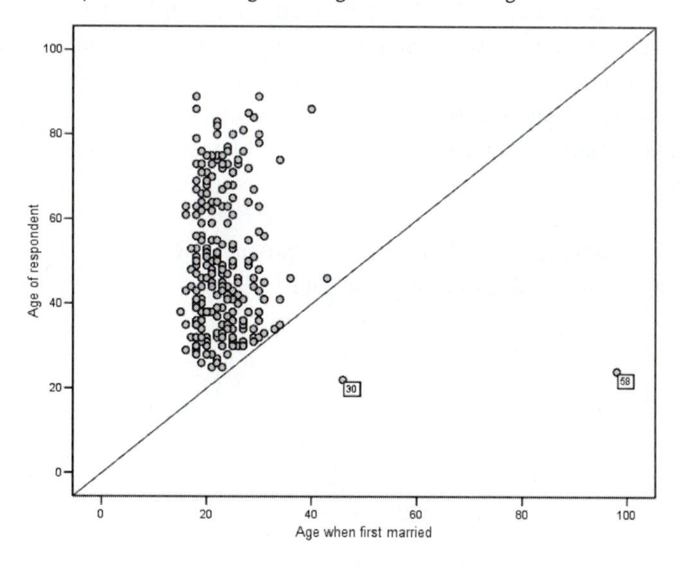

Tip: For large data files, the drawback to this approach is that it's tedious and prone to error. A better way is to create a new variable (see Chapter 5) that is the difference between the current age and the age at first marriage. Then use Data > Select Cases to select cases with negative values and Analyze > Reports > Case Summaries to list the pertinent information. Once you've remedied the age problem, you can create a new variable that is the sum of the age at first marriage and the duration of the first marriage. You can then find the difference between this sum and the current age. Reset the Select Cases criteria and use Case Summaries to list cases with offending values.

Is There Consistency?

For a survey, you often have questions that are conditional. For example, first you ask *Do you have a car?* If the answer is *Yes*, you ask insightful questions about the car. You can make Crosstabs tables of the responses to the main question with those to the subquestions. This often uncovers many inconsistencies that can plague you later when you have more car colors reported than cars owned by your respondents. You have to decide how to deal with these inconsistencies: Do you impute answers to the main question, or do you discard answers to subquestions? It's your decision.

Is There Agreement?

Often, you have pairs of variables that convey similar information in different ways. For example, you may have recorded both years of education and highest degree earned. Or, you may have created a new variable that groups age into categories, such as less than 25, 25 to 50, and older than 50. Compare the values of the two variables using crosstabulations. The table may be large, but it's easy to check the correspondence between the two variables. You can also identify problems by plotting the values of the two variables.

Tip: You are not checking the ability of SPSS Statistics to compute new variables. If there's a mistake, you're (almost always) at fault!

Are There Unusual Combinations of Values?

Outliers are cases that are far removed from the rest. They can wreak havoc in many statistical analyses. In the data-cleaning phase, your goal is to identify points that stick out so that you can make sure their values are correct. Identification of cases that are outliers is complicated by the fact that what constitutes an outlier depends on the variables that are being considered together. A case with "normal" values when two variables are considered individually may be noteworthy when the two variables are examined together. For example, if you consider weight and height together, a man who weighs 120 pounds and is six feet tall is an outlier, although the case need not be an outlier for the weight and height variables individually. For scale variables, you can plot pairs of variables and look for points that stick out like the one identified in Figure 4-10.

Figure 4-10
Unusual case

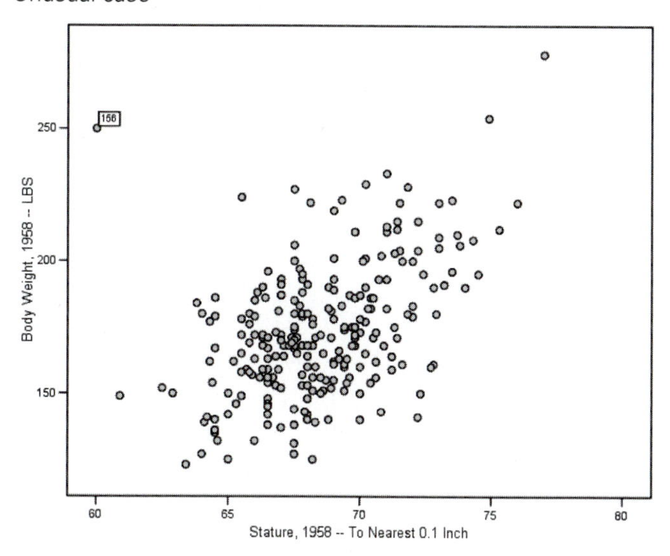

Tip: You can identify points in a scatterplot by specifying a variable in the Label Cases by text box in the Scatterplot dialog box. Double-click the plot to activate it in the Chart Editor. From the Elements menu, choose Data Label Mode or click on the Data Label Mode icon on the toolbar. This changes your cursor to a black box. Click the cursor over the point that you want identified by the value of the labeling variable. To go to that case in the Data Editor, right-click on the point, and then left-click. Make sure that the Data Editor is in Data View. To turn Data Label Mode off, click on the Data Label Mode icon on the toolbar.

Caution

The goal of this chapter was to indicate the importance of scrutinizing a data file prior to analysis. The outlined steps were only suggestions. There are many ways to look for "bad" values. The important point is that you should spare no effort in ensuring that your data file is healthy enough for analysis.

Chapter

5

Transforming Your Data

You've recorded birth dates and now you want to compute ages. Or you recorded individual scores on different tests and you want to average them. You may also want to transform the values of a variable by taking logs or some other mathematical function so that the distribution of the variable would better meet the assumptions needed for a statistical procedure. This chapter shows you how to perform the most common data transformations.

Before you compute new variables, check your data values using the procedures described in Chapter 4; otherwise, when you find a mistake, you'll have to redo everything.

Warning: This chapter won't teach you about the intricacies of the Transform menu. You'll have to use the Help system (and some trial and error) to become proficient. The goal of the chapter is to give you a sample of what can be done and to point you in the right direction.

Computing a New Variable

You can use SPSS Statistics to create a new variable that is based on the values of one or more existing variables. For example, you can compute the percentage of total family income earned by the wife. Or you can compute the body mass index as weight in kilograms divided by the square of height in meters. If you want to perform the same calculation for all of the cases in your data file, the transformation is called **unconditional**. If you want to perform different computations based on the values of one or more variables, the transformation is **conditional**. For example, if you compute an index differently for men and for women, the transformation is conditional. Both types of transformations can be performed in the Compute Variable dialog box.

One Size Fits All: Unconditional Transformation

When you choose Compute from the Transform menu, the Compute Variable dialog box opens. At the top left, you assign a new name to the variable that you will be computing. You do this by clicking in the Target Variable text box and typing. (In Figure 5-1, the new variable is *idealweight*.) You must follow the same rules for assigning variable names as you did when naming variables in the Data Editor.

Figure 5-1
Compute Variable dialog box with new variable and expression

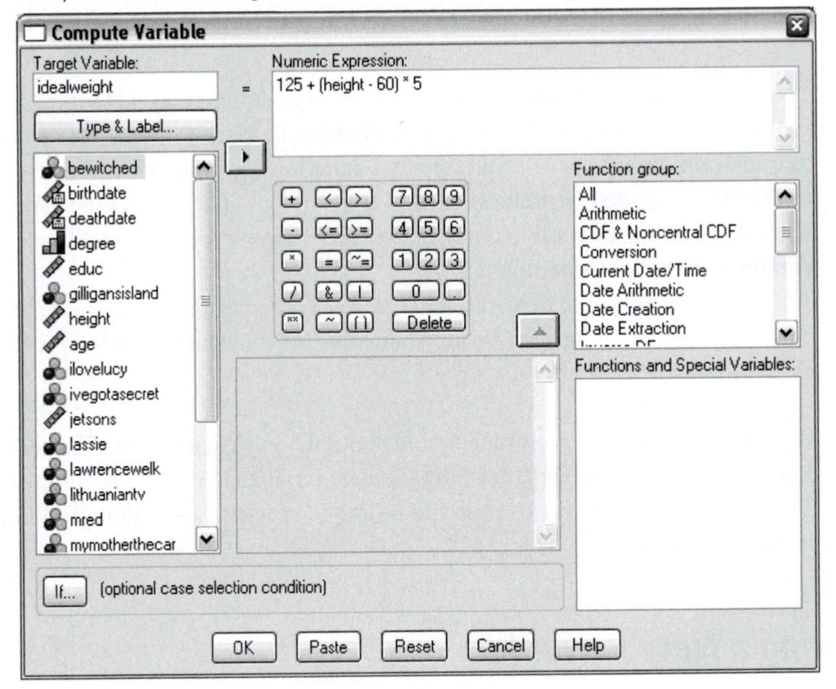

Warning: Although you can type the name of an existing variable and just replace its values, you shouldn't. Think of another variable name. If you reuse the same name and make a mistake specifying the transformation, you'll replace the values of the original variable with values that you don't want. If you don't catch the mistake right away and you save the data file, the original values of the variable are lost. SPSS Statistics will ask your permission to proceed if you try to use an existing variable name.

To specify the formula for the calculations that you want to perform, either type directly in the Numeric Expression text box (click in the box first) or use the calculator pad. Each time you want to refer to an existing variable, click it in the variable list and then click the arrow button. The variable name will appear in the formula at the blinking insertion point. Figure 5-1 shows the computation of a variable *idealweight*, which is 125 pounds, plus 5 pounds for every inch over 60 inches. (Using this formula, more of us can be ideal.) Once you click OK, the variable is added to your data file as the last variable.

Tip: Right-click on any button (except the numbers!) on the calculator pad or on any function for an explanation of what it means.

Using a Built-in Function

You'll notice that to the right of the calculator pad there is a list of functions that you can use in your calculations. Some of the functions, like square root and logarithms, should be familiar. Others are fairly exotic. Scroll through them and right-click on any function for a description. The built-in functions fall into seven main groups: arithmetic functions, statistical functions, string functions, date and time functions, distribution functions, random-variable functions, and missing-value functions.

If you want to use a function, click it when the blinking insertion point is placed where you want to insert the function into your formula, and then click the up arrow button. The function will appear in your formula, but it will have question marks for the arguments. (The **arguments** of a function are the numbers or strings that it operates on. In the expression SQRT(25), 25 is the sole argument of the SQRT function.) Enter a value for the argument, or double-click a variable to move it to the argument list. If there are more question-mark arguments, select them in turn and enter a value, move a variable, or somehow supply whatever suits the needs of the function.

Tip: For detailed information about any function and its arguments, from the Help menu, choose Topics, click the Index tab, and type the word functions. You can then select the type of function that you want.

Assigning Labels and Types

New variables, like old variables, need labels and types. If you don't click the Type & Label button below the new variable name (see Figure 5-1), SPSS Statistics will assign a variable type of numeric *sans* label.

Calculating Time: The Example

If you have dates in your data file, such as date of birth or date of admission to a hospital, you may want to calculate age or length of hospital stay. (Current medical trends suggest minutes might be the right measure for length of hospital stay.) You can do this in the Compute Variable dialog box, using special date functions. For example, if you have two all-encompassing date variables for a case, *birthdate* and *deathdate*, you can calculate age at death using the CTIME.DAYS function, as shown in Figure 5-2. The function CTIME.DAYS returns the time in days corresponding to a "time value," which, in the example, is computed as the difference between two variables in date format. The number of days is divided by 365.25 to convert to years.

Figure 5-2
Computing age at death

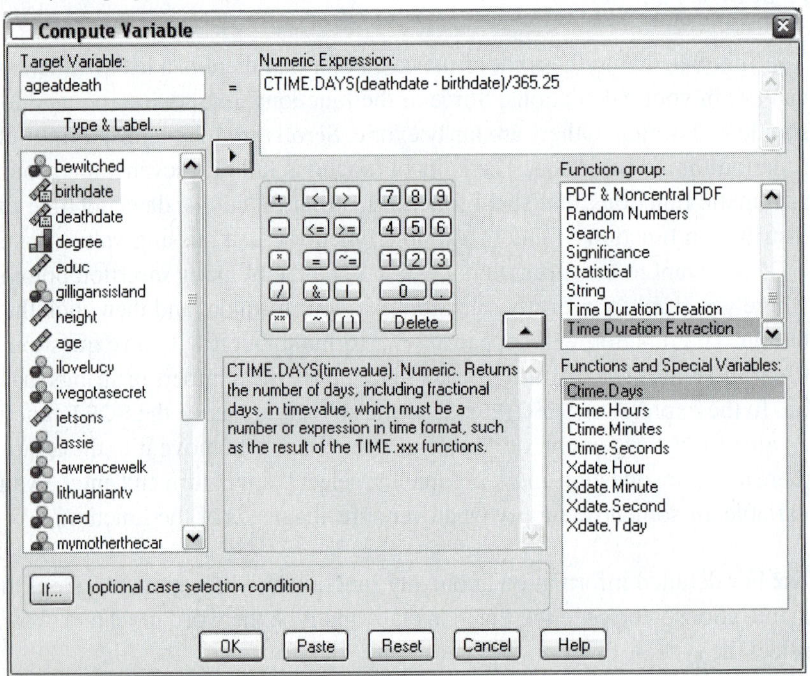

Warning: If you just subtract the two dates, the result will be the time in seconds between the two dates. Date and time functions are very beneficial, despite initial appearances.

If and Then: Conditional Transformations

If you want to use different formulas, depending on the values of one or more existing variables, you have to enter the formula and then click the button labeled If (shown at the bottom in Figure 5-2. This opens the dialog box shown in Figure 5-3).

Figure 5-3
Conditional specification for men

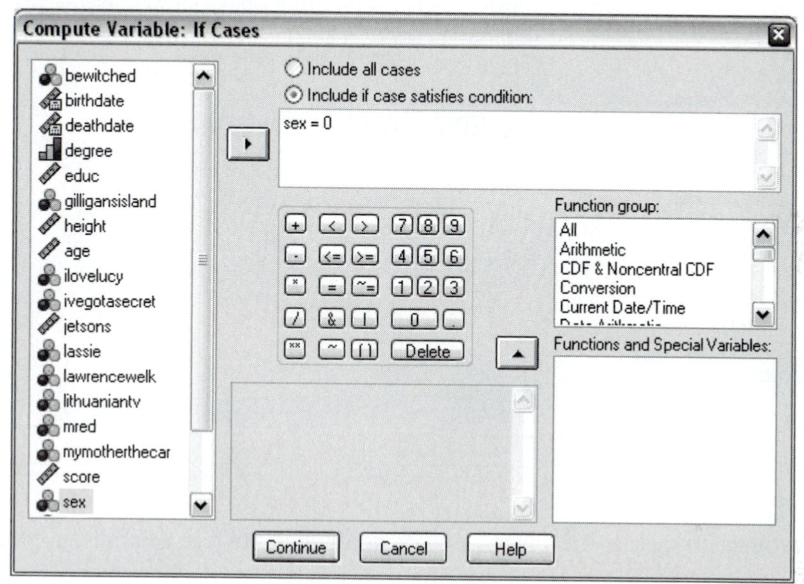

Using the same buttons as before (or by typing), you have to enter the condition that a case must satisfy for the computation to take place. For example, if you want to compute the ideal weight for men (*sex* = 0) as 125 pounds plus 5 pounds for every inch over 60 and for women (*sex* = 1) as 100 pounds plus 5 pounds for every inch over 60, you will have to use the Compute Variable function twice. First, enter the condition sex = 0, as shown in Figure 5-3, and the formula shown in Figure 5-1. Second, enter the condition sex = 1 and the formula adjusted for women. Figure 5-4 shows the dialog box when you've finished the specification for women and clicked the Continue button. Notice that the equation the condition must satisfy is displayed next to the If button.

Figure 5-4
Conditional calculation for women

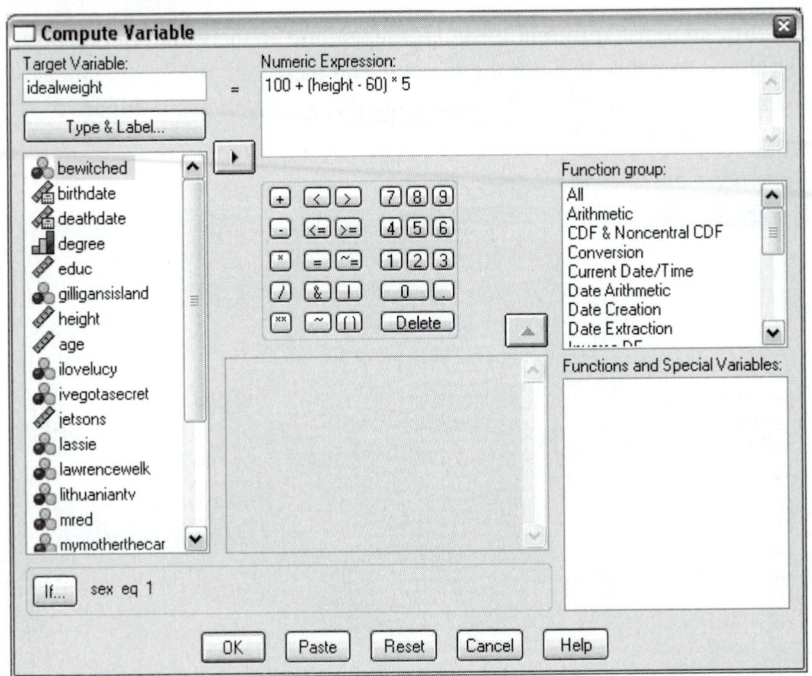

Warning: To raise a variable to a power, don't click * twice. You must use the ** button.

Visual Binning

When you have an ordered variable with many values, such as *age* or *income*, you may want to create a new variable that corresponds to ranges of values of the original variable. For example, you may want to create a new variable called *agegroup*, which has a code of 1 for people of age 25 or under, a code of 2 for people from age 26 through 50, and a code of 3 for people over 50.

The easiest way to group adjacent values for numeric variables is to use the Visual Binning facility. Choose Transform > Visual Binning to open the Visual Binning dialog box. (Figure 5-5 shows the dialog box with the selection of three variables to be binned.) Only variables that are numeric and either ordinal or scale appear in the list.

Figure 5-5
Visual Binning variable selection

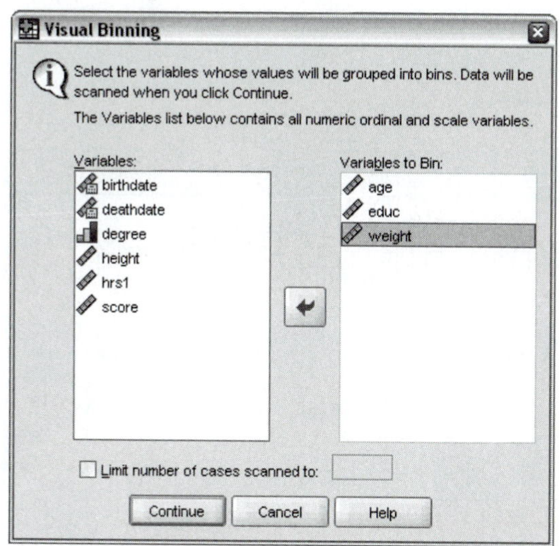

Tip: If you want to bin a numeric variable that doesn't appear on the list, change its type to either ordinal or scale.

Click Continue to open the dialog box shown in Figure 5-6.

Figure 5-6
Visual Binning specifications

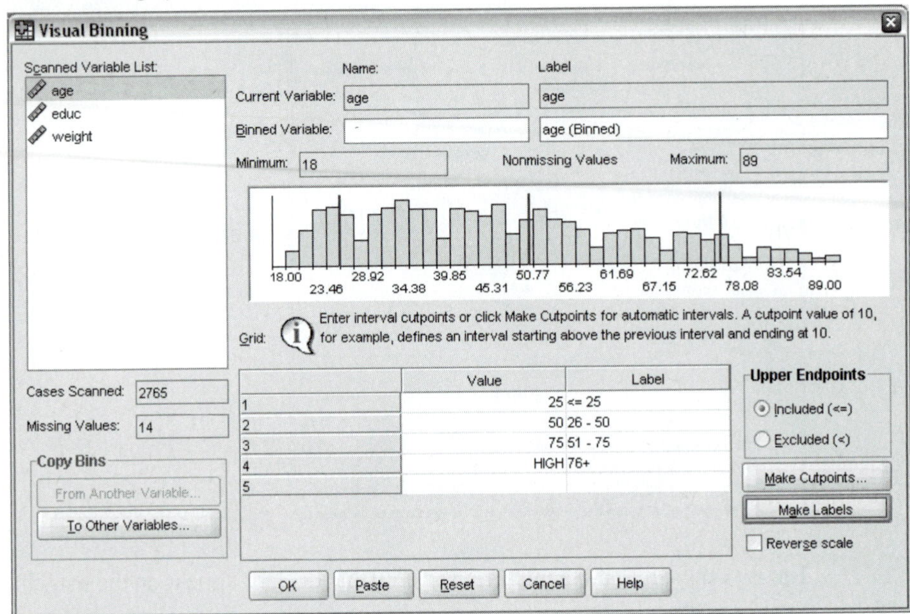

This dialog box is full of specifications. You have to give a name to the new variable and the upper bounds for each interval. You get to select whether the upper value is included in the interval by using the Upper Endpoints alternatives. For example, if you specify 50 as the upper bound of an interval, you get to choose whether 50 is in that interval or the next one.

There are several ways to specify intervals. You can enter numbers, or, by clicking the Make Cutpoints button, you can base the groupings on criteria such as percentiles, equal numbers of cases in groups, or standard deviations above and below the mean.

Once you've entered a cutpoint, a line appears at that point in the histogram scale. You can change the cutpoint by dragging its line in the histogram. Click the Help button for more information.

Warning: You must specify the upper boundary for the last interval that you want to create by using either an actual value or the keyword HIGH; otherwise, values that exceed the last cutpoint are set to system-missing. Go ahead and enter each of your cutpoints on a new row in the *Value* column. HIGH will jump ahead to the end of the list.

You can give labels to each of the categories of the new variable, or you can click the Make Labels button. Figure 5-6 shows the automatic labels generated by the system. If you want to reverse the order of the categories so that the lowest code is for the largest value, you can select the Reverse scale check box.

Changing the Coding Scheme

If you want to use the Visual Binning facility, your variables have to be numeric, and you must want to group only values that are adjacent. You can't create a new variable where the code of 1 is for people under age 25 or over age 75. If you want to create a new variable whose values can be any arbitrary rearrangement of the values of an existing variable, which can be either string or numeric, you have to choose Recode from the Transform menu. You're given a choice of whether you want to create new variables or alter the values of the existing variables. Be prudent; always create new variables.

Tip: If you're enamored with the name of a variable that's undergoing a transformation, go ahead and give the new variable an inferior name. After you've checked your transformation and know that it's right, you can go into the Data Editor in Variable View, delete the original variable, and then give the coveted name to the transformed variable. If you've made a mistake or changed your mind about the transformation, it's perfectly okay to reuse the transformed variable name. You don't want to keep incorrectly transformed variables in your data file. You just might end up using them.

When you choose Transform > Recode > Into Different Variables, the dialog box shown in Figure 5-7 opens. The objective is simple: Every original value has to be transformed into a value of the new variable. In the first dialog box, you select the name of the variable that will be recoded. Then, in the Output Variable Name text box, enter a name for the new variable. Click the Change button, and the new name appears after the arrow in the central list. In Figure 5-7, you are set up to recode the values of a variable called *degree* into a new variable called *degreecategories*. Finally, click the Old and New Values button to enter the recode specifications, as shown in Figure 5-8.

Figure 5-7
Recode into Different Variables dialog box

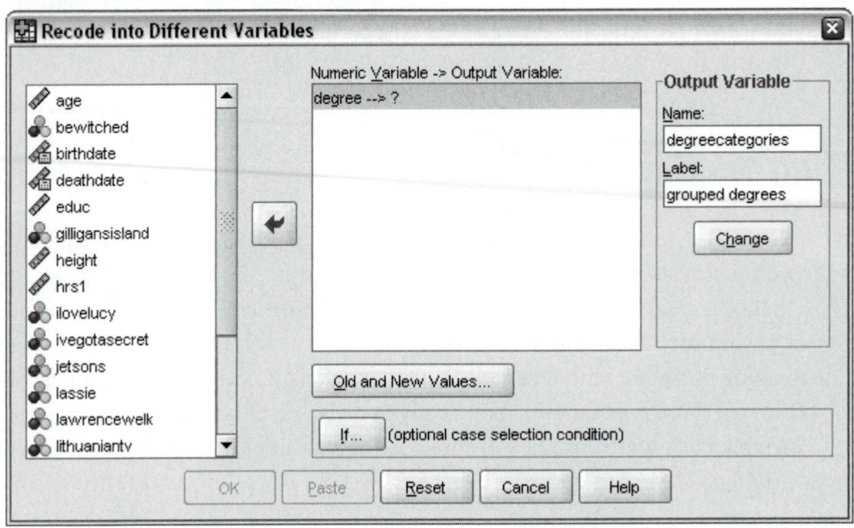

The dialog box in Figure 5-8 recodes the variable *degree* (with values of 0 = *less than high school*, 1 = *high school*, 2 = *junior college*, 3 = *bachelor's degree*, and 4 = *graduate degree*) into the new variable called *degreecategories*. This new variable will have a value of 1 for people for whom *degree* is 0 or 4, a value of 2 for people for whom *degree* is 1 or 3, and a value of 3 for people for whom *degree* is 2.

Figure 5-8
Old and New Values dialog box

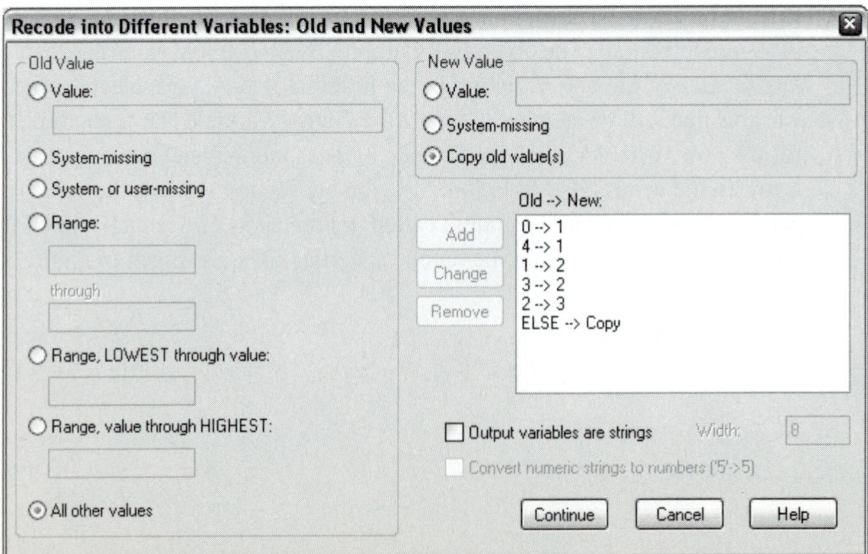

SPSS Statistics carries out the recode specifications in the order they are listed in the Old –> New list. When it finds a specification that applies to a case, it executes the specification and moves on to the next case. A case that starts with a value of 0 for *degree* will be recoded to 1 by the first specification in the list but will *not* be recoded again to 2 by the third specification.

Tip: Always specify all of the values even if you're leaving them unchanged. Select All other values and then Copy old value(s). Remember to click the Add button after entering each specification to move it into the Old –> New list; otherwise, it is ignored.

Checking the Recode

Always check to make sure that your transformation did what you wanted. The easiest way is to make a Crosstabs table of the original variable with the new variable containing recoded values. Figure 5-9 shows such a table for the *degree* and *degreecategories* variables. In the first row, you see that the values for *Less than high school* and *Graduate* are combined into a single category labeled 1. Similarly, in the second row, you see that *High school* and *Bachelor* are combined. Unless you carefully check the transformations that you specify with Recode, you may end up with major problems when the variable you think you're analyzing turns out to be something completely different.

Figure 5-9
Checking the recoded values

Count

		highest degree					
		Less than high school	High school	Junior college	Bachelor	Graduate	Total
grouping of degree	1	400	0	0	0	230	630
	2	0	1485	0	443	0	1928
	3	0	0	202	0	0	202
Total		400	1485	202	443	230	2760

Warning: After you've created a new variable with Recode, go to the Variable View in the Data Editor and set the missing values for each newly created variable.

Changing a String Variable to a Numeric Variable

Suppose that each person in your data file has the proper two-letter state identification for place of residence. Now you want to calculate certain statistics by state. SPSS Statistics won't allow string variables for some procedures, even when they are to be used only for grouping cases. The easiest solution is to assign consecutive integer codes to the states. Open the Automatic Recode dialog box (Transform > Automatic Recode). This is simple compared to some of the other transformation facilities. All you have to do is select a variable and indicate whether new numbers should be assigned in ascending or descending order, based on the sort order of the original variable's values. You also have to give the new variable a name, as shown in Figure 5-10.

The value labels for the new variable will be the original (pre-autorecode) values. This means that if the value 1 was assigned to the string *AK*, *AK* is automatically assigned as the value label for the value 1.

Figure 5-10
Automatic Recode dialog box

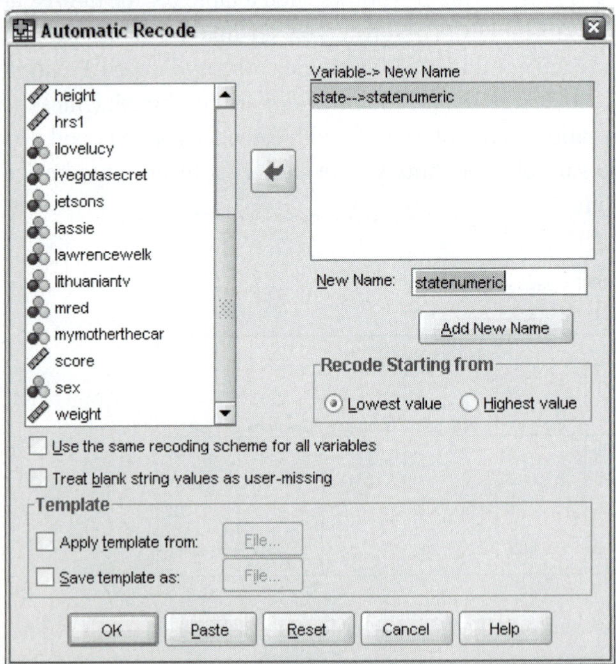

Tip: You can also use Automatic Recode for numeric variables. For example, if you coded the variable *dosage* as 0.25, 0.50, and 1.5, you can create a new variable *dosecategory*, which has the value of 1 for the dose of 0.25, the value of 2 for the dose of 0.50, and the value of 3 for the dose of 1.5.

Ranking Variables

If you have three cases with scores of 100, 50, and 76 on a test, you can replace the actual scores with ranks. If you assign the numerically lowest rank to the highest score, as usual, the new variable has values of 3, 1, and 2 for the three cases. If you assign the highest rank to the highest score, the new variable has values of 1, 3, and 2. To do this in SPSS Statistics, choose Transform > Rank Cases. This opens the dialog box shown in Figure 5-11. The Rank Cases option offers a myriad of choices, which, as always, are described in detail in the Help system. The basic idea is simple, however.

Select the variable or variables to be ranked. If you want to rank cases separately within groups of cases, say for males and females, move a grouping variable into the By list. You can assign ranks in either ascending or descending order.

SPSS Statistics prefixes the letter *R* to the original variable name for a ranked variable. If you don't like the new name, change it in the Data Editor.

Figure 5-11
Rank Cases dialog box

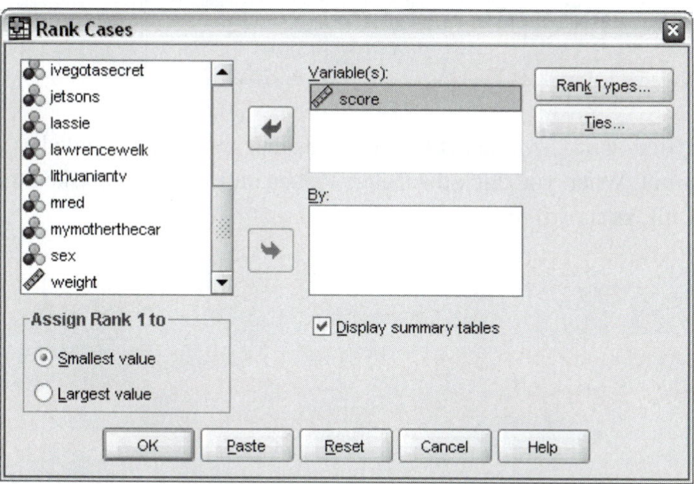

Warning: Only numeric variables can be ranked.

Counting Occurrences

You ask a person to indicate whether he or she watches each of 10 television shows. At some point, you want to know the total number of shows each person watches. You could go to the Compute Variable dialog box and add up the 0's and 1's, if that's how you coded *watch* and *don't watch*. A much simpler and more flexible strategy is to use the Count transformation. You specify a set of variables and the values they must have to be included in the count. The result for each case is the number of variables that satisfy the criterion for being counted. Figure 5-12 shows the Count dialog box.

Figure 5-12
Count dialog box

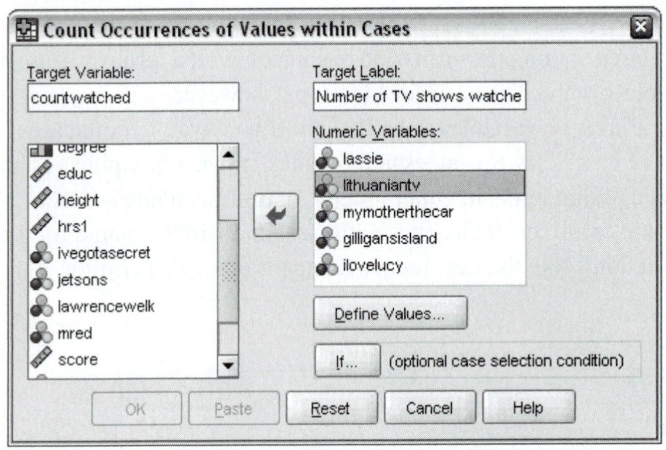

In the main dialog box, you select the variables that are to be evaluated as meeting the criterion or not. When you click the Define Values button, you are asked to specify the criterion itself, as shown in Figure 5-13.

Figure 5-13
Values to Count dialog box

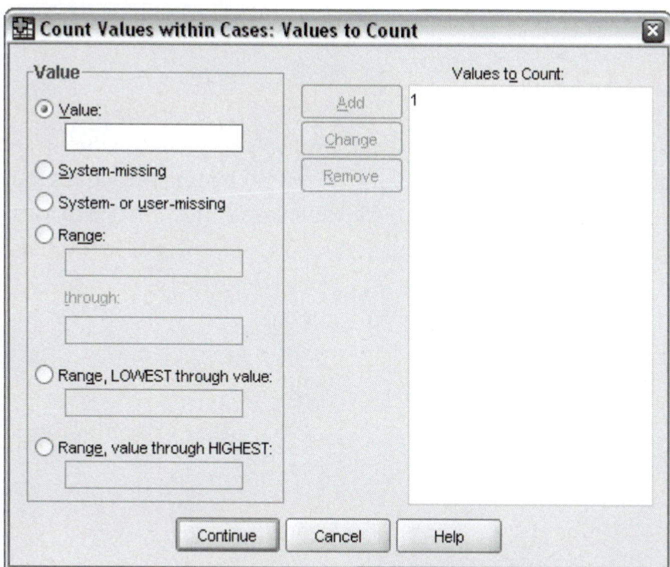

Here, you simply enter the values that you want to be counted and click the Add button after each one. Notice that you can also count missing values or ranges of values by choosing among the alternatives at the left.

Warning: Missing values are treated differently in formulas that you enter into the Compute Variable dialog box and in the statistical functions in the list. If you enter a formula into the Compute Variable dialog box and a case has a missing value for any variable in the equation, the result is a system-missing value. If you select a function, such as the mean, the results are computed using only cases with valid values. For the Count transformation, the count is over all variables with nonmissing values, unless you are counting missing-value codes.

Chapter

6

Describing Your Data

Once you have defined and cleaned your data file, you're ready to use simple statistical procedures and plots to examine your data. The goal is not to look for mistakes (although you may find some problems that you missed in data cleaning), but to look at how often different values occur, to determine what summary measures best describe the data, and to examine simple relationships between variables.

Even if your ultimate goal is to solve problems that have perplexed people in your field or organization for decades, using incredibly complex and innovative statistical analyses, you can learn much by starting simply. *Know thy data* is the golden rule of data analysis.

Examples

- How satisfied are people with their cable service providers?
- What is the average length of the commute to work in different areas of the country?
- How do people who voted in the last presidential election differ from people who were eligible to vote but did not?

In a Nutshell

All variables are not created equal. You must select appropriate summaries and displays based on the characteristics of the variables that you are summarizing, the properties of the statistics that you are computing, your audience, and your personal taste in graphical presentation techniques. (Don't have any yet? Just wait.) Some displays, like pie charts and bar charts, are familiar and don't require much explanation. There are also many specialized charts that are less familiar but are particularly suitable for summarizing statistical information about data.

Newspaper Reading: The Example

Who buys my product? Who uses my services? These are commonly asked questions. The answer may be very simple (*hospitals*) or considerably more complicated (*college-educated households with children under the age of 16 and household income in the top 10%*). Identifying people who buy particular products is important for targeting marketing efforts and for selling additional products to the group. As an example, consider the question *Who reads newspapers?* In this chapter, you'll use descriptive statistics and plots to guide you in expanding your publishing empire. The data are from the 2002 General Social Survey (GSS), a random sample of adults living in the United States (Davis and Smith, 2002).

Examining Tables and Charts of Counts

One of the simplest ways to summarize data is to count how often each of the values of a variable occurs. However, this is informative only for variables with a limited number of distinct values. For each variable in your data file, ask yourself the question *Does it make sense to look at each of the individual values?* If you've recorded household income in 2002 to the nearest dollar, would you really like to see how many people had each income? The prudent answer is *No*. On the other hand, if you asked people to name the state they would most like to live in, you might want to look at how many people chose each state.

Frequency Tables

A good starting point for the newspaper data is to count how often people report that they read a newspaper. Figure 6-1 is a frequency table of responses to the question *How often do you read a newspaper?*

Figure 6-1
Frequency table for newspaper reading

		Frequency	Percent	Valid Percent	Cumulative Percent
Valid	Everyday	369	13.3	40.6	40.6
	Few times a week	219	7.9	24.1	64.8
	Once a week	109	3.9	12.0	76.8
	Less than once a week	121	4.4	13.3	90.1
	Never	90	3.3	9.9	100.0
	Total	908	32.8	100.0	
Missing	NAP	1857	67.2		
Total		2765	100.0		

Each row of the table corresponds to one of the recorded answers. (If no one answers *Once a week*, then *Once a week* wouldn't appear in the frequency table.) From the column labeled *Frequency*, you see that 369 people gave the response *Everyday* and 90 people gave the response *Never*. The row labeled *Total* is the sum for all responses that aren't declared as missing values. Nine hundred and eight people gave a valid response to the question. (Notice the word *Valid* in the upper left corner.)

The row labeled *Missing* tells you how often the NAP missing-value code occurs. Your first reaction should be horror that 1,857 people failed to give a response. What if they were too engrossed in reading newspapers to answer questions?! Your analysis would be doomed. You'll be relieved to find out that the missing-value label *NAP* stands for *Not applicable*. It's used when people are not asked the question. (The GSS doesn't ask all people all questions. Different questionnaires ask different questions so that the interview doesn't become too long.) All of the people who were asked the newspaper question answered, since there are no cases with other missing values.

The next three columns contain percentages. The first column labeled simply *Percent* is the percentage of all cases in the data file with that value. Only 13% of the 2,765 people in the data file said they read a newspaper every day. You don't have to be a mathematical wizard to realize that this number is quite misleading, since even people who weren't asked the question are included. The next column, labeled *Valid Percent*, bases the percentage only on people who actually responded to the question. Almost 41% of people who were asked the question said they read the newspaper every day. You've got almost 60% of the population that you can convert to daily readers.

Warning: A large difference between the *Percent* and *Valid Percent* columns can signal big problems for your study. If the missing values result from people not being asked the question because that's the design of the study, you don't have to worry. If people weren't asked because the interviewer decided not to ask them or if they refused to answer, that's a different matter.

The *Cumulative Percent* column is the sum of the valid percentages for that row and all of the rows before it. It's useful only if the variable is measured at least on an ordinal scale. For example, the cumulative percentage for *Once a week* tells you that 76% of people read a newspaper at least once a week.

The valid data value that occurs most frequently is called the **mode**. For these data, *Everyday* is the modal category, since 369 of the respondents read the newspaper daily. The mode is not a particularly good summary measure, and if you report it, you should always indicate the percentage of cases with that value. For variables measured on a nominal scale, the mode is the only summary statistic that makes sense, but that isn't the case for this variable because there is a natural order to the responses.

Frequency Tables as Charts

You can display the numbers in a frequency table in a pie chart or bar chart. Even though prominent statisticians advise that one should "never use a pie chart," (van Belle, 2002; see also Tufte, 1983), periodical editors pay little heed, since charts are much more photogenic than frequency tables. Figure 6-2 is a pie chart of the frequency table of newspaper readership, with each slice topped with the percentage of cases. The size of the slice depends on the number of cases in each of the categories. Figure 6-3 is a bar chart of newspaper readership. The height of the bar depends on the number of cases in each of the categories.

Figure 6-2
Pie chart of newspaper reading

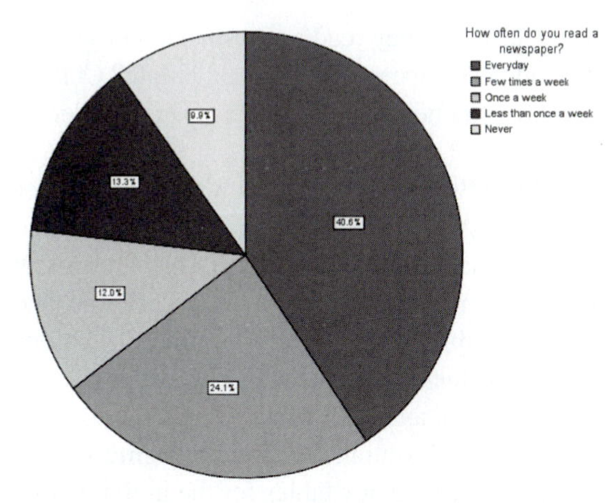

Figure 6-3
Bar chart of newspaper reading

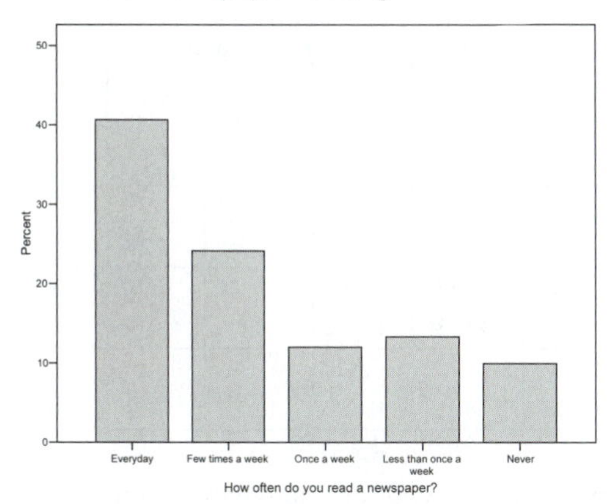

Warning: If you create a pie chart by choosing Descriptive Statistics > Frequencies, a slice for missing values is always included. Use Graphs > Legacy Dialogs > Pie or Graphs > Chart Builder and then click on a pie chart if you don't want to include a slice for missing values by default.

Examining Two-Way Tables of Counts

Now you know how often people claim they read a newspaper, but you still have no idea of what kind of people read newspapers. Are they old? Male? College educated? To find out who your readers are, you need to look at newspaper readership together with other variables.

Figure 6-4 is a crosstabulation, a two-way table of counts, for newspaper readership and gender. A crosstabulation gets its name, not from its disposition, but from the fact that the values of the two variables are considered together; they are crossed with each other. Gender is the row variable, since it defines the rows of the table, and frequency of reading the newspaper is the column variable, since it defines the columns. Each of the 10 unique combinations of the values of the two variables defines a cell of the table. From Figure 6-4, you see that 186 men and 183 women said they read a newspaper daily. The numbers in the *Total* row and column are called **marginals** because they are in the margins of the table. They are frequency tables for the individual variables.

Figure 6-4
Crosstabulation of newspaper reading and gender

			How often do you read the newspaper?					
			Everyday	Few times a week	Once a week	Less than once a week	Never	Total
	Male	Count	186	94	43	41	37	401
		Row %	46.4%	23.4%	10.7%	10.2%	9.2%	100.0%
	Female	Count	183	125	66	80	53	507
		Row %	36.1%	24.7%	13.0%	15.8%	10.5%	100.0%
Total		Count	369	219	109	121	90	908
		Row %	40.6%	24.1%	12.0%	13.3%	9.9%	100.0%

Tip: Don't be alarmed if the marginals in your crosstabulation aren't identical to the frequency tables for the individual variables. Only cases with valid values for both variables are in the crosstabulation, so if you have cases with missing values for one variable but not the other, they will be excluded from the crosstabulation. Respondents who tell you their gender but not how often they read a newspaper are included in the frequency table for gender but not in the crosstabulation of the two variables. Notice that only the 908 cases who answered the newspaper-reading question are included in the table.

Percentages

Although the counts in the cells are the basic elements of the table, they are usually not the best choice for reporting findings because they cannot be easily compared if there are different totals in the rows and columns of the table. For example, if you know that186 men and 183 women read the newspaper every day, you can conclude little about the relationship between the two variables unless you also know the total number of men and women in the sample.

For a crosstabulation, you can compute three different percentages:

- Row percentage: the cell count divided by the number of cases in the row times 100
- Column percentage: the cell count divided by the number of cases in the column times 100
- Total percentage: the cell count divided by the total number of cases in the table times 100

The three percentages convey different information, so it's important to choose the correct one for your problem. If one of the two variables in your table can be considered an independent variable and the other a dependent variable, make sure the percentages sum to 100 for each category of the independent variable.

Tip: A variable is termed **dependent** if its values are thought to depend on the values of another variable, called the **independent** variable. For example, income may depend on education, so education is the independent variable and income is the dependent variable.

Since, after conception, gender is always an independent variable and it defines the rows of the table, you want to calculate row percentages. They'll tell you what percentage of men and what percentage of women fall into each of the readership categories. This percentage is not affected by unequal numbers of males and females in your sample. From the row percentages in Figure 6-4, you see that 46% of men and 36% of women read a newspaper daily. That's what you want to know. Men in the sample are more likely to read a newspaper daily than are women.

For each category of newspaper readership, column percentages tell you the percentage of the category that are men and the percentage that are women. For example, in the *Everyday* category, the column percentages are 50.4% for men and 49.6% for women. That doesn't tell you that men and women are almost equally likely to read a newspaper every day. That tells you that of the people who read a newspaper every day, half are men and half are women. The column percentages depend on the number of men and women in the sample as well as how often they read a newspaper. If men and women have identical reading patterns but there are twice as many men in your survey as women, the column percentage for men will be twice as large as the column percentage for women. You can't draw any conclusions based on only the column percentages.

Tip: If you use row percentages, compare the percentages within a column. If you use column percentages, compare the percentages within a row.

Multiway Tables of Counts as Charts

You can plot the percentages in Figure 6-4 in a clustered bar chart like that shown in Figure 6-5. For each category of newspaper readership, there are separate (and not necessarily equal) bars for men and women. In the terminology of the Graph menu, gender is called the cluster variable. The values plotted are the percentage of all men and the percentage of all women who gave each response. You can easily see that men are more likely than women to read the newspaper each day. Although the same information is in the crosstabulation, it's easier to see in the bar chart.

Figure 6-5
Clustered bar chart

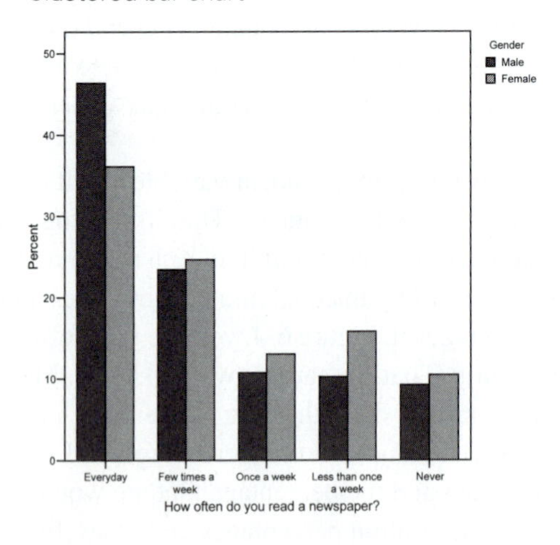

Tip: Always select percentage in the clustered bar chart dialog boxes; otherwise, you'll have a difficult time making comparisons within a cluster, since the height of the bars will depend on the number of cases in each subgroup. For example, you won't be able to tell if the bar for men who always read newspapers is higher because men are more likely to read a newspaper daily or because there are more men in the sample.

Control Variables

You can examine the relationship between gender and newspaper readership separately for each category of another variable, such as education (called the **control variable**). Figure 6-6 is a crosstabulation of gender and readership for each of three categories of highest degree. You see that the largest difference in daily reading between men and women is for college graduates (68% for men, 42% for women). The percentages are almost equal for those with less than a high school education.

Figure 6-6
Crosstabulation of newspaper reading by gender with degree as control

				How often do you read a newspaper?					
				Everyday	Few times a week	Once a week	Less than once a week	Never	Total
Degree categories	Less than high school	Male	Count	21	11	8	11	8	59
			Row %	35.6%	18.6%	13.6%	18.6%	13.6%	100.0%
		Female	Count	27	13	13	9	14	76
			Row %	35.5%	17.1%	17.1%	11.8%	18.4%	100.0%
	High school or junior college	Male	Count	100	69	29	23	25	246
			Row %	40.7%	28.0%	11.8%	9.3%	10.2%	100.0%
		Female	Count	108	76	40	59	32	315
			Row %	34.3%	24.1%	12.7%	18.7%	10.2%	100.0%
	Bachelor's or graduate	Male	Count	65	14	6	7	3	95
			Row %	68.4%	14.7%	6.3%	7.4%	3.2%	100.0%
		Female	Count	48	36	13	12	6	115
			Row %	41.7%	31.3%	11.3%	10.4%	5.2%	100.0%

As the number of variables in a crosstabulation increases, it becomes unwieldy to plot all of the categories of a variable. Instead, you can restrict your attention to particular responses. For example, if you want to see the relationship between education, gender, and newspaper reading, you can plot only the percentage of daily readers in each of the education and gender combinations.

In Figure 6-7, for each of five categories of education, you see the percentage of men and the percentage of women who read a newspaper daily. (The highest degree variable was recoded into three categories for the crosstabulation so that the chart would be smaller in size.) The horizontal line is at 40.6%, representing the percentage of people in the entire sample who read a newspaper daily. For women, there doesn't seem to be a strong relationship between education and newspaper reading. Male college graduates, however, are big daily newspaper readers (sports pages? *Wall Street Journal*?).

Figure 6-7
Clustered bar chart of gender and education with average line

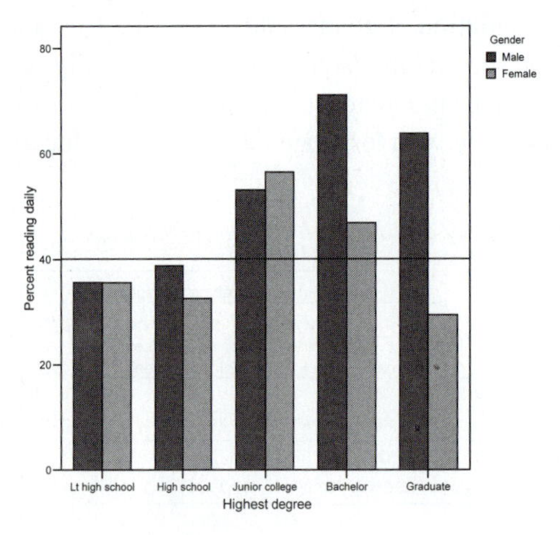

Summarize Scale Variables

If you want to look at variables such as age and income, you won't find frequency tables or crosstabulations useful. Imagine what a pie chart for age would look like; it would have many tiny slices. A bar chart would have too many bars to be useful. If the values of the variable can be ordered from smallest to largest, you can combine adjacent values and make a pie chart or bar chart of decades of age. You can't combine adjacent values if the variable is nominal, since adjacency of codes doesn't mean anything. If codes are assigned to states alphabetically, there's no reason to combine Alaska, Alabama, and Arkansas into a single category just because they all start with *A*.

Histograms

A histogram is a display that combines adjacent values for you and then draws a bar whose height depends on the number of cases that fall in that interval. From a histogram, you see the distribution of the values of a variable. You can tell whether there is a single peak or multiple peaks. You can also tell whether the distribution is approximately **symmetric**, meaning that the two sides are mirror images of each other. If a distribution has one tail that extends farther from the center than the other tail, the distribution is **skewed**. A distribution with a tail toward larger values is said to be skewed to the right; a distribution with a tail toward smaller values is said to be skewed to the left. Look for values that are far removed from the rest, called **outliers**. Always try to determine the reasons for outliers. Make sure that they are not data errors.

Figure 6-8 shows an age histogram for all people who answered the newspaper-reading question. The overall age distribution has a tail toward larger values because only adults over the age of 18 are included in the survey. After age 50, the number of people in the sample steadily decreases. Figure 6-9 shows a histogram of age, with cases subdivided by frequency of newspaper reading. Note that for daily readers, the age histogram is more symmetric. The bars between ages 35 and 75 are almost equal in height. That's because there are a lot of daily newspaper readers in the older age groups.

Figure 6-8
Age histogram for all cases questioned

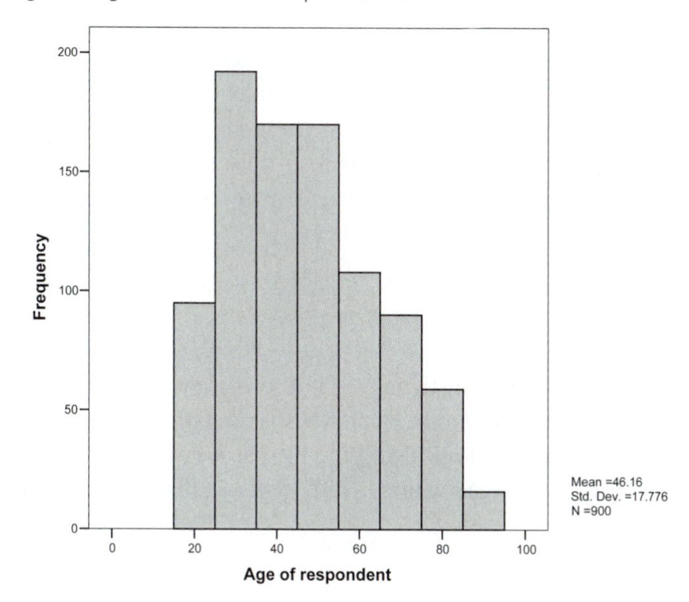

Figure 6-9
Age histogram by newspaper readership

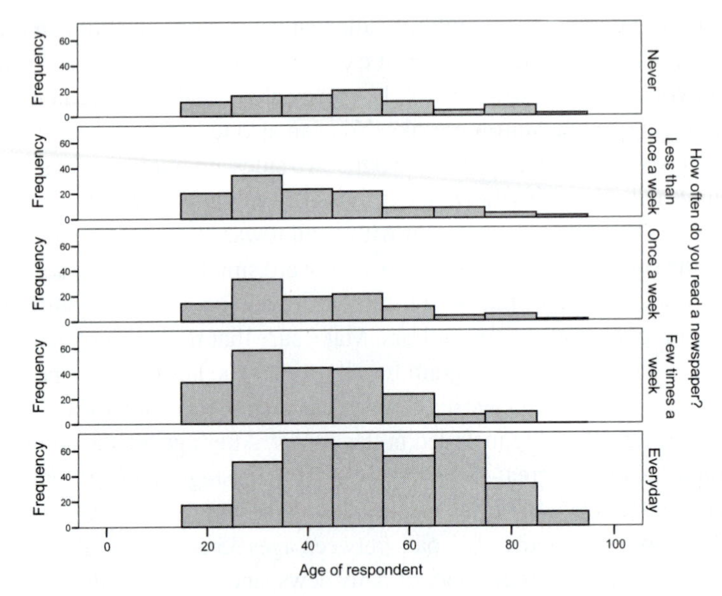

Warning: For integer data, you can get odd-looking histograms because even equal intervals may not include the same number of integer values. If you have outlying points, you may get a histogram with most values bunched together. In both cases, you should use the Chart Editor to specify convenient interval widths and starting points. If you use a small number of intervals, you are lumping observations and sacrificing detail; if you use a large number of intervals, you are seeing too much detail. The best way to tell what's "just right" is to try several different groupings and select the most informative one, the one with just enough detail.

Stem-and-Leaf Plots

A variant of the histogram that preserves more information about the actual values is the stem-and-leaf plot. In a histogram, you have a bar that represents just the cases with values in that interval. In a stem-and-leaf plot, the bar is composed of numbers that correspond to the actual observed values. Think of it as a histogram turned on its side, with a lot more detail.

Figure 6-10 is a stem-and-leaf plot of ages for people who never read the newspaper. Each observed value is subdivided into two components: the leading digit or digits (called the **stem**) and the trailing digit (called the **leaf**). For example, the age of 54 has a stem of 5 and a leaf of 4. If there are many cases, each stem is divided into two rows. The first row of each pair has cases with leaves of 0 through 4, while the second row has cases with leaves of 5 through 9. There can be additional rows at the beginning and the end of the stem-and-leaf plot that correspond to values that are far removed from the rest. The actual values would be displayed in parentheses.

Figure 6-10
Stem-and-leaf plot

```
AGE OF RESPONDENT Stem-and-Leaf Plot

 Frequency    Stem &  Leaf

    20.00        2 .  01122333444555566688
    15.00        3 .  002233355677799
    18.00        4 .  001233345555678999
    18.00        5 .  001112344455566799
     5.00        6 .  11158
     6.00        7 .  235778
     6.00        8 .  112479

 Stem width:  10
 Each leaf:       1 case(s)
```

At the bottom of the stem-and-leaf plot, you see the **stem width**. That's the value all stems have to be multiplied by if you want to get back to the original data values. Leaves have to be multiplied by one-tenth of the stem width. The sum of the two numbers is the original data value for that leaf. In this example, the stem width is 10, which means that a stem of 8 and a leaf of 4 represents an age of 84 $(8 \times 10 + 4 \times 1)$. For a three-digit number, such as blood pressure, the stem can be a multiple of 100 and the leaf a multiple of 10. A stem of 1 and a leaf of 5 correspond to a systolic blood pressure of 150. All systolic blood pressures in the 150's are represented by the same symbols: a stem of 1 and a leaf of 5. You can't distinguish between a systolic blood pressure of 151 and a systolic blood pressure of 159.

Tip: Stem-and-leaf plots are available only in the Analyze > Descriptive Statistics > Explore procedure. You can create separate stem-and-leaf plots for subgroups of cases by specifying the grouping variable in the Factor List. For example, you can get separate stem-and-leaf plots of age for each category of newspaper readership.

Variability and Central Tendency

Although frequency tables, bar charts, pie charts, and histograms are all useful displays, you also need summary measures that concisely describe the data. You want to be able to characterize readers and nonreaders of newspapers with a couple of well-chosen numbers. You want to answer questions like *What's a typical value?* and *How similar are the values?*

Warning: It's not always possible to adequately describe the values of a variable with one or two summary measures. For some distributions, there is no average or typical value.

The arithmetic mean and the variance (or the standard deviation) are the most commonly used measures of central tendency and of variability, respectively, although they are not the best for every situation.

- The **arithmetic mean** is calculated by summing the values of a variable for all observations and then dividing by the number of observations.
- The **variance** is calculated by finding the squared difference between an observation and the mean, summing it for all cases and then dividing by the number of observations minus 1.
- The **standard deviation** is calculated as the square root of the variance.

The **median** is another measure of central tendency. It's the middle value when observations are ordered from the smallest to the largest. For example, the median age for nine newspaper readers aged 31, 40, 55, 57, 60, 63, 63, 70, 79 is 60 years. If you have an even number of observations, the median is the average of the two numbers in the middle. The median of 20, 24, 26, and 80 is 25.

In Figure 6-11, you see descriptive statistics separately for each of the newspaper readership groups. The average age of people who read the newspaper every day is close to 53, compared to an average age of about 40 for people who read the newspaper less than once a week. People who never read the newspaper have an average age of 46. The pattern of average age is interesting. The most frequent and the least frequent readership groups have the largest average ages. The other three groups have similar averages.

Figure 6-11
Summary statistics for age

AGE OF RESPONDENT

		Mean	Median	Std. Deviation	Variance	Min	Max	Range	Std. Error of Mean	N
How often do you read the newspaper?	Everyday	52.60	51.00	17.78	316.23	19	89	70	.928	367
	Few times a week	40.75	39.00	15.51	240.49	19	84	65	1.053	217
	Once a week	41.72	40.50	16.56	274.13	20	87	67	1.593	108
	Less than once a week	40.35	36.00	16.03	257.04	19	89	70	1.464	120
	Never	45.99	45.00	18.17	330.13	20	89	69	1.937	88
	Total	46.16	44.00	17.78	316.00	19	89	70	.593	900

Tip: The first statistic you should look at is the number of cases in each group. It's easy to get carried away spinning theories about what's going on before realizing that some of the groups have only a handful of cases. If your groups are ordered and you have small numbers of cases in adjacent groups, you can combine adjacent groups so that you have larger numbers of cases in the groups.

Except for the *Less than once a week* group, the mean and median are fairly close in value. That's not always true. If the distribution of values is more or less symmetric, the mean and median are usually not too far apart. (For a perfectly symmetric distribution, the mean and median are equal.) If a distribution has a tail, on the other hand, the mean and median will differ in value. If the distribution has a tail toward larger values, the mean will be larger than the median. The opposite is true if the tail is toward smaller values. The reason for this is that numbers that are far removed from the rest can have a big effect on the mean but not on the median. Consider these five salaries in a company (in thousands): $50, $60, $65, $80, $500. The arithmetic mean is $151, the median is $65. The generous half-a-million dollar salary has a big effect on the mean, since that statistic is calculated using all of the actual salaries. The median is based on order, so the large value has a much smaller effect. In fact, the largest salary can be replaced by any number greater than $65 and the median remains unchanged.

Tip: Always make a histogram of your data values before you compute summary measures. Means and medians can be very misleading if your distribution doesn't have a single peak. For example, it's possible that two types of people may never read a newspaper: college students who get all the information that they deem newsworthy on the Web and feeble old people whose vision and alertness are impaired. The average age of the two groups combined may be around 50, but that's hardly a good indicator of the age of people who never read the newspaper.

From the variances in Figure 6-11, you see that age varies most in the *Never* group. The standard deviations are not too different among the groups. The smallest standard deviation is 15.5 years, the largest is 18.2 years. You should always scan the minimum and maximum values for a group to make sure that your valiant data-cleaning efforts were successful. The range is not a very good measure of variability because it depends on just two values, the minimum and the maximum.

The variance and standard deviation play a major role in describing data because of their importance in the normal distribution. (See "Normal Distribution" on p. 104 and Chapter 7 on hypothesis testing.) In many situations, they are not particularly good measures of variability because outlying observations exert a strong influence on their values. Percentiles (see "Percentiles" on p. 102) are often more useful in describing variability.

The **standard error** of the mean is calculated by dividing the standard deviation by the square root of the sample size in the group. It's a measure of how much you expect sample means to vary. You know that if you repeat the same survey or experiment using different people, you're going to get different results. The standard error is a measure of how much you expect the mean to vary when you take different samples of the same size from the same population. Using the standard error, you can calculate confidence intervals for the population value of the mean.

Tip: Since the mean can depend a lot on outlying observations, there are various modifications of the mean that attempt to diminish the effect of such outliers. These statistics, called **M-estimators**, are found in the Analyze > Descriptive Statistics > Explore procedure. They assign different weights to cases based on their distance from the mean or median. You can also calculate a **trimmed mean**, which excludes 5% of the largest and smallest observations from the computation of the mean.

You can examine the relationship between newspaper readership, age, and additional categorical variables by making separate tables of newspaper readership and age for each value (or combination of values) of the additional variable(s). For example, Figure 6-12 shows average ages when people are classified by both newspaper readership and highest degree attained in school. You see that for each category of education, people who read the newspaper every day are the oldest.

Figure 6-12
Summary statistics for age by education

			How often do you read a newspaper?					
			Everyday	Few times a week	Once a week	Less than once a week	Never	Total
Highest degree	Less than high school	Mean	60.57	43.13	47.62	49.60	56.77	53.16
		Median	69.00	34.50	42.00	46.50	53.50	53.00
		N	47	24	21	20	22	134
	High school or junior college	Mean	52.30	39.51	39.36	38.63	41.75	44.35
		Median	50.00	37.00	37.00	35.00	41.50	43.00
		N	208	144	69	81	56	558
	Bachelor's or graduate	Mean	49.81	43.22	43.89	37.95	48.00	46.57
		Median	50.50	42.00	42.50	36.00	49.00	45.00
		N	112	49	18	19	9	207
	Total	Mean	52.60	40.75	41.72	40.35	46.20	46.18
		Median	51.00	39.00	40.50	36.00	45.00	44.00
		N	367	217	108	120	87	899

Tip: Much of SPSS Statistics output is displayed in **pivot tables**. What that means to you is that you can easily switch rows and columns, change labels, and customize the table to your specifications. If you double-click a table, you switch into editing mode and can use the Pivot menu and the Pivoting Trays to move the entries. Try it. For more information on editing pivot tables, run the tutorial "Working with Output."

Plotting Pairs of Variables

You can make a bar chart of the means for each group, as shown in Figure 6-13, but a line chart of means is a better display. From the line chart in Figure 6-14, you can easily see what you've already found. At each education level, older people tend to read the newspaper daily. It's easier to see the relationship between newspaper readership and age in a line chart than in a table.

Figure 6-13
Bar chart of average age

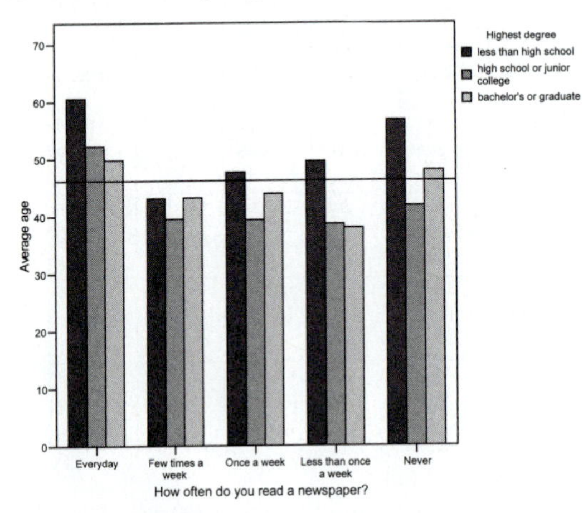

Figure 6-14
Clustered line charts of means

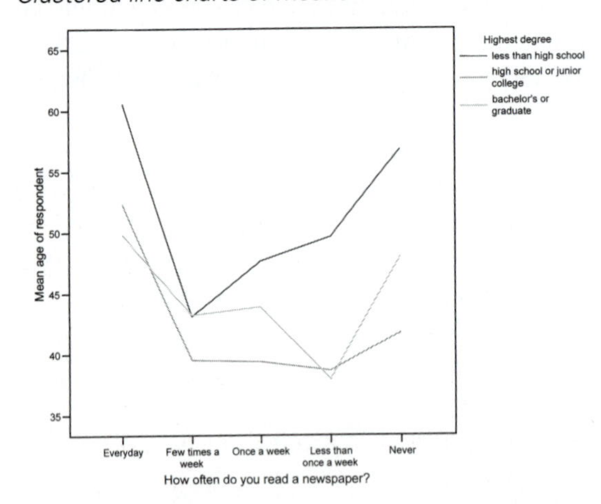

Warning: It's easy to manipulate the scale of a bar chart or line chart to make small differences appear large. If you start the vertical scale at 35 instead of 0, the differences between means will be magnified. Beware of this ploy when reading charts, and certainly don't do it yourself. Remember, this is a warning, not a tip!

Creating OLAP Cubes

When you classify your cases based on many categorical criteria, pivot tables can become large and unwieldy. It's hard to find just the number you want. If your goal is not to uncover relationships between variables but simply to report statistics (such as average age), you can create what's called an **OLAP cube**. (OLAP stands for *OnLine Analytical Processing*.) The advantage of an OLAP cube, such as the one in Figure 6-15, is that it can be manipulated interactively when you activate it in the SPSS Statistics Viewer.

Figure 6-15
OLAP cube of age

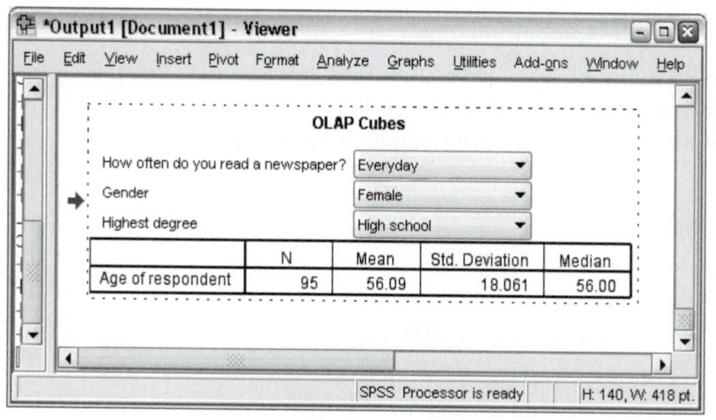

The cube in Figure 6-15 is already activated, as you can see by the cross-hatched border and the red arrow. You have three grouping variables: degree, gender, and newspaper readership. The table displays statistics for the particular combination of categories listed above it. Because the table is activated, each category is actually a pop-up menu, right there in the SPSS Statistics Viewer. If you switch the degree category from *High school* to *Less than high school*, the statistics displayed will change, right before your eyes.

Tip: OLAP cubes are obtained using Analyze > Reports > OLAP Cubes.

Percentiles

Percentiles are summary statistics based on cumulative counts. The idea behind these is very simple: you figure out values that split the distribution of the cases in various ways. For example, you calculate the age value below which 90% of the cases fall. Such values are called percentiles. The most popular percentile by far is the **median** or 50th percentile. Half of the cases have values below the median and half of the cases have values above the median. Many standardized tests give you percentile rankings as well as scores. That lets you determine where you stand in the pool of test-takers. A score at the 90th percentile is great, since only 10% of people scored higher than you did. A percentile of 15 is nothing to brag about, since 85% of people scored better than you did.

Figure 6-16 shows age percentiles for each category of newspaper reading. From the 75th percentile, you see that 25% of people who read the newspaper every day are older than 68 years and that 25% are younger than 39 years. Fifty percent of daily readers are between the ages of 39 and 68, compared to 50% of few-times-a-week readers who are between 27 and 50 years of age. Since the 25th, 50th, and 75th percentiles divide your sample into four equal groups, they are called **quartiles**. The difference between the 75th percentile and the 25th percentile is called the **interquartile range**. The interquartile range measures the spread of the observations. Small values tell you that observations don't vary much; large values indicate more spread.

Figure 6-16
Percentiles of age

		Age				
		How often do you read the newspaper?				
		Everyday	Few times a week	Once a week	Less than once a week	Never
Percentiles	5	25.00	20.00	22.00	21.05	22.00
	10	28.00	22.80	23.90	22.10	23.90
	25	39.00	27.00	26.25	28.25	30.50
	50	51.00	39.00	40.50	36.00	45.00
	75	68.00	50.00	52.75	50.00	55.75
	90	76.00	64.00	64.70	65.90	77.00
	95	81.60	72.20	75.00	75.75	81.55

Boxplots: Displaying Percentiles

When you use bar charts to compare means of groups, you ignore a lot of information about the distribution of the data. If you present histograms for each of the groups, you suffer from the opposite problem of too much information. A **boxplot** is a display that conveys much more information than a bar chart but less than the individual histograms. It's based on the percentiles of each group and is particularly useful for comparing several groups, although it won't show you gaps in the distribution of values and may not identify outliers that you might notice on a histogram.

Figure 6-17 shows boxplots of age for the five categories of newspaper readership. The horizontal line in the middle of each box is the category median, the 50th percentile. The bottom and top edges of the box are the 25th and the 75th percentiles. The length of the box is the interquartile range. The **whiskers** that extend from the top and bottom of the box represent the smallest and largest values that aren't outliers. Points outside the whiskers are identified as outliers (plotted with the symbol *O*) if they are between 1.5 and three box lengths from the edge of the box. They're identified as extremes (*) if they are more than three box-lengths away.

Figure 6-17
Boxplots of age

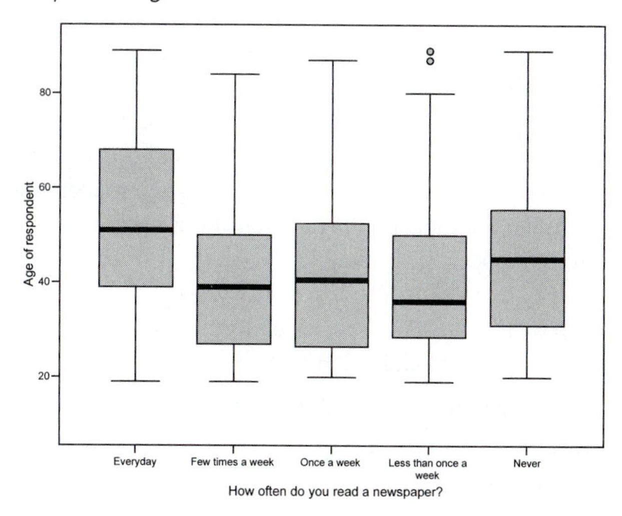

Tip: You can label outlying points with Label Cases By in the Explore dialog box (choose Analyze > Descriptive Statistics > Explore). If you don't select a labeling variable, the sequence number is used.

For most groups, you see that the median is pretty much in the center of the box. This means that the distribution of the variable is more or less symmetric. If the median is closer to the bottom of the box than to the top, the data are said to be **positively skewed**. This means that it has a tail toward larger values. If the median is closer to the top of the box than to the bottom, the opposite is true; the distribution is **negatively skewed**. The length of the tail is shown by the whiskers and the outlying and extreme points. For the age variable, the whisker extends toward larger ages. That's because the minimum age for inclusion is 18, while there is no upper limit set for age. (The General Social Survey sets all ages over 89 to 89, a very strange quirk that dates back to the prehistoric days of card sorters and punch cards!)

Warning: A special set of percentiles called Tukey's hinges are used for the boxplot display. They may differ a little from the other percentile estimates. Don't worry about it. The same percentile can be calculated in several slightly different ways, but its interpretation remains the same. For large sample sizes, most methods for calculating percentiles give similar answers.

Normal Distribution

Observed data distributions are often compared to a theoretical distribution called the **normal distribution**, since it plays an important role in statistical tests. Many variables such as weight, blood pressure, test scores, and running times in marathons are approximately normally distributed. Figure 6-18 is a histogram of height obtained from a sample of 240 men. Unlike the age distributions for newspaper readers, this distribution has a peak in the middle, and the number of observations decreases the farther away you go from the peak.

The curve superimposed on the data is a theoretical normal distribution with the same mean and variance. It is bell-shaped and symmetric.

- The mean and median are in the middle of the distribution and are equal.
- If you know the mean and standard deviation of a normal distribution, you know the exact distribution of values.

For example, 68% of the values in a normal distribution are within ±1 standard deviation of the mean, and 95% are within ±1.96 standard deviations.

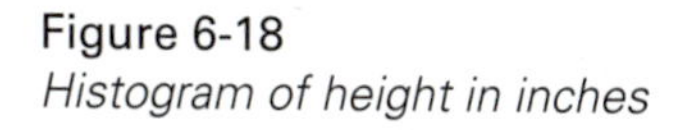

Figure 6-18
Histogram of height in inches

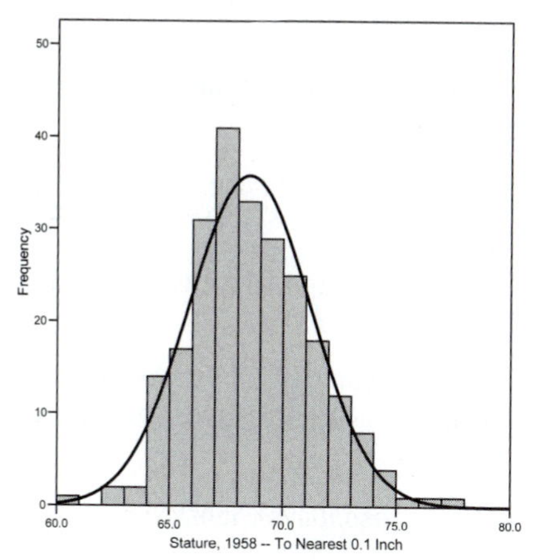

Because some statistical procedures require that data be a sample from a normal population, you can produce special plots that can be used for looking for deviations from normality. These are discussed in Chapter 8. If your data are not normally distributed, you can attempt to transform the data values to achieve normality. This is discussed in Chapter 5.

Standard Scores

Data values from any distribution can be expressed in standardized (z-score) form. The standard score tells you how many standard deviation units a particular case is from the mean. The mean of standard scores is 0, and the standard deviation is 1. A standard score of 0 tells you that a case has a value equal to the mean. A standard value of 2 indicates that a case is two standard deviations above the mean, whereas a score of –2 indicates that the case is two standard deviations below the mean. Standard scores are useful for comparing the values of a case on several variables. If you know that a woman has a standard score of –1.5 for a test of manual dexterity and a score of 2.2 for common sense, you know that she is less dexterous than most but blessed with common sense. If you know that manual dexterity scores and common sense scores are normally distributed, you have even more information because you know the percentage of cases above and below the standard scores.

Tip: Use the Analyze > Descriptive Statistics > Descriptives procedure to calculate standard scores and to automatically add them to your data file. Use Transform > Visual Bander to identify cases that are a certain number of standard deviations from the mean.

Navigating the Menus

The main procedures for describing data are found on the Analyze menu. Click Help in the main dialog box of each procedure for an overview of the procedure. Here's a short list of tasks and the best procedures for them.

- Making a frequency table: Analyze > Descriptive Statistics > Frequencies (your only choice for a frequency table).
- Identifying the five cases with the largest and the five cases with the smallest values: Analyze > Descriptive Statistics > Explore.
- Identifying any number of cases with large and small values: Sort the data by the values of the variable and list information using Analyze > Reports > Report Summaries in Rows.
- Making a stem-and-leaf plot: Analyze > Descriptive Statistics > Explore (your only choice).
- Making a boxplot: Analyze > Descriptive Statistics > Explore, or Graphs > Chart Builder.
- Making a pie chart: Graphs > Chart Builder (or Analyze > Descriptive Statistics > Frequencies if you don't mind a slice for missing values).
- Making a bar chart for a single variable: Graphs > Chart Builder (or Analyze > Descriptive Statistics > Frequencies).
- Calculating selected percentiles: Analyze > Descriptive Statistics > Frequencies. You can get fixed percentiles from Analyze > Descriptive Statistics > Explore.
- Calculating and saving standardized *z* scores: Analyze > Descriptive Statistics > Descriptives (your only choice).
- Calculating descriptive statistics for groups of cases: Analyze > Compare Means > Means if you want nice compact output; Analyze > Descriptive Statistics > Explore if you want fancy statistics and plots for each group.
- Testing normality, M-estimators, Levene test for equality of variance, spread-versus-level plots for transforming data: Analyze > Descriptive Statistics > Explore.
- Counts and percentages of the number of cases with all combinations of values of a set of grouping variables: Analyze > Descriptive Statistics > Crosstabs.
- Bar charts of many types: Graphs > Chart Builder.
- View individual cells of a multiway table: Analyze > Reports > OLAP Cubes.

Obtaining the Output

To produce the output in this chapter, open the file *gssdata.sav* and follow the instructions below. The last figure (Figure 6-18) requires the file *electric.sav.*

Several different procedures are available in the Graph menu for creating charts. Chart Builder has an easy to use interface based on selecting the desired chart and then dragging variable names to a prototype of the chart. Legacy dialogs require moving variable names to a more typical dialog box. Both are illustrated below.

Figure 6-1. From the Analyze menu choose Descriptive Statistics > Frequencies. Move *news* into the Variable(s) list. Click OK.

Figure 6-2. From the Graphs menu choose Legacy Dialogs. Select Pie chart and double click the chart. Move *news* into the Slice by? box. Click OK. Double-click the chart to activate it in the Chart Editor. From the Elements menu choose Show Data Labels. In the Properties dialog box, on the Data Value Labels tabs, select *Percent* in the Not Displayed list and click the arrow to the right of the list to move it into the Displayed list. Remove whatever you don't want from the Displayed list by selecting it and clicking the red X. Click Apply.

Figure 6-3. From the Graphs menu choose Legacy Dialogs. Select Simple Bar by double clicking the icon and move *news* into x-axis. In the Element Properties menu select Statistic: Percentage, and then click Apply.

Figure 6-4. From the Analyze menu choose Descriptive Statistics > Crosstabs. Move *sex* into the Row(s) list and *news* into the Column(s) list. Click the Cells button to open the Cell Display subdialog box. Select Observed in the Counts group and Row in the Percentages group, and Round cell counts in the Noninteger Weights group. Click Continue to close the Cell Display subdialog box, and then click OK to generate the table.

Figure 6-5. From the Graphs menu choose Legacy Dialogs > Bar. Select the Clustered icon and Summaries for groups of cases, and then click Define. Select *news* as the Category Axis variable and *sex* as the Define Clusters By variable. Select % of cases in the Bars Represent group. Click OK.

Figure 6-6. From the Analyze menu choose Descriptive Statistics > Crosstabs. Select *sex* as the row variable, *news* as the column variable, and move *degree3* into the Layer 1 of 1 box. Click the Cells button to open the Cell Display subdialog box. Select Observed in the Counts group, Row in the Percentages group, and Round cell counts in the Noninteger Weights group. Click Continue to close the Cell Display subdialog box, and then click OK to generate the table.

Figure 6-7. From the Graphs menu choose Legacy Dialogs > Bar. Select the Clustered icon and Summaries for groups of cases, and then click Define. Select *degree* as the Category Axis variable and *sex* as the Define Clusters By variable. Select Other statistic in the Bars Represent group. Move the variable *news* into the Variable box. Click the Change Statistic button. In the Statistic subdialog box, select Percentage inside, and enter the value 1 in both the Low and High text boxes. Click Continue to close the subdialog box, and then click OK to generate the clustered bar chart. Double-click the chart to activate it in the Chart Editor. From the Options menu select Y Axis Reference Line. Click the Reference Line tab. Enter 40.6 into the Y Axis Position box. Click Apply, and then click Close.

Figure 6-8. From the Data menu choose Select Cases. Select If condition is satisfied, and click the If button. In the If subdialog box, type missing(news) ne 1 into the text box. Click Continue, and then click OK. This selects people who were asked the newspaper-reading question and answered it. From the Graphs menu choose Legacy Dialog and then Histogram. Move *age* into the Variable box, and click OK. Double-click the chart to edit it in the Chart Editor. Click once on the chart to select it. From the Edit menu choose Properties. Click the Binning tab, set the Custom Value for Anchor to 15 and the Custom Interval width to 10. Click Apply, and then click Close.

Figure 6-9. From the Graphs menu choose Legacy Dialogs > Histogram. Move *age* into the Variable box, and *news* into the Panel by Rows box, and then click OK. Double-click the chart to edit it in the Chart Editor. Click once on the chart to select it. From the Edit menu choose Properties. Click the Binning tab, set the Custom Value for Anchor to 15 and the Custom Bin Size to an Interval width of 10. Click Apply, and then click Close.

Figure 6-10. From the Data menu choose Select Cases. Select All cases, and click OK. From the Analyze menu choose Descriptive Statistics > Explore. Select *age* as the Dependent variable and *news* as the Factor variable. Select Plots in the Display group. Click the Plots button. In the Plots subdialog box, select None in the Boxplots group and the Stem-and-leaf check box in the Descriptive group. Click Continue to close the subdialog box, and then click OK to generate stem-and-leaf plots for all categories. Figure 6-10 is the stem-and-leaf plot for the *Never* category.

Figure 6-11. From the Analyze menu choose Compare Means > Means. Select *age* as the Dependent variable and *news* as the Independent variable. Click the Options button. In the Options subdialog box, move all of the statistics except Mean from the Cell Statistics list into the Statistics source list. Now move Median, Standard Deviation, Variance, Minimum, Maximum, Range, Std. Error of Mean, and Number of Cases from the Statistics list into the Cell Statistics list. Click Continue to close the subdialog box, and then click OK in the main dialog box to generate the table.

Figure 6-12. From the Analyze menu choose Compare Means > Means. The *age* variable should still be in the Dependent List, and *news* should still be in the Independent List. In the Layer group, click the Next button. Move *degree3* into the Layer 2 of 2 Independent list. Click the Options button. In the Options subdialog box, move Standard Deviation, Variance, Minimum, Maximum, Range, and Std. Error of Mean into the Statistics source list. The Cell Statistics list should now display Mean, Median, and Number of Cases. Click Continue to close the subdialog box, and then click OK in the main dialog box to generate the table.

In the output Viewer, select the last generated item (labeled *Report*) and double-click the table to activate the Pivot Table Editor. From the Pivot menu choose Pivoting Trays. Click the *Statistics* icon in the Column dimension and drag it to the Row dimension as the last item. In the Row dimension, click the first icon (*How often do you read a newspaper?*) and drag it to the Column dimension. Click anywhere outside of the edited table to exit the Pivot Table Editor.

Figure 6-13. From the Graphs menu choose Legacy Dialogs > Bar. Select the Clustered icon and Summaries for groups of cases, and then click Define. In the Bars Represent group, select Other statistic, and then move *age* into the Variable box. This will automatically set the value to MEAN(age). Select *news* as the Category Axis variable, and select *degree3* for Define Clusters By. Click OK to generate the initial chart.

In the output Viewer, double-click the bar chart to open a Chart Editor window. From the Options menu choose Y Axis Reference Line. Enter the mean value of 46.18 into the Y Axis Position text box. Click Apply and then click Close.

Figure 6-14. From the Graphs menu choose Legacy Dialogs > Line. Select the Multiple icon and Summaries for groups of cases, and then click Define. In the Lines Represent group, select Other statistic and then move *age* into the Variable box. This will automatically set the value to MEAN(age). Select *news* as the Category Axis variable, and select *degree3* for Define Lines By. Click OK to generate the chart.

Figure 6-15. From the Analyze menu choose Reports > OLAP Cubes. Select *age* as the Summary Variable, and select *news*, *sex*, and *degree* as Grouping Variables. Click the Statistics button to open the Statistics subdialog box. Move Sum, Percent of Total Sum, and Percent of Total N from the Cell Statistics into the Statistics source list. Move Median from the Statistics list into the Cell Statistics list. Click Continue to close the subdialog box, and then click OK to generate initial OLAP Cubes.

In the output Viewer, select the last generated item (labeled *OLAP Cubes*) and double-click it to open the Pivot Table Editor. Change the first drop-down list item from Total to Everyday. Change the second drop-down list item from Total to Female. Change the third drop-down list item from Total to High School. The OLAP Cube should now be identical to the image displayed in Figure 6-15.

Figure 6-16. From the Analyze menu choose Descriptive Statistics > Explore. Select *age* as the Dependent variable and *news* as the Factor variable. Select Statistics in the Display group. Click Statistics. In the subdialog box, select *only* Percentiles. Click Continue, and then click OK in the main dialog box.

In the output Viewer, select the last generated table (labeled *Percentiles*) and double-click it to open the Pivot Table Editor. From the Pivot menu choose Pivoting Trays. Select the *Methods* label in the Row dimension and drag it to Layer. Select the *Percentiles* label in the Column dimension and drag it to Row. Select the *Dependent Variables* label in the Row dimension and drag it to Column. Select the *How often do you read a newspaper?* label in the Row dimension and drag it under the *Dependent Variables* label in Column. Click anywhere outside of the edited table to exit the Pivot Table Editor.

Figure 6-17. From the Graphs menu choose Legacy Dialogs > Boxplot. Select the Simple icon and Summaries for groups of cases, and then click Define. Move *age* into the Variable box, move *news* into the Category Axis box, and then click OK to generate the chart.

Figure 6-18. Open the SPSS Statistics data file *electric.sav.* From the Graphs menu choose Legacy Dialogs and then Histogram. Move *ht58* into the Variable box and select Display normal curve. Click OK to generate the chart.

Chapter

7

Testing Hypotheses

One of your missions in using SPSS Statistics may be to determine whether your results are *statistically significant*. You may have been warned that your assignment entails unearthing a mystical "*p* value" that will forever be entombed in a paper or report. Don't be embarrassed if you're not quite sure what's involved in this popular, yet mysterious, ritual. You are not alone! Because the basic idea is the same for all statistical procedures, this chapter outlines the game plan. The specifics of various tests are described in subsequent chapters. You may be tempted to skip this chapter and to head directly to those action-filled chapters that show you how to make SPSS Statistics spit *p* values. Don't do it. Calculating *p* values is easy. Understanding what they mean is the hard part. (For a more extensive discussion of hypothesis testing, see the *SPSS Statistics 17.0 Guide to Data Analysis*.)

In a Nutshell

Usually when you conduct an experiment or a survey, you want to draw conclusions about the population from which the sample is selected; for example, you want to draw conclusions about the opinions of all voters, based on a random sample of voters. To use statistical methods to draw conclusions about the population, you must design your study well and have a random sample of observations from the population of interest.

To test a hypothesis, you must:

- Formulate a null hypothesis and an alternative hypothesis.
- Assume that the null hypothesis is true.
- Calculate the probability of observing sample results as extreme as those you've observed if the null hypothesis is true (the observed significance level).
- Reject the null hypothesis if the observed significance level is small enough.

Setting the Stage

You can easily answer questions about the observations in your data file. You know with certainty how many men and how many women in your study say they are satisfied with your products or services or how many patients you've cured with the new treatment. Uncertainty arises when you want to draw conclusions about people or objects you haven't questioned or observed.

Defining Samples and Populations

The cases in your data file are known as a **sample**. You're no doubt familiar with the word *sample* because its common usage is similar to its usage in statistics. A sample of a new product at a grocery store is a small part of the available stock of that item. Similarly, in statistics, a sample is a part of a population, the totality of people or objects about which you want to draw conclusions.

Tip: Define your population of interest carefully. If you are studying hospitals, do you want to draw conclusions about a particular hospital? All hospitals in the region? All hospitals in a country? Statistical populations need not be people. They can be hamburgers sold by a particular chain, fish in rivers, or surfaces of planets.

Using Random Samples

It doesn't matter which slice of chocolate cake the solicitous salesperson offers you, since all product samples are pretty much the same (except for the frosting and size). Unfortunately, that's not usually the case when you take samples from other, less fattening populations. Samples of patients with the same disease can vary greatly on factors such as severity of disease and access to health care. Samples of voters can vary in terms of their candidate preferences. The statistical methods described in this book rely on the assumption that your data are a **random sample** from the population of interest. Random does not mean haphazard. Random means that all people or objects have the same chance of being included in your sample.

Creating a Good Experimental Design

Statistical methods don't rely on black magic. They rely on your following well-established principals of study design. If you don't, you can still use statistical software to generate all kinds of statistical tests—but they'll be meaningless. If you design a study poorly, most of the effort you spend analyzing it will be wasted. While it's always possible to reanalyze good data if you've made a mistake in the analysis, no amount of statistical analysis can salvage a poorly designed study.

You can't draw conclusions about the population if your sample is biased in some important way. A telephone survey conducted during a weekday afternoon will exclude people who work away from home. Don't expect to draw conclusions about the U.S. population from such a survey. Relying on volunteers for almost anything is also fraught with problems because people who volunteer are different from people who don't. Even worse is letting your marketing department determine who will be included in a survey of customer satisfaction.

Even if you've carefully selected a random sample, you won't be able draw correct conclusions if your observations are biased. Prejudices and hunches can easily influence the results of a study.

Some general principles of a well-designed study are:

- Phrase questions in such a way that they don't suggest any particular answer you might be seeking.
- Assign treatments or conditions randomly to subjects. If you don't use chance to assign who gets what treatment or condition, you may inadvertently let your prejudices creep in, and sicker patients may get the new treatment or poorer students may get the new computers.
- Use a control or placebo group. Compare new treatments or methods to existing treatments or methods, or to no treatment. Even extra attention or sugar pills can make people healthier or smarter.
- Observe all groups with the same diligence and objectivity.

Dealing with Missing Data

You may think that the topic of missing data concluded with declaring which codes in your data file are to be treated as missing. You're wrong. Specifying the codes for missing values is the easy part of dealing with missing values. The hard part is assessing the effect of missing data on your study as a whole, and on individual analyses that you carry out. You must confront the problem of missing data before you embark on testing hypotheses or building models. The most important question you must answer is *How are the values missing? Are they missing at random, or is there a reason, a non-ignorable reason, for their absence?*

If you ask people their incomes and randomly delete some of the values, the missing values are missing completely at random. In this situation, cases with missing values are indistinguishable from cases without missing values. Non-ignorable missing values occur when the likelihood of missing data is related to the variables you are studying. If people with high incomes are less (or more) likely to reveal their incomes to a questioner than people with low incomes, the missing values are called non-ignorable. Unless you take serious steps to deal with this problem, your analyses may be significantly impacted.

Warning: An even bigger problem sometimes occurs when you are missing cases in your data file because of refusals. For example, when you conduct a survey, some people you contact may refuse to participate. If people who refuse to be part of your survey or experiment are different from people who do participate (and you can't possibly be certain that they aren't), the results you draw based on cooperative people may not be true for the uncooperative people. That's potentially deadly for your analysis.

Understanding Missing-Data Strategies

In many of the statistical procedures in SPSS Statistics, you are offered different strategies for dealing with missing values. You can omit cases with missing values from an analysis, substitute the mean of a variable when the value is missing, or use partial information from a case. Unless you have just a few missing values and they are missing completely at random, none of these strategies is completely satisfactory, especially when analyzing several variables together. More sophisticated methods are necessary to properly deal with missing values (Little and Rubin, 2002).

The problems with simple missing-data strategies are fairly straightforward. If you completely eliminate cases with missing data from an analysis, your population changes. It becomes the population of all people or objects with complete information. The two populations may or may not be similar. You may also greatly reduce your sample size. If you replace the missing values with the mean of the variable for cases without missing values, you are distorting the variability of the data. If you cobble together a correlation matrix from correlation coefficients computed for different groups of cases, you don't really know what you have.

Tip: Always try to identify the patterns of the missing data for each variable. Compare the characteristics of subjects with missing values to those without missing values.

Testing a Hypothesis

When you conduct a survey or perform an experiment, you are interested in testing a hypothesis about the population from which you've selected your sample. You might want to know if the cure rate in the population is the same for two treatments, or if income is related to Internet use. You might think that because you're equipped with SPSS Statistics, the program will do all of the work for you. No such luck. Statistical hypothesis testing involves a series of steps, and SPSS Statistics is good for only one of them. And even then, it's up to you to point SPSS Statistics in the right direction.

Step 1: Specifying the Null Hypothesis and the Alternative Hypothesis

Every statistical test involves two hypotheses: the null hypothesis and the alternative hypothesis. There's nothing unsavory about a hypothesis, although the adjectives "null" and "alternative" might sound ominous. A **hypothesis** is a factual statement that may or may not be true. Men and women are equally likely to buy treadmills for their dogs; five-year survival rates are better for patients treated with this new drug than for patients treated with that old drug; the normal body temperature for human adults is 98.6. These are all hypotheses. Statistical hypotheses are stated in terms of population values of statistics, such as means. Population values of statistics are called **parameters**. They are the values you would get if you could calculate the statistic for all members of the population.

For every hypothesis, there is an opposite or contrary hypothesis that describes the state of the world if the original hypothesis is not true. For the hypothesis that men and women are equally likely to use the Internet, the contrary hypothesis is that men and women are not equally likely to use the Internet. A hypothesis and its corresponding contrary hypothesis, taken together, must cover all possible situations. It's often the case that the pair of hypotheses can be described as "something's going on" and "nothing's going on." Nothing's going on if men and women use the Internet equally; something's going on if they don't. Nothing's going on if two treatments have the same survival rates; something's going on if they don't. Statisticians call the nothing's-going-on hypothesis the null hypothesis and the something's-going-on hypothesis the alternative hypothesis.

The null hypothesis gets its less-than-flattering name from its role as the "no-difference" hypothesis, since no difference is usually not news. In most research situations, you set out to disprove the null hypothesis. When you conduct a survey or experiment, you are usually looking for that "something." You hypothesize that all people don't have the same attitudes, that your drug is better than the standard one, or that education increases income. You hope that you have gathered enough evidence so you can reject the null hypothesis and publish your results or make a memorable presentation.

Example: Educating Your Heart

As an example of hypothesis testing, consider data from the Western Electric Study (Paul et al., 1963). A sample of men initially without coronary heart disease was followed for 20 years to see who developed coronary heart disease. You're interested in determining whether there is a relationship between men developing heart disease and their education. The null hypothesis is that the average years of education are the same, in the population, for men who develop heart disease and for those who don't. The alternative hypothesis is that the two population means are not equal.

From Figure 7-1, you see that the average years of education for the 104 men in the sample who developed heart disease are 10.9. For the 108 men who did not develop heart disease, the average years of education are 12.4. The standard deviations in the two groups are fairly similar, around 2.5 years. Your goal is to determine what you can say about the population, based on the sample results.

Figure 7-1
Summary statistics for years of education

			N	Mean	Std. Deviation	Std. Error Mean
Years of education	Incidence of Coronary Heart Disease	No	108	12.40	2.75	.265
		Yes	104	10.89	2.59	.254

Prelude to the Remaining Steps

Once you have the two hypotheses specified, all you have to do is determine which one is true. You know one of them has to be true because together they cover all possible situations. Men with and without coronary heart disease either differ in average years of education or they don't. Unfortunately, even with SPSS Statistics squirming in your computer, you won't be able to select between the two hypotheses with complete certainty. That's because you don't have information available for everyone or everything in the population. If you did, you could draw correct conclusions all of the time by just calculating the results for your population. When you have only a random sample of data from one or more populations, you are forced to couch all of your conclusions in terms of probability statements.

You might wonder why you can't just simplify the process by assuming that what's true for the sample is also true for the population. The reason you can't do that is that a random sample from a population is not an exact miniature of the population.

Different random samples from the same population give different results. For *random* samples, you can determine how much results vary from sample to sample and how much they are expected to differ from the population value. If you don't have a random sample, you can't do that. That's why random samples are so important in statistics.

Step 2: Selecting the Appropriate Statistical Procedure

Different statistical procedures are used to test different types of hypotheses. Correlations, *t* tests, linear regression, analysis of variance, chi-square, and many other statistical tests you've heard about are used to test hypotheses. The hypotheses they test are different, and in later chapters, you'll determine which statistical test is appropriate for your situation. The selection of a procedure depends both on the type of hypothesis you want to test and on the characteristics of the data. The table on p. 126 gives you some idea about the procedures you might use for testing some simple hypotheses.

Two-Independent-Samples T Test

For the Western Electric data, you can use what's called a two-independent-samples *t* test to test the null hypothesis that two population means are equal. (Chapter 8 is devoted exclusively to this topic.)

Step 3: Checking Whether Your Data Meet the Required Assumptions

It may or may not be true that there is no such thing as a free lunch. It is certainly true that there is no such thing as a statistical test without assumptions. The assumptions differ from procedure to procedure and are described in subsequent chapters. Some statistical procedures go haywire when their assumptions are violated, while others remain fairly true to course. (A statistical procedure is termed **robust** if it can withstand violations of some of the underlying assumptions.) Some assumptions, such as independence of observations and normality, are common to many procedures.

Tip: Chapter 19 describes nonparametric tests that require less stringent assumptions about the distribution of the data.

Independence of Observations

The most important assumption that you have to make about your data is that the observations are independent of each other. This means that one observation isn't related in any way to another. If you are gathering data sequentially, it's possible that adjacent observations are in some way more similar (or, less frequently, more dissimilar) than non-adjacent observations. This is easy to imagine if you're working in a laboratory where equipment performance may change slightly with time. It can also occur when the *experimenter* changes with time. For example, a surgeon should improve with more experience, while a truancy counselor may get more or less effective with time, depending on whether they are growing experienced or weary. And, of course, multiple measurements of the same variable (such as blood pressure) for the same person are not independent. Multiple measurements like this are called **repeated measures** and require special methods of analysis.

Tip: If you've gathered data in sequence, always plot the data values against time to look for possible relationships with time.

If you take samples from several institutions, such as schools or hospitals, the observations within the same school or hospital are not independent because children attending the same school are probably more similar than children attending different schools. Patients treated in the same hospital are probably more similar than patients treated in different hospitals. Special statistical techniques for such data are discussed in Chapter 24.

Normality

In Chapter 6, you were introduced to the normal (bell-shaped) distribution. The assumption of normality figures prominently in hypothesis testing, although you yourself don't have to be normal (fortunately) to use statistical procedures. Many mathematical calculations are based on statistics (such as sample means) having particular distributions, such as the normal distribution, when repeated samples of the same size are taken from the same population. The importance of the normal distribution and exactly what has to be normally distributed differ from procedure to procedure. So does the importance of the normality assumption.

Tip: You can test whether a variable comes from a normal distribution using the tests for normality in the Analyze > Descriptive Statistics > Explore procedure.

Equal Variances

For many procedures that involve comparing the means of two or more groups, you have to make the assumption that the groups come from populations with equal variances. If you have fairly equal sample sizes in all groups, moderate violations of this assumption may not have much impact on your conclusions.

Transformations to the Rescue

If your data fail to satisfy the assumptions required for a particular procedure, you can attempt to transform the data values to better comply with the necessary assumptions. For example, you might find that taking logs or square roots of the data will make the distribution of a variable more normal. The Analyze > Descriptive Statistics > Explore procedure can be used to select a suitable transformation.

Warning: Transforming data brings its own set of problems. The transformed variable must then satisfy the required assumptions. The results become more difficult to explain because they are in logs or square roots of the original units of measurement.

Assumptions for the example. In this example, you have 108 men who did not develop coronary heart disease and 104 who did. As long as the observations are independent, you can go ahead and use the two-independent-samples *t* test without worrying about the shape of the distribution of education. That's because the sample size in each group is large. See Chapter 8 for further discussion.

Step 4: Assuming That the Null Hypothesis Is True

This is easily said, but like "I do" at a wedding, it has far-reaching consequences. Remember that all of the subsequent steps are based on this assumption.

Step 5: Calculating the Observed Significance Level

This is the step that SPSS Statistics will do for you if you've steered it to the correct procedure. You have to determine if your results are unusual if the null hypothesis is true. You must calculate the probability of getting results as extreme as those you've observed, assuming that the null hypothesis is true (the notorious *p* value!). In this example, you've observed a mean difference of 1.5 years of education between the two groups of men. You want to know how often you would expect to see a sample difference of at least 1.5 years if, in the population, the two groups have the same average years of education.

Figure 7-2
Independent-samples t test

		t-test for Equality of Means				
		t	df	Sig. (2-tailed)	Mean Difference	Std. Error Difference
Years of education	Equal variances assumed	4.091	210	.000	1.50	.368
Body mass index	Equal variances assumed	.284	238	.776	.114	.402

In the first row of Figure 7-2, in the column labeled *Sig. (2-tailed)*, you see that the observed significance level for the test of the null hypothesis that the average years of education are the same for men who develop coronary heart disease and those who don't is less than 0.0005. (SPSS Statistics rounds numbers, which is why it's displayed as 0.000.) Fewer than five samples in 10,000 would have a difference at least this large if the null hypothesis is true.

Figure 7-2 also contains a test of a second null hypothesis, namely that the population average body mass index (weight in kilograms divided by the square of height in meters) is the same for the two groups. For that null hypothesis, the observed significance level is 0.776. The probability of observing a difference of this magnitude when the null hypothesis is true is quite large. Almost 8 out of 10 samples would have a difference at least this large if the null hypothesis is true.

Warning: SPSS Statistics will often display significance levels of 0.000. That doesn't mean that the observed significance level is 0. It means that the observed significance level is less than 0.0005. Don't ever report an observed significance level as 0. An observed significance level of 0 would mean that it is *impossible* to get your results if the null hypothesis is true. To see the significance level to more decimal places than shown, activate the Pivot Table Editor and double-click the cell. Make sure the column width is large enough to display the number. You'll see the probability in scientific notation. The negative number after the letter *e* tells you how many places to move the decimal to the left.

Step 6: Deciding Whether to Reject the Null Hypothesis

If the p value is small enough (usually less than 0.05), reject the null hypothesis. Traditionally, 0.05 is used as the threshold for "small enough," although a more stringent criterion of 0.01 is also used. These criteria are called the **significance levels** or **alpha levels** for a statistical significance test. If your observed significance level is less than 0.05, your results are said to be "statistically significant" at the 5% level. If your observed significance level is less than 0.01, your results are said to be statistically significant at the 1% level.

There's nothing sacred about 0.05 as the cutoff for an unlikely event. There's remarkably little difference, after all, between an observed significance level of 0.049 and one of 0.051. Both indicate that the observed results are unlikely if the null hypothesis is true. You should always evaluate the actual strength of the evidence against the null hypothesis instead of relying on a magical number for what is statistically significant and what is not.

Because computer programs calculate the actual observed significance levels, you should report them. Knowing that $p = 0.003$ is more informative than knowing that $p < 0.05$. If you're planning to publish your results, that's what most journals require.

Tip: Make sure that you know which observed significance level corresponds to which test because SPSS Statistics sometimes shows several significance levels in the same pivot table.

Cheers for Education!

Because the observed significance level for education in Figure 7-2 is quite small, you can reject the null hypothesis that the two population means are equal. The average years of education are significantly higher for the men who didn't develop heart disease than for those who did. On the other hand, you see that the average body mass index is not significantly different between the two groups.

Warning: Despite everything stated above, *statistical significance* is overrated. Too often, it's a poor substitute for careful thought, common sense, and good research practices. Use it wisely.

Tails on the Test

Sometimes an alternative hypothesis states a direction for the anticipated difference. For example, if you are interested in the effects of massage therapy on test scores, you may have a preconceived notion that massage therapy can only help scores or leave them unchanged. You exclude the possibility that massage therapy may relax students so much that they fail to study and actually get worse scores. If you can specify the direction of the difference prior to conducting the study and analyzing the data, you can adjust the observed significance level to take that into account. You can use what's called a **one-tailed test**, meaning that you reject the null hypothesis based on probability calculations for only the anticipated direction. The observed significance level is smaller than for an alternative hypothesis that doesn't indicate the direction of the difference. That means you're more likely to reject the null hypothesis when it is false. Most tests in SPSS Statistics are for alternative hypotheses that don't specify the direction of the difference. That's why they're labeled as **two-tailed**.

Warning: You should use one-tailed tests with extreme caution because it's difficult to know the direction of the difference with certainty. In medicine, many treatments thought only to benefit patients have turned out to be harmful. Don't declare that you're using a one-tailed test simply to make your observed results more unlikely.

Calculating Confidence Intervals

The observed significance level doesn't tell you anything about the magnitude of the difference in the population. You can have very small observed significance levels for differences that are small and for differences that are large, depending on the number of cases in the samples and the observed variability of the data. In addition to the observed significance level, you should calculate what's called a **confidence interval**. A confidence interval is a range of values that you have reason to believe includes the unknown population value. A confidence interval can be calculated for the population value of any statistic whose distribution is known.

From Figure 7-3, you see that the 95% confidence interval for the population difference in years of education between the two groups is from 0.78 to 2.23 years. For body mass index, the 95% confidence interval is from –0.68 to 0.91. For each variable, you can be 95% confident that the true population difference is within this interval. Note that the value of 0 is included in the 95% confidence interval for body mass index because you can't reject the null hypothesis that the difference between the two population means is 0. It isn't included in the 95% confidence interval for the difference in education because you could reject the null hypothesis that the difference between the two population means is 0.

Figure 7-3
Confidence intervals

		t-test for Equality of Means			
		Mean Difference	Std. Error Difference	95% Confidence Interval of the Difference	
				Lower	Upper
Years of education	Equal variances assumed	1.50	.368	.779	2.229
Body mass index	Equal variances assumed	.114	.402	-.678	.907

Confidence intervals are constructed so that they include the unknown population values with a designated likelihood (for example, 95%). For the same dataset, as you increase the confidence level, say from 95% to 99%, the interval becomes wider. That's because the only way you can be more confident that a confidence interval contains the unknown population value is to make the interval wider. The relationship between hypothesis tests and confidence intervals is discussed in Chapter 8 on *t* tests.

Warning: A particular confidence interval either does or does not contain the unknown population value, but you don't know which assertion is true. You do know that 95% of your 95% confidence intervals include the unknown true population value. If you knew the population value, of course, you would know with certainty whether it was in a given confidence interval, and you wouldn't need to bother with hypothesis testing.

Reporting Your Results Correctly

Because hypothesis testing depends on probability statements and not on certain knowledge, the conclusions you can correctly draw are limited. If the observed p value is smaller than 0.05, you can say that you have rejected the null hypothesis. You can also say that if the null hypothesis is true, you would expect to see results as extreme as the ones you've observed fewer than five times out of 100. If your observed p value is greater than 0.05, you can say that you failed to reject the null hypothesis. Your observed results are not all that unlikely if the null hypothesis is true.

Tip: Read the article "Statistical Methods in Psychology Journals: Guidelines and Explanations" found at *www.spss.com/research/wilkinson/Publications/apasig.pdf* for many useful guidelines for reporting results and for insightful commentary on the use of statistics in research (Wilkinson, 1999).

Warnings on Interpreting Results of Hypothesis Testing

The results of hypothesis testing must be reported carefully. *You can never conclude with certainty that the null hypothesis is either true or false.* It is easy to make pretentious-sounding claims that are not statistically justified.

- Never conclude from a hypothesis test that differences in one variable cause differences in the other unless you are analyzing a carefully controlled experiment. For example, you can't conclude that by staying in school you will decrease your risk of heart disease. It may be that earning more will let you join a health club and exercise. Or it may be that younger people are better educated and also less likely to develop heart disease.
- Never equate statistical significance with practical significance. When you reject the null hypothesis, it is not necessarily the case that the difference or association you found is important or noteworthy. For large samples, even very small differences in means may be *statistically significant.* For example, if you find that a particular treatment prolongs life by one week compared to another, even if you've collected enough data to make the difference statistically significant, it is of little practical importance. That's why you should always examine the actual observed differences or the magnitudes of measures of association and focus only on those that are both statistically significant and practically meaningful. Always report the actual difference or coefficient, as well as its observed significance level.
- Never equate failure to reject the null hypothesis with the null hypothesis being true. Your failure to reject the null hypothesis doesn't mean it is true. It could simply mean that you haven't gathered enough evidence to reject it. If your sample size is small, you may fail to reject the null hypothesis even when the population difference is large. That's why it's important, before you embark on a study, to determine how big a sample size you need in order to detect what you consider to be an important difference.
- Never use the phrase "accept the null hypothesis" because it implies that you believe the null hypothesis is true. You can legitimately reject the null hypothesis, but you can't accept it. (That sounds unfair, but it is so, nonetheless.)

- Never equate rejection of the null hypothesis with certainty that the null hypothesis is false.
- Never assign probabilities to the null hypothesis or alternative hypothesis being true or false. The null hypothesis is either true or it is false. You can't know which. The alternative hypothesis is either true or it is false. Again, you can't know which.
- Never claim that the observed significance level is the probability that the null hypothesis is true. The *p* value tells you the probability of seeing results at least as extreme as the ones you've observed if the null hypothesis is true. The null hypothesis is either true or it is not.
- Never claim that the *p* value is the "probability that the results are due to chance." You can't talk about the probability of the results being due to chance unless you know that the null hypothesis is true.

To Err Is Statistical

Whenever you test a hypothesis, you can be wrong in two different ways. You can reject the null hypothesis when it is true (this is a Type 1 error), or you can fail to reject the null hypothesis when it is false (this is a Type 2 error). Type 1 and Type 2 errors are closely related; the more stringent you set your criterion for rejecting the null hypothesis, the greater the chance that you fail to detect a true difference when it really does exist. You control a Type 1 error with your decision as to how small the observed significance level has to be for you to say you've rejected the null hypothesis.

Your ability to reject the null hypothesis when it is false depends on four factors:

- How large the "effect" (for example, the difference between the groups) is in the population
- How much variability there is within the groups
- How large your sample size is
- How small the significance level is at which you want to say that you reject the null hypothesis

Before you conduct a survey or an experiment, you want to make sure that your sample size will be large enough to detect differences that you think are important. If you collect your data from a sample that is too small, you may not be able to find even large differences that exist in the population. This issue is termed **statistical power**. Detailed discussion of statistical power is beyond the scope of this book.

Commonly Used Tests for Popular Hypotheses

The chapters that follow describe statistical tests for frequently tested null hypotheses. If there's a particular hypothesis you want to test, use the following chart to find the right chapter. In these chapters, you'll find recommendations for alternative tests if your data don't meet the required assumptions.

Hypothesis	Statistical procedure and chapter
A sample comes from a population with a particular mean.	One-sample *t* test (Chapter 8)
Two independent population means are equal.	Two-independent-samples *t* test (Chapter 8)
Two related (paired) population means are equal.	Paired-samples *t* test (Chapter 8)
Two or more independent population means are equal.	One-way analysis of variance (Chapter 9)
Two categorical variables are independent.	Chi-square test (Chapter 10)
Two or more proportions are equal.	Chi-square test (Chapter 10)
There is no linear relationship between two variables.	Bivariate (Pearson) correlation (Chapter 11)

Obtaining the Output

To produce the output in this chapter, open the file *electric.sav* and follow the instructions below.

Figures 7-1 to 7-3. From the Analyze menu choose Compare Means > Independent-Samples T Test. Select *eduyr* and *bmi* as Test Variables and *chd* as the Grouping Variable. Click Define Groups and specify the value 0 for Group 1 and 1 for Group 2. Click Continue to close the subdialog box, and then click OK in the main dialog box to generate the figures.

Chapter

8

T Tests

One of the most frequently asked questions in data analysis is whether two means are significantly different. Do people who exercise have different average serum cholesterol levels from those who don't? Do autistic children have different average manual dexterity scores than children in the population as a whole? Does enrollment in a truancy reduction program decrease unexcused absences? Although often not explicitly stated, the question refers not to the sample means but to the underlying population means. You don't need fancy statistical tests to tell you whether two sample means are equal. You know with certainty what the results are for the sample.

The statistical procedure that is most often used to test hypotheses about two population means is the *t* test. (Tests that require less restrictive assumptions about the data are discussed in Chapter 19.) There are three different flavors of *t* tests: one-sample, two-independent-samples, and paired-samples (or matched-cases). On the Analyze > Compare Means menu, SPSS Statistics offers a separate procedure for each. You must choose the appropriate *t* test, based on the design of your study.

In a Nutshell

You are testing the null hypothesis that two population means are equal. The alternative hypothesis is that they are not equal. There are three different ways to go about this, depending on how the data were obtained. The plot line in all three of them is the same:

- Compute the observed difference and its standard error from your sample.
- Calculate a *t* value by dividing the observed difference by its standard error.

- Use the *t* distribution to calculate how often you would expect a *t* value at least as large as the one you observed if the two population means are equal.
- If the observed significance level is small enough, reject the null hypothesis that the two population means are equal.

Deciding Which T Test to Use

Your study design dictates which of the three types of *t* tests you need to analyze your data. The data file has to be arranged in a particular format for each type of *t* test. The statistical assumptions differ somewhat as well. Neither the one-sample *t* test nor the paired-samples *t* test requires any assumption about the population variances, but the two-sample *t* test does.

Tip: When reporting the results of a *t* test, make sure to include the actual means, differences, and standard errors. Don't give just a *t* value and the observed significance level.

One-Sample T Test

If you have a single sample of data and want to know whether it might be from a population with a known mean, you have what's termed a **one-sample** design, which can be analyzed with a one-sample *t* test.

Examples

- You want to know whether CEOs have the same average score on a personality inventory as the population on which it was normed. You administer the test to a random sample of CEOs. The population value is assumed to be known in advance. You don't estimate it from your data.
- You're suspicious of the claim that the normal body temperature is 98.6 degrees (Fahrenheit). You want to test the null hypothesis that the average body temperature for human adults is the long-assumed value of 98.6, against the alternative hypothesis that it is not. The value 98.6 is not estimated from the data; it is a known constant. You take a single random sample of 1,000 adult men and women and obtain their temperatures.

- You think that 40 hours no longer defines the traditional workweek. You want to test the null hypothesis that the average workweek is 40 hours, against the alternative that it is not. You ask a random sample of 500 full-time employees how many hours they worked last week.
- You want to know whether the average IQ score for children diagnosed with schizophrenia differs from 100, the average for the population of all children. You administer an IQ test to a random sample of 700 schizophrenic children. Your null hypothesis is that the population value for the average IQ score for schizophrenic children is 100, and the alternative hypothesis is that it is not.

Data Arrangement

For the one-sample *t* test, you have one variable that contains the values for each case. For example, to test the null hypothesis that a random sample of children comes from a population with an average IQ score of 100, your data file looks like that shown in Figure 8-1.

Figure 8-1
One-sample t test data arrangement

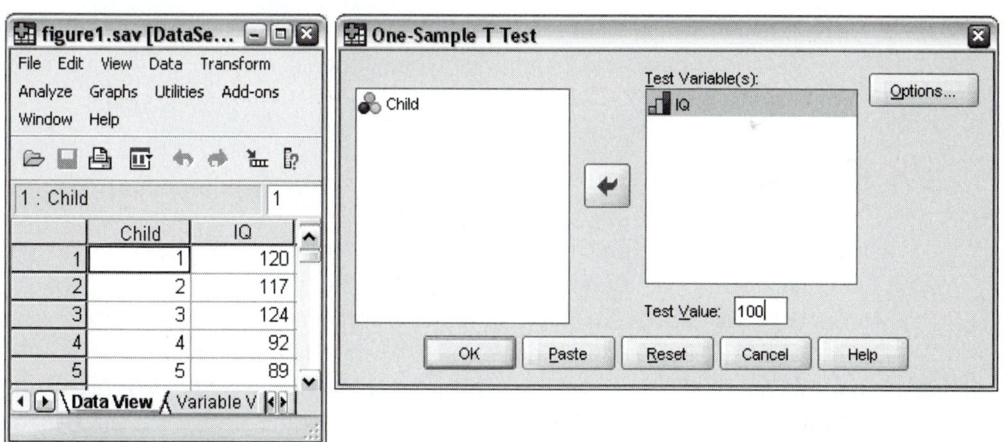

The population value against which you will be testing does not appear in the data file. It is specified in the Test Value text box in the One-Sample T Test dialog box, as shown on the right side of Figure 8-1.

Paired-Samples T Test

You use a **paired-samples** (also known as **matched-cases**) *t* test if you want to test whether two population means are equal, and you have two measurements from pairs of people or objects that are similar in some important way. For example, you've observed the same person before and after treatment or you have personality measures for each CEO and their non-CEO sibling. Each "case" in this data file represents a pair of observations.

Examples

- You are interested in determining whether self-reported weights and actual weights differ. You ask a random sample of 200 people how much they weigh and then you weigh them on a scale. You want to compare the means of the two related sets of weights.
- You want to test the null hypothesis that husbands and wives have the same average years of education. You take a random sample of married couples and compare their average years of education.
- You want to compare two methods for teaching reading. You take a random sample of 50 pairs of twins and assign each member of a pair to one of the two methods. You compare average reading scores after completion of the program.

Data Arrangement

In a paired-samples design, both members of a pair must be on the same data record. Different variable names are used to distinguish the two members of a pair. For example, if, for each child, you have scores before and after an intervention and want to test whether the population values for the average before and after scores are equal, your data file should look like that in Figure 8-2. The Paired-Samples T Test dialog box is shown on the right side of Figure 8-2.

Figure 8-2
Paired-Samples T Test dialog box

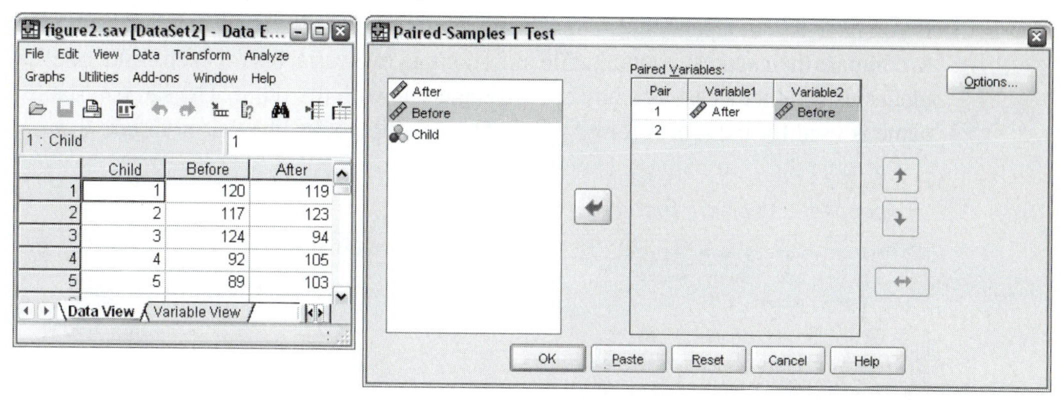

Warning: When you click the first variable of a pair, it doesn't move to the list box; instead, it moves to the lower left box labeled *Current Selections*. Only when you click a second variable and move it into Current Selections can you move the pair into the Paired Variables list.

Two-Independent-Samples T Test

If you have two independent groups of subjects, such as CEOs and non-CEOs, men and women, or people who received a treatment and people who didn't, and you want to test whether they come from populations with the same mean for the variable of interest, you have a **two-independent-samples** design. In an independent-samples design, there is no relationship between people or objects in the two groups. The *t* test you use is called an independent-samples *t* test.

Examples

- You want to test the null hypothesis that, in the U.S. population, the average hours spent watching TV per day is the same for males and females.
- You want to compare two teaching methods. One group of students is taught by one method, while the other group is taught by the other method. At the end of the course, you want to test the null hypothesis that the population values for the average scores are equal.
- You want to test the null hypothesis that people who report their incomes in a survey have the same average years of education as people who refuse.

Data Arrangement

If you have two independent groups of subjects (for example, boys and girls) and want to compare their scores, your data file must contain two variables for each child: one that identifies whether a case is a boy or a girl, and one with the score. The same variable name is used for the scores for all cases. The data file looks like that in Figure 8-3.

Figure 8-3
Independent-Samples T Test dialog box

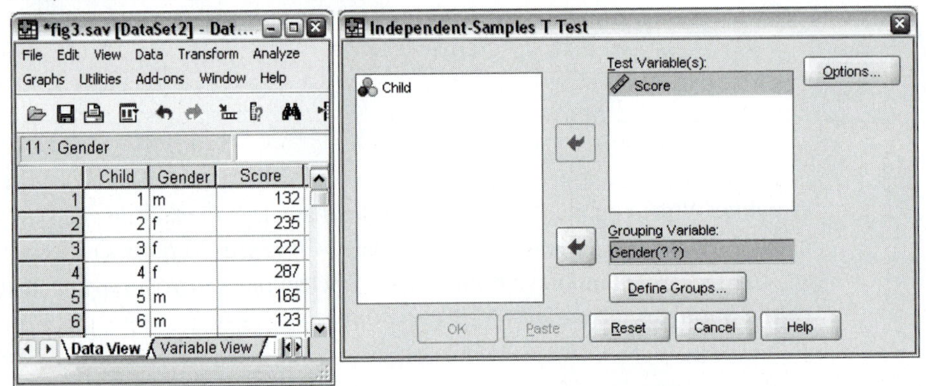

To run the two-independent-samples *t* test, you have to tell SPSS Statistics which variable defines the groups. That's the variable *Gender*, which is moved into the Grouping Variable box. Notice the two question marks after the variable name. They will disappear after you use the Define Groups dialog box (shown in Figure 8-4) to tell SPSS Statistics which values of the variable should be used to form the two groups.

Tip: Right-click the variable name in the Grouping Variable box and choose Variable Information from the pop-up menu. Now you can check the codes and value labels you've defined for that variable.

Figure 8-4
Define Groups dialog box

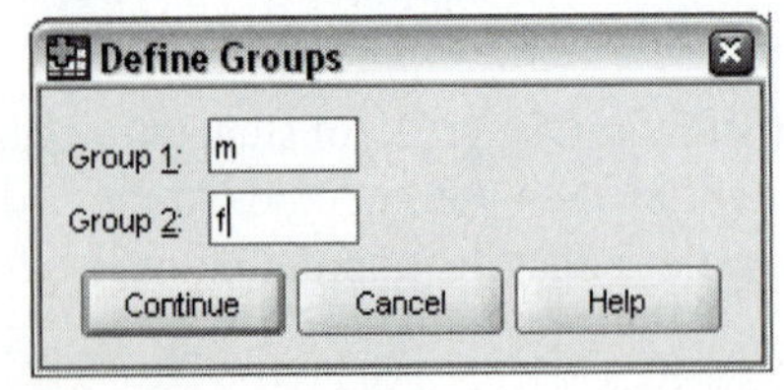

Warning: In the Define Groups dialog box, you must enter the actual values that you entered into the Data Editor, not the value labels. In Figure 8-4, the actual data values are *m* and *f*. (If you used the codes of *1* for male and *2* for female and assigned them value labels of *m* and *f*, then you enter the values *1* and *2*, not the labels *m* and *f*, into the Define Groups dialog box.)

Analyzing Truancy Data: The Example

Truancy (unexcused absence from school) is an important social problem, because children who fail to attend school are much more likely not to graduate and to be involved in criminal activities. Nicely et al. (2002) evaluated the effect of a truancy reduction program (TRP) in grade schools in Nashville, Tenn. For each of 299 students, they recorded the percentage of total school days with unexcused absences in the year prior to enrollment in the TRP program and the same percentage in the year following enrollment in the TRP program.

Figure 8-5 shows an excerpt from an abridged data file called *truancy.sav*. For each student, you have an identification number, grade, gender, the percentage of unexcused absences before enrollment in TRP (*prepct*), the percentage of unexcused absences after enrollment in TRP (*postpct*), and the difference between the before and after percentages (*diffpct*).

Figure 8-5
Data Editor with all variables

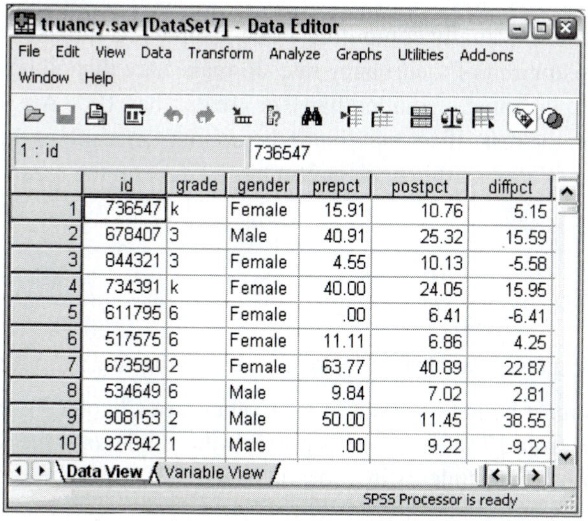

truancy.sav [DataSet7] - Data Editor

File Edit View Data Transform Analyze Graphs Utilities Add-ons Window Help

1 : id 736547

	id	grade	gender	prepct	postpct	diffpct
1	736547	k	Female	15.91	10.76	5.15
2	678407	3	Male	40.91	25.32	15.59
3	844321	3	Female	4.55	10.13	-5.58
4	734391	k	Female	40.00	24.05	15.95
5	611795	6	Female	.00	6.41	-6.41
6	517575	6	Female	11.11	6.86	4.25
7	673590	2	Female	63.77	40.89	22.87
8	534649	6	Male	9.84	7.02	2.81
9	908153	2	Male	50.00	11.45	38.55
10	927942	1	Male	.00	9.22	-9.22

Data View | Variable View

SPSS Processor is ready

One-Sample T Test

Consider whether the observed truancy rate before intervention (the percentage of school days missed because of truancy) differs from an assumed nationwide truancy rate of 8%. You have one sample of data (students enrolled in the TRP program) and you want to compare their results to a fixed, specified-in-advance population value.

The null hypothesis is that the sample comes from a population with an average truancy rate of 8%. (Another way of stating the null hypothesis is that the difference in the population means between your population and the nation as a whole is 0.) The alternative hypothesis is that your sample doesn't come from a population with a truancy rate of 8%.

From Figure 8-6, you see that, for the 299 students in this sample, the average truancy rate is 14.2%. You know that even if the sample is selected from a population in which the true rate is 8%, you don't expect your sample to have an observed rate of exactly 8%. Samples from the same population vary. What you want to determine is whether it's plausible for a sample of 299 students to have an observed truancy rate of 14.2% if the population value is 8%.

Figure 8-6
Descriptive statistics for pre-intervention truancy

	N	Mean	Std. Deviation	Std. Error Mean
Percent truant days pre intervention	299	14.20	13.07	.76

Tip: Before you embark on actually computing a one-sample *t* test, make certain checks. Look at the histogram of the truancy rates to make sure that all of the values make sense. Are there percentages smaller than 0 or greater than 100? Are there values that are really far from the rest? If so, make sure they're not the result of errors. If you have a small number of cases, outliers can have a large effect on the mean and the standard deviation.

Checking the Assumptions

To use the one-sample *t* test, you have to make certain assumptions about the data:

- The observations must be independent of each other (see Chapter 7). In this data file, students came from 17 schools, so it's possible that students in the same school may be more similar than students in different schools. If that's the case, the estimated significance level may be smaller than it should be, since you don't have as much information as the sample size indicates. (If you have 10 students from 10 different schools, that's more information than having 10 students from the same school because it's plausible that students in the same school are more similar than students from different schools.) That's something to keep in mind. Independence is one of the most important of the assumptions that you have to make when analyzing data.

- In the population, the distribution of the variable must be normal, or the sample size must be large enough so that it doesn't matter. The assumption of normally distributed data is required for many statistical tests. The importance of the assumption differs, depending on the statistical test. In the case of a one-sample *t* test, the following guidelines are suggested (Moore, 1995): If the number of cases is fewer than 15, the data should be approximately normally distributed; if the number of cases is between 15 and 40, the data should not have outliers or be very skewed; for samples of 40 or more, even markedly skewed distributions are acceptable. Because you have close to 300 observations, there's little need to worry about the assumption of normality.

Tip: If you have reason to believe that the assumptions required for the *t* test are violated in an important way, you can analyze the data using one of the nonparametric tests described in Chapter 19.

Testing the Hypothesis

To test the null hypothesis that the sample comes from a population with a mean of 8%:

- Compute the difference between the observed sample mean and the hypothesized population value. From Figure 8-6, the difference is 6.2% (14.2% – 8%).

- Compute the standard error of the difference. This is a measure of how much you expect sample means, based on the same number of cases from the same population, to vary. The hypothetical population value is a constant and doesn't contribute to the variability of the differences, so the standard error of the difference is just the standard error of the mean. Based on the standard deviation in Figure 8-6, the standard error equals

$$s_{\bar{x}} = \frac{13.07}{\sqrt{299}} = 0.756$$

Figure 8-7
One-sample t test

	Test Value = 8					
					95% Confidence Interval of the Difference	
	t	df	Sig. (2-tailed)	Mean Difference	Lower	Upper
Percent truant days pre intervention	8.207	298	.000	6.20	4.72	7.69

- Compute the *t* statistic by dividing the observed difference by the standard error of the difference.

$$t = \frac{14.204 - 8}{0.756} = 8.21$$

- Use the *t* distribution to determine if the observed *t* statistic is unlikely if the null hypothesis is true. To calculate the observed significance level for a *t* statistic, you have to take into account both how large the actual *t* value is and how many degrees of freedom it has. For a one-sample *t* test, the degrees of freedom is one fewer than the number of cases. From Figure 8-7, you see that the observed significance level is less than 0.0005. Your observed results are very unlikely if the true rate is 8%, so you reject the null hypothesis. Your sample probably comes from a population with a mean larger than 8%.

Tip: To obtain observed significance levels for an alternative hypothesis that specifies direction, often known as a one-sided or one-tailed test, divide the observed two-tailed significance level by two. Be very cautious about using one-sided tests.

Examining the Confidence Interval

If you look at the 95% confidence interval for the population difference, you see that it ranges from 4.7% to 7.7%. You don't know whether the true population difference is in this particular interval, but you know that 95% of the time, 95% confidence intervals include the true population values. Note that the value of 0 is not included in the confidence interval. If your observed significance level had been larger than 0.05, 0 would have been included in the 95% confidence interval.

Tip: There is a close relationship between hypothesis testing and confidence intervals. You can reject the null hypothesis that your sample comes from a population with any value outside of the 95% confidence interval. The observed significance level for the hypothesis test will be less than 0.05.

Paired-Samples T Test

You've seen that your students have a higher truancy rate than the country as a whole. Now the question is whether there is a statistically significant difference in the truancy rates before and after the truancy reduction programs. For each student, you have two values for unexcused absences. One is for the year before the student enrolled in the program; the other is for the year in which the student was enrolled in the program. Because there are two measurements for each subject, a before and an after, you want to use a paired-samples t test to test the null hypothesis that average before and after rates are equal in the population.

Tip: The reason for doing a paired-samples design is to make the two groups as comparable as possible on characteristics other than the one being studied. By studying the same students before and after intervention, you control for differences in gender, socioeconomic status, family supervision, and so on. Unless you have pairs of observations that are quite similar to each other, pairing has little effect and may, in fact, hurt your chances of rejecting the null hypothesis when it is false.

Before running the paired-samples t test procedure, look at the histogram of the differences shown in Figure 8-8. You see that the shape of the distribution is symmetric (not too far from normal). Many of the cases cluster around 0, indicating that the difference in the before and after scores is small for these students.

Figure 8-8
Histogram of differences

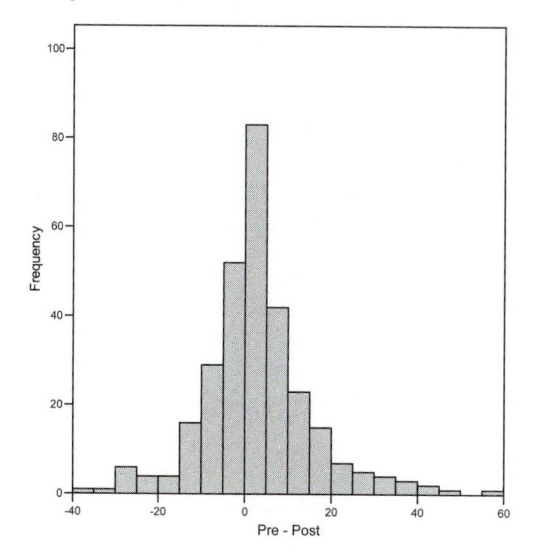

Checking the Assumptions

The same assumptions about the distributions of the data are required for the one-sample t test and the paired-samples t test. The observations should be independent; if the sample size is small, the distribution of differences should be approximately normal. Note that the assumptions are about the differences, not the original observations. That's because a paired-samples t test is nothing more than a one-sample t test on the differences. If you calculate the differences between the pre- and post-values and use the one-sample t test with a population value of 0, you'll get exactly the same statistic as using the paired-samples t test.

Testing the Hypothesis

From Figure 8-9, you see that the average truancy rate before intervention is 14.2% and the average truancy rate after intervention is 11.4%. That's a difference of about 2.8%.

Figure 8-9
Paired-samples t test descriptive statistics

		Mean	N	Std. Deviation	Std. Error Mean
Pair 1	Percent truant days pre intervention	14.20	299	13.07	.76
	Percent truant days post intervention	11.44	299	11.18	.65

To see how often you would expect to see a difference of at least 2.8% when the null hypothesis of no difference is true, look at Figure 8-10. The t statistic, 3.8, is computed by dividing the average difference (2.77%) by the standard error of the mean difference (0.73). The degrees of freedom is the number of pairs minus one. The observed significance level is less than 0.0005, so you can reject the null hypothesis that the pre-intervention and post-intervention truancy rates are equal in the population. Intervention appears to have reduced the truancy rate.

Figure 8-10
Paired-samples t test results

		Paired Differences					t	df	Sig. (2-tailed)
		Mean	Std. Deviation	Std. Error Mean	95% Confidence Interval of the Difference				
					Lower	Upper			
Pair 1	Percent truant days pre intervention - Percent truant days post intervention	2.77	12.69	.73	1.32	4.21	3.77	298	.000

Warning: The conclusions you can draw about the effectiveness of truancy reduction programs from a study like this are limited. Even if you restrict your conclusions to the schools from which these children are a sample, there are many problems. Because you are looking at differences in truancy rates between adjacent years, you aren't controlling for possible increases or decreases in truancy that occur as children grow older. For example, if truancy increases with age, the effect of the truancy reduction program may be larger than it appears. There is also potential bias in the determination of what is considered an "excused" absence.

The 95% confidence interval for the population change is from 1.3% to 4.2%. It appears that if the program has an effect, it is not a very large one. On average, assuming a 180-day school year, students in the truancy reduction program attended school five more days after the program than before. The 95% confidence interval for the number of days "saved" is from 2.3 days to 7.6 days.

A paired-samples design is effective only if you have pairs of similar cases. If your pairing does not result in a positive correlation coefficient between the two measurements of close to 0.5, you may lose power (your computer stays on, but your ability to the reject the null hypothesis when it is false fizzles) by analyzing the data as a paired-samples design (van Belle, 2002). From Figure 8-11, you see that the correlation coefficient between the pre- and post-intervention rates is close to 0.5, so pairing was probably effective.

Figure 8-11
Correlation coefficient for pre- and post-intervention

		N	Correlation	Sig.
Pair 1	Percent truant days pre intervention & Percent truant days post intervention	299	.461	.000

Warning: Although well-intentioned, paired designs often run into trouble. If you give a subject the same test before and after an intervention, the practice effect, instead of the intervention, may be responsible for any observed change. You must also make sure that there is no carryover effect; that is, the effect of one intervention must be completely gone before you impose another.

Two-Independent-Samples T Test

You've seen that the intervention seems to have had a small, although statistically significant, effect. One of the questions that remains is whether the effect is similar for boys and girls. Three hypotheses are of interest: Is the average truancy rate the same for boys and girls prior to intervention? Is the average truancy rate the same for boys and girls after intervention? Is the change in truancy rates before and after intervention the same for boys and girls?

Figure 8-12 shows summary statistics for the two groups for all three variables. Boys had somewhat larger average truancy scores prior to intervention than did girls. The average scores after intervention were similar for the two groups. The difference between the average pre- and post-intervention is larger for boys. You must determine whether these observed differences are large enough for you to conclude that, in the population, boys and girls differ in average truancy rates. You can use the two-independent-samples *t* test to test all three hypotheses.

Figure 8-12
Independent-samples t tests for boys and girls

			N	Mean	Std. Deviation	Std. Error Mean
Percent truant days pre intervention	Gender	Male	147	15.35	13.82	1.14
		Female	152	13.10	12.25	.99
Percent truant days post intervention	Gender	Male	147	11.36	10.95	.90
		Female	152	11.51	11.44	.93
Pre - Post	Gender	Male	147	3.98	13.56	1.12
		Female	152	1.59	11.72	.95

Checking the Assumptions

You must assume that all observations are independent. If the sample sizes in the groups are small, the data must come from populations that have normal distributions. If the sum of the sample sizes in the two groups is greater than 40, you don't have to worry about the assumption of normality. The two-independent-samples *t* test also requires assumptions about the variances in the two groups. If the two samples come from populations with the same variance, you should use the "pooled" or equal-variance *t* test. If the variances are markedly different, you should use the separate-variance *t* test. Both of these are shown in Figure 8-13.

You can test the null hypothesis that the population variances in the two groups are equal using the Levene test, shown in Figure 8-13. If the observed significance level is small (in the column labeled *Sig.* under *Levene's Test*), you reject the null hypothesis that the population variances are equal. For this example, you can reject the null hypothesis that the pre-intervention truancy variances are equal in the two groups. For the other two variables, you can't reject the null hypothesis that the variances are equal.

Figure 8-13
Two-independent-samples t test

		Levene's Test for Equality of Variances		t-test for Equality of Means						
									95% Confidence Interval of the Difference	
		F	Sig.	t	df	Sig. (2-tailed)	Mean Difference	Std. Error Difference	Lower	Upper
Percent truant days pre intervention	Equal variances assumed	5.25	.023	1.49	297	.138	2.25	1.51	-.72	5.22
	Equal variances not assumed			1.49	290.23	.139	2.25	1.51	-.73	5.22
Percent truant days post intervention	Equal variances assumed	.12	.727	-.12	297	.906	-.15	1.30	-2.70	2.40
	Equal variances not assumed			-.12	296.97	.906	-.15	1.29	-2.70	2.40
Pre - Post	Equal variances assumed	1.68	.196	1.64	297	.102	2.40	1.46	-.48	5.28
	Equal variances not assumed			1.63	287.91	.103	2.40	1.47	-.49	5.29

Testing the Hypothesis

In the two-independent-samples *t* test, the *t* statistic is computed the same as for the other two *t* tests. It is the ratio of the difference between the two sample means divided by the standard error of the difference. The standard error of the difference is computed differently, depending on whether the two variances are assumed to be equal or not. That's why you see two sets of *t* values in Figure 8-13. In this example, the two *t* values and confidence intervals based on them are very similar. That will always be the case when the sample size in the two groups is almost the same.

The degrees of freedom for the *t* statistic also depends on whether you assume that the two variances are equal. If the variances are assumed to be equal, the degrees of freedom is two fewer than the sum of the number of cases in the two groups. If you don't assume that the variances are equal, the degrees of freedom is calculated from the actual variances and the sample sizes in the groups. The result is usually not an integer.

From the column labeled *Sig. (2-tailed)*, you can't reject any of the three hypotheses of interest. The observed results are not incompatible with the null hypothesis that boys and girls are equally truant before and after the program and that intervention affects both groups similarly. Note that the value 0 is included in all three of the 95% confidence intervals.

Warning: When you compare two independent groups, one of which has a factor of interest and the other that does not, you must be very careful about drawing conclusions. For example, if you compare people enrolled in a weight-loss program to people who are not, you cannot attribute observed differences to the program unless the people have been randomly assigned to the two programs.

Obtaining the Output

To produce the output in this chapter, open the file *truancy.sav* and follow the instructions below.

Figures 8-6 and 8-7. From the Analyze menu choose Compare Means > One-Sample T Test. Select *prepct* as the Test Variable. Type 8 in the Test Value text box, and click OK.

Figure 8-8. From the Graphs menu choose Legacy Dialogs > Histogram. Select *diffpct* as the Variable, and click OK.

Figures 8-9, 8-10, and 8-11. From the Analyze menu choose Compare Means > Paired-Samples T Test. Select *prepct* and *postpct* as the Paired Variables, and click OK. In the output Viewer, the table used in Figure 8-9 is labeled *Paired Samples Statistics*; Figure 8-10 is *Paired Samples Test*; Figure 8-11 is *Paired Samples Correlations*.

Figures 8-12 and 8-13. From the Analyze menu choose Compare Means > Independent-Samples T Test. Select *prepct*, *postpct*, and *diffpct* as the Test Variables and *gender* as the Grouping Variable. Select the Grouping Variable box and click Define Groups. Specify *f* for Group 1 and *m* for Group 2. Click Continue to close the subdialog box, and then click OK. In the output Viewer, the table used in Figure 8-12 is labeled *Group Statistics*, and Figure 8-13 is labeled *Independent Samples Test*.

Chapter

One-Way Analysis of Variance

Although many worthwhile hypotheses can be tested by comparing the means from two groups, more complicated experiments and questions necessitate testing hypotheses about more than two means. One-way analysis of variance is used to test the null hypothesis that several independent population means are equal. You can also identify which pairs of means are significantly different from each other using multiple comparison procedures or by examining linear combinations of the group means.

Examples

- You want to prescribe the smallest dosage of a drug that is effective for treating fever. You select five possible dosages and administer them to people with high fevers. You record the time in minutes until the temperature returns to normal. Do average times for fever reduction differ for the five dosages?
- How do promotions affect product sales? As a marketing manager, you assign one of three display types to each of 100 stores and record the change in dollar sales for your product from the previous week. Are all three display types equally effective?
- You have four possible ways of filling containers on a production line. You randomly assign 50 workers to each process and determine the weight of the resulting boxes. You want to know whether there is a relationship between the final weight of the boxes and the method used to fill them.

In a Nutshell

You have several independent groups of cases, and you want to test whether they come from populations with the same mean for a variable of interest. The null hypothesis is that all population means are equal. The alternative hypothesis is (surprise!) that they are not all equal. There's a known statistical relationship between the variability of means and the variability of individual observations from the same population. If your sample means vary more than you would expect if the null hypothesis is true, you reject the null hypothesis that the samples come from populations with the same mean. The statistical procedure for doing this is called **analysis of variance** because you are examining the variability of the sample means in order to draw conclusions about the population means.

Anorexia: The Example

Hand et al. (1994) present data for three groups of anorexic girls who were treated with either cognitive behavioral therapy, family therapy, or no therapy at all (the "control group"). For each girl, they report weights before and after treatment and the change in weight. Positive values indicate that a girl has gained weight and negative values, that she has lost weight. The null hypothesis you want to test is that in the population, all three treatments result in the same average weight change. The alternative hypothesis is that they do not.

Arranging the Data

Figure 9-1 shows part of the data file *anorexia.sav*. For each girl, you have five values: an ID number, the type of therapy received, the pre-treatment weight, the post-treatment weight, and the difference between the post- and pre-treatment weights.

Figure 9-1
Data file for analysis of variance

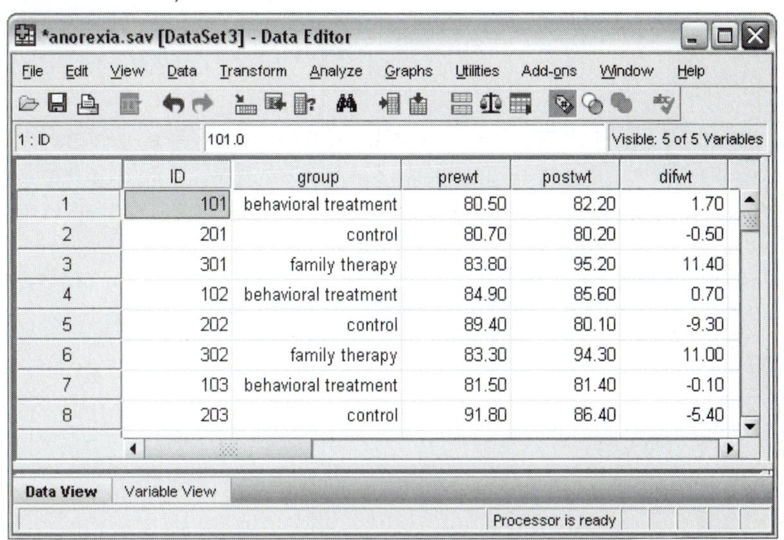

	ID	group	prewt	postwt	difwt
1	101	behavioral treatment	80.50	82.20	1.70
2	201	control	80.70	80.20	-0.50
3	301	family therapy	83.80	95.20	11.40
4	102	behavioral treatment	84.90	85.60	0.70
5	202	control	89.40	80.10	-9.30
6	302	family therapy	83.30	94.30	11.00
7	103	behavioral treatment	81.50	81.40	-0.10
8	203	control	91.80	86.40	-5.40

Tip: If you want to restrict the analysis of variance to only a subset of the groups defined by the grouping variable, you have to use Select Cases on the Data menu or set the codes for groups that you don't want to include as *missing*.

Examining the Data

As always, the first step is to examine the data values. Because the data come from three different groups, you want to look at the values separately for each group. In particular, you want to make sure that there are no very large or small values that distort the group means.

Figure 9-2 is a boxplot of weight change. Notice that the median weight change is largest for the family therapy group. It's close to 0 for both of the other groups. The lengths of the boxes, a measure of the spread of the data values, is smallest for the behavioral treatment group. However, the behavioral treatment group has quite a few values identified as outliers, meaning that they are more than 1.5 box lengths from the end of the box. If you look at a stem-and-leaf plot of the values for the behavioral therapy group, you'll notice that there are no weight losses between 6.1 and 11.6 pounds. That's an interesting finding, suggesting that behavioral treatment may have either little or no effect, or a large effect. If that's the case, the mean may not be a good summary measure for weight change.

Figure 9-2
Boxplot of weight change

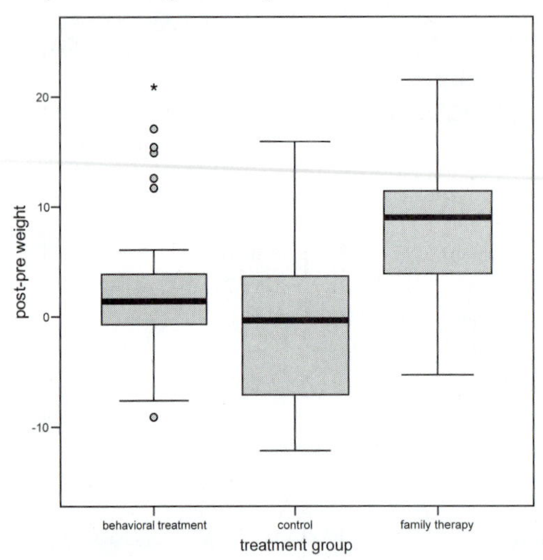

Descriptive statistics for each of the groups are shown in Figure 9-3. The average weight gain is 7.26 pounds for the family therapy group, 3.01 pounds for the behavioral treatment group, and –0.45 pounds for the control group (the untreated anorexic girls in the control group lost almost half a pound). The 95% confidence intervals for the behavioral treatment and family therapy groups don't include the value 0 and have positive endpoints, so you can reject the null hypothesis that there was no change in weight for these two groups. However, for the control group, the value 0 is in the confidence interval, so you can't reject it as a plausible population value for average weight loss.

Figure 9-3
Descriptive statistics for weight change

post-pre weight

	N	Mean	Std. Deviation	Std. Error	95% Confidence Interval for Mean Lower Bound	95% Confidence Interval for Mean Upper Bound	Min	Max
Behavioral treatment	29	3.01	7.31	1.36	.23	5.79	-9.10	20.90
Control	26	-.45	7.99	1.57	-3.68	2.78	-12.20	15.90
Family therapy	17	7.26	7.16	1.74	3.58	10.94	-5.30	21.50
Total	72	2.76	7.98	.94	.89	4.64	-12.20	21.50

The confidence intervals for the three groups are plotted in Figure 9-4. Notice that the confidence intervals for the family therapy group and the control group do not overlap. The control group and behavioral therapy confidence intervals do overlap.

Figure 9-4
Plot of confidence intervals

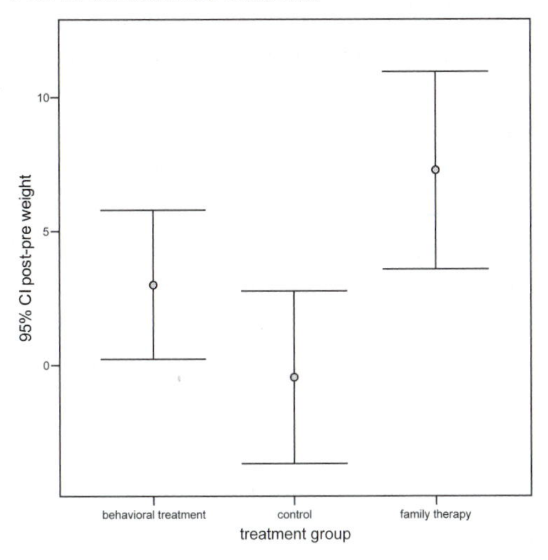

Warning: Two means can be significantly different from each other and yet have confidence intervals that overlap. Confidence intervals can overlap by as much as 29% when the means are significantly different (van Belle, 2002).

Checking the Assumptions

To use analysis of variance, the observations must be independent random samples from normal populations with equal variances.

Independence

As always, the assumption of independence is important. If you know the order in which data values were obtained, you can plot the values of each group in sequence to look for a relationship between successive values. If you use analysis of variance when observations are correlated with each other, the observed significance level may be smaller than it should be.

Equal Variances

The importance of the assumption of equal variances in all of the groups depends on whether they all have roughly equal sample sizes. If they do, violation of the equality-of-variance assumption has little effect on the observed significance levels. If the groups have markedly different sample sizes and the smaller groups have the larger variances, you increase your chance of rejecting the null hypothesis when it is true. If the larger groups have the smaller variances, you increase your chance of failing to reject the null hypothesis when it is false. Moore (1995) suggests that as long as the ratio of the largest to the smallest variances is less than 4:1, the results will be approximately correct.

You can test the null hypothesis that several population variances are equal using the Levene test, as shown in Figure 9-5. For this example, the observed significance level is large, so you don't reject the null hypothesis of equal group variances. SPSS Statistics offers several tests that don't require the assumption of equal variances. These are discussed later.

Figure 9-5
Levene test

post-pre weight

Levene Statistic	df1	df2	Sig.
.314	2	69	.731

Tip: If you violate the required assumptions, your observed significance level may be either too small or too large. Unless you've masterminded a ghastly violation, look at the observed significance level and determine how much it can change without impacting your conclusions very much. If your original p value is quite small or large, even reasonably large changes in it won't alter your conclusions. So, even if your observed p value isn't exactly right, you can live with it. (You can also compute the Kruskal-Wallis test described in Chapter 19 and see whether the results agree if you make less stringent assumptions about the origins of the data.)

Normality

The easiest way to check normality is to subtract the group mean from each observation and then look at the distribution of the differences from the group means. These differences are called **residuals**. You have to subtract the group means because the data may be coming from populations with different means. You can look at a histogram of the residuals or at special plots and tests for normality.

Figure 9-6 is a **normal quantile plot** of the weight changes after group means are subtracted from the individual weight changes. Each observed weight difference is plotted against the expected weight difference if the data are from a normal distribution. If the sample is exactly normally distributed, all the points should fall on the straight line. It's unlikely you'll ever see that unless you hallucinate about statistical analyses. Instead, the points should more or less cluster around the line.

Figure 9-6
Q-Q plot of residual weights

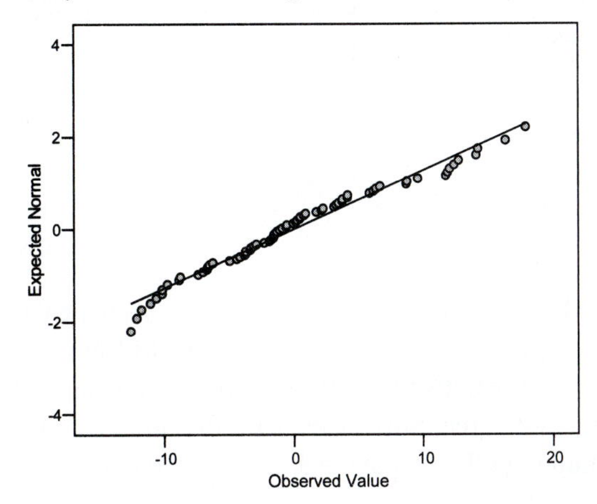

Tip: The easiest way to compute residuals is to abandon the one-way analysis of variance procedure and run Analyze > General Linear Model > Univariate. In the Univariate dialog box, click Save, and then select any of the residuals listed in the Save dialog box.

In Figure 9-6, the small observed values fall quite a bit below the predicted normal line, indicating that the tail toward small values is shorter than it would be if the distribution were normal. You can also look at a plot of the differences between the observed and predicted values, as shown in Figure 9-7. Ideally, you want the values to fall randomly about the zero line.

Figure 9-7
Detrended normal plot

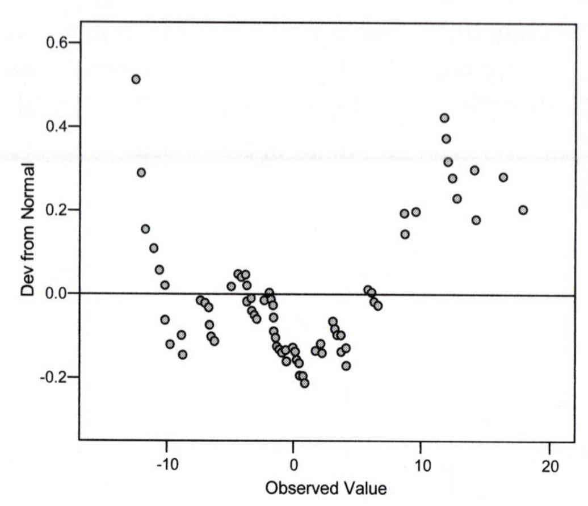

You can also compute, regardless of how you are attired, formal statistical tests that the variables come from normal populations. Figure 9-8 shows results from two commonly used tests of normality. They are sensitive to different types of deviations from normality, so the results need not agree. Small values for the observed significance level lead you to reject the null hypothesis that the data come from normal populations.

Figure 9-8
Tests of normality

	Kolmogorov-Smirnov[1]			Shapiro-Wilk		
	Statistic	df	Sig.	Statistic	df	Sig.
Residual for difwt	.091	72	.200*	.967	72	.057

*. This is a lower bound of the true significance.

1. Lilliefors Significance Correction

Warning: If your sample size is large, even small deviations from normality that won't affect your analysis may cause you to reject the null hypothesis that the samples come from normal populations. Make your decisions based on the diagnostic plots.

Testing the Hypothesis

The components you need to test the null hypothesis that all population means are equal are shown in Figure 9-9. The critical pieces are in the last two columns: the *F* statistic and its associated significance probability. If the observed significance level is small, you can reject the null hypothesis that all population means are equal. In this example, you reject the null hypothesis that the three treatments result in the same average weight change.

Figure 9-9
Analysis-of-variance table

post-pre weight

	Sum of Squares	df	Mean Square	F	Sig.
Between Groups	614.644	2	307.322	5.422	.006
Within Groups	3910.742	69	56.677		
Total	4525.386	71			

Figure 9-9 contains two estimates of variability:

- The between-groups mean square tells you how much individual observations should vary if the null hypothesis is true. This is computed from the variability of the sample means.
- The within-groups mean square tells you how much the observations within the groups really do vary.

To calculate the *F* statistic, you find the ratio of the two estimates of variance. You divide the between-groups mean square by the within-groups mean square. Large values of the *F* ratio provide evidence against the null hypothesis. The distribution of the *F* statistic depends on the number of groups that you are comparing and the number of cases in the groups. The between-groups degrees of freedom is one fewer than the number of groups, and the within-groups degrees of freedom is the number of cases minus the number of groups.

Tip: To calculate eta squared (the proportion of variance in the dependent variable "explained" by the grouping variable), use the Means procedure, and in the Options dialog box, select Anova table and eta.

Welch and Brown-Forsythe Tests

If the variances in the groups are very unequal and the sample sizes in the groups are quite different, you can use statistical procedures that don't require that all population variances are equal. Two such tests are shown in Figure 9-10. The results from these tests agree with the analysis-of-variance results. The three treatments differ in average weight loss.

Figure 9-10
Robust equality of means tests

post-pre weight

	Statistic[1]	df1	df2	Sig.
Welch	5.355	2	41.136	.009
Brown-Forsythe	5.497	2	62.959	.006

1. Asymptotically F distributed.

Warning: Unless the girls have been randomly assigned to the three treatment groups, you can't compare the effectiveness of the three treatments. For example, families that select family therapy may differ from families that won't agree to such an intervention, and differences in such family characteristics may affect outcome.

Analyzing Change

The variable that you analyzed is the difference between two weights: the weight before treatment and the weight after treatment. Intuitively, this seems a better endpoint to analyze than the final weight in each group because obviously the groups didn't start out with exactly the same initial weights, although you hope the groups were not too different.

There are several ways you can analyze change data. You can use analysis of covariance (a special type of analysis of variance, described in Chapter 22) to adjust statistically for initial weight differences. Or, you can analyze the initial weights and the final weights separately for the groups, as shown in Figure 9-11.

Figure 9-11
ANOVA of pre- and post-treatment weights

		Sum of Squares	df	Mean Square	F	Sig.
Pre-treatment weight	Between Groups	32.569	2	16.285	.599	.552
	Within Groups	1874.346	69	27.164		
	Total	1906.915	71			
Post-treatment weight	Between Groups	918.987	2	459.493	8.651	.000
	Within Groups	3665.058	69	53.117		
	Total	4584.044	71			

From the observed significance levels in Figure 9-11, you cannot reject the null hypothesis that the initial weights are equal in all groups. You can reject the null hypothesis that final weights are different. That's what you want to see. If you found significant differences in pre-treatment weights, the analysis would be much more problematic.

Tip: When you obtain multiple measurements for an individual, don't analyze just the first and the last. You can learn a lot about a problem by looking at how the values change with time. You'll have to use specialized statistical techniques for analyzing **repeated measures** data. For more information, see Chapter 23 and Chapter 24.

Pinpointing the Differences

If you reject the null hypothesis that all population means are equal based on the analysis of variance, you still don't know which pairs of means are statistically different from each other. There are many possibilities. It could be that all three population means differ from each other, or that two don't differ from each other but differ from the third. As the number of means increases, so does the number of pairs of means that may be different.

Your first thought may be to compare all possible pairs using the independent-samples *t* test. The drawback to this approach is that if you don't adjust the usual significance level for the fact that you're making multiple comparisons, you increase the chance that you'll call differences statistically significant when they're not. For example, if you compare 15 pairs of means, all of which are from the same population, and use $p < 0.05$ as your criterion for significance, your probability of finding at least one pair of significantly different means is more than a half.

Tip: If you're just exploring a dataset or investigating preliminary hypotheses, you may be willing to take the risk of an increased chance of calling differences significant when they're really not. If you set up very stringent criteria for detecting differences, you may overlook promising directions for further research.

Atoning for Many Comparisons

To sidestep the problem of rejecting the null hypothesis too often when it is true, you can use specially designed **multiple comparison procedures** that protect you from erroneously calling differences significant. You pay a price; you are less likely to find true differences between pairs of means when they do exist, because multiple comparison procedures insist on larger differences between means as the number of comparisons increases than does the ordinary *t* test.

SPSS Statistics offers you a bewildering choice of multiple comparison procedures. It's almost like choosing a new car, albeit cheaper. Don't take different procedures for test drives on the same dataset because that cancels your insurance against errors.

Click the Help button in the One-Way ANOVA Post Hoc dialog box to get descriptions of the multiple comparison procedures. If you're really brave or really bored, examine the formulas in the online algorithms document mentioned in Chapter 2. They all have different properties and protect you in various ways from calling too many differences statistically significant. Some of them are called multiple range tests because they identify groups of means that don't differ from one another. There are also different tests, depending on whether you are willing to assume that the variances in all groups are equal.

The most straightforward multiple comparison test is the Bonferroni procedure. All it does is multiply the usual observed significance level by the number of comparisons you're making. So if the usual (called *unprotected*) observed significance level is 0.01 and you are comparing eight pairs of means, the Bonferroni corrected significance level is 0.08. The Bonferroni procedure is recommended for a small number of pairwise comparisons between means and for contrasts among means (see "Contrasts: Testing Linear Combinations of Means" on p. 156).

Tip: Try to narrow down the comparisons that you are interested in before you start analyzing the data. With multiple comparison procedures, the fewer comparisons you actually make, the more likely you are to find true differences when they exist. If you want to make comparisons against only a control group, there are special procedures for just that.

Comparing All Possible Means

Figure 9-12 shows all possible comparisons between the three treatment groups using the Tukey multiple comparison procedure. With its comforting name, Tukey's honestly significant difference (HSD) test has weathered the onslaught of new tests well. It is one of the oldest and is still in favor for all pairwise comparisons of means.

The differences between pairs of means are in the column labeled *Mean Difference (I-J)*. All you have to do is figure out which groups are (*I*) and (*J*). The columns to the left of the group labels tell you which group is *I* and which group is *J*. For example, the comparison of behavioral treatment and the control has behavioral treatment as (*I*) and control as (*J*). The (*J*) means are always sorted from smallest to largest, based on their mean difference with the group (*I*) mean. Each combination of means appears twice in the table with different signs. (Don't worry, this is treated as one pair for the adjustment for the number of comparisons made.) The standard error of the mean difference is in the column after the difference.

Figure 9-12
Tukey's honestly significant difference (HSD)

Dependent Variable: post-pre weight
Tukey HSD

				Mean Difference (I-J)	Std. Error	Sig.	95% Confidence Interval	
							Lower Bound	Upper Bound
(I) treatment group	behavioral treatment	(J) treatment group	control	3.46	2.03	.212	-1.41	8.33
			family therapy	-4.26	2.30	.161	-9.77	1.25
	control	(J) treatment group	behavioral treatment	-3.46	2.03	.212	-8.33	1.41
			family therapy	-7.71*	2.35	.005	-13.34	-2.09
	family therapy	(J) treatment group	behavioral treatment	4.26	2.30	.161	-1.25	9.77
			control	7.71*	2.35	.005	2.09	13.34

*. The mean difference is significant at the .05 level.

The column labeled *Sig.* is the Tukey corrected significance value. (If you didn't correct for multiple comparisons, the significance levels would be about half as large as the ones reported.) Each of the 95% confidence intervals for the mean difference is also corrected for the fact that you are making many comparisons. It's wider than it would be if you didn't adjust for the number of comparisons.

From Figure 9-12, you see that only two groups are significantly different from each other. The family therapy group is significantly different from the control group. It's marked with an asterisk in the table.

Warning: You'll always find twice as many asterisks as there are significantly different pairs of means because the same pair appears twice.

Identifying Homogenous Subsets

For some multiple comparison tests, including Tukey's HSD, your output may also appear in a table like that shown in Figure 9-13. Instead of asterisks that denote which groups are different, the table shows which groups are similar. Each column of the table shows group means that are not significantly different from each other. The first subset indicates that the control and behavioral treatment groups are homogenous (that is, not different from each other). The second subset shows that behavioral treatment and family therapy are not different from each other. From this test, you conclude that family therapy differs from the control because the two means don't appear together in a column.

Figure 9-13
Homogeneous subsets

Tukey HSD[1,2]

		N	Subset for alpha = .05	
			1	2
treatment group	control	26	-.4500	
	behavioral treatment	29	3.0069	3.0069
	family therapy	17		7.2647
	Sig.		.275	.144

Means for groups in homogeneous subsets are displayed.

1. Uses Harmonic Mean Sample Size = 22.767.
2. The group sizes are unequal. The harmonic mean of the group sizes is used. Type I error levels are not guaranteed.

Warning: If you have unequal sample sizes in the groups, the homogeneous subset table uses a harmonic average of group sizes to determine a common standard error for all pairs of means. That's why the significance levels for the same comparisons differ between Figure 9-12 and Figure 9-13. Figure 9-12 is based on the actual sample sizes in the groups. Figure 9-13 is based on the average. You're usually better off using the actual sample sizes.

Contrasts: Testing Linear Combinations of Means

When you are comparing two population means, you are testing the null hypothesis that $\mu_1 - \mu_2 = 0$, where μ_1 is the population value of the mean for the first group and μ_2 is the population value for the second group.

This is a simple example of a **linear contrast of means**. A linear contrast of means has the general form $a_1\mu_1 + a_2\mu_2 + \ldots + a_n\mu_n = 0$, where a_1, a_2, and so on, are coefficients that you select. For each contrast, these coefficients must sum to 0. For example, if you want to test the null hypothesis that the population mean for the control is equal to the average of the population means for behavioral therapy and family therapy, you can set up the following contrast:

$$(1 \times \mu_{Control}) - \left(\frac{1}{2} \times \mu_{Behavioral}\right) - \left(\frac{1}{2} \times \mu_{Family}\right) = 0$$

Tip: You can specify the same contrast in many ways. For example, the contrast for the control against the treatments could be

$$(2 \times \mu_{Control}) - (1 \times \mu_{Behavioral}) - (1 \times \mu_{Family}) .$$

The value of the contrast will change. The observed significance level will not change.

Setting Up Contrasts in SPSS Statistics

Figure 9-14 shows the dialog box for the linear contrast of the control to the average of the two treatments. Integer coefficients are used to avoid having to fuss with decimal points. For each group, in order, you have to enter a coefficient and then click Add. If you don't want to include a group in a contrast, you must specify a value of 0 for the coefficient for that group. The dialog box shows you the sum of the coefficients under the Coefficients list. The groups are coded 1 = *behavioral treatment*, 2 = *control*, and 3 = *family therapy*; therefore, the coefficients are entered in the sequence 1, –2, and 1.

Figure 9-14
Contrasts dialog box in One-Way ANOVA procedure

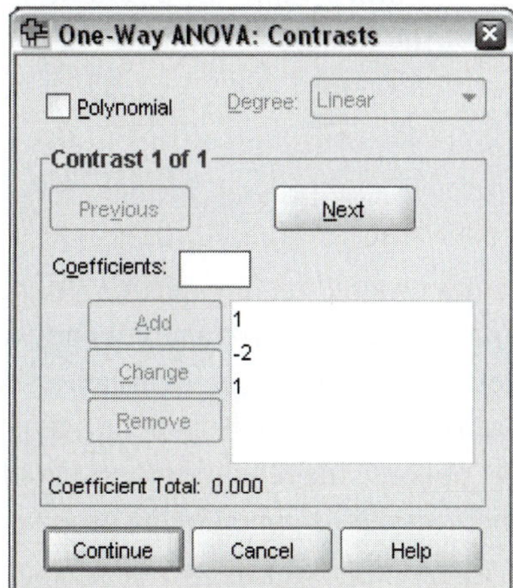

Testing the Contrast

You see the coefficients for the contrast that you are testing in Figure 9-15. Make sure that this is the contrast you intended to enter.

Figure 9-15
Contrast coefficients

		treatment group		
		behavioral treatment	control	family therapy
Contrast	1	1	-2	1

The actual test results are shown in Figure 9-16. For each contrast, you see two tests: one based on the assumption that all groups come from populations with the same variance and the other without this assumption. The value of the contrast is the coefficients applied to the sample means. For this example, the contrast value is

$$1(\overline{X}_{Behavioral}) - 2(\overline{X}_{Control}) + 1(\overline{X}_{Family}) = 3.01 - 2(-0.45) + 7.26 = 11.17$$

Figure 9-16
Test of contrast

				Value of Contrast	Std. Error	t	df	Sig. (2-tailed)
post-pre weight	Assume equal variances	Contrast	1	11.17	3.74	2.98	69	.004
	Does not assume equal	Contrast	1	11.17	3.83	2.92	47.38	.005

The *t* value is the ratio of the contrast to its standard error. Based on the observed significance level, you can reject the null hypothesis that the population mean for the control group is equal to the average of the population means for the two treatment groups.

Polynomial Contrasts

In this example, the treatment groups are not ordered on any meaningful scale. If the grouping variable represents an underlying continuum, such as frequency of reading a newspaper or dosage of a drug, you can test hypotheses about the relationship between the grouping variable and the dependent variable. You can test the null hypothesis that the relationship is linear, quadratic, cubic, or some higher-order function of the actual values of the grouping variable. The numeric codes assigned to the grouping value are used for the scale.

Figure 9-17 is a plot of years of education and average ages for people in five categories of newspaper reading: every day, a few times a week, once a week, less than once a week, and never. In this example, frequency of reading a newspaper is coded 1 through 5. That's an ordinal scale because the distances between adjacent groups has no mathematical meaning. You see that education decreases somewhat linearly with decreasing frequency of reading a newspaper. The relationship between age and frequency of reading is more complicated; it looks like a U-shaped curve.

Figure 9-17
Plot of cell means for education and for age

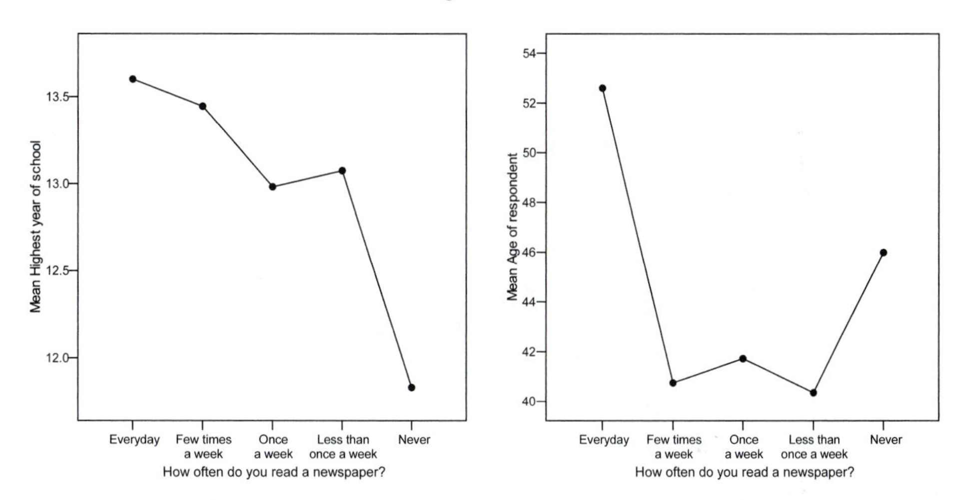

Tip: If there is an underlying scale to the grouping variable, such as actual dosages of a drug or the number of marriages, make sure you use that number for the group code if you want to look at the relationship between the level of the grouping variable and the mean of the dependent variable.

Linear Contrast

A linear contrast is the simplest polynomial contrast. With a linear contrast, you test the null hypothesis that the average highest year of school is linearly related to the frequency of reading a newspaper. Figure 9-18 is an analysis-of-variance table for the null hypothesis that the average years of education are the same for the five categories of newspaper readership. From the F statistic and associated significance level in the row labeled *Between Groups (Combined)*, you can reject the null hypothesis. To test whether the relationship between the average years of education and the newspaper-reading category is linear, look at the rows labeled *Linear Term.*

Figure 9-18
Linear contrast

Highest year of school

			Sum of Squares	df	Mean Square	F	Sig.
Between Groups	(Combined)		242.626	4	60.656	6.544	.000
	Linear Term	Unweighted	221.294	1	221.294	23.875	.000
		Weighted	195.923	1	195.923	21.137	.000
		Deviation	46.703	3	15.568	1.680	.170
Within Groups			8332.856	899	9.269		
Total			8575.482	903			

If the sample sizes in the groups are not the same, this row is further subdivided into three pieces:

- The *Unweighted* row is a test of the null hypothesis that there is no linear relationship when the same weight is assigned to all sample means, regardless of how many cases there are in each group.
- The *Weighted* row is a test of the same null hypothesis, using the actual sample sizes in each of the groups.
- The *Deviation* row is a test of the null hypothesis that the linear term is sufficient to explain the observed variation in the group means.

In this example, you can reject the null hypothesis that there is no linear relationship between the average years of education and the frequency of reading a newspaper, since the observed significance level for the unweighted and weighted tests is less than 0.0005. You don't reject the null hypothesis that the linear polynomial is enough to explain the relationship, since the observed significance level is 0.17.

Tip: Use the unweighted test for the polynomial if all means are equally important and the sample sizes in each group are adequate.

If you want to specify a polynomial contrast for average age for the data shown in Figure 9-17, a linear relationship isn't a reasonable summary. You'll have to investigate higher-order polynomial terms, such as the quadratic and cubic. The output looks similar except that there are additional terms after the linear term.

Obtaining the Output

To produce the output in this chapter, open the file *anorexia.sav* and follow the instructions below.

Figure 9-2. From the Graphs menu choose Legacy Dialogs > Boxplot. Select the Simple icon and Summaries for groups of cases, and then click Define. Select *difwt* for the Variable and *group* for the Category Axis variable. Click OK.

Figures 9-3 and 9-9. From the Analyze menu choose Compare Means > One-Way ANOVA. Select *difwt* for the Dependent List and *group* for the Factor variable. Click Options. In the subdialog box, select Descriptive in the Statistics group. Click Continue, and then click OK.

Figure 9-4. From the Graphs menu choose Legacy Dialogs > Error Bar. Select the Simple icon and Summaries for groups of cases, and then click Define. Select *difwt* for the Variable and *group* for the Category Axis variable. Accept all of the other default values in the dialog box. Click OK.

Figure 9-5. From the Analyze menu choose Compare Means > One-Way ANOVA. Select *difwt* for the Dependent List and *group* for the Factor variable. Click Options. In the subdialog box, select Homogeneity of variance test in the Statistics group. Click Continue, and then click OK.

Figures 9-6, 9-7, and 9-8. From the Analyze menu choose Descriptive Statistics > Explore. Select *res_1* for the Dependent List but do not include any other variables in the other lists. Click Plots. In the subdialog box, select Normality plots with tests. Click Continue, and then click OK.

Figure 9-10. Click the Dialog Recall button on the SPSS Statistics toolbar and select One-Way ANOVA. Do not change the variable assignments in the main dialog box. Click Options. In the subdialog box, select both Brown-Forsythe and Welch in the Statistics group. Click Continue, and then click OK.

Figure 9-11. Click the Dialog Recall button on the SPSS Statistics toolbar and select One-Way ANOVA. Click Reset to clear all prior settings. Select *prewt* and *postwt* for the Dependent List and *group* for the Factor variable. Click OK.

Figures 9-12 and 9-13. Click the Dialog Recall button on the SPSS Statistics toolbar and select One-Way ANOVA. Click Reset to clear all prior settings. Select *difwt* for the Dependent List and *group* for the Factor variable. Click Post Hoc. In the subdialog box, select Tukey in the Equal Variances Assumed group. Click Continue, and then click OK.

Figures 9-15 and 9-16. Click the Dialog Recall button on the SPSS Statistics toolbar and select One-Way ANOVA. Click Post Hoc and deselect Tukey in the Equal Variances Assumed group. Click Continue. In the One-Way ANOVA dialog box, click Contrasts. In the subdialog box, type the value 1 in the Coefficients text box, and click Add. Type the value –2 and click Add. Type the value 1 and click Add. Click Continue, and then click OK. In the output Viewer, Figure 9-15 is labeled *Contrast Coefficients*; Figure 9-16 is the table labeled *Contrast Tests*.

Figure 9-17. From the File menu choose Open > Data. Select *gssdata.sav* and click OK. From the Graphs menu choose Legacy Dialogs > Line. Select the Simple icon and Summaries for groups of cases, and then click Define. Select Other statistic in the Line Represents group. Select *educ* for the Variable (it will display as the summary function *MEAN(educ)*). Select *news* for the Category Axis variable. Click OK.

Click the Dialog Recall button on the SPSS Statistics toolbar and select Define Simple Line: Summaries for Groups of Cases. Select *MEAN(educ)* for the Variable box and move it back into the source variable list. Select *age* for the Variable (it will display as the summary function *MEAN(age)*). Click OK. The two charts will appear separately in the output Viewer.

To add the line markers seen in Figure 9-17, double-click each chart to open the Chart Editor. From the Elements menu choose Add Markers. Click the Apply button, and then click Close. The markers will now appear on the chart. From the File menu choose Close to return to the output Viewer. Repeat the same steps with the second chart.

Figure 9-18. From the Analyze menu choose Compare Means > One-Way ANOVA. Select *educ* for the Dependent List and *news* for the Factor variable. Click Contrasts. Select Polynomial in the subdialog box, leaving the Degree drop-down list set to Linear. Click Continue, and then click OK.

Chapter

10

Crosstabulation

Well-planned and carefully labeled tables are often the most critical component of a good presentation or paper. When a table has counts of the number of cases with particular combinations of values of the two variables, the table is known as a **crosstabulation** (or simply a **crosstab**). The observed counts and percentages in a crosstabulation describe the relationship between the two variables in your sample. But you must take additional steps if you want to draw conclusions about the relationship of the variables in the population. You can use the chi-square test to test the null hypothesis that two categorical variables are independent. You can summarize the strength of the relationship between the variables by using specially constructed statistics called **measures of association**.

Examples

- You ask a random sample of people how satisfied they are with their jobs. You want to know what characteristics of the employee and the job are related to satisfaction levels. (Test that two variables are independent.)
- You want to know whether four types of toys are equally likely to be defective at the time of purchase. (Test that several proportions are equal.)
- You want to describe how strongly various characteristics of a school are related to high school students' satisfaction with their education (measures of association).
- You want to know if people view a product more favorably after seeing a demonstration of the product (McNemar test).
- You want to determine if there is a relationship between a defendant's race and receiving the death penalty and if the relationship depends on the race of the victim. (Test that the odds ratio is 1.)

In a Nutshell

You classify cases based on values for two or more categorical variables (for example, type of health insurance coverage and satisfaction with healthcare). Each combination of values is called a cell. To test whether the two variables that make up the rows and columns are independent, you calculate how many cases you expect in each cell if the variables are independent, and you compare these expected values to those actually observed using the chi-square statistic. If your observed results are unlikely if the null hypothesis of independence is true, you reject the null hypothesis.

You can measure how strongly the row and column variables are related by computing measures of association. There are many different measures, and they define association in different ways. In selecting a measure of association, you should consider the scale on which the variables are measured, the type of association you want to detect, and the ease of interpretation of the measure.

You can study the relationship between a dichotomous (two-category) risk factor and a dichotomous outcome (for example, family history of a disease and development of the disease), controlling for other variables (for example, gender) by computing special measures based on odds.

Chi-Square Test: Are Two Variables Independent?

If you think that two variables are related, the null hypothesis that you want to test is that they are *not* related. Another way of stating the null hypothesis is that the two variables are **independent**. Independence has a very precise meaning in this situation. It means that the probability that a case falls into a particular cell of a table is the product of the probability that a case falls into that row and the probability that a case falls into that column.

Warning: The word *independent* as used here has nothing to do with dependent and independent variables. It refers to the absence of a relationship between two variables.

Is It in the Stars?

As an example of testing whether two variables are independent, look at Figure 10-1, a crosstabulation of astrological sign and perception of life's excitement, based on data from the General Social Survey. (Zodiacal signs are grouped into the usual *triplicities*, according to dominant element.*) If you're willing to allow data to influence your view about astrology, you can use the data in Figure 10-1 to test the null hypothesis that birth triplicity is unrelated to how you perceive life. From the row percentages, you see that the percentage of people who find life exciting is not exactly the same in the four astrological groups, although it's fairly similar. About 46% of the fire and water groups, 45% of the earth group, and 47% of the air group find life exciting. You see that the differences for the other two categories of life excitement are similarly small.

Warning: The chi-square test requires that all observations be independent. This means that each case can appear in only one cell of the table. For example, if you apply two different treatments to the same patients and classify them both times as improved or not improved, you can't analyze the data with the chi-square test of independence.

Figure 10-1
Birth triplicity and perception of life

			How's life?			
			Exciting	Routine	Dull	Total
Dominant element	Fire	Count	2788	2931	306	6025
		Expected Count	2776.3	2940.4	308.3	6025
		Row %	46.3%	48.6%	5.1%	100.0%
	Earth	Count	2690	2956	323	5969
		Expected Count	2750.5	2913.1	305.4	5969
		Row %	45.1%	49.5%	5.4%	100.0%
	Air	Count	2915	3002	309	6226
		Expected Count	2868.9	3038.5	318.6	6226
		Row %	46.8%	48.2%	5.0%	100.0%
	Water	Count	2738	2900	298	5936
		Expected Count	2735.3	2897.0	303.7	5936
		Row %	46.1%	48.9%	5.0%	100.0%
Total		Count	11131	11789	1236	24156
		Expected Count	11131.0	11789.0	1236.0	24156
		Row %	46.1%	48.8%	5.1%	100.0%

* Thanks to Dr. Bruce Stephenson of the Adler Planetarium for incisive and expensive consultation on matters astrological, astronomical, and oenological.

Computing Expected Values

You use the chi-square test to determine if your observed results are unlikely if the two variables are independent in the population. Two variables are independent if knowing the value of one variable tells you nothing about the value of the other variable. The triplicity of a person's birth and his or her perception of life are independent if the probability of any triplicity/perception-of-life combination is the product of the probability of that triplicity times the probability of that perception of life. For example, under the independence assumption, the probability of being born in an Earth sign and finding life exciting is

$$\text{Probability(Earth sign)} \times \text{Probability(life exciting)}$$

$$P = \frac{5{,}969}{24{,}156} \times \frac{11{,}131}{24{,}156} = 0.114$$

If the null hypothesis is true, you expect to find in your table 2,750 excited people born in the Earth signs. You see this expected value in the row labeled *Expected Count* in Figure 10-1.

The chi-square test is based on comparing these two counts: the observed number of cases in a cell and the expected number of cases in a cell if the two variables are independent. The Pearson chi-square statistic is

$$\chi^2 = \sum_{\text{all cells}} \frac{(\text{observed} - \text{expected})^2}{\text{expected}}$$

Tip: By examining the differences between observed and expected values in the cells (the residuals), you can see where the independence model fails. You can examine actual residuals and residuals standardized by estimates of their variability to help you pinpoint departures from independence by requesting them in the Cells dialog box of the Analyze > Descriptive Statistics > Crosstabs procedure.

Determining the Observed Significance Level

From the calculated chi-square value, you can estimate how often in a sample you would expect to see a chi-square value at least as large as the one you observed if the independence hypothesis is true in the population. If the observed significance level is small enough, you reject the null hypothesis that the two variables are independent. The value of chi-square depends on the number of rows and columns in the table. The degrees of freedom for the chi-square statistic is calculated by finding the product of one fewer than the number of rows and one fewer than the number of columns. (The degrees of freedom is the number of cells in a table that can be arbitrarily filled when the row and column totals are fixed.) In this example, the degrees of freedom is 6.

From Figure 10-2, you see that the observed significance level for the Pearson chi-square is 0.59, so you cannot reject the null hypothesis that birth triplicity and perception of life are independent. That means all readers should find this book equally exciting. The sample size in this example is very large, so even small differences among the triplicities should have been detected.

Warning: A conservative rule for use of the chi-square test requires that the expected values in each cell be greater than 1 and that most cells have expected values greater than 5. After SPSS Statistics displays the pivot table with the statistics, it displays the number of cells with expected values less than 5 and the minimum expected count. If more than 20% of your cells have expected values less than 5, you should combine categories, if that makes sense for your table, so that most expected values are greater than 5.

Figure 10-2
Tests of independence for triplicity and perception of life

	Value	df	Asymp. Sig. (2-sided)
Pearson Chi-Square	4.651[1]	6	.589
Continuity Correction			
Likelihood Ratio	4.641	6	.591
Linear-by-Linear Association	.332	1	.564
N of Valid Cases	24156		

1. 0 cells (.0%) have expected count less than 5. The minimum expected count is 303.73.

Examining Additional Statistics

SPSS Statistics displays several statistics in addition to the Pearson chi-square when you ask for a chi-square test, as shown in Figure 10-2.

- The likelihood-ratio chi-square has a different mathematical basis than the Pearson chi-square, but, for large sample sizes, it is close in value to the Pearson chi-square. It is seldom that these two statistics will lead you to different conclusions.
- The linear-by-linear association statistic is also known as the Mantel-Haenszel chi-square. It is based on the Pearson correlation coefficient. It tests whether there is a linear association between the two variables. You should not use this statistic for nominal variables. For ordinal variables, the Mantel-Haenszel test is more likely to detect a linear association between the variables than is the Pearson chi-square test; it is more powerful.
- A continuity-corrected chi-square value is shown for tables with two rows and two columns. Some statisticians claim that this leads to a better estimate of the observed significance level, but the claim is disputed.
- Fisher's exact test is calculated if any expected value in a 2-by-2 table is less than 5. You get exact probabilities of obtaining the observed table or one more extreme if the two variables are independent and the marginals are fixed. That is, the number of cases in the rows and columns of the table is determined in advance by the researcher.

Warning: The Mantel-Haenszel test is calculated by using the actual values of the row and column variables, so if you coded three unevenly spaced dosages of a drug as 1, 2, and 3, those values are used for the computations.

Are Proportions Equal?

A special case of the chi-square test for independence is the test that several proportions are equal. For example, you want to test whether the percentage of people who report themselves to be very happy has changed during the time that the General Social Survey has been conducted. Figure 10-3 is a crosstabulation of the percentage of people who said they were very happy for each of the decades. You see that the highest value is for the earliest years. Almost 35% of people questioned in the 1970s claimed they were very happy, compared to 31% in this millennium.

Calculating the Chi-Square Statistic

If the null hypothesis is true, you expect 32.1% of people to be very happy in each decade, the overall *Very happy* rate. You calculate the expected number in each decade by multiplying the total number of people questioned in each decade by 32.1%. The expected number of *Not very happy* people is 67.9% multiplied by the number of people in each decade. These values are shown in Figure 10-3. The chi-square statistic is calculated in the usual fashion.

Figure 10-3
Happiness with time

			Decade				Total
			1972-1979	1980-1989	1990-1999	2000-2002	
General happiness	Very happy	Count	3637	4475	4053	1296	13461
		Expected Count	3403.4	4516.7	4211.5	1329.4	13461.0
		Column %	34.3%	31.8%	30.9%	31.3%	32.1%
	Not very happy	Count	6977	9611	9081	2850	28519
		Expected Count	7210.6	9569.3	8922.5	2816.6	28519.0
		Column %	65.7%	68.2%	69.1%	68.7%	67.9%
Total		Count	10614	14086	13134	4146	41980
		Expected Count	10614.0	14086.0	13134.0	4146.0	41980.0
		Column %	100.0%	100.0%	100.0%	100.0%	100.0%

From Figure 10-4, you see that the observed significance level for the chi-square statistic is less than 0.0005, leading you to reject the null hypothesis that in each decade people are equally likely to describe themselves as very happy. Notice that the difference between years is not very large; the largest percentage is 34.3% for the 1970s, while the smallest is 30.9% for the 1990s. The sample sizes in each group are very large, so even small differences are statistically significant, although they may have little practical implication.

Figure 10-4
Chi-square tests

	Value	df	Asymp. Sig. (2-sided)
Pearson Chi-Square	34.180[1]	3	.000
Continuity Correction			
Likelihood Ratio	33.974	3	.000
Linear-by-Linear Association	25.746	1	.000
N of Valid Cases	41980		

1. 0 cells (.0%) have expected count less than 5. The minimum expected count is 1329.43.

Introducing a Control Variable

To see whether both men and women experienced changes in happiness during this time period, you can compute the chi-square statistic separately for men and for women, as shown in Figure 10-5.

Tip: You get separate chi-square tests for every combination of values of the control variable(s) moved into the layer box.

Figure 10-5
Chi-square tests by gender

			Value	df	Asymp. Sig. (2-sided)
Gender	Male	Pearson Chi-Square	3.677	3	.298
		Continuity Correction			
		Likelihood Ratio	3.668	3	.300
		Linear-by-Linear Association	.901	1	.343
		N of Valid Cases	18442		
	Female	Pearson Chi-Square	42.987	3	.000
		Continuity Correction			
		Likelihood Ratio	42.712	3	.000
		Linear-by-Linear Association	35.904	1	.000
		N of Valid Cases	23538		

You see that for men, you can't reject the null hypothesis that happiness has not changed with time. You can reject the null hypothesis for women. From the line plot in Figure 10-6, you see that in the sample, happiness decreases with time for women but not for men.

Figure 10-6
Line chart of percentage very happy with time for men and women

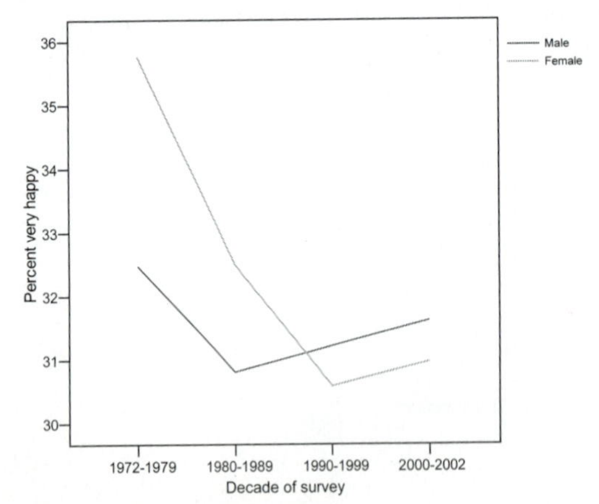

Measuring Change: McNemar Test

The chi-square test can also be used to test hypotheses about change when the same people or objects are observed at two different times. For example, Figure 10-7 is a crosstabulation of whether a person voted in 1996 and whether he or she voted in 2000. An interesting question is whether people were more likely to vote in one of the years than the other. The cases on the diagonal of the table don't provide any information because they behaved similarly in both elections. You have to look at the off-diagonal cells, which correspond to people who voted in one election but not the other. If the null hypothesis that likelihood of voting did not change is true, a case should be equally likely to fall into either of the two off-diagonal cells.

The binomial distribution is used to calculate the exact probability of observing a split between the two off-diagonal cells at least as unequal as the one observed, if cases in the population are equally likely to fall into either off-diagonal cell. This test is called the **McNemar test**.

Figure 10-7
Crosstabulation of voting behavior

Count

		Vote in1996 election		Total
		Yes	No	
Vote in 2000 election	Yes	1539	151	1690
	No	187	502	689
Total		1726	653	2379

The McNemar test can be calculated for a square table of any size to test whether the upper half and the lower half of a square table are symmetric. This test is labeled as the McNemar-Bowker test for square tables with more than two rows and columns. From Figure 10-8, you see that you can't reject the null hypothesis that people who voted in only one of the two elections were equally likely to vote in either.

Figure 10-8
McNemar test for changes in voting

	Value	df	Asymp. Sig. (2-sided)	Exact Sig. (2-sided)
McNemar Test				.057[1]
N of Valid Cases	2379			

1. Binomial distribution used.

Warning: Because the same person is asked whether he or she voted in 1996 and whether he or she voted in 2000, you can't make a table in which the rows are years and the columns are whether he or she voted. Each case would appear twice in such a table.

How Strongly Are Two Variables Related?

If you reject the null hypothesis that two variables are independent, you may want to describe the nature and strength of the relationship between the two variables. There are many statistical indexes that you can use to quantify the strength of the relationship between two variables in a cross-classification. No single measure adequately summarizes all possible types of association. Measures vary in the way they define perfect and intermediate association and in the way they are interpreted. Some measures are used only when the categories of the variables can be ordered from lowest to highest on some scale.

Warning: Don't compute a large number of measures and then report the most impressive as if it were the only one examined.

You can test the null hypothesis that a particular measure of association is 0, based on an approximate *t* statistic shown in the output. If the observed significance level is small enough, you reject the null hypothesis that the measure is 0.

The standard errors of some of the measures of association are labeled *Asymptotic* in the output because they are correct only for large sample sizes. The *t* value is labeled *Approximate* for the same reason. You can compute confidence intervals for measures of association by using the standard error labeled *Asymp Std Error 1*. This estimate of standard error, unlike the one used for computing the *t* value, is not based on the assumption that the null hypothesis is true; however, it is also correct only for large sample sizes.

Tip: Measures of association should be calculated with as detailed data as possible. Don't combine categories with small numbers of cases, as was suggested above for the chi-square test of independence.

Measures of Association for Nominal Variables

If you have variables measured on nominal scales, you're limited in what you can say about their association. It doesn't make sense to say that preferred car color increases with region of the country. For nominal variables, you can't say anything about the "direction" of the association.

Measures of association for nominal variables are of two types: those based on the chi-square statistic and those based on the proportional reduction in error (PRE).

Chi-Square-Based Measures

Because the first step in many analyses is testing independence with a chi-square test, it's not surprising that some measures of association are based on the chi-square statistic. These measures don't require the categories of the variables to have any meaningful order. The major drawback to the chi-square-based measures is that they are very difficult to interpret because they are not based on any kind of intuitive measure of association.

Tip: You can't just compare chi-square statistics because the value of chi-square depends on the sample size as well as the discrepancy from independence. If you increase your sample size by a factor of 5, you multiply the chi-square value by 5, assuming the relationship between the two variables is the same. You can get large values for chi-square when the sample size is large, even if the discrepancies between the observed and expected counts are relatively small.

Because the only restriction on the value of the chi-square statistic is that it has to be 0 or greater, measures of association based on the chi-square statistic attempt to normalize it so that the values fall in a range of 0 to 1. That makes it possible to compare the measures for tables of different dimensions and different sample sizes.

Phi Coefficient

You calculate the phi coefficient by dividing the chi-square by the sample size (*N*) and then taking the square root:

$$\phi = \sqrt{\frac{\chi^2}{N}}$$

The phi coefficient (ϕ) can be greater than 1 in value. For a 2-by-2 table, SPSS Statistics attaches a minus sign, if necessary, so that the phi coefficient is equal to the Pearson correlation coefficient.

Cramér's V

Cramér's V is bounded by 0 and 1 and can attain the maximum of 1 for a table of any size. Cramér's V is:

$$V = \sqrt{\frac{\chi^2}{N(k-1)}}$$

where k is the smaller of the number of rows and columns in the table.

Contingency Coefficient

The contingency coefficient (C) is always between 0 and 1, but the upper limit can't always be attained. For example, in a table with four rows and four columns, the maximum value of the contingency coefficient is about 0.87.

$$C = \sqrt{\frac{\chi^2}{\chi^2 + N}}$$

An Example

Consider the newspaper-reading data from Chapter 6. The crosstabulation of highest degree attained and frequency of reading the newspaper is shown in Figure 10-9. Notice that the frequency of the newspaper-reading variable has been recoded so that the values go from least frequent to most frequent.

Tip: Whenever you compute measures of association for ordinal variables, you want to make sure that they are both coded in the same way: from smallest to largest or largest to smallest. Positive coefficients will mean that the values of the variables increase together; negative coefficients will mean that as values of one variable increase, the values of the other variable decrease.

Figure 10-9
Crosstabulation of degree and newspaper readership

			How often do you read a newspaper?					
			Never	Less than once a week	Once a week	Few times a week	Everyday	Total
Highest degree	Lt high school	Count	22	20	21	24	48	135
		Row percent	16.3%	14.8%	15.6%	17.8%	35.6%	100.0%
	High school	Count	56	72	66	134	178	506
		Row percent	11.1%	14.2%	13.0%	26.5%	35.2%	100.0%
	Junior college	Count	1	10	3	11	30	55
		Row percent	1.8%	18.2%	5.5%	20.0%	54.5%	100.0%
	Bachelor	Count	4	13	12	31	80	140
		Row percent	2.9%	9.3%	8.6%	22.1%	57.1%	100.0%
	Graduate	Count	5	6	7	19	33	70
		Row percent	7.1%	8.6%	10.0%	27.1%	47.1%	100.0%
Total		Count	88	121	109	219	369	906
		Row percent	9.7%	13.4%	12.0%	24.2%	40.7%	100.0%

Chi-square and the chi-square-based measures are shown in Figure 10-10. Cramér's *V* is the smallest, and phi is the largest. On a scale of 0 to 1, the coefficients are quite small, although they are significantly different from 0. The observed significance level for the chi-square-based measures is the same as for the Pearson chi-square test.

Figure 10-10
Chi-square-based measures

	Value	df	Asymp. Sig. (2-sided)
Pearson Chi-Square	49.146[1]	16	.000

1. 0 cells (.0%) have expected count less than 5. The minimum expected count is 5.34.

		Value	Approx. Sig.
Nominal by Nominal	Phi	.233	.000
	Cramer's V	.116	.000
	Contingency Coefficient	.227	.000
N of Valid Cases		906	

Proportional Reduction in Error Measures

Proportional reduction in error (PRE) measures are easier-to-interpret alternatives to chi-square-based measures. You compare two error rates: the error when one variable is used to predict the values of another and the error when no additional information is available. The general form of a PRE measure is

$$\text{PRE coefficient} = \frac{\text{Error without } X - \text{Error with } X}{\text{Error without } X}$$

where X is the variable that may supply additional information. Different prediction rules are used by different PRE measures.

Coefficient Lambda

A simple PRE coefficient is lambda (λ). It's based on the rule that for each category of the independent variable, you predict the category of the dependent variable that occurs most frequently (the modal category). Lambda measures association in a very specific way. It measures the reduction in error when values of one variable are used to predict values of the other, using the modal categories. If this particular type of association is absent, lambda is 0. Measures that define association differently need not be 0.

- Lambda always ranges between 0 and 1.
- A value of 0 means that the additional variable is of no help in predicting the dependent variable. Lambda is 0 when two variables are independent, but a lambda of 0 doesn't necessarily mean that variables are independent.
- A value of 1 means that you can predict the dependent variable perfectly by knowing the value of the independent variable. That happens when only one value of the dependent variable occurs for each value of the independent variable.
- You can compute three different lambdas for a table: one with the row variable as dependent, one with the column variable as dependent, and a symmetric lambda, which is based on predicting the row and column variables in turn.

Consider the data in Figure 10-11, a crosstabulation of depth of hypnosis and success in treatment of migraine headaches by suggestion (Cedercreutz, 1978). You want to know if depth of hypnosis helps you to predict success in treatment.

Figure 10-11

Depth of hypnosis and outcome

			Outcome of treatment			Total
			Cured	Better	No change	
Depth of hypnosis	Deep	Count	13	5	0	18
		Row %	72.2%	27.8%	.0%	100.0%
	Medium	Count	10	26	17	53
		Row %	18.9%	49.1%	32.1%	100.0%
	Light	Count	0	1	28	29
		Row %	.0%	3.4%	96.6%	100.0%
Total		Count	23	32	45	100
		Row %	23.0%	32.0%	45.0%	100.0%

To compute lambda, you first compute the two errors:

- **Error without depth of hypnosis.** The best guess of the results of treatment when no other information is available is the modal category—the outcome category with the largest number of observations. In this example, it's *No change*. If you predict *No change* for everyone, you will misclassify 55 people out of the total of 100.

- **Error using depth of hypnosis.** For each level of hypnosis, you predict the outcome that occurs most frequently. For deep hypnosis, you predict *Cured*; for medium hypnosis, you predict *Better*; and for light hypnosis, you predict *No change*. You'll be wrong for 5 of the deep hypnosis patients, 27 of the medium hypnosis patients, and 1 of the light hypnosis patients. The total number of errors is $5 + 27 + 1 = 33$.

  You can now calculate lambda as

$$\lambda = \frac{55 - 33}{55} = 0.40$$

You can reduce your prediction error by 40% by including information about the depth of hypnosis. That's the value for lambda with the outcome of treatment as the dependent variable in Figure 10-12.

Figure 10-12
Lambda statistics for depth of hypnosis and outcome

			Value	Asymp. Std. Error[1]	Approx. T[2]	Approx. Sig.
Nominal by Nominal	Lambda	Symmetric	.353	.113	2.753	.006
		Depth of Hypnosis Dependent	.298	.147	1.723	.085
		Outcome of Treatment Dependent	.400	.105	3.076	.002

1. Not assuming the null hypothesis.
2. Using the asymptotic standard error assuming the null hypothesis.

Warning: If you divide the value of the measure by the standard error labeled *Asymp. Std. Error*, you won't get the *t* value. The *t* value has as its denominator a standard error that is computed based on the assumption that the null hypothesis is true.

Goodman and Kruskal's Tau

When you computed lambda, you made the same prediction for all cases in a particular row or column. Goodman and Kruskal's tau (τ) is based on making predictions in the same proportion as the marginal totals. Within each category of the independent variable, you randomly predict the outcome in the same proportion as observed for that category.

- **Error without depth of hypnosis.** Predict *Cured* for 23% of the cases, *Better* for 32% of the cases, and *No change* for the remaining 45%. Using this rule, you correctly classify 23% of the 23 *Cured* patients, 32% of the 32 *Better* patients, and 45% of the 45 *No change* patients. That's a total of 35.78 patients classified correctly. That means you misclassify about 64 people.

- **Error with depth of hypnosis.** For deep hypnosis, predict *Cured* 72% of the time and *Better* 28% of the time. For medium depth, predict *Cured* 19% of the time, *Better* 49% of the time, and *No change* 32% of the time. For light hypnosis, predict *Better* 3% of the time and *No change* 97% of the time. This results in correct classification for about 58 cases and incorrect classification for about 42.

The proportional reduction in error is

$$\tau(\text{migraine} \mid \text{hypnosis}) = \frac{64 - 42}{64} = 0.34$$

By incorporating depth of hypnosis, you have reduced your error of prediction by about 34%. This is the value for Goodman and Kruskal's tau shown in Figure 10-13.

Just as with lambda, there are different versions for Goodman and Kruskal's tau, depending on which variable is considered to be the dependent variable.

Figure 10-13
Goodman and Kruskal's tau

			Value	Asymp. Std. Error[1]	Approx. T[2]	Approx. Sig.
Nominal by Nominal	Goodman and Kruskal tau	Symmetric				
		Depth of Hypnosis Dependent	.294	.063		.000[3]
		Outcome of Treatment Dependent	.345	.049		.000[3]

1. Not assuming the null hypothesis.
2. Using the asymptotic standard error assuming the null hypothesis.
3. Based on chi-square approximation

Ordinal Measures

Proportional reduction in error measures can be computed for variables measured on any scale because they don't use any order information. You can interchange rows or columns in the table, and you'll get the same values for PRE measures. If your variables are measured on an ordered scale (such as from lowest to highest), you can compute measures that use this additional information. These measures can indicate direction because it makes sense to talk about values of one variable increasing as values of the other variable increase or decrease.

Warning: If you have ordinal variables whose values are strings (for example, *small*, *medium*, *large*, *xlarge*), be sure to assign ordered numeric codes before computing measures of association that use the values of the data. SPSS Statistics orders string variables by their alphabetical sort order. That results in the ordering *large*, *medium*, *small*, *xlarge*.

Measures Based on Examining Pairs

Several measures of association for ordinal variables are based on classifying all possible pairs of cases as either concordant, discordant, or tied.

- A pair of cases is **concordant** if the case with the larger value of the first variable also has the larger value for the second variable.

- A pair of cases is **discordant** if the case with the larger value of the first variable has the smaller value for the second variable.
- If the two cases have identical values on one or on both variables, they are said to be **tied**.

For example, if you are studying education and income, two cases are concordant if the case with more education also has more income. The pair is discordant if the case with more education has less income. They are tied if either income or education is the same. If the majority of pairs is concordant, the variables have a positive association; if the majority of pairs is discordant, the association is negative. If concordant and discordant pairs are equally likely, an association of this type doesn't exist.

All of the ordinal measures of association calculated by SPSS Statistics have the same numerator—the difference between the number of concordant and discordant pairs. The measures vary in the way the difference is normalized. You want the measure to range from –1 to +1 for all tables. Because cases in the same row and in the same column of a table are tied, you can't just divide by the total number of pairs in the table.

Gamma

Gamma (γ) is the simplest and most popular of the ordinal measures. It's defined as

$$\gamma = \frac{P - Q}{P + Q}$$

where P is the number of cases concordant on the two variables and Q is the number of discordant cases.

The gamma statistic ignores the number of cases with ties and includes only untied cases in the computation. Gamma can be thought of as the difference between the probability that a pair of cases is concordant and the probability that a pair of cases is discordant, assuming the absence of ties. You can also think of gamma as a proportional reduction in error measure. Its absolute value is the reduction in error between guessing concordant or discordant based on which occurs more often in the data, and guessing based on the fair toss of a coin. Gamma equals 1 if all observations are concentrated in the upper-left to lower-right diagonal of the table.

Warning: If observations are independent, gamma is 0, but a gamma of 0 implies independence only for a 2-by-2 table.

Kendall's Tau-b and Kendall's Tau-c

Two measures that take ties into account are Kendall's tau-*b* and tau-*c*. Kendall's tau-*b* includes, in the denominator, pairs that are tied on one variable but not on both.

$$\tau_b = \frac{P - Q}{\sqrt{(P + Q + T_X) \times (P + Q + T_Y)}}$$

where T_X is the number of cases with ties only on variable *X* and T_Y is the number of cases with ties only on variable *Y*. The disadvantage of tau-*b* is that it can have the values of +1 and –1 only if the table is square.

A measure that can attain or nearly attain +1 and –1 for any table is tau-*c*.

$$\tau_c = \frac{2m(P - Q)}{N^2(m - 1)}$$

where *m* is the smaller of the number of rows and columns and *N* is the total sample size.

From the measures of association in Figure 10-14 for the crosstabulation of highest degree and newspaper readership, you see that the values of these statistics are greater than 0. The relationship between the two variables is positive; as the values for highest degree increase, the values for frequency of reading a newspaper increase.

Figure 10-14
Kendall's tau-b, Kendall's tau-c, and gamma

		Value	Asymp. Std. Error[1]	Approx. T[2]	Approx. Sig.
Ordinal by Ordinal	Kendall's tau-b	.153	.028	5.486	.000
	Kendall's tau-c	.130	.024	5.486	.000
	Gamma	.226	.041	5.486	.000
N of Valid Cases		906			

1. Not assuming the null hypothesis.
2. Using the asymptotic standard error assuming the null hypothesis.

You see that the three measures are roughly similar in magnitude, although gamma is larger than the other two. That's because there are a lot of cases with tied values, and gamma just ignores them.

Asymmetric Ordinal Measures: Somers' d

All of the *ordinal* measures discussed so far are symmetric, since they don't distinguish between independent and dependent variables. They just quantify the strength and the positive or negative direction of the association. If you want to consider one of your variables as dependent and the other as independent, you can modify gamma by including the number of pairs not tied on the independent variable (X) in the denominator. Somers' d is

$$D_Y = \frac{P - Q}{P + Q + T_Y}$$

where T_Y is the number of pairs of cases tied on the dependent variable only. The value of Somers' d will always be less than the value of gamma. Somers' d can be interpreted as the excess of concordant pairs over discordant pairs among pairs not tied on the independent variable. A symmetric version of Somers' d uses, in the denominator, the average value of the denominators of the two asymmetric versions. From Figure 10-15 for the newspaper data, Somers' d, with newspaper readership as the dependent variable, is 0.16; the symmetric version is 0.15.

Figure 10-15
Somers' d

			Value	Asymp. Std. Error[1]	Approx. T[2]	Approx. Sig.
Ordinal by Ordinal	Somers' d	Symmetric	.152	.028	5.486	.000
		Highest degree Dependent	.142	.026	5.486	.000
		How often do you read a newspaper? Dependent	.164	.030	5.486	.000

1. Not assuming the null hypothesis.
2. Using the asymptotic standard error assuming the null hypothesis.

Tip: The approximate t values are always the same for gamma, Kendall's tau-b and tau-c, and all versions of Somers' d. This means that if you reject the null hypothesis that one of them is 0, you can reject the null hypothesis that the others are 0.

Eta Coefficient

If the dependent variable has values that are ordered on an interval scale (for example, number of children or years of education) while the independent variable has a limited number of categories, you can calculate the proportion of variability in the dependent variable that is explained by the values of the independent variable. Eta squared is an asymmetric measure that doesn't require the independent variable to have any ordering. It ranges in value from 0 to 1.

Measures Based on Correlation Coefficients

If both variables are measured on an ordered scale, you can compute the usual Pearson correlation coefficient between the two variables. You can also calculate the Pearson correlation coefficient using ranks instead of the original values. This is called the **Spearman correlation**. For the newspaper data from Figure 10-16, the Spearman correlation coefficient is 0.18, indicating a weak linear association between the categories of the two variables.

Figure 10-16
Correlation-based measures

		Value	Asymp. Std. Error[1]	Approx. T[2]	Approx. Sig.
Interval by Interval	Pearson's R	.177	.031	5.408	.000[3]
Ordinal by Ordinal	Spearman Correlation	.179	.032	5.483	.000[3]
N of Valid Cases		906			

1. Not assuming the null hypothesis.
2. Using the asymptotic standard error assuming the null hypothesis.
3. Based on normal approximation.

Measuring Agreement

When you have the same person or object rated or diagnosed by two observers, you can measure the degree of agreement between the two raters using Cohen's kappa coefficient. The kappa coefficient doesn't require that the categories be ordered. Figure 10-17 shows two observers' ratings of the classroom styles of 72 teachers (Bishop, Fienberg, and Holland, 1975). The percentages in the table are total percentages that express the number of cases in each cell as a percentage of the total number of cases in the table.

Figure 10-17
Crosstabulation of two raters

			Teacher2			
			Authoritarian	Democratic	Permissive	Total
Teacher1	Authoritarian	Count	17	4	8	29
		% of Total	23.6%	5.6%	11.1%	40.3%
	Democratic	Count	5	12	0	17
		% of Total	6.9%	16.7%	.0%	23.6%
	Permissive	Count	10	3	13	26
		% of Total	13.9%	4.2%	18.1%	36.1%
Total		Count	32	19	21	72
		% of Total	44.4%	26.4%	29.2%	100.0%

The simplest measure that comes to mind is the percentage of teachers that received the same rankings from the two judges. You just sum the percentages for the diagonal cells. Over 58% of the ratings of the two judges were identical. The problem with this simple measure is that it doesn't take into account chance agreement. Even if two judges were assigning ratings without actually observing the teachers, they would be expected to agree, by chance alone, a certain percentage of the time.

To compute chance agreement, for each judge you find the percentage of teachers assigned to each of the three categories, and then you calculate the chance agreement for each type of teacher by multiplying the two percentages. For the authoritarian cell, you expect chance agreement to be $(29/72) \times (32/72) = 0.179$, or 17.9%. Chance agreement for the democratic cell is $(17/72) \times (19/72) = 0.062$, and for the permissive cell, it's $(26/72) \times (21/72) = 0.105$. The sum of these is 34.6%, the percentage of all teachers that you would expect to be classified the same merely by chance.

Warning: The two variables that contain the ratings must have the same coding scheme; otherwise, SPSS Statistics will not compute kappa. SPSS Statistics also will not compute kappa if one of the raters did not use one or more of the codes, resulting in a nonsquare table or if the raters agree perfectly because the denominator of the statistic is not defined.

Cohen's Kappa

The difference between the observed proportion of cases in which the raters agree and that expected by chance is $0.583 - 0.346 = 0.237$. Cohen's kappa (κ) normalizes this difference by dividing it by the maximum difference possible, given the number of rankings in each margin. For this example, the largest possible non-chance agreement is $1.0 - 0.346 = 0.654$. Kappa is

$$\kappa = \frac{\text{observed} - \text{chance}}{1.0 - \text{chance}} = \frac{0.583 - 0.346}{1.0 - 0.346} = 0.362$$

Large values of Cohen's kappa indicate high degrees of agreement. According to J. L. Fleiss (1981), values exceeding 0.75 suggest strong agreement above chance, values in the range of 0.40 to 0.75 indicate fair levels of agreement above chance, and values below 0.40 are indicative of poor agreement above chance levels. You can test the null hypothesis that the population value for kappa is 0, based on the observed significance level in Figure 10-18. In this example, you reject the null hypothesis that kappa has a value of 0 in the population.

Figure 10-18
Cohen's kappa

		Value	Asymp. Std. Error[1]	Approx. T[2]	Approx. Sig.
Measure of Agreement	Kappa	.362	.091	4.329	.000
N of Valid Cases		72			

1. Not assuming the null hypothesis.
2. Using the asymptotic standard error assuming the null hypothesis.

Warning: Although kappa is a frequently reported statistic, it has serious shortcomings. Its value depends on the marginal distribution of the variables. You can't compare kappas for tables in which marginal distributions are different.

Measuring Risk in 2-by-2 Tables

Special measures are used to study the relationship between a dichotomous independent variable, sometimes called a **risk factor**, and the occurrence of an event (for example, smoking and lung cancer, race and the death penalty, alcohol and auto accidents, and treatment and cure).

Data can come from three different experimental designs:

- A **cohort** study, in which you follow two groups (one with the risk factor and the other without) and record how often the event of interest occurs in each group.
- A **cross-sectional** study, in which you take a random sample of individuals and count how many fall into each of the four cells of the table.
- A **case-control** study, in which you examine a group of individuals who have experienced the event of interest and a group of individuals who have not. For each of the individuals, you record whether the risk factor was present or not.

Ideally, you want to determine how much more or less likely you are to experience the event of interest if the risk factor is present. For example, you want to say that a smoker is four times more likely to develop lung cancer than a nonsmoker; that someone who has been drinking is 10 times more likely to be involved in an auto accident than someone who has not; or that someone who uses SPSS Statistics is 20 times more likely to achieve fame and glory than someone who doesn't. What you can actually say depends on the design of the study.

Two measures that are frequently used to describe the relationship between a risk factor and an event are the **relative risk ratio** and the **odds ratio**. Relative risk can only be computed for cohort and cross-sectional studies, while odds ratios can be computed for all three types of studies. For rare events, the odds ratio is an approximation to the relative risk ratio.

Measuring the Relative Risk

Relative risk is the ratio of the probability that an event occurs for someone with the risk factor to the probability that the event occurs for someone without the risk factor. If the relative risk ratio or the odds ratio is close to 1, the risk factor and the event are independent. If the ratio is greater than 1, people with the risk factor are more likely to experience the event than people without the risk factor. If the ratio is less than 1, people with the risk factor are less likely to experience the event than people without the risk factor.

Monitoring Deliveries

As an example of relative risk computation, consider data relating fetal monitoring and infant death. (The data are from Schlesselman, 1982.) Figure 10-19 shows the results of a historical cohort study of 15,846 deliveries in which 7,182 babies had fetal monitoring and 8,664 babies did not have fetal monitoring. Of the babies that were not monitored, 48 died. Of the monitored babies, 23 died.

Figure 10-19
Crosstabulation of fetal monitoring and outcome

			Status		
			dead	alive	Total
Monitor	Not monitored	Count	48	8616	8664
		Row %	.55%	99.45%	100.0%
	Monitored	Count	23	7159	7182
		Row %	.32%	99.68%	100.0%
Total		Count	71	15775	15846
		Row %	.45%	99.55%	100.0%

The relative risk ratio is the probability of a monitored baby dying divided by the probability of an unmonitored baby dying. From the row percentages, you see that 0.32% of the monitored babies and 0.55% of the unmonitored babies died. The relative risk ratio is 1.730. In Figure 10-20, the relative risk ratio for a baby dying is labeled *For cohort Status = dead.* That means you're comparing the probabilities of the event with the value label *Dead.*

Relative risks always appear in a pivot table together with the odds ratio. SPSS Statistics can't determine how you collected the data or which of the two events you are interested in. That's why it computes measures for all situations. Make sure you pluck the right numbers from the tables.

Tip: Specify the outcome variable for the columns of your table and the risk factor for the rows. The value of the risk factor that indicates that the risk factor is present should be in the first row. Don't be confused by the labeling in the rows labeled *For cohort.* The variable and value labels refer to the value that defines the event in the relative risk computation for a cohort study. The row for *Status = dead* gives the relative risk ratio for *dead* as the outcome; the row labeled *Status = alive* gives the relative risk ratio for *alive* as the outcome. It's up to you to recognize that you're interested in the risk of ending up dead.

Figure 10-20
Risk estimates

	Value	95% Confidence Interval	
		Lower	Upper
Odds Ratio for Monitor (not monitored / monitored)	1.734	1.054	2.853
For cohort Status = dead	1.730	1.053	2.841
For cohort Status = alive	.998	.996	1.000
N of Valid Cases	15846		

Calculating the Odds Ratio

If your data are from a case-control study, you can't calculate the relative risk ratio because you can't calculate the probability of someone with and without the risk factor experiencing the event. For a case-control study, the odds ratio is used to measure the relationship between the event and the risk factor.

The odds ratio is the ratio of two odds: the odds that a case has the risk factor and the odds that a control has the risk factor. (The odds is the number of cases with the risk factor divided by the number of cases without the risk factor.) For the fetal monitoring example:

- The odds that a dead infant was not monitored are $48/23$ (or 2.0870).
- The odds that a live infant was not monitored are $8616/7159$ (or 1.2035).
- The odds ratio is $2.0870/1.2035$, or 1.734, as shown in the first row of Figure 10-20.

If the event of interest occurs infrequently and if the total sample size is large, the odds ratio from a case-control study can be used as an estimate of relative risk. In this example, because death is a rare event, the odds ratio and the relative risk ratio are very close in value. For a cohort study, the odds ratio is the ratio of the relative risks for the two outcomes, as you can see in Figure 10-20.

If the event and the risk factor are independent, you expect the odds ratio to be close to 1. If the odds ratio is greater than 1, there is a positive association between the risk factor and the event; for odds ratios less than 1, there is a negative association. A negative association means that presence of the factor decreases the likelihood of the event occurring.

Figure 10-20 also displays 95% confidence intervals for the odds ratio and for the relative risk. You are 95% confident that the true relative risk ratio is between 1.05 and 2.84. That's a much wider range than you would like. If the population value for the relative risk ratio is only slightly larger than 1, monitoring may not be cost-effective. If the true value is around 2.5, monitoring may be worthwhile. The value of 1 is not included in the 95% confidence interval, so you can reject the null hypothesis that the relative risk ratio is 1.

Warning: Rejecting the null hypothesis that the odds ratio is 1 doesn't mean that the risk factor causes the event. It is just associated with it.

Tests of Conditional Independence

You can also base the test of the null hypothesis that the odds ratio is 1 on the two statistics shown in Figure 10-21. If their observed significance level is small, you reject the null hypothesis that the risk factor and the event are independent or, equivalently, that the odds ratio is 1. In this example, you reject the null hypothesis that the odds of dying are the same for infants who are monitored during delivery and for those who are not.

Figure 10-21
Tests of conditional independence

	Chi-Squared	df	Asymp. Sig. (2-sided)
Cochran's	4.811	1	.028
Mantel-Haenszel	4.301	1	.038

Under the conditional independence assumption, Cochran's statistic is asymptotically distributed as a 1 df chi-squared distribution, only if the number of strata is fixed, while the Mantel-Haenszel statistic is always asymptotically distributed as a 1 df chi-squared distribution. Note that the continuity correction is removed from the Mantel-Haenszel statistic when the sum of the differences between the observed and the expected is 0.

Tip: When you have only two variables, Cochran's test of conditional independence is equivalent to Pearson's chi-square. The Mantel-Haenszel test is the continuity-corrected chi-square.

Stratifying the Cases

You can introduce a control or stratification variable into the computation of the odds statistics by specifying it as a layer variable in the Crosstabs dialog box. For example, the women in the monitoring study were assigned to one of three groups based on how risky the delivery was predicted to be. Figure 10-22 shows the counts in each of the cells.

Figure 10-22
Crosstabs of risk categories and outcome

Count

				Status		
				Dead	Alive	Total
Risk	High	Monitor	No	29	112	141
			Yes	10	84	94
		Total		39	196	235
	Medium	Monitor	No	7	331	338
			Yes	4	267	271
		Total		11	598	609
	Low	Monitor	No	12	8173	8185
			Yes	9	6808	6817
		Total		21	14981	15002

Separate risk estimates for each of the three groups are shown in Figure 10-23. You see that the relative risk ratio ranges from 1.933 for high-risk pregnancies to 1.110 for low-risk pregnancies. That's almost a twofold difference.

Figure 10-23
Risk estimates for Figure 10-22

				95% Confidence Interval	
			Value	Lower	Upper
Riskcat	High	Odds Ratio for Monitor (not monitored / monitored)	2.175	1.005	4.709
		For cohort Status = dead	1.933	.990	3.777
		For cohort Status = alive	.889	.797	.991
		N of Valid Cases	235		
	Medium	Odds Ratio for Monitor (not monitored / monitored)	1.412	.409	4.873
		For cohort Status = dead	1.403	.415	4.743
		For cohort Status = alive	.994	.973	1.015
		N of Valid Cases	609		
	Low	Odds Ratio for Monitor (not monitored / monitored)	1.111	.468	2.637
		For cohort Status = dead	1.110	.468	2.634
		For cohort Status = alive	1.000	.999	1.001
		N of Valid Cases	15002		

Testing Hypotheses about the Odds Ratios

When you have separate odds ratios for several strata, you can test null hypotheses about the odds ratios in the strata, as well as estimate a pooled value for it.

Are the Odds Ratios 1 in All Strata?

Both the Cochran and the Mantel-Haenszel tests of conditional independence shown in Figure 10-24 are used to test the null hypothesis that the risk factor and the event are independent in all strata. The assumptions for each of them are given in the footnote that SPSS Statistics has thoughtfully displayed below the figure.

You can't reject the null hypothesis that monitoring and outcome are independent at each level of risk of the delivery. That's a different conclusion than you reached before when all of the cases were pooled. The reason for the difference is that the distribution of high-, medium-, and low-risk cases in the monitored and unmonitored groups is slightly different. When you take into account the imbalance of the types of cases in the two groups, you can't reject the null hypothesis that the two variables are independent, controlling for the risk of delivery.

Figure 10-24
Tests of conditional independence

	Chi-Squared	df	Asymp. Sig. (2-sided)
Cochran's	3.160	1	.075
Mantel-Haenszel	2.720	1	.099

Under the conditional independence assumption, Cochran's statistic is asymptotically distributed as a 1 df chi-squared distribution, only if the number of strata is fixed, while the Mantel-Haenszel statistic is always asymptotically distributed as a 1 df chi-squared distribution. Note that the continuity correction is removed from the Mantel-Haenszel statistic when the sum of the differences between the observed and the expected is 0.

Warning: The tests of conditional independence do not perform well if the pattern of the association differs across the strata, particularly if they are opposite. You may fail to reject the null hypothesis when there is an association that is not consistent across strata.

Are the Odds Ratios Equal in All Strata?

Tests for the equality of the odds ratios across the three strata are shown in Figure 10-25. The Breslow-Day test is a more general test for the equality of odds ratios because its null hypothesis of equal odds ratios does not constrain them to all equal 1. The Tarone test is a modification of the Breslow-Day test that performs better under most circumstances. Based on the large observed significance level, you can't reject the null hypothesis that the odds ratios are equal in all three strata.

Figure 10-25
Test for equality of odds ratios

	Chi-Squared	df	Asymp. Sig. (2-sided)
Breslow-Day	1.340	2	.512
Tarone's	1.340	2	.512

What Is the Estimate of the Common Odds?

If you don't reject the null hypothesis that all odds ratios are equal, you can combine them into an estimate of a common odds ratio. Figure 10-26 contains the Mantel-Haenszel estimate of the common odds ratio and its logarithm, which is labeled *ln(Estimate)*. If you do reject the null hypothesis that the odds ratios are equal across the strata, it doesn't make sense to obtain a pooled estimate.

Figure 10-26
Mantel-Haenszel estimate of common odds ratio

Statistics	Estimate			1.594
	ln(Estimate)			.466
	Std. Error of ln(Estimate)			.264
	Asymp. Sig. (2-sided)			.077
	Asymp. 95% Confidence Interval	Common Odds Ratio	Lower Bound	.950
			Upper Bound	2.675
		ln(Common Odds Ratio)	Lower Bound	-.051
			Upper Bound	.984

The Mantel-Haenszel common odds ratio estimate is asymptotically normally distributed under the common odds ratio of 1.000 assumption. So is the natural log of the estimate.

Warning: The Breslow-Day statistic requires large sample sizes in each stratum. The Mantel-Haenszel test requires merely a large overall sample size. Also, it should not be used if the association between the risk factor and the outcome changes markedly over the strata.

The pooled estimate of the odds ratio is 1.6, and the 95% confidence interval is from 0.95 to 2.68. Because the value of 1 is included, you cannot reject the null hypothesis that the odds ratio is 1. The observed significance level for the test of the null hypothesis that the common odds ratio is 1 is 0.077.

Tip: You can specify a constant other than 1 for the test of the common odds ratio. The asymptotic significance level is then, for the null hypothesis, that the population common odds ratio is the specified value.

Megatip: Entering Tables Directly

If you have a table with counts and want to compute tests and measures of association for it, you can enter the cell counts directly into SPSS Statistics. For example, Figure 10-27 shows how you might enter the table in Figure 10-11 (the hypnosis table) into the Data Editor.

Figure 10-27
Data Editor for depth of hypnosis

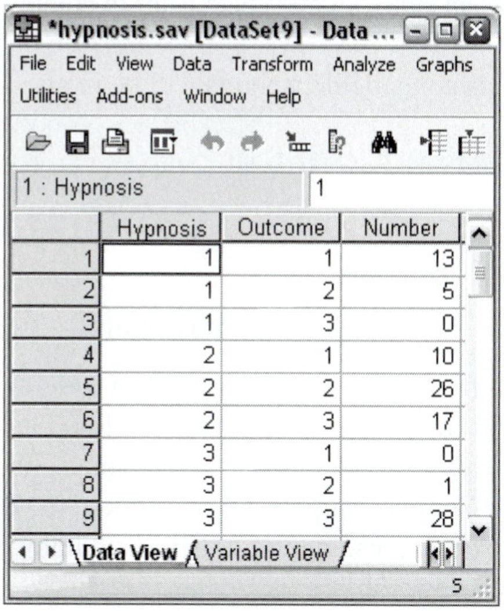

*hypnosis.sav [DataSet9] - Data ...
File Edit View Data Transform Analyze Graphs Utilities Add-ons Window Help
1 : Hypnosis 1

	Hypnosis	Outcome	Number
1	1	1	13
2	1	2	5
3	1	3	0
4	2	1	10
5	2	2	26
6	2	3	17
7	3	1	0
8	3	2	1
9	3	3	28

Data View Variable View

For each cell in the table, you enter the value of the row variable, the value of the column variable, and the count of the number of cases. Then, from the Data menu, choose Weight Cases. The resulting Weight Cases dialog box is shown in Figure 10-28.

Figure 10-28
Weight Cases dialog box

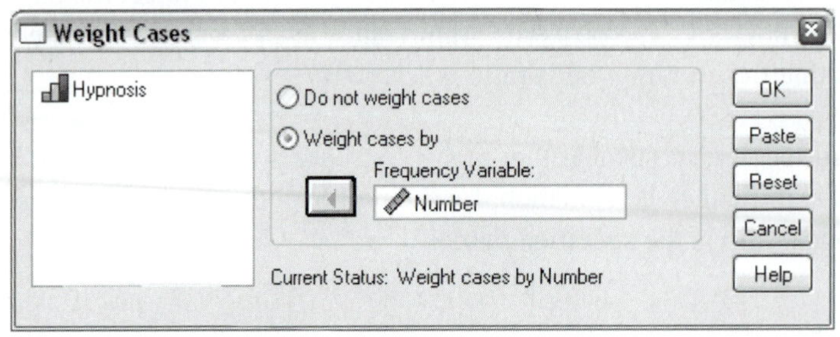

Move the name of the variable that contains the cell counts into the Frequency Variable field. That's all there is to it. For purposes of crosstabulating hypnosis and outcome, the 9-case data file in Figure 10-27 is fully equivalent to the 100-case data file underlying Figure 10-11. What's more, a table with cell counts in the thousands would be just as easy to enter into the Data Editor. You would enter larger values for *Number* into the third column, but you wouldn't have to enter any more cases. If you specify *hypnosis* as the row variable and *outcome* as the column variable in Crosstabs, you will get Figure 10-11 exactly. You can select statistics and cell entries to compute measures of association and tests for the table.

When you save a file with weights applied, the weights will be in effect the next time you open it.

Obtaining the Output

To produce the output in this chapter, follow the instructions below.

Figures 10-1 and 10-2. Open the data file *aggregatedgss.sav*. From the Analyze menu choose Descriptive Statistics > Crosstabs. Move *triplicity* into the Row(s) list, and move *life* into the Column(s) list. Click Cells to open the Cell Display subdialog box. Select Observed and Expected in the Counts group, and select Row in the Percentages group. Click Continue, and then click Statistics. In the Statistics subdialog box, select Chi-square. Click Continue, and then click OK.

Figures 10-3 and 10-4. From the Analyze menu choose Descriptive Statistics > Crosstabs. Click Reset. Move *happy* into the Row(s) list, and move *decade* into the Column(s) list. Click Cells to open the Cell Display subdialog box. Select Observed and Expected in the Counts group, and select Column in the Percentages group. Click Continue, and then click Statistics. In the Statistics subdialog box, select Chi-square. Click Continue, and then click OK. Edit labels in the Pivot Table Editor to reproduce the figures exactly.

Figure 10-5. Proceed as for Figures 10-3 and 10-4, but move *sex* into the Layer 1 of 1 list.

Figure 10-6. From the Graphs menu choose Legacy Dialogs > Line. Select the Multiple icon and Summaries for groups of cases, and then click Define. In the dialog box that appears, move *decade* into the Category Axis box, and move *sex* into the Define Lines by box. Select Other statistic, move *happy* into the Variable list, and then click Change Statistic. In the Statistic subdialog box, select Percentage inside and type 1 into both the Low and High text boxes. Click Continue, and then click OK.

Figures 10-7 and 10-8. Open the data file *gssdata.sav*. From the Analyze menu choose Descriptive Statistics > Crosstabs. Move *vote00* into the Row(s) list and *vote96* into the Column(s) list. Click Statistics, and in the Statistics subdialog box, select McNemar. Click Continue, and then click OK.

Figures 10-9 and 10-10. From the Analyze menu choose Descriptive Statistics > Crosstabs. Click Reset. Move *degree* into the Row(s) list, and move *newsreordered* into the Column(s) list. Click Cells to open the Cell Display subdialog box, and then select Observed in the Counts group and Row in the Percentages group. Click Continue, and then click Statistics. In the Statistics subdialog box, select Contingency coefficient and Phi and Cramér's V. Click Continue, and then click OK.

Figures 10-11 to 10-13. Open the data file *hypnosis.sav*. From the Analyze menu choose Descriptive Statistics > Crosstabs. Move *hypnosis* into the Row(s) list, and move *outcome* into the Column(s) list. Click Cells to open the Cell Display subdialog box, and then select Observed in the Counts group and Row in the Percentages group. Click Continue, and then click Statistics. In the Statistics subdialog box, select Lambda. Click Continue, and then click OK.

Figures 10-14 to 10-16. Open the data file *gssdata.sav*. From the Analyze menu choose Descriptive Statistics > Crosstabs. Move *degree* into the Row(s) list, and move *newsreordered* into the Column(s) list. Click Statistics to open the Statistics subdialog box. Select Gamma, Somers' d, Kendall's tau-b, Kendall's tau-c, and Correlations. Click Continue, and then click OK.

Figures 10-17 and 10-18. Open the data file *kappa.sav.* From the Analyze menu choose Descriptive Statistics > Crosstabs. Move *teacher1* into the Row(s) list, and move *teacher2* into the Column(s) list. Click Cells, and in the Cell Display subdialog box, select Observed in the Counts group and Total in the Percentages group. Click Continue, and then click Statistics. In the Statistics subdialog box, select Kappa. Click Continue, and then click OK.

Figures 10-19 to 10-26. Open the data file *relrisk.sav.* From the Analyze menu choose Descriptive Statistics > Crosstabs. For Figures 10-19 to 10-21, move *monitor* into the Row(s) list, and move *status* into the Column(s) list. Click Cells to open the Cell Display subdialog box. Select Observed in the Counts group and Row in the Percentages group. Click Continue, and then click Statistics. In the Statistics subdialog box, select Risk and Cochran's and Mantel-Haenszel statistics. Click Continue, and then click OK.

Figures 10-22 to 10-26. Reopen the Crosstabs dialog box and move *riskcat* into the Layer 1 of 1 list. Click Cells to open the Cell Display subdialog box. Deselect Row in the Percentages group. Click Continue, and then click OK.

Chapter

11

Correlation

The term *correlation* is firmly entrenched in our vocabularies. Everyone knows that smoking is correlated with cancer, that voting behavior is correlated with education, and that knowing the right people is correlated with success. In everyday usage, *correlation* is just a vague term indicating some type of relationship. In statistics, **correlation** has a precise definition. In this chapter, you'll use the Bivariate Correlations procedure to calculate correlation coefficients and test the null hypothesis that the correlation coefficient is 0 in the population. You'll also use the Partial Correlations procedure to calculate the correlation coefficient between two variables while controlling for the linear effects of other variables.

Examples

- You want to determine if there is a relationship between sales and dollars spent on advertising. For a particular product each month, you record the dollars spent on advertising and the total sales revenue.
- You want to determine if body fat changes with age. You take a random sample of adults and measure their body fat and record their age.
- You want to study the relationship between education and salary in a particular job category. Because people have differing years of job experience, you want to determine the correlation coefficient between education and salary, controlling for the linear effects of work experience.

In a Nutshell

The **Pearson correlation coefficient** is a measure of the linear association between two variables. It ranges in value from –1 to +1. The absolute value of the Pearson correlation coefficient tells you the strength of the linear relationship. If the sign of the coefficient is positive, the values of the two variables increase together. If the sign of the correlation coefficient is negative, as values of one variable increase, the values of the other variable decrease. Many different relationships between pairs of variables can have the same correlation coefficient, so it's important to plot the data before calculating a correlation coefficient.

The **Spearman correlation coefficient** is calculated by replacing data values with ranks. You can use this nonparametric correlation coefficient if your data don't meet the normality assumptions needed for hypothesis tests about the population correlation coefficient or if there are outlying observations to which the Pearson correlation coefficient is sensitive.

The **partial correlation coefficient** quantifies the strength of the linear relationship between two variables while controlling for the linear effects of one or more variables. You can use it to eliminate spurious correlations and to uncover hidden relationships.

Body Fat: The Example

Body measurements are related in various ways. Many of us have personal data to support the observation that pounds on the scale translate to inches everywhere: on the abdomen, thighs, arms, and hips. Johnson (1996)* presents a dataset with body fat (as a percentage), age, weight, height, and 10 body circumference measurements for 252 men. In this chapter, you'll examine relationships between circumference measurements and body fat percentage as estimated from submerging individuals under water, having them exhale, and then measuring water displacement.

* I thank Roger Johnson and Garth Fisher for making the data available.

Plotting the Data

The first step in examining the relationship between two variables is to plot them. Figure 11-1 is a scatterplot of weight in pounds, on the horizontal, or *x,* axis, and the percentage of that weight that is fat, on the vertical, or *y,* axis. Each point in the scatterplot represents the values of body fat percentage and total weight for an individual. The circled point is for a man who weighs 219 pounds and has a body fat percentage of 47.5.

Figure 11-1
Percentage of body fat and weight

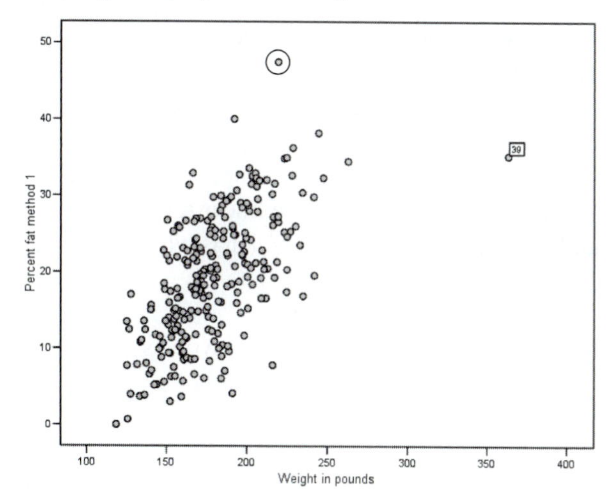

From the plot, you can determine whether there is a pattern to the points and whether there are any strange data points. You see that there is one data point in the upper right part of the plot, corresponding to a case ID number of 39. The point represents a large individual who weighs 375 pounds. Notice that the point doesn't fit the pattern of the rest of the data points. The body fat percentage is too low for the observed weight. There is also a very lean man with an estimated body fat percentage of 0. Outlying points are of concern because they can have a large effect on the value of the correlation coefficient that is used to measure the strength of the linear relationship between two variables.

Warning: Don't ignore strange data points. The next step in this analysis should be to check the values for unusual cases. If they are not the result of data entry or recording error, you should determine if there was a problem in obtaining the measurement. It may be that the water displacement method used to estimate body fat percentage was not suitable for a large person. There may not have been enough water to displace. It's also possible that the formula for estimating body fat percentage from density may not be accurate for large weights or for small weights. If there isn't a good reason for excluding a point, you must keep a watchful eye on it throughout the analysis.

In Figure 11-1, you see that, for the most part, as weight increases, so does the percentage of body fat. You can draw a straight line to summarize the relationship between the two variables. When points cluster around a straight line, the variables are said to be linearly related. If the values of the two variables increase together, the variables are positively related. If the values of one variable increase as the values of the other variable decrease, the variables are negatively related.

Figure 11-1 is a plot of a positive linear relationship, while Figure 11-2, a plot of body fat percentage and density, shows a negative linear relationship. (Body fat percentage is calculated as a function of density, which is why the two variables are so closely related.)

Warning: Whenever the computation of two variables involves common variables, the variables will be related.

Figure 11-2
Percentage of body fat and body density

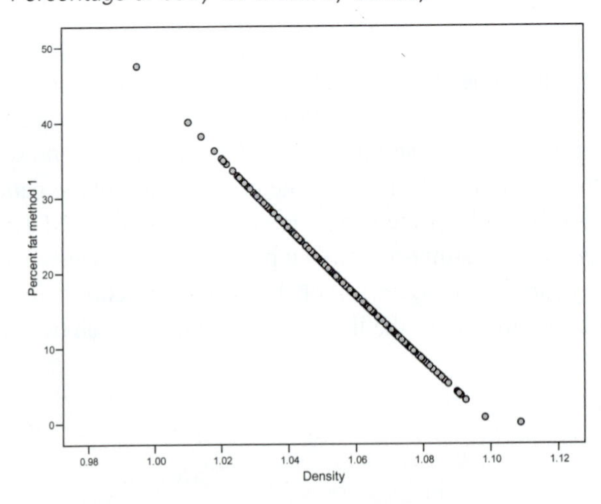

The two plots differ, not only in whether the two variables have a positive or negative association but also in how closely the points cluster around a straight line. In Figure 11-1, the points are loosely spread around a straight line, while in Figure 11-2, the points cluster closely around a straight line. If you look at Figure 11-3, the scatterplot of body fat and height, you see yet another type of relationship—an absence of a relationship. Body fat and height aren't related in any discernible way. The line in the plot is the average body fat percentage for all of the men. The data points randomly scatter around the mean.

Figure 11-3
Scatterplot of body fat percentage and height

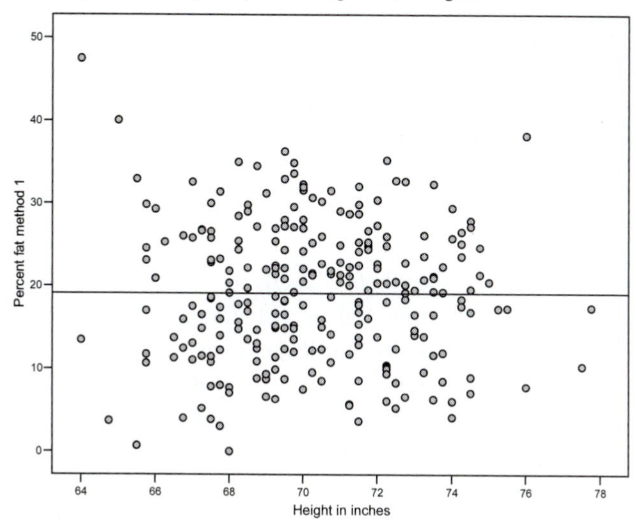

Examining the Scatterplot Matrix

Although you can make plots of individual pairs of variables, relationships are easier to see if you arrange the plots in a single display. For example, Figure 11-4 is a scatterplot matrix of body fat, height, weight, and density.

Figure 11-4
Scatterplot matrix

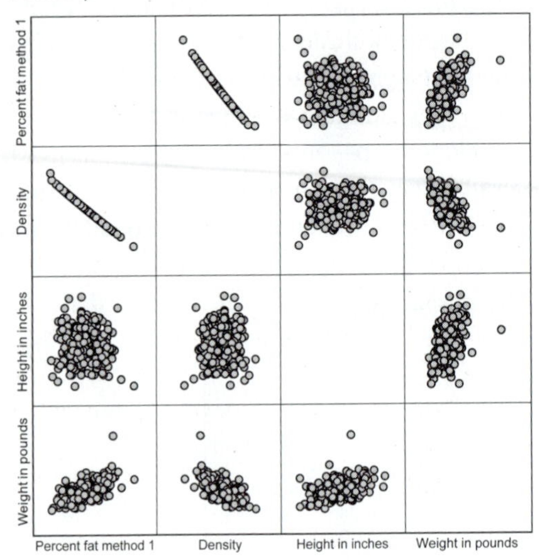

To tell what's in each plot, look at the labels beneath and to the left of the display. They define the variables plotted on the horizontal (*x*) and vertical (*y*) axes. For example, the first row of the matrix has body fat on the vertical axis for all three of the plots. In the first non-empty plot, density is on the horizontal axis; in the second, height; and in the third, weight. Figures 11-1, 11-2, and 11-3 are all in the first row of the matrix. Each pair of variables appears twice in the matrix, each variable taking a turn on the horizontal and vertical axis.

Using the Pearson Correlation Coefficient

Examining plots is the best way to gauge the nature and strength of the relationship between two variables. However, you need an objective measure to replace subjective descriptions like *strong*, *weak*, *I can't make up my mind*, and *none*. If the variables are related in a linear fashion, you can quantify the strength of the relationship between the two variables with the Pearson correlation coefficient.

The Pearson correlation coefficient:

- Is a measure, in absolute value, of the strength of the linear relationship between two variables.
- Ranges in value from –1 to +1. A value of either +1 or –1 means that you can perfectly predict the values of one variable from the values of the other.
- Is 0 when there is no linear relationship between two variables.
- Is positive if the values of the two variables increase together.
- Is negative if the values of one variable increase as the values of the other variable decrease.
- Is symmetric, since the correlation of x and y is the same as the correlation of y and x.
- Is unaffected by linear transformations, such as adding a constant to all numbers or dividing all numbers by a constant.
- Is the proportion, when squared, of the variance of one variable that can be explained by the other.

Warning: Never compute correlation coefficients for nominal variables, even if they are nicely coded with numbers. A correlation between social security number and income is meaningless.

Correlation Matrix

In Figure 11-5, you see the correlation coefficients for all pairs of the three variables. The correlation between height and weight is 0.49, between body fat and height is –0.02, and between body fat and weight is 0.61.

Figure 11-5
Correlation matrix

		Percent fat method 1	Weight in pounds	Height in inches
Percent fat method 1	Pearson Correlation	1	.613**	-.025
	Sig. (2-tailed)	.	.000	.694
	N	252	252	252
Weight in pounds	Pearson Correlation	.613**	1	.487**
	Sig. (2-tailed)	.000	.	.000
	N	252	252	252
Height in inches	Pearson Correlation	-.025	.487**	1
	Sig. (2-tailed)	.694	.000	.
	N	252	252	252

**. Correlation is significant at the 0.01 level (2-tailed).

On the diagonal, you have coefficients of 1 because any variable is perfectly related to itself. The upper and lower parts of the matrix are identical because the correlation coefficient is a symmetric measure. The correlation coefficient between body fat and height is the same as the correlation coefficient between height and body fat.

Warning: Two variables may have a correlation coefficient of 0 and have a strong nonlinear relationship, perhaps following a U-shaped or an S-shaped curve.

Testing Hypotheses about Correlation Coefficients

You can use the correlation coefficient as just a descriptive summary of the relationship between two variables in your dataset. If your data are a random sample from a population, you can also test hypotheses about the relationship of the variables in the underlying population.

The statistical test that the population correlation coefficient (denoted as ρ) is 0 requires that your data come from a population in which the two variables jointly have a bivariate normal distribution. The observations must also be independent. Bivariate normality is not an easy assumption to test. It is always the case that if the two variables have a joint normal distribution, the distribution of each of the variables individually is normal. If you want to test the null hypothesis that the sample comes from a population with a known nonzero value for the correlation coefficient, you have to use special procedures described in Hays (1994).

Warning: Don't compute correlation coefficients between many pairs of variables just to see if something is significant. Remember that even if the variables are not correlated in the population, you still expect to find five significant sample coefficients in every 100, if you use a 5% cutoff for determining statistical significance. If you are examining many correlation coefficients, consider using the Bonferroni adjustment, which entails multiplying the observed significance level by the number of coefficients examined.

The entry labeled *Sig. (2-tailed)* in Figure 11-5 is the observed significance level for the test of the null hypothesis that the population value for the correlation coefficient is 0. If the observed significance level is small, you can reject the null hypothesis that there is no relationship between the two variables in the population. In Figure 11-5, you can reject the null hypothesis that the population correlation coefficient is 0 for all pairs of variables except height and body fat. Coefficients that have observed significance levels smaller than 0.05 are flagged with a single asterisk (*) in the table, and coefficients with observed significance levels less than 0.01 are flagged with double asterisks (**).

Tip: Always look at the magnitude of the correlation coefficient as well as the observed significance level. For large sample sizes, even very small correlation coefficients will have small observed significance levels. Statistically significant doesn't mean important or useful.

Pitfalls of Correlation Coefficients

The correlation coefficient may well lead the list of most popular statistics because it seems to capture the essence of the relationship between two variables with a single number between –1 and +1. Nothing could be further from the truth, since so many different relationships can result in the same correlation coefficient. Figure 11-6 shows four different plots all with a correlation coefficient of 0.82. (The first three are from Anscombe, 1973).

Figure 11-6
Four plots with the same correlation coefficient

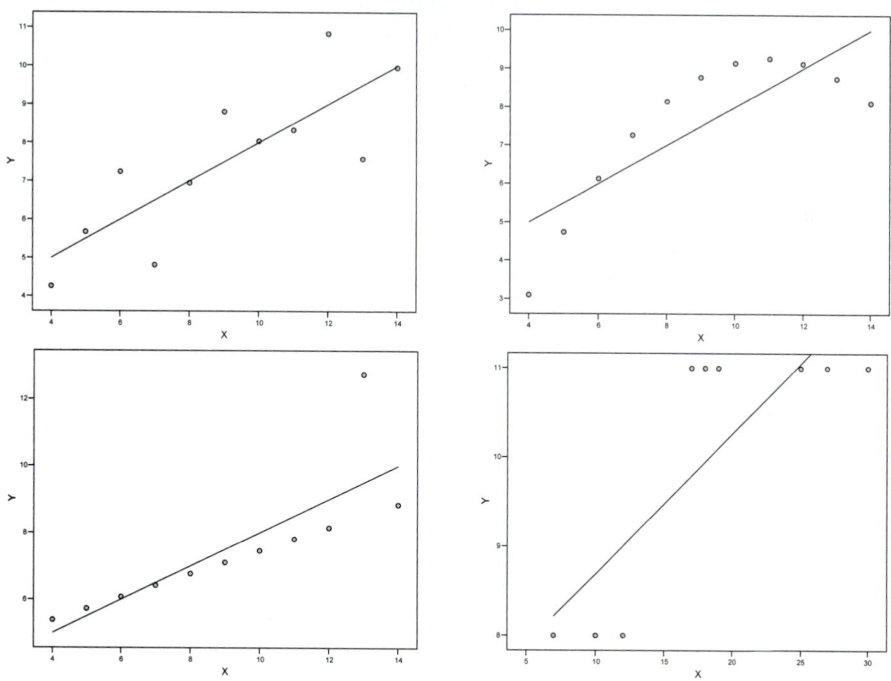

The correlation coefficient is an appropriate summary measure for the first plot only, since a straight line is a reasonable summary of the relationship between those two variables. In the second plot, the relationship between the two variables is not linear, so it doesn't make sense to describe how tightly the points cluster around a straight line. In the third plot, you see that the perfect relationship between the two variables is distorted by a single point. In the fourth plot, there appear to be two subgroups of cases in which there is no linear relationship between the two variables. However, if you combine the two subgroups on a single plot, there appears to be a relationship between the two variables. The punch line is clear: If you don't plot your data, you can't tell whether a correlation coefficient is a good summary of the relationship between the two variables.

The value of a correlation coefficient also depends on the range of values for which observations are taken. It's possible that even if there is a linear relationship between two variables, you won't detect it if you consider a small range of values of the variables. For example, height may be a poor predictor of weight if you restrict your range of heights to those over six feet.

Warning: A correlation coefficient doesn't tell you anything about the cause-and-effect nature of the relationship between the variables. Just because two variables are correlated doesn't mean that one causes another. (It has long been known in Europe that there is a positive correlation between the number of storks in an area and the birthrate. That's because rural areas, where storks live, have higher birthrates, not because of any baby-delivering properties of storks.)

Comparing Two Indexes

Often, there are several different tests or measuring devices that purport to measure the same quantity. They may differ in accuracy, cost, and, perhaps, even in pain inflicted. Sometimes you may have a "gold standard" that is known to be accurate. In such situations, you are faced with the task of comparing the results from two alternative methods to determine if they can be used interchangeably.

The body fat data file contains two estimates of body fat: one based on a formula proposed by Siri (1956) and one by Brozek et al. (1963). Figure 11-7 is a plot of the body fat percentages for the two methods. The correlation coefficient is 1, which might lead unsophisticated analysts to conclude that the two methods are equivalent. This is an unwarranted conclusion. Whenever you have two variables that purport to measure the same thing, you have to compare their actual values; you can't just rely on the correlation coefficient between the two variables.

Figure 11-7
Scatterplot of two methods for measuring percentage of body fat

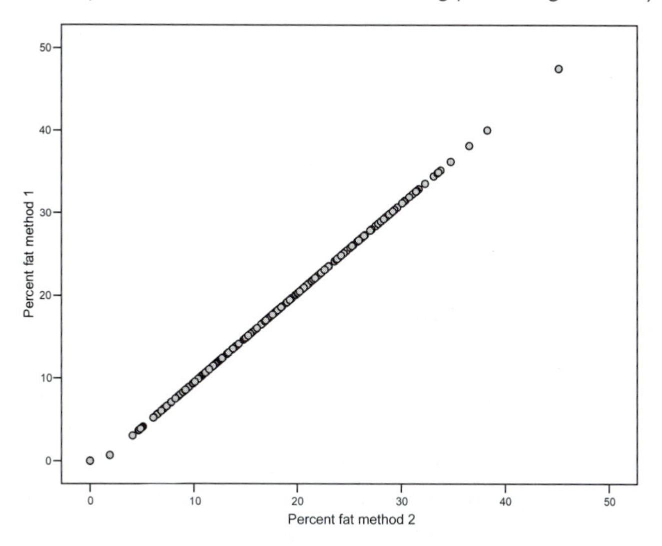

If you look at the descriptive statistics shown in Figure 11-8, you see that the means and standard deviations of the two measurements are not equal. The first method has both a larger mean and a larger standard deviation.

Figure 11-8
Descriptive statistics for two measures of body fat

		Mean	N	Std. Deviation	Std. Error Mean
Pair 1	Percent fat method 1	19.1566	252	8.38077	.52794
	Percent fat method 2	18.9356	252	7.74922	.48815

Based on the paired *t* test shown in Figure 11-9, you can reject the null hypothesis that the population values for the two means are equal. Two variables that have a large correlation coefficient can differ in means and standard deviations. That's why the correlation coefficient alone is not a good measure of agreement between two methods. Two methods are equivalent if the correlation coefficient is 1 *and* the means and standard deviations are equal. (For further discussion, see van Belle, 2002.)

Figure 11-9
Paired t test for two measures of body fat

		Paired Differences			t	df	Sig. (2-tailed)
		Mean	Std. Deviation	Std. Error Mean			
Pair 1	Percent fat method 1 - Percent fat method 2	.22	.64	.04	5.51	251	.000

Tip: Plot the difference between the two measurements against the sum of the two measurements. You can then see how the difference varies over the range of values of the measurement (Bland and Altman, 1995).

Basing Correlation Coefficients on Ranks

The Pearson correlation coefficients require the assumption of normality. For ordinal data or for interval data that do not satisfy the normality assumption, you can calculate a correlation coefficient that is based on ranks. The Spearman rank correlation coefficient, if there are no ties, is the Pearson correlation coefficient when the data values are replaced by ranks. (If there are ties, some adjustments are made.)

Like the Pearson correlation coefficient, the rank correlation ranges between –1 and +1, where –1 and +1 now indicate a perfect linear relationship between the ranks of the two variables. The interpretation is the same, except that the relationship is between the ranks, not the actual values. For you to use the Spearman correlation coefficient, the actual values of the two variables don't have to be linearly related, but their ranks do.

Tip: Correlation coefficients based on ranks are less sensitive to outliers. If the Pearson and Spearman coefficients are quite different in value, you should determine why. It may be that you have an outlying point that has a lot of effect on the Pearson correlation coefficient.

Using Partial Correlation Coefficients

Whenever you examine the relationship between two variables, you have to be concerned about the effect of other variables on the relationship of interest. For example, daily ice cream consumption and the number of murders in a large city are usually positively correlated. That's because ice cream consumption and murders are both related to a third variable—temperature. The correlation between ice cream consumption and murders is said to be spurious. If you are interested in studying the relationship between ice cream consumption and murder, you must control for the effects of temperature. You'd like to hold temperature constant and find the correlation between ice cream consumption and murder. In statistical jargon, you want to compute the correlation between ice cream consumption and murder, "controlling for" the effect of temperature. That's what a partial correlation coefficient does. Properly used, partial correlation coefficients can help you uncover spurious relationships, identify intervening variables, and detect hidden relationships.

A partial correlation coefficient measures the linear association between two variables while adjusting for the linear effects of one or more additional variables. If you control for the effects on one variable, the partial correlation coefficient is called a **first-order coefficient**; if you control for two variables, it's a **second-order coefficient**. Sometimes the ordinary correlation coefficient is called a **zero-order correlation** because it has no control variables.

The properties of a partial correlation coefficient are like those of the ordinary correlation coefficient: it ranges in value from –1 to +1. The absolute value tells you the strength of the relationship, and the sign tells you the direction.

Warning: The partial correlation coefficient should be used only when the relationships between pairs of variables are linear.

Calculating Partial Correlation Coefficients

In the body fat example, you expect the various circumferences to be related to each other because, on average, thin people have smaller circumferences for most body parts and heavier people have larger circumferences for most body parts. To study body "construction," you may want to know what the relationship is between two circumferences when weight is held constant. For example, are abdominal circumference and knee circumference linearly related for people who weigh the same?

In the first column of Figure 11-10, you see correlation coefficients between abdominal circumference and various other body circumferences. You see that chest, hip, knee, and neck circumferences are positively correlated with abdominal circumference. That's not surprising. Hot fudge sundaes of old are stored everywhere.

Figure 11-10
Partial correlations with abdominal circumference

Control	None	Weight	Height	Height and weight
Chestcirc	0.92 **	0.59 **	0.91 **	0.40 **
Hipcirc	0.87 **	0.25 **	0.88 **	0.09
Kneecirc	0.74 **	-0.09	0.76 **	0.04
Neckcirc	0.75 **	0.06	0.75 **	-0.05

Warning: These coefficients were calculated with SPSS Statistics. SPSS Statistics cannot, however, produce a table in this format.

Controlling for Weight

The column labeled *Weight* in Figure 11-10 contains the partial correlation coefficients between abdominal circumference and the four other circumference measures when the linear effect of weight is eliminated. That's an estimate of the strength of the linear relationship between the pairs of measures if all people have the same weight. All of the partial correlation coefficients are smaller in value than the ordinary correlation coefficients. That's because much of the observed correlation between abdominal circumference and the other four circumferences can be attributed to differences in weight.

The correlations between knee and abdominal circumferences and that between neck and abdominal circumferences are very close to 0. In fact, you can't reject the null hypothesis that, in the population, their value is 0. If people are of the same weight, measuring their necks and knees won't tell you much about their abdomens. The partial correlation coefficients for chest and hip circumference, controlling for weight, are also smaller than the zero-order correlation coefficients, but they are still reasonably large and significantly different from 0. For a fixed weight, as hip and chest size increase, so does abdominal circumference.

Tip: A single asterisk (*) indicates that you can reject the null hypothesis that the population value of the coefficient is 0, using the 0.05 significance level. Two asterisks (**) tell you that you can reject the null hypothesis, using the 0.01 significance level.

Controlling for Height

To see what happens if you control for height, look at the column labeled *Height* in Figure 11-10. The coefficients haven't changed much compared to the ordinary correlation coefficients in the column labeled *None*. If you hold height constant, all four of the other circumference measurements are still linearly related to abdominal circumference. That's because height alone is not a good linear predictor of circumferences. In fact, you saw in Figure 11-5 that the correlation coefficient between height and body fat percentage is very close to 0.

Controlling for Height and Weight

The last column of Figure 11-10 shows you what happens to the relationships between the circumference measurements when both height and weight are held constant. The coefficients for both chest and hip circumference are much smaller than when weight alone was held constant. Both height and weight affect the relationship of chest and hip circumference with abdominal circumference.

When height and weight are held constant, there is no linear relationship of abdominal circumference with knee circumference, and the relationship of abdominal circumference with neck circumference is small and negative.

Testing Hypotheses about Partial Correlation Coefficients

To test hypotheses about the population partial correlation coefficient, you have to assume that the observations are independent and that the data come from what's known as a multivariate normal distribution. The observed significance level depends on the absolute value of the coefficient, the sample size, and the order of the coefficient.

Figure 11-11 shows the second-order partial correlation coefficients and their two-tailed observed significance levels for four of the circumference measurements when you control for both height and weight. You see that only hip circumference is related to the other circumferences when height and weight are held constant. If you look at zero-order correlations between the pairs of circumferences, they are all significantly different from 0.

Figure 11-11
Partial correlation coefficients and observed significance

				chestcirc	hipcirc	kneecirc	neckcirc
Control Variables	weight & height	chestcirc	Correlation	1.000	-.289	-.104	.093
			Significance (2-tailed)	.	.000	.102	.144
			df	0	248	248	248
		hipcirc	Correlation	-.289	1.000	.183	-.315
			Significance (2-tailed)	.000	.	.004	.000
			df	248	0	248	248
		kneecirc	Correlation	-.104	.183	1.000	-.096
			Significance (2-tailed)	.102	.004	.	.129
			df	248	248	0	248
		neckcirc	Correlation	.093	-.315	-.096	1.000
			Significance (2-tailed)	.144	.000	.129	.
			df	248	248	248	0

Steps for Calculating the Partial Correlation Coefficient

The idea behind partial correlation coefficients is fairly straightforward. It's based on building a linear regression model that is used to predict values of one variable from another (see Chapter 13). To calculate the partial correlation coefficient between abdominal circumference and chest circumference while controlling for weight, you would follow these steps:

- Calculate a linear regression model that predicts the values of abdominal circumference from weight, the control variable.
- Calculate the differences between the observed and predicted values of abdominal circumference (the residuals for abdominal circumference).
- Calculate a linear regression model to predict the values of chest circumference from weight.
- Calculate the differences between the observed and predicted values of chest circumference (the residuals for chest circumference).
- Calculate the (zero-order) correlation coefficient between the two sets of residuals.

Fortunately, you don't have to do any of this because the Partial Correlations procedure will do it all. If you want to control for more than one variable, say weight and height together, you must use both of the variables together to predict the values of abdominal circumference and chest circumference.

Warning: If you calculate the partial correlation coefficient from the residuals using the Pearson correlation procedure, the degrees of freedom is not correct. The order of the coefficient must be subtracted from the reported degrees of freedom.

Uses of the Partial Correlation Coefficient

Partial correlation coefficients are used to detect spurious correlations between two variables. A spurious correlation is one in which the correlation between two variables results solely from the fact that one of the variables is correlated with a third variable that is the true predictor. In the body fat dataset, you saw that the observed relationship between neck circumference and abdominal circumference vanished when the effect of weight was removed.

You can also find hidden relationships using partial correlation coefficients. For example, if you can't find a relationship between two variables that you expect to be related, it's possible that one or more additional variables are suppressing the expected relationship. For example, it may be that *A* is not correlated with *B* because *A* is negatively related to *C*, which is positively related to *B*.

As an example, suppose a marketing research company wants to examine the relationship between the need for transmission-rebuilding kits and the intent to purchase such a kit. Initial examination of the data reveals almost no correlation between the two variables ($r = 0.01$). However, the data show a negative relationship ($r = -0.5$) between income and need to buy and a positive relationship ($r = 0.6$) between income and intent to buy. If you control for the effect of income using a partial correlation, the first-order partial between need and intent, controlling for income, is 0.45. Income hid the relationship between need and intent to buy.

Missing Values

In statistical procedures that involve many variables, SPSS Statistics offers two possible treatments for missing values: pairwise deletion and listwise deletion. In listwise deletion, a case that has a missing value for any variable in the analysis is dropped from the analysis. For example, if you are calculating correlation coefficients between 10 variables, a case would be dropped if it had a missing value for any of the 10 variables. If missing values are scattered randomly throughout the variables, this may result in a greatly reduced sample size, since many cases are eliminated.

Tip: If one or two variables account for most of the missing values, consider eliminating those variables from your analysis.

In pairwise missing-value treatment, the computation of each correlation coefficient is based on all cases that have valid values for that pair of variables. For example, if a case has valid values for variables 1, 3, and 5 only, it is used in computations involving only variable pairs 1 and 3, 1 and 5, and 3 and 5. Often, pairwise deletion is a bad strategy for dealing with missing data (Little and Rubin, 2002).

Many problems can arise with correlation matrices calculated using pairwise missing-value treatment. One of them is inconsistency. There are some relationships between correlation coefficients that are clearly impossible but may seem to occur when different cases are used to estimate different coefficients. For example, if age and weight and if age and height have large positive correlations, it is impossible for height and weight to have a large negative correlation in the same sample. However, if the same cases are not used to estimate all three coefficients, such an anomaly can occur.

Warning: If you use pairwise deletion, you may encounter problems in procedures that do calculations based on the correlation matrix. You may get a message that a correlation matrix is "not positive definite" or that it is "singular."

There is no single sample size that is associated with a matrix constructed using pairwise missing-value treatment, since each coefficient can be based on a different number of cases. Significance levels obtained from analyses based on pairwise matrices must be interpreted with caution.

Megatip: Identifying Points in a Scatterplot

In the Chart Editor, you can label points by the value of the labeling variable you selected in the Scatterplot dialog box, by the sequential case number used by SPSS Statistics to identify data points (labeled *Case Number*), and by the values of the variables plotted on the axes. You can label all points or selected points. Whatever you want to do, the first step is to double-click the chart to activate the Chart Editor.

Labeling Individual Points with the Default Variable Value

Click the Data Label Mode button in the Chart Editor (that's the black square with a black cross in the middle), or from the Chart Editor choose Elements > Data Label Mode. This changes your cursor to the same symbol. You cannot edit the chart when you are in Data Label Mode. All you can do is identify points.

Click the point you want to identify. A value will appear. If you specified a labeling variable in the Scatterplot dialog box, that's the value. If you did not specify a labeling variable in the Scatterplot dialog box, the label is the case number. If you click a point that has already been identified, the label will go away. To turn off Data Label Mode, click the Data Label Mode button on the toolbar. To change the value or values being displayed, follow the directions in "Changing the Variables Used to Identify Points" below. If several cases are represented by the same point on the plot, a dialog box will open that allows you to choose the case label you want to display.

Tip: When in Data Label Mode, you can locate a case in the Data Editor by right-clicking the point and then right-clicking on the Go To Case pop-up menu. This takes you to the Data Editor, with the point highlighted (if you are in Data View).

Labeling All Points with the Default Variable Value

From the Elements menu, choose Show Data Labels. All of the points are now labeled with the default values of the labeling variable. If you specified a labeling variable in the Scatterplot dialog box, that's the value. If you did not specify a labeling variable in the Scatterplot dialog box, the label is the case number. To turn off all of the labels, choose Elements > Hide Data Labels.

Changing the Variables Used to Identify Points

From the Elements menu, choose Show Data Labels. This will label all of the points with the value labels of the variable specified in the Scatterplot dialog box. It will also open the Properties dialog box. In the Properties dialog box, click the Data Value Labels tab. Variable names used to label the points appear in the Displayed list. To add more variables to be used for labeling (a point can be labeled with several values), select and move one or more of the variables from the Not Displayed list into the Displayed list by clicking the green arrow. To remove a variable from the Displayed list, select it and click the red X. Click Apply, and then click Close. The selections you make in the Properties dialog box determine which variables are used as labels in Data Label Mode as well.

Obtaining the Output

To produce the output in this chapter, open the file *bodyfat.sav* and follow the instructions below. The Graphs can also be produced by the Chart Builder, which is on the Graphs menu.

Figures 11-1, 11-2, and 11-3. From the Graphs menu, choose Legacy Dialogs > Scatter/Dot, select the Simple Scatter icon, and click Define. For Figure 11-1, move *fatpct1* into the Y Axis box and *weight* into the X Axis box, and then click OK. For Figure 11-2, move *density* into the X Axis box, and click OK. For Figure 11-3, move *height* into the X Axis box, and click OK. To show the reference line for the mean in Figure 11-3, double-click the chart to open it in the Chart Editor. From the Options menu, choose Y Axis Reference Line. Type the number 19.1 into the Y Axis Position box, click Apply, and then click Close.

Figure 11-4. From the Graphs menu, choose Legacy Dialogs > Scatter/Dot; then select the Matrix Scatter icon and click Define. Move *fatpct1*, *density*, *height*, and *weight* into the Matrix Variables list. Click OK.

Figure 11-5. From the Analyze menu, choose Correlate > Bivariate. Move *fatpct1*, *weight*, and *height* into the Variables box. Select Pearson, and then click OK.

Figure 11-6. Cannot be produced from the data file.

Figure 11-7. From the Graphs menu, choose Legacy Dialogs > Scatter/Dot. Select the Simple Scatter icon and click Define. Move *fatpct1* into the Y Axis box and *fatpct2* into the X Axis box. Click OK.

Figures 11-8 and 11-9. From the Analyze menu, choose Compare Means > Paired-Samples T Test. Move *fatpct1* and *fatpct2* into the Paired Variables list. Click OK.

Figures 11-10 and 11-11. From the Analyze menu, choose Correlate > Partial. Move *chestcirc*, *hipcirc*, *kneecirc*, and *neckcirc* into the Variables list. To get zero-order coefficients, click Options and select Zero-order correlations; then click Continue. To get partials controlling for weight, move *weight* into the Controlling For list, and then click OK. To get partials controlling for height, move *height* into the Controlling For list. To get partials controlling for height and weight, move both variables into the Controlling For list. This produces Figure 11-11. Select the right numbers from all of this output and use a word processor to create Figure 11-10.

Chapter

12

Bivariate Linear Regression

A correlation coefficient tells you how closely points cluster around a straight line, but it leaves you in suspense as to the actual identity of the line. If you want to predict the values of one variable from another, you need more information about the mysterious line. (Fortunately, for a line, this just means knowing the slope and intercept.)

This chapter outlines the basics of bivariate linear regression, which is used when you want to predict the values of a dependent variable from the values of a single independent variable. If you have multiple independent variables, you want to use multiple linear regression, which is discussed in Chapter 13.

Warning: If you didn't reject the null hypothesis that the population correlation coefficient is 0, you won't be able to fit a regression line that has a slope significantly different from 0.

Examples

- How are final salaries at a previous job related to beginning salaries for a new job?
- Can you predict bone density from a laboratory assay?
- How much can you reduce your abdominal circumference by exercising an hour a day?

In a Nutshell

You want to predict values of a dependent variable from values of an independent variable. You select the line that best fits the data points, using the criterion that the sum of squared differences between the points and the line is as small as possible. This line is known as the **least-squares regression line**. To test hypotheses about the population regression line, you must make certain assumptions about the origins of the data. You can test these assumptions using plots that are based on the differences between the observed and the predicted values of the dependent variable. The correlation coefficient between the observed and predicted values tells you how well the line fits the data.

Predicting Body Fat: The Example

Body fat percentage is one of the myriad of magical numbers that characterize the contemporary human machine. Its importance is ever increasing, with even the government urging overweight citizens to shed fat. (Will that translate to less pork in government?) Body fat percentage is not an easy measurement to obtain, since it involves subtracting the weight of bones, water, and other essential nonfatty parts from the total weight. Given the difficulties involved in removing these parts from a living human, body fat is measured indirectly, using calipers, passing electric currents through the body, or weighing people under water. The last method requires repeated total submersion and exhaling air from the lungs. This is not exactly a fun procedure that can be carried out in the privacy of your home.

Since body fat percentage is linearly related to the circumference of various body regions, such as the abdomen, chest, hips, and thighs, you can use statistical methods to try to predict body fat from measurements that require only a tape measure and an honest eye. If you find a good model, total submersion can be relegated to its rightful place in religious rites.

Plotting the Values

Figure 12-1 is a scatterplot of body fat percentage and abdominal circumference, measured in centimeters, for the sample of 252 men described in Chapter 11. The least-squares regression line has been drawn through the data points. The correlation coefficient measures how closely the points cluster around this very line.

Figure 12-1
Scatterplot of body fat percentage and abdominal circumference

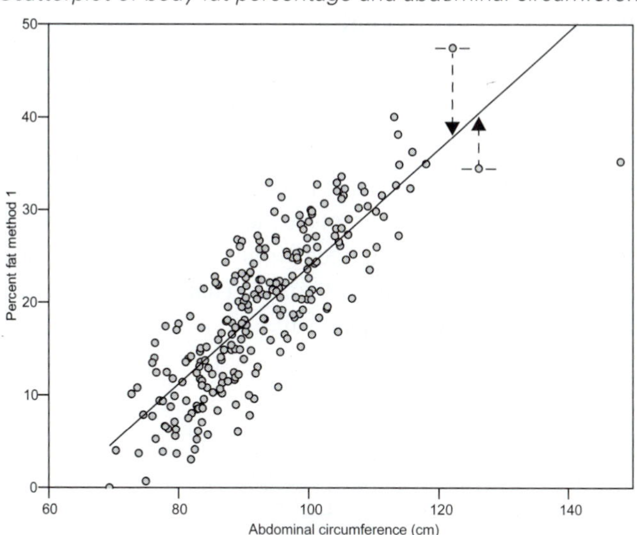

Warning: There is an outlying point (case number 39) that has elicited the appropriate anguish in the previous chapter. Read about it there. Watch for it during the analyses.

Calculating the Least-Squares Regression Line

The general form of the line to predict body fat percentage from abdominal circumference is

$$\text{PredFat} = B_0 + (B_1 \times \text{AbdomCirc})$$

Here, B_0 is the intercept and B_1 is the slope. The slope tells you how much body fat percentage changes for a one-centimeter change in abdominal circumference. The intercept is the value of predicted body fat percentage when the abdominal circumference is 0. (The intercept is not an interesting number in most regression analyses. Body fat percentage is probably not a major concern for a man without an abdomen.)

Warning: SPSS Statistics has an option for fitting a model without an intercept term (called *regression through the origin* in statistical literature), but don't use it without very good reason. The meaning of some of the statistics from regression through the origin are different from those described here. It may sound like a good thing to do if you know that the line must pass through the origin, but it's almost always a bad idea.

The equation for the least-squares line in Figure 12-1 is

$$\text{PredFat} = -39.425 + (0.633 \times \text{AbdomCirc})$$

The coefficients are in the column labeled *B* in Figure 12-2. The standardized regression coefficient (labeled *Beta*) is the slope of the regression line you obtain when you standardize both the independent and dependent variable to have a mean of 0 and a standard deviation of 1. The coefficients define the line that has the smallest sum of squared vertical distances from the data points to the line. Any other line that you draw through the points has a larger sum of squared vertical distances. The dashed lines in Figure 12-1 show these distances for several points. That's why the regression line is sometimes admiringly called the least-squares line.

Figure 12-2
Regression coefficients for predicting body fat percentage

			Unstandardized Coefficients		Standardized Coefficients		
			B	Std. Error	Beta	t	Sig.
Model	1	(Constant)	-39.425	2.658		-14.832	.000
		Abdominal circumference (cm)	.633	.029	.814	22.187	.000

Warning: The coefficients of the regression line depend on which variable is the dependent variable and which variable is the independent variable. You'll get a different set of coefficients if you try to predict abdominal circumference from body fat percentage.

Predicting Body Fat

You can use the least-squares regression line to predict values of body fat percentage for men with abdominal circumferences in the same range as the ones in the data. For example, for a man with an abdominal circumference of 90 centimeters, the predicted body fat percentage is

$$\text{PredFat} = -39.425 + (0.633 \times 90) = 0.175 = 17.5\%$$

For men in the data file, you can compare the actual and predicted body fat values. The difference between the actual and predicted values in called a **residual**. For a man with a 90-centimeter abdomen and an observed body fat percentage of 20, the residual is

$$\text{residual} = \text{observed} - \text{predicted} = 20\% - 17.5\% = 2.5\%$$

Tip: You can read the predicted value from Figure 12-1 by finding 90 centimeters on the horizontal axis and seeing what value of body fat corresponds to 90 centimeters on the regression line. The residual for a point is the vertical distance from the point to the regression line.

Determining How Well the Line Fits

From the coefficients of a linear regression line, you can't tell how well the line fits the data. You can get a slope of 0.63 when all of the points fall right on the regression line or when they are scattered around the line as in Figure 12-1. A simple way to summarize how well an estimated regression line fits the observed data is to calculate the correlation coefficient between the observed and predicted values. That's the number in the column labeled *R* in Figure 12-3. The square of the correlation coefficient, labeled *R Square*, is the proportion of the variability in body fat percentage that is explained by differences in abdominal circumference. In this sample, about 66% of the variability in body fat percentages is explained by abdominal girth.

Figure 12-3
Regression goodness-of-fit statistics

		R	R Square	Adjusted R Square	Std. Error of the Estimate
Model	1	.814[1]	.663	.662	4.87352

1. Predictors: (Constant), Abdominal circumference (cm)

In a regression analysis with a single independent variable, the correlation coefficient between the observed and predicted values is the same as the absolute value of the correlation coefficient between the dependent and independent variables. The standardized regression coefficient in Figure 12-2 is also equal to the correlation coefficient.

Warning: A model always fits the sample from which it is calculated better than it will fit another sample. *Adjusted R Square* is an attempt to correct for this.

Testing Hypotheses about the Population Regression Line

If your data are a random sample from some population, you may want to draw conclusions about the population regression line. As is always the case, you have to make some assumptions before you can test regression hypotheses. All observations must be independent. In the population, for each value of the independent variable, there must be a normal distribution of values of the dependent variable. All of these distributions must have the same variance. The means of all of the population distributions must fall on a straight line.

Figure 12-4 is a schematic summary of the requisite assumptions. The line in the figure is the population regression line that relates abdominal circumference and body fat percentage. For each circumference value, there is a distribution of values of body fat. That's reasonable if you think about it. You don't expect all men with the same circumference to have exactly the same value for body fat percentage. You expect a distribution of values. To use statistical tests, you have to assume that the distributions are normal and have the same variance. You'll see how to use residuals to check these assumptions in "Searching for Violations of Assumptions" on p. 225.

Figure 12-4
Regression assumptions

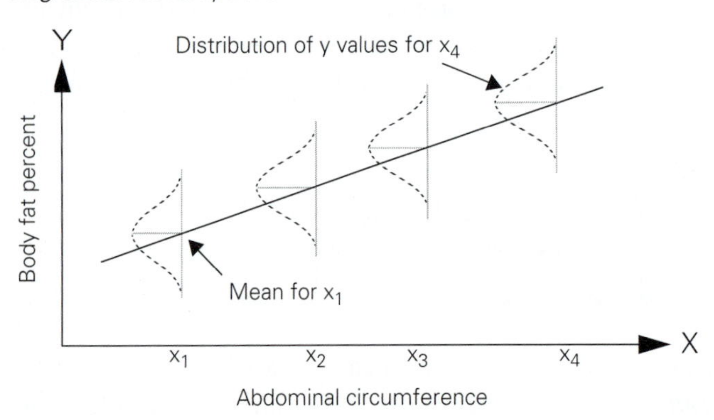

Tip: Sometimes you can transform a nonlinear model to a linear one using transformations. For example, for the model $y = a \times c^x$, you can take logarithms of both sides of the equation and arrive at a linear model for the relationship of *log* y and x.

Testing for a Slope of Zero

The first, and often only, null hypothesis that you want to test is that the population slope is 0. This null hypothesis says that there is no linear relationship between the dependent variable and the independent variable. You can use the t value in Figure 12-5 and its observed significance level to test that the population slope is 0. The t value is the ratio of the coefficient to its standard error. From the small observed significance level, you can reject the null hypothesis that the slope is 0. There does appear to be a linear relationship between the two variables.

Testing that the population slope is 0 is equivalent to testing that the population correlation coefficient between the two variables is 0. If you reject the null hypothesis that the population slope is 0, you reject the null hypothesis that the population correlation coefficient between the two variables is 0.

Warning: Rejecting the null hypothesis that the slope is 0 doesn't tell you that you have a good equation for prediction. You have to look at the actual value of R^2 to determine whether the prediction is good enough for your needs. If you're calculating body fat to help you decide whether you should exercise more, this model is probably good enough. If you're calculating body fat so that you can undergo some delicate surgery that requires a very svelte figure for survival, the estimate may not be good enough.

You can also calculate confidence intervals for the value of the population slope. From Figure 12-5, you see that the 95% confidence interval for the slope is from 0.58 to 0.69.

Figure 12-5
Confidence interval for the slope

			Unstandardized Coefficients		Standardized Coefficients			95% Confidence Interval for B	
			B	Std. Error	Beta	t	Sig.	Lower Bound	Upper Bound
Model	1	(Constant)	-39.425	2.658		-14.832	.000	-44.661	-34.190
		Abdominal circumference (cm)	.633	.029	.814	22.187	.000	.577	.689

Tip: If you want to test the null hypothesis that the population slope has a particular value (for example, 1), determine if that value is included in the 95% confidence interval for the population slope. If it's not in there, you can reject the null hypotheses, using a significance level of 0.05.

Predicting Future Values

From a regression equation, you can make two types of predictions. You can predict the body fat percentage for a particular man with a specified abdominal girth, or you can predict the average body fat percentage for all men with a specified girth. As you saw, the predicted body fat percentage is 17.5 for a man with a girth of 90 centimeters. That's also the predicted mean value for all men with a girth of 90 centimeters.

Although the predicted value is the same in both situations, the standard error of the prediction is not. There is more variability associated with the estimate for an individual prediction than for the mean prediction. That's not unexpected. Even if you were able to predict perfectly the mean value, you wouldn't necessarily be able to make a perfect prediction for a particular individual. Not all men with the same abdominal circumference have the same body fat percentage; instead, there is a normal distribution of values.

Figure 12-6 is a plot of the 95% prediction intervals for both the mean prediction and the prediction for an individual case. The outer curves are for the individual prediction. The tight curves in the center are for the mean prediction. The curves are narrowest at the average value of the independent variable and spread out from there.

For the mean prediction, the prediction interval is simply the confidence interval for the population mean value of the dependent variable for each value of the independent variable. For an individual prediction, the prediction interval is a range of values that you expect to include the value for a future case. For an individual man with an abdomen of 90 centimeters, the 95% prediction interval for body fat percentage ranges from 7.92 to 27.16. For the mean, body fat percentage of such men is considerably narrower, from 16.92 to 18.16.

Figure 12-6
Prediction intervals

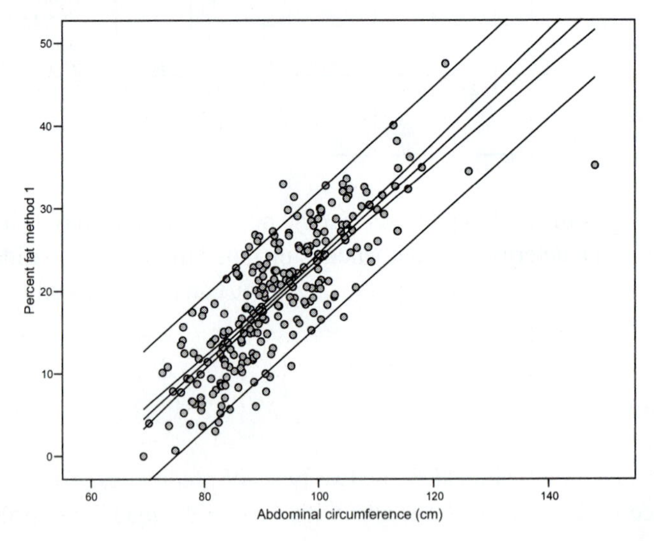

Warning: Don't use the regression line to make predictions outside of the observed range of values of the independent variable. The relationship between the two variables may be different outside of the observed range.

Searching for Violations of Assumptions

You can use the residuals you calculate from the regression equation to look for violations of the assumptions that are necessary for hypothesis testing. Besides the ordinary (unstandardized) residual, SPSS Statistics calculates four other closely related residuals:

- **Standardized residuals** are the ordinary residuals divided by the standard deviation of the residuals. The mean of the standardized residuals is 0, and the standard deviation is 1. Dividing residuals by their standard error makes it easier to judge their size, since most standardized residuals should be less than 3 in absolute value.
- **Studentized residuals** are the ordinary residuals divided by an estimate of the standard deviation that varies from point to point.
- **Deleted residuals** are the difference between the observed and predicted values when a case is excluded from the computation of the regression statistics.
- **Studentized deleted residuals** are the deleted residuals divided by their standard error that varies from point to point.

Tip: Studentized residuals are the most sensitive to outliers, so don't hurt their feelings; use them in plots.

Is the Relationship Linear?

You can evaluate the assumption that the relationship of the two variables is linear by looking at a scatterplot of the data. The points should cluster around a straight line. You can also check this assumption by plotting residuals against the values of the independent variable and against the values of the predicted variable. In both of these plots, you shouldn't see any relationship between the two variables; instead, the points should be randomly scattered in a band around 0. Figure 12-7 is a plot of the Studentized residuals against the predicted values. Except for case 39 (the outlier), the points appear to be randomly scattered.

Figure 12-7
Plot of Studentized residuals and predicted values

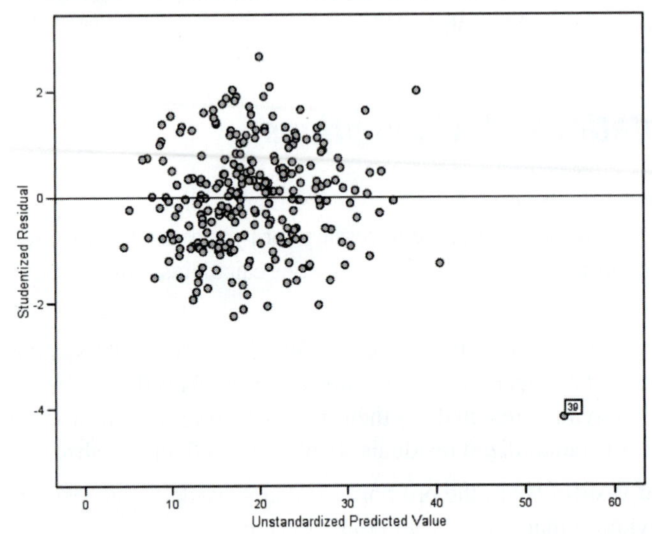

Tip: It's a good idea to plot residuals against other independent variables that you think might be related to the dependent variable. If you see a nonrandom pattern, you may want to include additional individual variables in your regression. (See Chapter 13.)

Are the Errors Independent?

Whenever the data are collected sequentially, you should plot residuals against the sequence variable, even if you think time is irrelevant. Experimental equipment and experimenters can both deteriorate (or improve) with time.

If the body fat data were collected in a known sequence, you could plot the sequence number against the residuals, as shown in Figure 12-8. If you see patterns in the points, such as a clustering of positive and negative residuals, or residuals increasing or decreasing with time, you should be concerned.

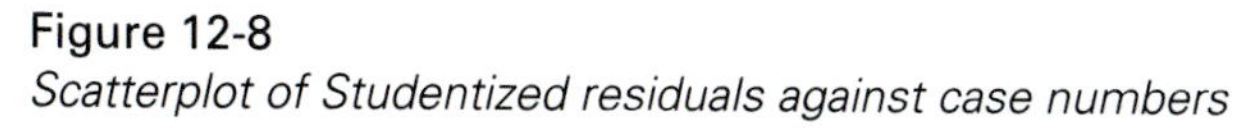

Figure 12-8
Scatterplot of Studentized residuals against case numbers

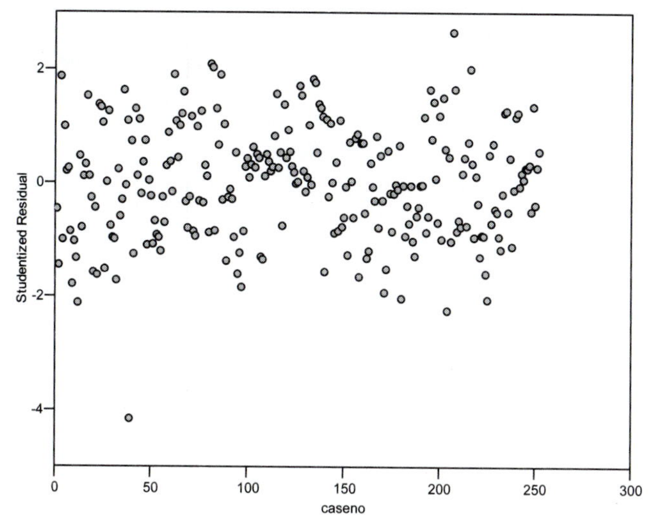

Tip: You can use the Durbin-Watson statistic to test whether adjacent residuals are correlated. The possible values of the statistic range from 0 to 4. If the residuals are not correlated with each other, the statistic is close to 2. Values less than 2 indicate positive correlation of the residuals, and values greater than 2 indicate negative correlation. You have to consult special tables of the Durbin-Watson statistic to test the null hypothesis that the residuals are independent.

Are the Variances Equal?

To check whether the variances are constant, plot the residuals against the predicted values. If the spread of the residuals increases or decreases during the course of the plot, you should question the assumption of constant variance for all values of the independent variable. The spread of the residuals in Figure 12-7, a plot of Studentized residuals against predicted values, appears fairly constant over the range of the predicted value. Figure 12-9 is a plot in which the spread of the residuals increases with values of the independent variable.

Figure 12-9
Plot of residuals with nonconstant variance

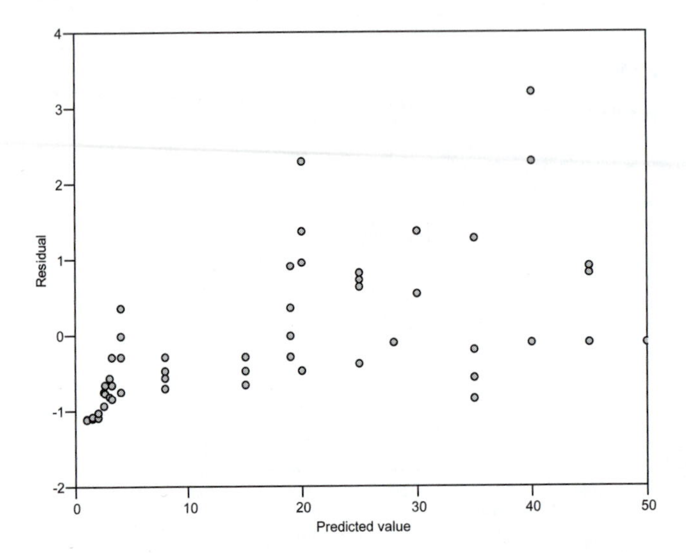

Warning: If you plot the residuals against the observed values of the dependent variable, you don't expect a random scatter of residuals. The slope of the line will always be $1 - R^2$.

Are the Residuals Normally Distributed?

You can check the normality assumption by examining the distribution of residuals. Figure 12-10 is a histogram of the Studentized residuals, and Figure 12-11 is a normal probability plot. Except for the one large residual, there's no reason to believe that the assumption of normality is violated. The normal probability plot shows a reasonably good fit to the normal distribution.

Tip: To change the width of the bars in a histogram, double-click the histogram to activate it in the Chart Editor. From the Edit menu, choose Select X Axis. Click the Histogram Options tab, select Custom in the Bin Sizes group, and then specify the bin widths or the number of bins.

Figure 12-10
Histogram of Studentized residuals

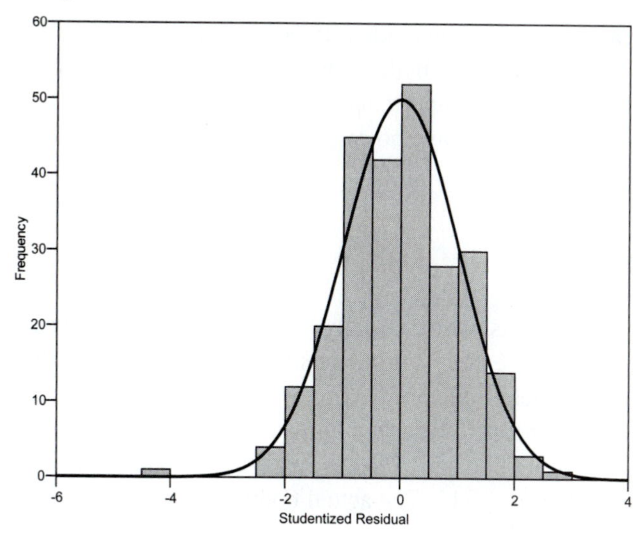

Figure 12-11
Normal probability plot of Studentized residuals

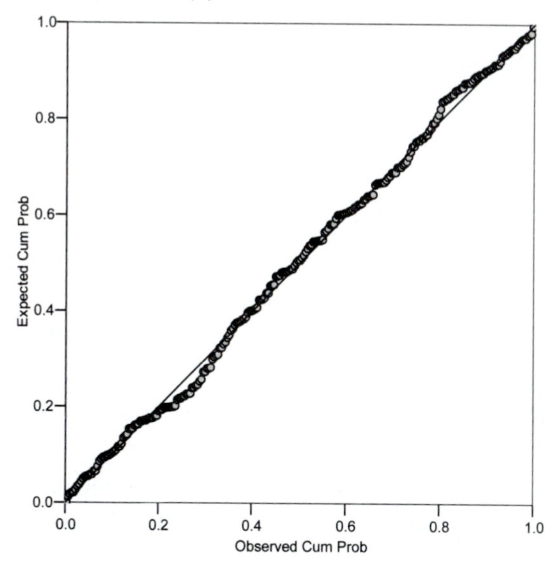

Looking for Unusual Points

Whenever you fit a model, you want to identify points that are different from the others in important ways. In a regression model, you want to look for three types of points: points that have large residuals, points that have unusual combinations of values for the independent variables (high-leverage points), and points that affect the values of the coefficients (influential points).

Are There Large Residuals?

You can easily identify points with unusually large positive or negative residuals. You can find them on stem-and-leaf plots of the residuals, or you can spot them in plots of the residuals against predicted values or values of the independent variable. Cases with absolute values of the standardized residual greater than 3 raise concerns.

There's only one case in this data file that has a large value for the standardized residual, as shown in Figure 12-12. The actual body fat percentage is 35.2, while the predicted fat value is 54.3. That's a residual of –19.1%. The standardized value for the residual is –3.9, indicating that it is almost four standard deviations away from the mean of 0.

Figure 12-12
Case with standardized residual greater than 3

		Std. Residual	Percent fat method 1	Predicted Value	Residual
Case Number	39	-3.922	35.20	54.3125	-19.11352

That's the same very heavy person with a body fat percentage value that's out of line with the rest. It's unclear whether there were problems with the body fat measurement or whether the formula for estimating body fat percentage from density may not be accurate for someone of this size. Of course, it's possible that the point is correct.

Are There High-Leverage Points?

A high-leverage point is a case with unusual values for the independent variable. Because there is only one independent variable in this example, it's easy to identify the case with the different value of the independent variable. It's the very heavy individual. Cases with unusual values for the independent variables can have a large effect on the regression coefficients.

Are There Influential Points?

An influential point is a case that has a large impact on the regression equation. Unlike influential friends who are an asset to us, influential points are bad for a regression equation. You don't want to have a model whose coefficients are heavily dependent on the values for one or two cases. Special statistics that depend on the values of both the residual and the leverage are needed to identify influential points. Cook's distance (not related to any abdominal measures) is the most frequently used measure of influence. It's based on the changes in the regression coefficients when a case is excluded from the computation of the regression equation.

Figure 12-13 is a scatterplot of Cook's distance against the case numbers. There is one case with a large value for Cook's distance, and, as expected, it is case 39. Cook's distances that are greater than 1 are usually of concern.

Figure 12-13
Scatterplot of Cook's distance

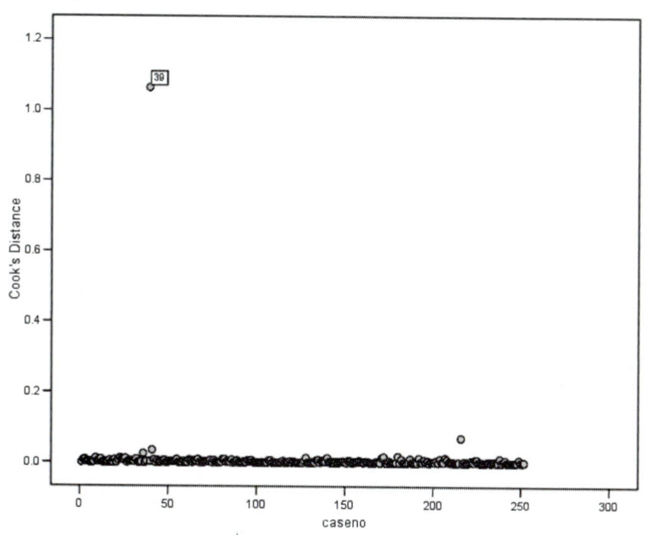

Tip: If you plot the diagnostic statistics against the case numbers, you can use Data Label Mode in the Chart Editor to identify the influential points.

You can also look at the actual changes in the regression coefficients. Figure 12-14 shows how much the regression coefficient for the abdominal circumference changes when individual cases are removed from the model. The cases listed are those that result in the five largest positive changes and the five largest negative changes. The column labeled *Value* is the actual change in the coefficient (called *DfBeta*). If you eliminate case 39 from your database, the regression coefficient for abdominal circumference would change by –0.041. It would go from the observed value of 0.63 to 0.67.

Tip: You can also save the standardized change in the coefficient (standardized DfBeta). Examine points for which the standardized change is larger than

$$\frac{2}{\sqrt{N}}$$

Figure 12-14
Cases that result in large changes for the regression coefficient

			Case Number	Value
DFBETA abdomcirc	Highest	1	216	.01009
		2	36	.00566
		3	172	.00449
		4	192	.00416
		5	182	.00365
	Lowest	1	39	-.04086
		2	41	-.00718
		3	180	-.00407
		4	238	-.00398
		5	128	-.00377

Dealing with Violations of the Assumptions

As is usually the case in life, it's easier to identify a problem than to fix it. If you uncover serious violations of the assumptions, you can pursue one of two strategies: you can formulate an alternative model, such as weighted least squares, or you can transform the variables so that you can use the original model. Transformations usually involve taking logs, square roots, or reciprocals of the original values.

Coaxing a Nonlinear Relationship to Linearity

You can try to make the relationship between your variables linear by transforming either the independent or the dependent variable, or both. The regression assumptions are for the dependent variable, so if you tinker with the scale of the independent variable, you don't have to worry about unraveling other assumptions. If you transform the dependent variable, you change its distribution. The distribution of the transformed variable must satisfy the assumptions of the analysis. For example, if you take logs of body fat percentage, then it is the logs that must be normally distributed with constant variance.

The choice of transformation depends on the actual data. If a relationship appears nearly linear for part of the data but is curved for the rest, taking the log of the dependent variable may result in a better linear fit. Figure 12-15 is a before-and-after plot of a nonlinear relationship that is transformed to linearity by taking logs of the dependent variable.

Other transformations that may diminish the curvature are $-1/Y$ and $\sqrt{Y}$. You have to see what works for your data.

Figure 12-15
Before and after transformation plots

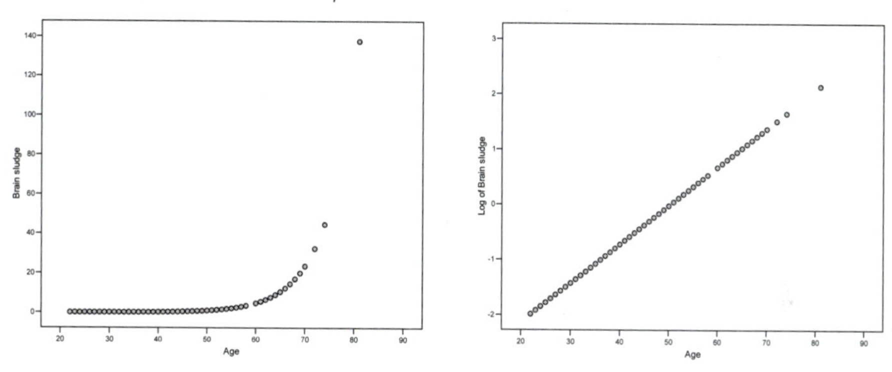

Coping with Non-Normality

If the distribution of residuals has a tail toward larger values, a log transformation may cause it to retract. If the tail is toward smaller values, taking square roots may solve the problem. Regression hypothesis tests are usually not very sensitive to departures from normality.

Tip: Worry about non-normality last. Fixing other problems may resolve the normality issue as well.

Stabilizing the Variance

If the variance of the residuals isn't constant, you can try a variety of remedial measures:

- If the variance is proportional to the mean of the dependent variable and all values are positive, take the square root.
- If the standard deviation is proportional to the mean and all values are positive, try the logarithmic transformation.
- If the standard deviation is proportional to the square of the mean, use the reciprocal.
- If the dependent variable is a proportion or rate, try the arcsine transformation.

Adding More Circumferences: Multiple Linear Regression

You saw that, for this sample, you could explain about 66% of the variability in body fat percentage from just the abdominal circumference. You're no doubt very curious about whether you could get an even better prediction if you included several circumferences in a model. Figure 12-16 shows summary statistics when abdominal, wrist, and hip circumferences are used to predict body fat percentage. With this model, you can explain 72% of the variability in body fat. That's somewhat larger than the 66% from abdominal circumference alone.

Warning: Adding independent variables always improves the fit of the model in the sample. Additional variables do not necessarily improve the fit of the model in the population.

Figure 12-16
Model summary multiple regression

		R	R Square	Adjusted R Square	Std. Error of the Estimate
Model	1	.849[1]	.722	.718	4.44929

1. Predictors: (Constant), Wrist circumference (cm), Abdominal circumference (cm), Hip circumference (cm)

The coefficients for predicting body fat percentage from the three circumferences are shown in Figure 12-17. You see that the coefficients are positive for abdominal circumference but negative for hip and wrist circumferences. One possible explanation for the negative coefficient for hip circumference is the high correlation between abdominal and hip circumferences, Their contribution to the model is shared. The negative coefficient for wrist circumference may tell you how big a person's bones are. Bones have to be subtracted from the total weight to get at body fat percentage.

Figure 12-17

Multiple regression coefficients

			Unstandardized Coefficients		Standardized Coefficients		
			B	Std. Error	Beta	t	Sig.
Model	1	(Constant)	1.551	6.212		.250	.803
		Abdominal circumference (cm)	.928	.055	1.195	17.026	.000
		Hip circumference (cm)	-.329	.083	-.282	-3.970	.000
		Wrist circumference (cm)	-1.944	.394	-.217	-4.932	.000

Tip: Read Chapter 13 to learn how to fit a linear regression model with more than one independent variable.

Obtaining the Output

To produce the output in this chapter, open the file *bodyfat.sav* and follow the instructions below.

Figure 12-1. From the Graphs menu, choose Legacy Dialogs > Scatter/Dot, select the Simple Scatter icon, and click Define. Move *fatpct1* into the Y Axis box and *abdomcirc* into the X Axis box. Click OK. In the output Viewer, double-click the chart to activate the Chart Editor. From the Elements menu, choose Fit Line at Total. In the Properties dialog box, select Linear. Click Apply, and then click Close. To match the plot exactly, you'll have to use the Chart Editor to set the number of decimal places for the axes.

Figures 12-2, 12-3, 12-5, and 12-12. From the Analyze menu, choose Regression > Linear. Move *fatpct1* into Dependent box and *abdomcirc* into the Independent(s) list, and click Statistics. In the Statistics subdialog box, select Estimates, Confidence intervals, Model fit, and Casewise diagnostics, and click Continue. Click Save to display the Save subdialog box, and then select Unstandardized in the Predicted Values group, Studentized in the Residuals group, Cook's in the Distances group, and DfBeta(s) and Standardized DfBeta(s) in the Influence Statistics group. Click Continue, and then click OK.

Figure 12-6. Double-click to activate Figure 12-1 in the Chart Editor. From the Elements menu, choose Fit Line at Total. In the Properties dialog box, select Linear, and then select Mean from the Confidence Intervals group. Click Apply. Now select Individual from the Confidence Intervals group and click Apply again. Click Close. You should now have the regression line and two sets of prediction intervals on the plot.

Figure 12-7. From the Graphs menu, choose Legacy Dialogs > Scatter/Dot, select the Simple Scatter icon, and click Define. Click Reset to clear the boxes for Y Axis and X Axis, if necessary. Move *sre_1* into the Y Axis box, *pre_1* into the X axis box, and *caseno* into the Label Cases By box. Click OK. To see the value of the Studentized residual for the outlying point, double-click the chart to activate the Chart Editor, and then choose Data Label Mode from the Elements menu. Place the pointer over the outlying point and click. (The variables with underscores won't be available to you if you did not save them in the Regression Save subdialog box.)

Figure 12-8. From the Graphs menu, choose Legacy Dialogs > Scatter/Dot, select the Simple Scatter icon, and click Reset to clear the boxes for Y Axis and X Axis, if necessary. Move *sre_1* into the Y Axis box and *caseno* into the X Axis box. Click OK. Double-click the chart to activate the Chart Editor. From the Options menu, choose Y Axis Reference Line. In the Properties dialog box, type 0 in the Y Axis Position box. Click Apply, and then click Close.

Figures 12-9 and 12-15. These figures cannot be produced from the data file.

Figure 12-10. From the Graphs menu, choose Legacy Dialogs > Histogram. Move *sre_1* into the Variable box. Select Display normal curve, and click OK. If you want the chart to look exactly like Figure 12-10, you'll have to edit it in the Chart Editor.

Figure 12-11. From the Analyze menu, choose Descriptive Statistics > Q-Q. Move *sre_1* into the Variables list, and click OK.

Figure 12-13. Follow the same instructions as Figure 12-7 but move *coo_1* into the Y Axis box and *caseno* into the X Axis box. Click OK.

Figure 12-14. From the Analyze menu, choose Descriptive Statistics > Explore. Move *dfb1_1* into the Dependent List. Click the Statistics button and select Outliers. Click Continue, and then click OK.

Figures 12-16 and 12-17. Follow the same instructions as for Figure 12-2, adding *hipcirc* and *wristcirc* to the list of independent variables.

Chapter

13

Multiple Linear Regression

In the real world, relationships among variables are usually complex and difficult to unravel. For physical phenomena, such as the time it takes dense objects to fall from various heights, there's an underlying mathematical equation that exactly relates the values of the two variables. Unfortunately, nature seems to have overlooked (or hidden well) the exact equations that govern the relationship between variables, such as sales and marketing dollars, test scores and class attendance, or blood pressure and exercise. However, even in these situations, you can use mathematical models to organize and simplify the observed relationships so that you can better understand the processes at work. But always keep in mind the caveat that "all models are wrong, but some are useful" (Box, 1979).

In this chapter, you'll use multiple linear regression to model the relationship between a single dependent variable and a set of independent variables. Questions you can answer with a regression analysis include:

- Can the values of the dependent variable be predicted from the values of the independent variables?
- Which variables are linearly related to the dependent variable?
- Can a subset of independent variables be identified that are useful for predicting the dependent variable?

Examples

- What's the relationship between systolic blood pressure and age, weight, height, cigarette smoking, and physical activity?
- Can you predict a salesperson's total dollar sales for the year based on years of experience, age, sales territory, and education?
- What characteristics of a student and his or her undergraduate college are related to scores on the Graduate Record Exam?

In a Nutshell

You predict the values of a dependent variable as a linear combination of the values of one or more independent (predictor) variables. The coefficients of the model are chosen so that the sum of squared differences between the observed and predicted values (in other words, the sum of squared residuals) is as small as possible. If you have a large number of independent variables, you can identify a subset of independent variables that are useful for predicting the dependent variable.

To test hypotheses about the population values of the regression coefficients, you have to make certain assumptions. You examine residuals to look for violations of the assumptions. You examine the data to look for observations that may have more than their fair share of impact on the final equation or are poorly predicted by the model.

Reading Scores: The Example

Some schools produce students that score well on standardized tests and some schools don't. Some hospitals have good results for a particular surgical procedure while others don't. The question that everyone wants to answer is: *Why?* Unfortunately, building a regression model that relates outcomes to variables such as teacher qualifications or the volume of surgeries performed can't answer the real questions. You can't draw cause-and-effect conclusions from such a model. Models, however, are useful for identifying variables that are related to the outcome. Careful thought and a good regression model can suggest avenues to explore.

Tip: Before collecting your own data and building your model, make sure to examine models that others have reported for similar problems. That's often a good starting point for your own analysis.

Formulating the Problem

The most crucial step in fitting any type of model to data is understanding the question you want to answer and its relationship to the data you have available or will collect. Even the best statistical software can't tell you that the data you are analyzing do not shed light on the problem that you are interested in because you neglected to include important independent variables or because the dependent variable you selected is a poor indicator of what you're really interested in. Building a good model is as much of an art as it is a science.

Selecting the Variables

You have to decide how to define the dependent variable and decide which independent variables to consider. (In this chapter, to match the labels used in the SPSS Statistics dialog boxes, the outcome variable is called a **dependent variable**, and the predictor variables are called **independent variables**. When the values of the independent variables are not under the control of the experimenter, the dependent variable is sometimes called the **criterion**, and the independent variables are called the **predictors**.) Sometimes the decisions are easy. If you're looking at predictors of height, the dependent variable is well defined. However, you still need to carefully establish the protocol for obtaining even such a straightforward measurement. Will heights be self-reported? Will standing straight be encouraged?

If you're concerned with the performance of institutions, selecting the dependent variable to study is often considerably more complicated. You can use a large number of criteria to evaluate a school: standardized test scores, graduation rates, and attendance figures, to name a few. Your conclusions may be quite different, depending on how you specify the dependent variable.

You encounter similar problems with selecting the independent variables for study. For many interesting problems, there's an excess of independent variables. For the school performance model, independent variables can be characteristics of the students, of the faculty, of the family, or of the neighborhood, and that's just the start. Unfortunately, independent variables are usually correlated with each other and that makes unraveling the individual contributions of the variables very difficult.

If you neglect to include important variables, your model will be seriously impaired. For example, if you are trying to model the post-surgery length of stay in the hospital and you neglect to include information about whether the surgery was planned or was an emergency, you may be overlooking the best predictor of length of stay.

Tip: Whenever you build a statistical model, you are faced with the question of how large a sample you need to obtain reasonably good estimates of the coefficients of the independent variables. That's a difficult question to answer. Harrell (2001) cites studies suggesting sample sizes of 10 to 20 cases for each independent variable. If your sample size is not large enough, you have the very real risk of overfitting the model; that is, your model will not validate when applied to a different sample. If you're analyzing large surveys, such as the General Social Survey, these concerns are minimal. With large sample sizes, you want to make sure that results that are statistically significant are also substantively important, since even very small coefficients may lead you to reject the null hypothesis.

Selecting the Mathematical Model

Model formulation also involves selecting the type of mathematical model to fit to the data. Some models, such as those in the life sciences or physical sciences, are fairly well established. You may need to fit sums of exponentials or logistic curves. If you know the mathematical form for the model that you want to fit and it is not intrinsically linear, you should choose the SPSS Statistics procedures that can estimate that type of model. If prior research doesn't dictate a particular type of model, you want to fit the simplest model that is consistent with your data.

One of the simplest models that you can build to predict the values of a dependent variable from those of one or more independent variables is to assume a linear relationship between the variables. If you have a single independent variable, the analysis is termed a **bivariate linear regression** (see Chapter 12). When you have two or more independent variables, the analysis is called **multiple linear regression**.

Multiple Linear Regression Model

As an example of linear regression, we'll look at the relationship between percentage of students scoring at grade level on a reading test and various characteristics of their schools: attendance, class size, percentage of low-income students, and percentage of limited-English students. The data for this example are from a sample of Chicago high schools.

The general formula for a linear model for predicting the values of a dependent variable (Y) from one or more independent variables (X) is

$$\hat{Y} = B_0 + B_1X_1 + B_2X_2 + \ldots + B_nX_n$$

where $\hat{Y}$ is the predicted value and X_i is the ith independent variable.

For example, if you want to predict test scores for schools, based on attendance rates, percentage of students who have limited English proficiency, and average class size, the model is

$$\text{pred. score} = B_0 + B_1(\text{attendance}) + B_2(\text{limited English}) + B_3(\text{class size})$$

Before You Start

Before beginning to build a regression model, you want to look at the distribution of the individual variables and at relationships between pairs of variables. A thorough initial examination may identify problems in the data as well as guide subsequent analysis.

Examining Descriptive Statistics

Figure 13-1 shows very basic descriptive statistics for each of the variables. This is an inadequate look at the variables but a convenient summary of the number of cases with valid values for each of the variables (in the column labeled *N*). You can determine whether there are variables that have a disproportionate share of missing values. (Income is always a good bet, although it's not a problem here because these data are obtained from school records.) The last row, *Valid N (listwise)*, is particularly important. It tells you that many cases have complete information for all of the variables you are considering for the analysis.

Figure 13-1
Descriptive statistics and sample sizes

	N	Minimum	Maximum	Mean	Std. Deviation
Attendance %	81	73.7	95.8	86.565	4.6625
Average class size	82	4.80	27.20	17.8610	4.46365
Limited English %	82	.0	22.3	4.170	6.4632
Low income %	82	23.1	98.6	80.928	15.1344
Reading at national norm %	81	.0	100.0	30.164	22.2573
Valid N (listwise)	80				

Tip: Use the Descriptives procedure to determine how many cases have valid values for each variable in the data file. If you specify listwise variable deletion, the Regression procedure calculates descriptive statistics only for cases that are used in building the model.

If you have a small number of cases (less than 5%) with missing values, you can probably exclude those cases and proceed with the analysis. If you have a larger proportion of cases with missing values, you have to select a strategy for dealing with missing values. If you have randomly missing values for an independent variable that is not related to other independent variables, you may be able to use mean substitution, although this results in covariances that are too small. Otherwise, you may need more complicated imputation strategies.

Examining Correlations

You should also examine the correlation matrix for all pairs of independent variables. Very large correlations among independent variables can cause computational problems as well as increase the difficulty of interpreting your results. If you know which variables are highly correlated with each other, you'll find it easier to understand your results. However, even if you don't find pairs of variables that are highly correlated with each other, it is possible that there are strong linear relationships when more than two independent variables are considered together.

Figure 13-2 shows the correlation coefficients for the variables in this example. Notice that the two variables most strongly correlated with the percentage scoring at norm, the attendance percentage and the low-income percentage, are also correlated with each other.

Figure 13-2
Correlation coefficients

		Reading at national norm %	Attendance %	Average class size	Limited English %	Low income %
Pearson Correlation	Reading at national norm %	1.000	.721	.511	-.123	-.714
	Attendance %	.721	1.000	.593	-.010	-.518
	Average class size	.511	.593	1.000	.036	-.440
	Limited English %	-.123	-.010	.036	1.000	.103
	Low income %	-.714	-.518	-.440	.103	1.000

Plotting the Variables

In a bivariate regression model, you can plot the two variables and quickly see the relationship between them. When you have more than one independent variable, the process becomes more complicated. There's not a single plot that shows how all of the variables are related. Instead, you're forced to look at the variables two at a time. Since the variables are interrelated, you can't tell whether a strangeness you observe on one plot is the result of relationships with other variables that are not in the plot. Before you proceed further, you should:

- Plot the dependent variable against each of the independent variables and determine whether a line is a reasonable summary measure. Figure 13-18 on p. 263 is a plot of two variables that are strongly related but not in a linear fashion. You may need to transform variables to achieve linearity, or you may need to include additional terms, such as squares of the values in the regression equation. Note also whether the variance of the dependent variable is fairly similar over the range of values of the independent variables.
- Plot the values of the independent variables against each other so that you know how the variables are related. Eliminate any variables that are duplicates. You can use a scatterplot matrix to plot all possible pairs of variables. In all of the plots, look for points that stand out from the rest. See if there is a reason for such points. For example, some of the high schools are in correctional facilities, so the relationship between scores and attendance may be different from the more traditional schools.

Tip: The most likely reasons for a strange point are data entry or coding errors. Check that first.

Estimating the Coefficients of the Model

Figure 13-3 shows the regression coefficients that are used to predict percentage reading at norm from four independent variables. The coefficients are estimated using the method of ordinary least squares. They result in the smallest sum of squared differences between the observed and predicted values of the dependent variable.

Based on the coefficients in Figure 13-3, the equation to predict the percentage of students reading at grade level in a particular school is

$$\begin{aligned}\text{pred. score} = & -110.568 + (2.205 \times \text{attendance}) + (0.212 \times \text{class size}) \\ & - (0.259 \times \text{limited English}) - (0.652 \times \text{low income})\end{aligned}$$

Figure 13-3
Regression coefficients

			Unstandardized Coefficients		Standardized Coefficients
			B	Std. Error	Beta
Model	1	(Constant)	-110.568	37.558	
		Attendance %	2.205	.411	.463
		Average class size	.212	.416	.042
		Limited English %	-.259	.230	-.074
		Low income %	-.652	.114	-.448

For a school with an attendance percentage of 86, average class size of 19.7 students, low-income percentage of 91.1, and limited-English percentage of 0, the predicted percentage reading at norm is 23.8.

$$\text{pred. score} = -110.568 + (2.205 \times 86) + (0.212 \times 19.7) - (0.259 \times 0) - (0.652 \times 91.1) = 23.8$$

The regression coefficient for an independent variable tells you how much the value of the dependent variable changes for a one-unit change in the independent variable when all of the other independent variables are held constant. The intercept is the predicted value of the dependent variable when all independent variables have the value of 0.

The difference between the observed and the predicted value is called the **residual**. For the previously described school, the observed percentage reading at norm is 27.1. The residual is

$$\text{residual} = \text{observed score} - \text{predicted score} = 27.1 - 23.8 = 3.3$$

Examining How Well the Model Predicts

Once you have the regression coefficients, the first thing you want to know is whether the model predicts the observed values well. You can't tell from the size of the coefficients because the magnitude of the coefficient depends primarily on the units in which each variable is measured. An intuitive way to tell how well the model fits the data is to compare the observed and predicted values of the dependent variable. Figure 13-4 shows some measures based on the correlation between observed and predicted values.

Figure 13-4
Summary of model

		R	R Square	Adjusted R Square	Std. Error of the Estimate
Model	1	.827[1]	.684	.667	12.8596

1. Predictors: (Constant), Low income %, Limited English %, Average class size, Attendance %

Some commonly used statistics for this purpose are the following:

- **Multiple *R*** is the correlation coefficient between the observed and predicted values. It ranges in value from 0 to 1.
- ***R* square** (R^2) is the proportion of variability in the dependent variable that is attributable to the regression equation. It is the square of multiple *R*. For example, an R^2 of 0.7 tells you that 70% of the observed variability in the percentage of students reading at grade level is attributable to differences in attendance, class size, low-income percentage, and limited-English percentage. This is a very important statistic in regression analysis.
- **Adjusted *R* square** corrects for R^2 being an overly optimistic estimate of how well the model fits the population and decreases R^2 accordingly. A model always fits the sample on which it is based better than it will fit the population, especially if the sample size is small. Adjusted R^2 is also useful for comparing models with different numbers of independent variables because it adjusts for the increase expected in sample R^2 when additional variables are included in a model, even if those variables have no predictive power in the population.
- **Standard error of the estimate** is the standard deviation of the residuals. It doesn't have an upper bound but depends on the actual data values. For a successful regression model, the standard error of the estimate should be considerably smaller than the standard deviation of the dependent variable. That is, observations should vary less about the regression line than about the mean. When the sample size is large, the standard error of the estimate approximates the standard error of an individual prediction.

Warning: The value of R^2 depends on the actual values of the independent variables in your sample. If you have a random sample of data, R^2 should indicate roughly how well the model fits in the population. If you have restricted the range of the independent variables, the sample R^2 may be small, even if the population value is large.

Assumptions for Testing Hypotheses

You can compute multiple R^2 in any situation, although its interpretation changes if a constant is not included in the model. Multiple R^2 tells you how well the model fits your data. If you want to draw conclusions about the population from which your data originate, you encounter all of the usual problems. You know you can't assume that what's true for the sample is true for the population. Different samples from the same population will result in regression equations with different values for the regression coefficients and different multiple *R*s. You have to figure out if your coefficients and multiple *R* are unlikely if there is no relationship in the population between the dependent variable and the independent variables.

If your data are a sample from some underlying population (such as all high schools in Chicago), and if you want to draw conclusions about the population based on the sample results, your data must satisfy the following assumptions:

- All observations are independent of one another. (Reading scores in one school don't affect scores in another school.)
- In the population, there is a linear relationship between the dependent variable and the independent variables.
- For each combination of values of the independent variables in the population, there is a normal distribution of values of the dependent variable. All of these normal distributions have the same variance. (For example, for any combination of values of attendance, class size, low-income percentage, and limited-English percentage, the percentage scoring at grade level is assumed to be normally distributed.)

There are many statistical tools that you can use to examine your data for violations of these assumptions. They are discussed later in this chapter.

Regression Hypotheses

For a particular regression equation, you want to test two types of hypotheses: hypotheses about how well the overall regression model predicts the dependent variable, and hypotheses about the regression coefficients for individual variables or groups of variables. You want to know whether the population value for multiple *R* is greater than 0 and whether the population values for the coefficients for individual variables are different from 0.

Tests about the Equation

You can test the null hypothesis that the population value for multiple R is 0 by using the F test shown in Figure 13-5. This is called the overall F test for the model. The analysis-of-variance (ANOVA) table in Figure 13-5 divides the observed variability in the dependent variable into two parts: the sum of squares explained by the regression and the residual sum of squares (the variability not explained by the regression). The total sum of squares is the sum of these two numbers. If you divide the regression sum of squares by the total sum of squares, you will get multiple R^2.

The observed significance level for the F statistic tells you how often you would expect to see a sample value for multiple R of 0.83 or larger when the true population value is 0. Because the observed significance level is less than 0.0005, you reject the null hypothesis. The test that multiple R is 0 is equivalent to testing that the population values of all of the regression coefficients in the equation except the constant are 0. In regression analysis, you are usually interested in the contributions of the individual variables, so the overall test is important only if it is not significant, leaving you unable to reject the null hypothesis that there is no linear relationship between the dependent variable and all of the independent variables.

Warning: You may find, particularly if your dataset is small, that some coefficients are significantly different from 0 while the overall test of the regression is not. Consider removing some of the nonsignificant variables from the equation.

Figure 13-5
Analysis-of-variance table

			Sum of Squares	df	Mean Square	F	Sig.
Model	1	Regression	26836.909	4	6709.227	40.571	.000[1]
		Residual	12402.723	75	165.370		
		Total	39239.632	79			

1. Predictors: (Constant), Low income %, Limited English %, Average class size, Attendance %

Tests about Individual Coefficients

Even if you reject the null hypothesis that the population value for multiple R is 0, that doesn't mean that all of the variables in the equation have regression coefficients that are significantly different from 0. To test whether a particular coefficient is 0, look first at the column labeled t in Figure 13-6. This column is computed by dividing each coefficient by its standard error. The observed significance level for testing the null hypothesis that in the population the value of the coefficient is 0 is in the column labeled *Sig.* In this example, you can't reject the null hypothesis that the coefficients for class size and limited English are 0. Confidence intervals for the population values of the regression coefficients are constructed in the usual way from the estimated coefficient and its standard error. They are shown in the last two columns of Figure 13-6.

Figure 13-6
Tests of individual coefficients

			Unstandardized Coefficients				95% Confidence Interval for B	
			B	Std. Error	t	Sig.	Lower Bound	Upper Bound
Model	1	(Constant)	-110.568	37.558	-2.944	.004	-185.387	-35.749
		Attendance %	2.205	.411	5.370	.000	1.387	3.022
		Average class size	.212	.416	.509	.612	-.617	1.041
		Limited English %	-.259	.230	-1.127	.263	-.717	.199
		Low income %	-.652	.114	-5.739	.000	-.878	-.426

The value of the regression coefficient for a particular independent variable depends on what other variables are in the equation. Figure 13-7, for example, shows the regression coefficients when attendance rate is removed from the equation. Notice how the values for all of the other coefficients are changed. Unlike the previous model, you can reject the null hypothesis that the population value for the coefficient for class size is 0. Notice that the coefficient for class size is positive. This means that the larger the class size, the larger the score when low income and limited English are in the equation. One possible explanation is that special-needs children, who are in small classrooms, may have affected both the scores and the average class size. This is a variable that should be examined further.

Figure 13-7
Regression equation without attendance rate

			Unstandardized Coefficients		Standardized Coefficients		
			B	Std. Error	Beta	t	Sig.
Model	1	(Constant)	80.420	15.067		5.338	.000
		Average class size	1.217	.427	.241	2.848	.006
		Limited English %	-.317	.263	-.092	-1.205	.232
		Low income %	-.875	.124	-.598	-7.043	.000

Warning: Don't pull a variable from a regression equation and make absolute statements about the importance of that variable in predicting the dependent variable.

The importance of a variable depends on the company it's keeping in the equation. An unimportant variable in one regression model may be an important variable in another regression model that predicts the same dependent variable. Any statement that you make about a coefficient depends on the other variables in the regression model. That's the curse of having correlated independent variables.

Standardized Coefficients

The magnitude of a regression coefficient isn't necessarily related to how good a predictor the variable is, since the size of the coefficient depends in large part on the units of measurement for the variable. The coefficient for a variable measured in feet will be 12 times as large as the coefficient for the same variable measured in inches. If you have an equation with variables such as income in dollars, highest year of education, and number of children, you can't compare the size of the regression coefficients.

One way to make the coefficients easier to compare is to compute **standardized coefficients** (also known as beta coefficients). Standardized coefficients are the coefficients you get if you standardize both the dependent variable and each of the independent variables to have a mean of 0 and a standard deviation of 1. This manipulation doesn't do anything for the problem that the value of a coefficient depends on what other variables are in the model, but it may be useful for comparing the coefficients in a particular model. Standardized coefficients are shown in Figure 13-8.

Tip: When you have one independent variable, the standardized regression coefficient is equal to the correlation coefficient with the dependent variable. When you have two or more correlated independent variables, the standardized coefficients are no longer correlation coefficients and are not restricted to be between –1 and +1.

Figure 13-8
Standardized coefficients

			Unstandardized Coefficients		Standardized Coefficients		
			B	Std. Error	Beta	t	Sig.
Model	1	(Constant)	-110.568	37.558		-2.944	.004
		Attendance %	2.205	.411	.463	5.370	.000
		Average class size	.212	.416	.042	.509	.612
		Limited English %	-.259	.230	-.074	-1.127	.263
		Low income %	-.652	.114	-.448	-5.739	.000

Tip: There may be several explanations for why a variable you think should be related to the dependent variable has a coefficient that's not significantly different from 0. If the variable is correlated with one or more variables in the model, the information it contributes may already be there. (Remove the related variables and see if the situation changes.) If the sample size is small, you may not be able to detect the relationship. Or if the range of values of the variable is limited, there may not be much of a relationship between the variables over the limited range. If you look at the relationship between education and income for only those people earning more than $100,000, you may not see much of a relationship.

Part and Partial Coefficients

Two other statistics that are used to assess the contribution of an independent variable to an existing model are the part and partial correlation coefficients (see Figure 13-9). Both can range in value from –1 to +1.

- The **part correlation coefficient** is the signed square root of the change in multiple R^2 when the variable is added to the equation. A large absolute value for the part coefficient tells you that the variable provides unique information about the dependent variable that is not available from the other independent variables in the equation. The part correlation coefficient is also the correlation between the dependent variable and the independent variable when the linear effects of the other independent variable have been removed from the independent variable. In other words, it is the correlation coefficient between the dependent variable and the residuals of an equation in which the other independent variables in the model are used to predict the independent variable.
- The **partial correlation coefficient** is the correlation coefficient between the independent variable and the dependent variable when the linear effects of the other independent variables have been removed from both the dependent variable and the independent variable. The square of the partial correlation coefficient tells you what proportion of the unexplained variance in the dependent variable is explained when that variable is entered into the model.

The relative ranking of the variables will be the same whether you use part or partial coefficients.

Figure 13-9
Correlation coefficients

			Correlations		
			Zero-order	Partial	Part
Model	1	Attendance %	.721	.527	.349
		Average class size	.511	.059	.033
		Limited English %	-.123	-.129	-.073
		Low income %	-.714	-.552	-.373

Warning: Just because variables are related does not mean that one causes the other. Don't infer that the reading scores at a school will increase if you manage to increase the attendance rate. For example, it may be that attendance rates are large in schools that have good teachers and facilities for learning. Increasing attendance at schools with poor teachers may have no effect on reading scores. When data come from an observational study, you can't infer cause and effect. It's only if you set the values of the independent variable and watch the effects on the dependent variable that you can establish cause and effect.

Including Categorical Variables

All of the independent variables considered so far were scale variables that needed no special preparation. If you want to include a nominal or ordinal variable as an independent variable in a regression, you must take special steps.

Two-category independent variable. If you want to include a dichotomous categorical variable (such as yes or no or male or female), you can go ahead without any special preparations. It's easiest to interpret the coefficient if you code the categories as 0 and 1. Then the coefficient for the variable is the effect of the category labeled 1. For example, if you want to include a variable that indicates whether a school has a selective enrollment policy or has to take all comers, you can create a variable *Selective* that has the value of 1 if the school has a selective enrollment policy and a value of 0 if the school does not. You can then interpret the coefficient as the effect of being in the category labeled 1. That is, the value of the coefficient is the difference in predicted values for the group coded 1 as compared to the group coded 0 when all other independent variables are held constant. If the coefficient for *Selective* is statistically significant, you can conclude that in that model, the type of school is probably related to the outcome studied.

More-than-two-category independent variable. If you have a nominal or ordinal independent variable with more than two categories, you must create a set of independent variables to represent the variable. The number of variables that you create is one less than the number of categories. For example, if type of school has four categories (General, Selective, Military, and Magnet), you must create three new variables. There are several coding schemes that can be used, but the easiest is to code the variables as 0 and 1 "dummy variables." (That describes not the crafter of the variables but the lack of meaning of the values assigned.) The three dummy variables can be coded as

Selective	= 1 if selective school,	= 0 otherwise.
Military	= 1 if military school,	= 0 otherwise.
Magnet	= 1 if magnet school,	= 0 otherwise.

The category General is not coded separately. If the other three variables are coded as 0, then you know the category is General. It is called the **reference category**. The coefficients of the variables are interpreted as the difference between that category and the reference category. You must enter all of the variables that represent the categorical variable into the model. Unfortunately, SPSS Statistics won't treat the set of dummy variables as a single variable, so you'll get separate observed significance levels for each dummy variable (see the Tip on p. 253).

Comparing Two Models: Change in R-Square

To compare two models, one of which includes more independent variables than the other, you can calculate how much R^2 changes when the additional variables are added to the model. For example, to see how much class size and school size contribute to the model for reading score achievement, you can compare two models: one with only attendance, low-income percentage, and percentage with limited English and the other with the previous three variables as well as class size and school size.

Tip: Removing variables from a complete model or adding them to an incomplete model results in the same change in R^2. If you add the variables, the change is positive. If you remove them, the change is negative. The observed significance level for the change is the same.

In Figure 13-10, you see that the R^2 statistic is 0.683 for the model with three variables and 0.693 for the model with all five variables. This is not much of a change, especially since you know that as you add variables, the sample R^2 always increases, even if the fit doesn't improve in the population. To test the null hypothesis that the population change in R^2 is 0, a statistic called the F change is computed. Its observed significance level is in the column labeled *Sig. F Change*.

Figure 13-10
Change statistics

					Change Statistics				
		R	R Square	Adjusted R Square	R Square Change	F Change	df1	df2	Sig. F Change
Model	1	.826[1]	.683	.670	.683	54.540	3	76	.000
	2	.832[2]	.693	.672	.010	1.217	2	74	.302

1. Predictors: (Constant), Low income %, Limited English %, Attendance %
2. Predictors: (Constant), Low income %, Limited English %, Attendance %, Total enrollment, Average class size

The first value for *Sig. F Change* corresponds to the test of the hypothesis that R^2 equals 0 for the first model. (That can be thought of as the change between a model with no independent variables and the first model.) Based on the small observed significance level, you reject the null hypothesis that adding attendance, percentage with low income, and percentage with limited English didn't change the population value for R^2. The next line is a test of the null hypothesis that the population change in R^2 associated with adding class size and total enrollment to the model with attendance, percentage with low income, and percentage with limited English is 0. Since the observed significance level is large $(p = 0.302)$, you can't reject the null hypothesis that the change in R^2 is 0 for the population. Class size and total enrollment added little, if anything, to the previous model.

Tip: To get an overall test of the significance of a multicategory variable, look at the change statistics when all of the pieces of the variable are in the model and when they are all removed.

Many Paths Lead to the Same Hypothesis

Hypotheses in regression analysis can be formulated and tested in different ways, but most of them can be stated in terms of the change in R^2. It's important to recognize what hypotheses are equivalent so that you don't keep repeating the same finding in different ways in an article or report.

- Significance tests of R^2 and of R are the same.
- Testing the null hypothesis that a coefficient is 0, based on the t statistic, is equivalent to testing whether the change in R^2 when that variable is added or removed is 0.

- A test of the null hypothesis that a set of coefficients is 0 is equivalent to the test that the change in R^2 when the variables are added is 0.
- The overall *F* test in the analysis-of-variance table is equivalent to the test that R^2 is 0 when all variables are in the model. It is also equivalent to testing that all coefficients for variables in the model are 0.
- Significance tests of standardized and unstandardized coefficients and part and partial coefficients for the same variable are identical.
- The *F* change for a single variable is equivalent to the *t* test for the coefficient of the variable that is being entered or removed. (If you square the *t* statistic, you will get the value of *F* change.)
- The *F* change for removing variables from a complete model and the *F* change for adding them to an incomplete model are the same. If you add the variables, the change in R^2 is positive. If you remove the variables, the change is negative. The observed significance level of *F* change is the same.

Using an Automated Method for Building a Model

If you are interested in just testing hypotheses about regression coefficients and you have enough cases to estimate all of the coefficients, there may be no real need for a model with a small number of independent variables. However, if you have a large number of independent variables and want to build a more parsimonious model, you can try to eliminate variables that don't contribute to the prediction.

It's tempting to let the computer screen the independent variables and identify those that seem to be related to the dependent variable. SPSS Statistics includes three different methods that you can use to build models.

Warning: The automated methods that are described in this section do not produce the best model in any statistical sense. These methods are criticized for depending on chance associations among variables observed in the sample. They don't solve problems caused by correlated independent variables. Instead, they make arbitrary choices. Automated algorithms should never substitute for good judgment. Understand your variables before exposing them to an automated model-building algorithm. Treat with a good degree of skepticism any model that the automated methods produce. The models that result from automated methods are suggestions, not final answers. Save your sincere belief until you can validate the model on another dataset (see Harrell, 2001).

Examining the Methods

All of the model-building methods available in SPSS Statistics are based on examining how much multiple R^2 changes as variables are entered or removed from an equation. You set the criteria to be used for entering and removing variables.

- **Forward variable selection.** This method starts with a model that has only the constant in it. Next, you include the variable that causes the largest increase in multiple R^2, provided that it meets the criteria you've set for entry. You then look at all of the variables that are not yet in the model and choose the variable that increases multiple R^2 the most and meets the entry criteria. You keep doing this until there are no more variables that meet the criteria for entry.
- **Backward elimination.** This method starts with a model that includes all possible independent variables. Then you remove the variable that causes the least change in multiple R^2, provided that it meets the preset criterion for removing a variable. At each step, you evaluate all of the variables in the model and remove the one that results in the smallest change in R^2. You stop when there are no more variables that meet the criterion for removal.
- **Stepwise variable selection.** This is the most commonly used method for variable selection. It is just like forward selection except that after a variable is entered, all variables already in the model are examined to see if any of them meet the criteria for removal. It's possible for a variable that has entered the model to be removed at a later step. (This can happen when independent variables are correlated with one another.)

Warning: The SPSS Statistics variable selection algorithm does not enter all pieces of a dummy-coded variable together. If only certain pieces are present in the model, the reference category must then be regarded as all of the excluded groups. You may want to force the nonselected categories into the model.

Setting Entry and Removal Criteria

For all three of the variable selection methods, you must establish the criteria for entering and removing variables. For forward selection, you need only an entry criterion; for backward elimination, you need only a removal criterion; and for stepwise variable selection, you need both. Entry and removal criteria can be either a p value or the actual F change that a variable must have to enter or to be removed.

The two criteria are not necessarily equivalent because a fixed *F* value has different significance levels, depending on the number of variables currently in the equation. For large sample sizes, the significance level and the corresponding *F* value are comparable.

Warning: The actual significance levels associated with variable entry and removal are difficult to determine, since many variables are being examined at each step. Strictly speaking, the *p* values are correct only when you conduct a single hypothesis test. Just think of the entry and removal criteria as guidelines, not exact significance levels.

Stepwise Selection: An Example

As an example of variable selection, let the stepwise algorithm build a model using the independent variables of attendance rate, low-income percentage, limited-English percentage, and class size and using the default criterion of $p = 0.05$ for a variable to enter and $p = 0.10$ for a variable to be removed.

Tip: For the stepwise method, make sure your criterion for a variable to enter is more stringent than the criterion for a variable to be removed; otherwise, the same variable could be entered and removed forever. SPSS Statistics warns you if you slip up.

Following the Action

This is how the variables are selected:

- **Step 1.** Find the variable with the largest absolute value for the correlation coefficient with percentage reading at norm. Check whether the observed significance level for that correlation is less than 0.05 (the entry criterion). Attendance rate has the largest correlation with the reading scores, and the observed significance level is less than 0.05, so attendance rate enters the model.
- **Step 2.** Examine all of the variables not in the model. Figure 13-11 shows statistics for variables not in the model after Step 1. Select the variable with the smallest observed significance level less than 0.05, which turns out to be low-income percentage, and enter it into the model. This results in the second model.

Figure 13-11
Excluded variables for Model 1

			Beta In	t	Sig.	Partial Correlation	Collinearity Statistics
							Tolerance
Model	1	Average class size	.129[1]	1.326	.189	.149	.648
		Limited English %	-.116[1]	-1.487	.141	-.167	1.000
		Low income %	-.465[1]	-6.148	.000	-.574	.731

1. Predictors in the Model: (Constant), Attendance %

- **Step 3.** Examine the variables in the model to see if either of them meets the removal criterion ($p = 0.1$ in this example), and if so, remove the variable. (This won't happen with two variables in the model.) Keep repeating this step until no variables meet the removal criterion.
- **Step 4.** Examine all of the variables not in the model. Select the variable with the smallest observed significance level less than or equal to the entry criterion, and enter it into the model. If no variable meets the entry criterion, STOP; otherwise, keep entering and removing variables until there are no more variables that meet the entry or removal criteria.

From Figure 13-12, you see that the observed significance level for both of the variables that are not in Model 2—average class size and limited-English percentage—is greater than 0.05, so variable selection stops after attendance and low income are entered.

Figure 13-12
Excluded variables for Model 2

			Beta In	t	Sig.	Partial Correlation	Collinearity Statistics
							Tolerance
Model	2	Average class size	.035[1]	.423	.673	.048	.624
		Limited English %	-.071[1]	-1.097	.276	-.125	.987
		Low income %					.

1. Predictors in the Model: (Constant), Attendance %, Low income %

Tip: If you want to see the values of criteria, such as Mallow's CP, the Akaike Information Criterion (AIC), the Amemiya's Prediction Criterion (PC), or the Schwarz Bayesian Criterion (BIC), in building your model, you have to use syntax. Set up the model as you normally would using Analyze > Regression > Linear, but do not click the OK button. Instead, click Paste, add the keyword SELECTION to the /STATISTICS subcommand, and execute the syntax using the Run menu (or icon).

Summarizing the Steps

Figure 13-13 is a summary of the activity. In the first model, attendance percentage enters the equation because it is linearly related to test scores most strongly, based on the change in R^2. In the second step, Model 2, low-income percentage is added to the first model. Variable selection stops after low-income percentage is added.

Figure 13-13
Variables entered

		Variables Entered	Variables Removed	Method
Model	1	Attendance %	.	Stepwise (Criteria: Probability-of-F-to-enter <= .050, Probability-of-F-to-remove >= .100).
	2	Low income %	.	Stepwise (Criteria: Probability-of-F-to-enter <= .050, Probability-of-F-to-remove >= .100).

The changes in R^2 are shown in Figure 13-14. When only attendance is in the model, multiple R^2 is 52%. The addition of percentage of students from low-income families increases multiple R^2 to 68%. You know that the change is significantly different from 0 because you specified that the observed significance level had to be less than 0.05 for a variable to enter.

Figure 13-14
Changes in R^2

					Change Statistics				
		R	R Square	Adjusted R Square	R Square Change	F Change	df1	df2	Sig. F Change
Model	1	.721[1]	.520	.513	.520	84.378	1	78	.000
	2	.823[2]	.678	.669	.158	37.800	1	77	.000

1. Predictors: (Constant), Attendance %
2. Predictors: (Constant), Attendance %, Low income %

The coefficients are shown in Figure 13-15. Again, notice that the coefficient for attendance changes when low-income percentage enters the model. This is because the two variables are correlated.

Figure 13-15
Regression coefficients at each step

			Unstandardized Coefficients		Standardized Coefficients		
			B	Std. Error	Beta	t	Sig.
Model	1	(Constant)	-266.630	32.384		-8.233	.000
		Attendance %	3.430	.373	.721	9.186	.000
		Low income %					
	2	(Constant)	-112.591	36.609		-3.075	.003
		Attendance %	2.283	.360	.480	6.344	.000
		Low income %	-.677	.110	-.465	-6.148	.000

Calculating Predicted Values

You can use a regression equation to make two types of predictions. You can predict the percentage scoring at grade level for a new school, based on its attendance rate and low-income percentage. Or you can predict the mean percentage scoring at grade level for all schools with the same attendance rates and low-income rates. The predicted value is the same for both situations. What differs is the standard deviation of the predicted value. As you would expect, there's less variability when predicting the average for all schools than when predicting the score for an individual school. Even if the regression assumptions are met perfectly, you don't expect all schools with the same values for the independent variables to have the same observed reading score. There is a distribution of values.

You can save the following variables, among others, in your data file:

- **Unstandardized predicted values** are the predicted values of the dependent variable obtained from the regression equation.
- **Standardized predicted values** are the unstandardized predicted values in *z* score form. That is, they have a mean of 0 and a standard deviation of 1 for cases used to build the model if listwise missing value treatment is used.
- **Adjusted predicted values** are the predicted values for a case when it is excluded from the computation of the regression model.
- **Confidence interval for the population mean** value for the dependent variable for the observed values of the independent variables.

- **Prediction intervals for a future value** of the dependent variable for the observed values of the independent variables.
- **Standard error of the mean prediction** is a measure of the variability of the prediction of the mean value. It's used in computing the confidence interval for the population mean value.

Warning: A regression equation should be used only to make predictions for values of the independent variable that are in the same range as the values actually observed in the sample. You don't know if the model applies for other ranges of values of the independent variable.

After the Model Is Selected

Once you've selected a model that makes substantive and statistical sense, you're still not done. You have to check for violations of the regression assumptions and make sure that there are no strange data points that are influencing your results. You don't want a regression equation that changes dramatically if one of the schools is removed from the analysis. SPSS Statistics offers a vast array of statistics that are useful for checking assumptions and gauging the influence of individual points.

Tip: If possible, it's always wise to try your model on a set of data that wasn't used to estimate the regression model. That's because a model almost always fits the sample from which it was derived better than it will fit another sample from the same population. If your dataset is large enough, you can randomly split it into two parts: one for building the model and one for testing the model. Obtain predicted values for the cases that weren't used to build the model and calculate the square of the correlation coefficient between the observed and predicted values. If the resulting R^2 is markedly different from the original, you may have included more independent variables in the model than your sample sizes can support; that is, you have overfit the model. However, split-sample methods require large sample sizes—Harrell (2001) recommends at least 100 cases—and don't validate the final model that is based on all cases. Methods based on bootstrapping are better ways of validating the model.

Checking for Violations of Regression Assumptions

To test hypotheses about the population regression, you have to make the assumptions outlined in "Assumptions for Testing Hypotheses" on p. 246. Once you have selected a regression model, you can use the residuals (the differences between the observed and predicted values) to check for violations of the assumptions. If the assumptions are met, the residuals are approximately an independent sample from a normal distribution with a mean of 0 and a constant variance for all combinations of values of the independent variables.

Calculating Residuals

SPSS Statistics computes several different types of residuals that are useful for checking the model:

- **Unstandardized residuals** are just the difference between the observed and predicted values of the dependent variable. The actual values for the residuals depend on the scale on which a variable is measured. You can't tell whether an unstandardized residual is large or small without more information.
- **Standardized residuals** are the ordinary residuals recomputed to have a standard deviation of 1. That makes it easier to detect residuals that are large in absolute value. (You expect most standardized residuals to be between –2 and +2.)
- **Studentized residuals** are calculated by dividing each residual by an estimate of its standard deviation. The estimate varies from point to point depending on the values of the independent variables. The Studentized residual more precisely reflects differences in the true error variance from point to point. Usually, standardized and Studentized residuals are close in value, but not always.
- **Deleted residuals** are the difference between the observed and predicted values when a case is excluded from the regression model; that is, the regression model is computed without the case. Deleted residuals are used to identify cases that have large effects on the regression equation.
- **Studentized deleted residuals** are the deleted residuals divided by their standard deviation. That makes it easier to look for unusual values.

Warning: SPSS Statistics gives variable names to all of the regression statistics that are calculated for each case. Naming of the variables is done by sequence, not by model. For example, if you save only the predicted variable for the first model and save both the predicted variable and the residual for the second model, the variables will be called *pred_1*, *pred_2*, and *res_1*. *Pred_1* is for the first model and *pred_2* and *res_1* are for the second model.

Examining the Residuals

Once you've generated variables containing residuals, you have many ways to use them to tap into a wealth of information about the model.

- **Identify additional predictors.** Plot the residuals against values of independent variables that are not in the equation. If you see a relationship between the two variables, consider including the variable in the regression model.
- **Check normality.** Make histograms or stem-and-leaf plots of the residuals. The distribution should be approximately normal. Figure 13-16 is a histogram of the residuals from the final regression equation for the reading score data. The distribution is symmetric, and normality can't be ruled out. The statistical tests used in regression are not affected much by moderate departures from normality.

Figure 13-16
Histogram of Studentized residuals

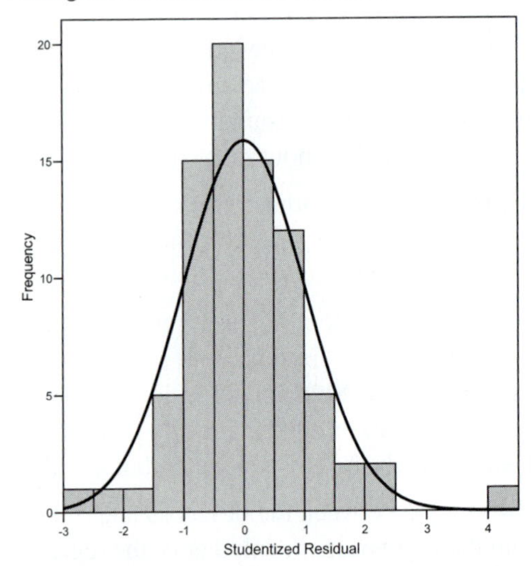

- **Check linearity.** Plot the residuals against the predicted values of the dependent variable and against each of the independent variables. You should see a horizontal band of residuals, as shown in Figure 13-17. If you see a pattern, such as that shown in Figure 13-18, the relationship between the dependent variable and the independent variables may not be linear.

Figure 13-17
Residuals plotted against predicted values

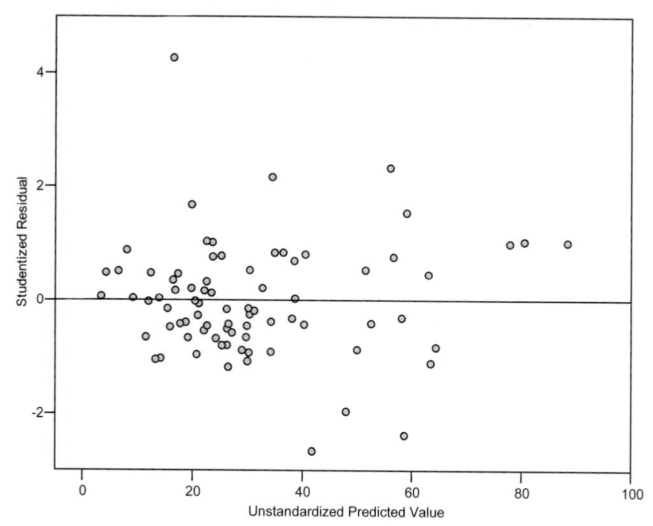

Figure 13-18
Nonrandom residuals

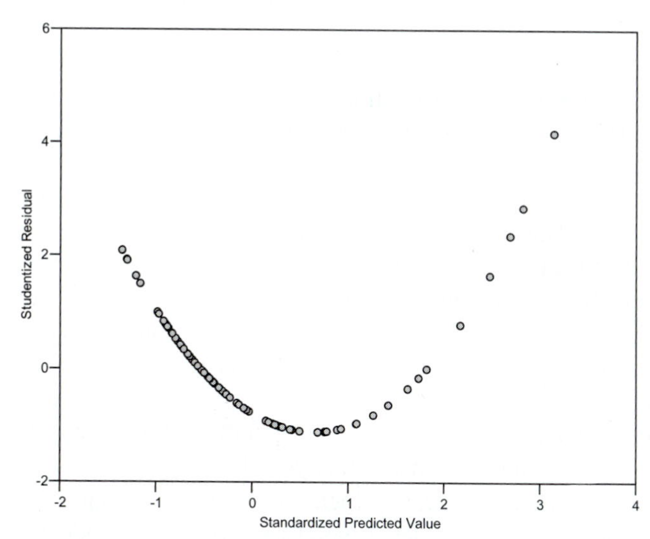

- **Check equality of variance.** If the spread of the residuals changes over the range of the independent variable, as shown in Figure 13-19, the assumption of equal variances may be violated. Notice that in Figure 13-17, the spread of the residuals seems to increase with increasing predicted values.

Figure 13-19
Nonconstant residuals

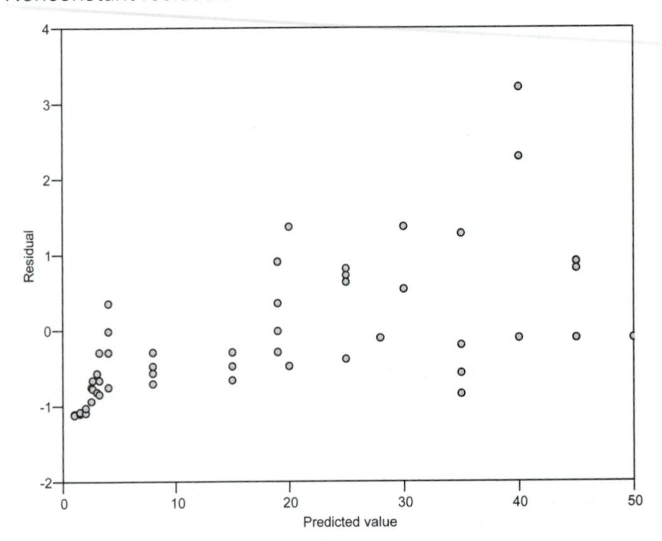

- **Check independence.** If your observations were collected in a sequence, you should plot the residuals against the order in which they were obtained. If you see a pattern, such as small or large residuals being next to each other, the observations may not be independent. The Durbin-Watson statistic is a test for correlation of adjacent residuals. The values of the statistic range from 0 to 4. If the residuals are not correlated with each other, the value is close to 2. Values less than 2 mean that adjacent residuals are positively correlated. Values greater than 2 mean that adjacent residuals are negatively correlated. You must consult special tables to obtain significance levels for the Durbin-Watson test.

If you find evidence that the assumptions necessary for regression are violated, you can attempt to transform the data. This means that you change the scale on which the variables are measured. For example, take logarithms, squares, or square roots of either the dependent or independent variables. If you transform the dependent variable, you change its distribution. It is the transformed distribution that must satisfy the assumptions of the analysis.

Tip: If you use SPSS Statistics procedures that automatically create new variables, delete unnecessary ones before you save the data file. Label the ones you're keeping so that you will know what model they came from. You don't want 30 sets of diagnostic statistics from regression in your data file.

Looking for Unusual Observations

In life, it's an asset to be considered "influential." In model building, being influential is bad. An influential point is one that has a large effect on the results of an analysis. You don't want a model that depends largely on a particular data point. You want a regression equation that remains fairly constant, regardless of the actual schools included in the sample.

Tip: You can compute a sequence number for your cases by choosing Transform > Compute and then typing Sequence = $casenum. (You don't have to enter a credit card number to use variables with a dollar sign in front of them. They're special variables created by SPSS Statistics.)

Identifying Large Residuals

Cases with large standardized residuals, called outliers, may be influential in determining the coefficients of the regression equation. You should always examine them to see if there's anything unusual about them, other than the fact that the model doesn't fit them well.

Figure 13-20 is a casewise listing of schools with standardized residuals greater than 2.5 in absolute value. You see that there is one school that has an observed reading score much greater than predicted. Several explanations are possible. First, you should make sure that all of the data values are correct. There's no need to spin elaborate theories about exceptional schools if the problem is in data entry or recording. If the data values are correct, see if there are other simple explanations. For example, is this a high school located in a correctional facility? How many students actually took the test?

A possible explanation for this outlier is that, compared to the other schools, very few students actually took the test. This suggests that you might want to include the number or percentage of students tested in each school as an independent variable. Average test scores based on small numbers of students will usually vary more than test scores based on larger numbers of students from the same population.

Figure 13-20
Casewise listing of standardized residuals greater than 2.5 in absolute value

		schoolname	Std. Residual	Reading at national norm %	Predicted Value	Residual
Case Number	1	Flower Vocational	4.228	70.6	16.423	54.1773
	6	North Lawndale Charter	-2.615	8.2	41.706	-33.5063

Identifying Unusual Values of Independent Variables

It's important to identify cases that have unusual values for the independent variables. First, you want to check the values to make sure that they are correct. If they are correct, you want to see if they are having a large effect on the regression coefficients. Two closely related statistics for identifying unusual observations in terms of the independent variables are leverage and Mahalanobis distances.

- **Leverage** is the distance between the values of the independent variables for a case and the average for all cases. SPSS Statistics calculates centered leverages.
- **Mahalanobis distances** is the leverage multiplied by one less than the number of cases.

From the leverage plot in Figure 13-21, you see that there is one case that particularly stands out from the rest. That's case number 63. It has an attendance percentage of 95, compared to an overall average of 87%; the low-income percentage is 23, compared to the average of 81%.

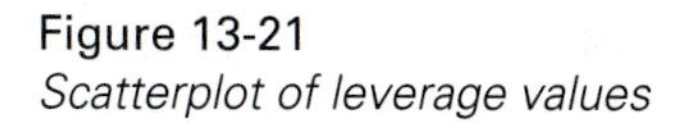

Figure 13-21
Scatterplot of leverage values

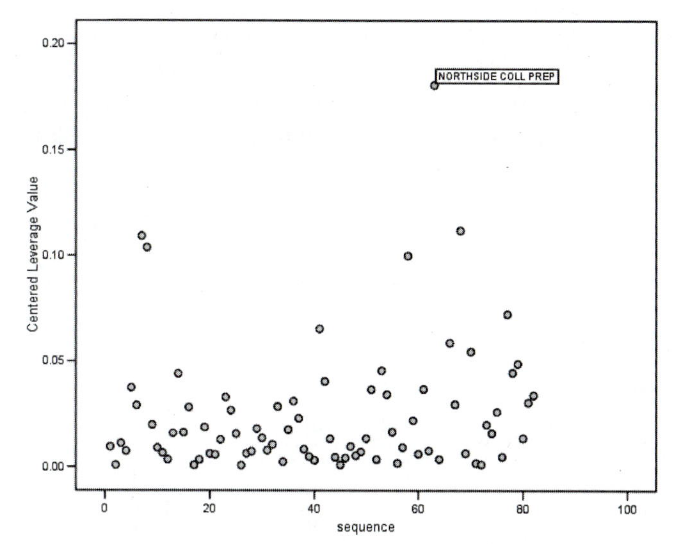

Tip: Various rules of thumb have been suggested for leverage values. One says values less than 0.2 are *safe*, values above 0.5 are *dangerous*, and the rest are *risky*. Another rule says if $p > 6$ and $(n - p) > 12$, use $(3 \times p)/n$ as a cutoff, where p is the number of variables. As a general rule, it's a good idea to plot all of the values and examine the ones that are different from the rest.

Looking for Influential Points

You can't tell if a case has a large effect on the estimates of the regression coefficients from just the residuals or the leverage values. If a case has a large residual but is average in terms of the independent variables, the case may not have much effect on the estimates of the regression coefficients. Similarly, even if a case has unusual values for the independent variables and even if the residual is not unusual, the case may have a large effect on the regression coefficients. You should examine the residual, the leverage, and Cook's distance in your search for influential points.

SPSS Statistics calculates many different statistics that are useful for identifying influential points:

- **Changes in all coefficients.** Cook's distance is a summary measure that is based on how much regression coefficients for all variables change when a case is removed from the regression. Large values for Cook's distance, certainly values greater than 1, are troublesome. You notice several cases with large Cook's distances in Figure 13-22.

Figure 13-22
Cook's distances

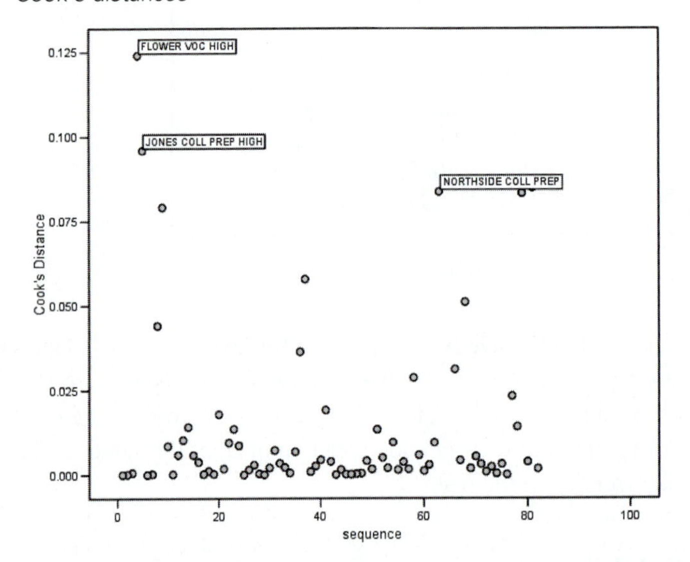

- **Changes in individual coefficients.** You can see the effect on each coefficient of deleting each case from the regression equation. The actual change is called **DfBeta**, and a standardized form is also available. Figure 13-23 is a plot of Standardized DfBeta for low-income percentage for the Chicago schools. You see that most of the schools have values that cluster in a band around 0, meaning that they have little effect on the values of the coefficient. You might want to look at the points that are somewhat further removed from the rest.

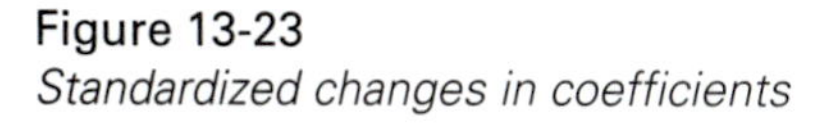

Figure 13-23
Standardized changes in coefficients

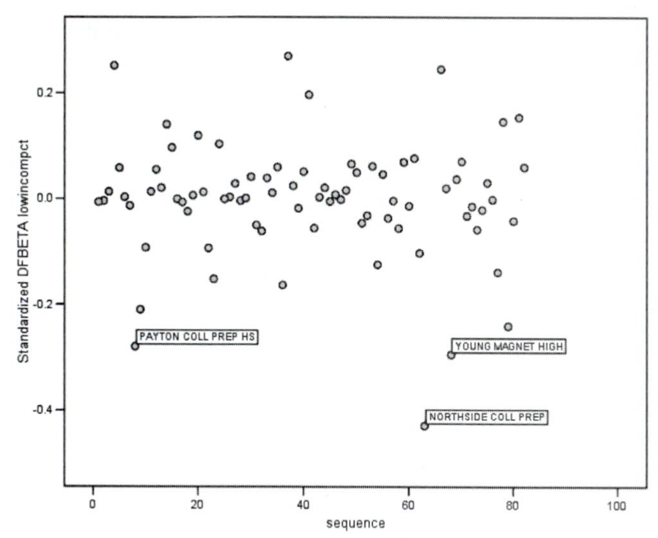

- **Changes in predicted values.** You can also examine the change in the predicted value when a case is deleted from the computation of the model. Select DfFit for the unstandardized value and Standardized DfFit for a standardized value. Again, if you plot these values, you would expect them to cluster in a horizontal band around 0. An individual case shouldn't change predicted values much.
- **Changes in residuals.** The residual for a case when it is not included in the computation of the model is called a **deleted residual**. The standardized version is called a **Studentized deleted residual**. If you plot the ordinary residual against the deleted residual, you expect points to cluster around a straight line because omitting a point shouldn't change the predicted value and the residual by much. If you find points that are far removed from the line, you should examine them to determine what causes them to be unusual.

Warning: If you use a variable selection method and all variables are not in the final model, and if you request SdfBeta or DfBeta to be saved, the program computes as many new variables as there are independent variables in the independent variable list. For variables not in the final equation, the values for all cases are system-missing.

Partial Regression Plots

A very useful plot for checking linearity and identifying influential observations is the partial regression plot. There is a separate plot for each independent variable. The partial regression plot for a particular independent variable is a plot of two sets of residuals: the residuals when the dependent variable is predicted from all of the variables except that variable, and the residuals when that independent variable is predicted from all of the other independent variables.

Figure 13-24 is a partial regression plot for attendance rate. If attendance is linearly related to test scores, the plot should show a linear relationship, and it does. If the points fall along a curve, you have reason to doubt that the relationship between the two variables is linear. You may need to transform the variable or include additional terms, such as the variable squared. Points that are far removed from the rest are influential points for determining the regression coefficient for the variable.

Figure 13-24
Partial regression plot

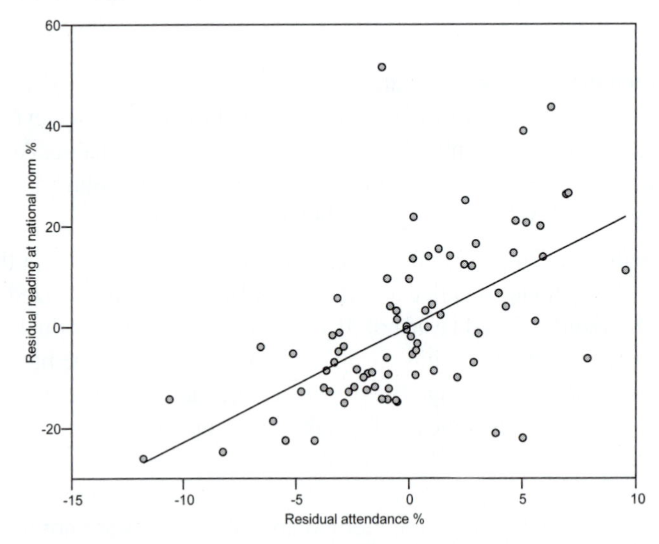

The partial regression plot has the following noteworthy features: The slope of the line is equal to the coefficient for the variable in the multiple regression, the correlation between the two variables is the partial correlation coefficient, and the residuals from the least-squares line are equal to the residuals from the final multiple regression equation, which includes all of the independent variables.

Tip: When you build a regression model, the default treatment of cases with missing values is to exclude any case that has a missing value for any independent variables listed in the dialog box. This is not a good strategy unless the percentage of excluded cases is small (that is, less than 5%). If you settle on a final model that includes a subset of independent variables, you should rerun the model, including only those variables that are in the final model. The effect of this is to bring back into the analysis cases that were excluded because of missing values for variables that are not included in the final model. If you have many cases with missing values, you may have difficulties interpreting any model that you fit. If schools that perform poorly didn't submit their reading scores, for example, you couldn't draw any conclusions about reading scores and the other independent variables.

Looking for Collinearity

Collinearity refers to the situation in which there is a high multiple correlation when one of the independent variables is regressed on the others. The problem with collinear variables is that they provide very similar information, and it is difficult to separate the effects of the individual variables. SPSS Statistics has diagnostics that you can use to detect collinear data and to assess the effect of the collinearity on the estimated coefficients.

Tip: Suspect collinearity if you have regression coefficients with the wrong signs, large standard errors for coefficients, or significant overall regression but no significant variables. The best solution for such problems is to create new variables that are based on the sets of collinear variables.

- **Tolerance.** The tolerance is $1 - R^2$, where R^2 is obtained from predicting the variable from all of the other independent variables in the model. Tolerance values close to 1 tell you that the variable is not linearly related to the other independent variables. Tolerance values close to 0 tell you that the variable is strongly related to the other independent variables in the model. (By default, SPSS Statistics will not allow variables with tolerances less than 0.0001 to enter a regression model because small values can lead to computational problems.)
- **Variance inflation factor.** The variance inflation factor is the reciprocal of the tolerance. It's called a variance inflation factor because it measures the increase in the variances of the coefficients due to the correlations of the independent variables. Figure 13-25 shows the tolerances and VIFs for the variables in the initial model with four independent variables. The tolerances are large, indicating that the variables are not highly correlated with each other.

- **Condition index.** A measure that compares the magnitude of the eigenvalues is the condition index. It is the square root of the ratio of the largest eigenvalue to each successive eigenvalue. A condition index greater than 15 indicates a possible problem, and an index greater than 30 suggests a serious problem with collinearity. (See Belsley, Kuh, and Welsch, 1980.)
- **Variance proportions.** The variances of each of the regression coefficients, including the constant, can be decomposed into a sum of components associated with each of the eigenvalues. If a high proportion of the variance of two or more coefficients is associated with the same eigenvalue, there is evidence for near-dependency.

Figure 13-25
Collinearity statistics

			Collinearity Statistics	
			Tolerance	VIF
Model	1	Attendance %	.566	1.766
		Average class size	.621	1.611
		Limited English %	.981	1.019
		Low income %	.693	1.443

Warning: The condition index and variance proportions depend on the scaling of the independent variables. For example, if you just subtract off their mean, you might draw completely different conclusions about collinearity problems than you would from the original data matrix.

A Final Comment

Rarely are assumptions not violated one way or another in regression analysis (or any other statistical procedure). This is not, however, a justification for ignoring the assumptions. Producing regressions without considering possible violations of the necessary assumptions can lead to results that are difficult to interpret and apply. Significance levels, confidence intervals, and other results are sensitive to certain types of violations and cannot be interpreted in the usual fashion if serious violations exist. By carefully examining residuals, you are in a much better position to pursue analyses that solve the problems you are investigating. Even if everything is not perfect, you can at least gauge the potential for serious difficulties.

Obtaining the Output

To produce the output in this chapter, open the data file *school.sav* and follow the instructions below.

Figure 13-1. From the Analyze menu choose Descriptive Statistics > Descriptives. Move *attendpct*, *classize*, *limitenglpct*, *lowincompct*, and *tapreading* into the Variable(s) list. Click OK.

Figures 13-2 to 13-6, 13-8, 13-9, and 13-25. From the Analyze menu choose Regression > Linear. Move *tapreading* into the Dependent box, and move *attendpct*, *classize*, *limitenglpct*, and *lowincompct* into the Independent(s) list. Click Statistics. In the Statistics subdialog box, select Estimates, Confidence intervals, Model fit, R squared change, Descriptives, Part and partial correlations, and Collinearity diagnostics. Click Continue, and then click OK.

Figure 13-7. Repeat the instructions in the previous example but omit *attendpct* from the Variable(s) list.

Figure 13-10. From the Analyze menu choose Regression > Linear. Click Reset. Move *tapreading* into the Dependent box, and move *attendpct*, *limitenglpct*, and *lowincompct* into the Independent(s) list. Click the Next button. In the Block 2 of 2 group, move *classize* and *enroll* into the Independent(s) list. Click Statistics. In the Statistics subdialog box, select Model fit and R squared change. Click Continue, and then click OK.

Figures 13-11 to 13-15, 13-20, and 13-24. From the Analyze menu choose Regression > Linear. Click Reset. Move *tapreading* into the Dependent box, and move *attendpct*, *classize*, *limitenglpct*, and *lowincompct* into the Independent(s) list. From the Method drop-down list choose Stepwise. Move *schoolname* into the Case Labels box. Click Statistics. In the Statistics subdialog box, select R squared change and Casewise diagnostics. Next, select Outliers outside, and type 2.5 in the Standard Deviations text box. Click Continue, and then click Save. In the Save subdialog box, select Unstandardized in the Predicted Values group, Studentized in the Residuals group, Cook's and Leverage values in the Distances group, and Standardized DfBeta(s) in the Influence Statistics group. Click Continue, and then click Plots. In the Plots subdialog box, select Produce all partial plots. Click Continue, and then click OK.

Figure 13-16. From the Graphs menu choose Legacy Dialogs > Histogram. Move *SRE_1* into the Variable box. Select Display normal curve, and click OK.

Figure 13-17. From the Graphs menu choose Legacy Dialogs > Scatter/Dot. Select the Simple Scatter icon, and click Define. Move *SRE_1* into the Y Axis box, move *PRE_1* into the X Axis box, and click OK.

Figures 13-18 and 13-19 were not obtained from these data.

Figure 13-21. From the Graphs menu choose Legacy Dialogs > Scatter/Dot, select the Simple Scatter icon, and click Define. Move *LEV_1* into the Y Axis box, move *sequence* into the X Axis box, and click OK.

Figure 13-22. From the Graphs menu Legacy Dialogs > Scatter/Dot. Select the Simple Scatter icon, and click Define. Move *COO_1* into the Y Axis box, move *sequence* into the X Axis box, and click OK.

Figure 13-23. From the Graphs menu choose Legacy Dialogs > Scatter/Dot. Select the Simple Scatter icon, and click Define. Move *SDB4_1* into the Y Axis box, move *sequence* into the X Axis box, and click OK. Double-click the chart to open it in the Chart Editor. From the Options menu choose Y Axis Reference Line. In the Properties dialog box, type the value 0 into the Y Axis Position text box, and then click Apply and Close.

Chapter

14

Discriminant Analysis

Gazing into crystal balls is not the exclusive domain of soothsayers. Judges, college admissions counselors, bankers, and many other professionals must foretell outcomes such as parole violation, success in college, and credit worthiness. An intuitive strategy is to compare the characteristics of a potential student or credit applicant to those of cases whose success or failure is already known. You reach a prediction based on evaluating similarities and differences. Often, this is done subjectively, using only experience and wisdom. However, as problems grow more complex and the consequences of bad decisions become more severe, you need a more objective procedure for predicting outcomes.

Linear discriminant analysis is a statistical technique often used to examine whether two or more mutually exclusive groups can be distinguished from each other based on linear combinations of values of predictor variables and to determine which variables contribute to the separation. (**Mutually exclusive** means that a case can belong to only one group.) To use discriminant analysis, you must have information for a set of cases whose group membership you know. For example, you must have people who were found to be good credit risks and people who were found to be bad credit risks. For each of them, you might have information about yearly income, age, marital status, and net worth.

When you conduct a linear discriminant analysis, the questions you want to answer are:

- What are the linear combinations of predictors whose average values best separate the groups?
- Can you substantively interpret the linear combinations?

- Are the linear combinations useful in actually classifying the cases into groups?
- Is there a smaller subset of predictor variables that captures the differences between the groups?

Tip: Don't confuse discriminant analysis and cluster analysis. If you know what groups the cases belong to, you need discriminant analysis. If you want to form groups based on characteristics of the cases, you want cluster analysis.

Examples

- Predict which banks will fail, based on assets, debts, and other financial information.
- Predict which students should be placed into an advanced mathematics class on the basis of SAT scores, grades in algebra, and class rank.
- Predict which patients will survive after aneurysm surgery, based on the length of surgery, age, location of aneurysm, and type of surgical procedure.
- Predict which of three car models a person will buy, based on miles driven in a year, age, income, and family size.

In a Nutshell

You have two or more mutually exclusive groups of cases. For each case, you have information about what group it belongs to and variables that may be useful for distinguishing the groups. You calculate discriminant scores that are linear combinations of predictor variables. The scores are chosen so that their values are similar for cases in the same group and different for cases in different groups. For example, if you have three groups of people classified as drivers of sedans, sports cars, and SUVs, you want to have sedan drivers having similar scores, sports-car aficionados having similar scores, and SUV drivers having similar scores.

You evaluate the success of your analysis by using the discriminant scores to actually assign cases to the groups. You compute the percentage of cases correctly classified by the discriminant functions.

Predicting Internet Use: The Example

How do people who use the Internet differ from those who don't? Income? Age? Education? Children? If you are a marketing type, identifying Internet users may be important for targeting Internet-related products and services. The actual characteristics may not matter much, provided the prediction is good. (You don't want to waste your resources interrupting dinners or offering exciting prizes to people who are unlikely to buy your product.) If you are a social scientist, you may be more interested in identifying the dimensions on which Internet users and non-users differ, rather than your ability to actually predict whether a particular individual uses the Internet.

As an example of discriminant analysis, consider data from the Library Association Survey, a national random telephone survey of over 3,000 people (Rodger et al., 2000). One of the goals of the survey was to study the relationship between Internet use and library use. Because you have a random sample of people who identified themselves as Internet users and non-users, you can build a model for predicting Internet usage. The predictor variables you'll use are age in years, education in years, gender (coded 1 = *male*, 0 = *other*), suburban (1 = *suburb*), kids (1 = *children under 17*), work (1 = *employed full or part time*), and income in thousands of dollars. Respondents were asked to choose a range of income values, and the midpoint of the range is used in the analysis.

Warning: If your groups are formed from scores on some variable (with high scores and low scores defining the groups), consider using a statistical procedure that works with the actual scores, since you lose information by creating categories.

Before You Start

To prepare for your analysis, do the following:

- Look for outliers in the predictor variables because discriminant analysis is very sensitive to outliers that affect the means of the groups.
- Plot pairs of independent variables to see if the relationships among them are approximately linear.
- If you have nominal predictor variables, transform them to a set of dummy variables.

Examining Missing Values

Before you start any analysis, you want to determine whether missing values are a potential problem. For discriminant analysis, values can be missing for either the grouping variable (Internet use) or any of the predictor variables. From Figure 14-1, you see that Internet usage is missing for only one case. This means that you don't have to worry about people who refuse to divulge whether they use the Internet. For more sensitive information, like cheating on taxes or speeding, you have to worry about both the people who fail to divulge the information and the truthfulness of the information given. If your groups are not correctly identified, you may have serious problems interpreting any results.

From Figure 14-1, you see that there are 780 people who are missing information on at least one predictor variable (indiscriminately labeled *discriminating variable*). That's more than 25% of all of the survey respondents. This is a problem you can't ignore. You have to look at each of the potential predictor variables to see if there is a single variable that accounts for the large number of cases with missing values or whether all variables contribute to the problem.

Figure 14-1
Case processing summary

			N	Percent
Unweighted Cases	Valid		2316	74.8
	Excluded	Missing or out-of-range group codes	1	.0
		At least one missing discriminating variable	780	25.2
		Both missing or out-of-range group codes and at least one missing discriminating variable	0	.0
		Total	781	25.2
	Total		3097	100.0

The results in Figure 14-2 are not particularly surprising. Almost 25% of the cases have missing values for income. The 71 cases with missing values for age are of little import compared to those who understandably refused to give information about income to a stranger on the telephone.

Figure 14-2
Distribution of missing values

	Cases					
	Included		Excluded		Total	
	N	Percent	N	Percent	N	Percent
Age * Internet use	3026	97.7%	71	2.3%	3097	100.0%
Gender * Internet use	3096	100.0%	1	.0%	3097	100.0%
Income * Internet use	2344	75.7%	753	24.3%	3097	100.0%
Children less than 17 years old * Internet use	3077	99.4%	20	.6%	3097	100.0%
Suburban * Internet use	3096	100.0%	1	.0%	3097	100.0%
Employed * Internet use	3096	100.0%	1	.0%	3097	100.0%
Years of education * Internet use	3052	98.5%	45	1.5%	3097	100.0%

From the graphs in Figure 14-3, you see that failure to divulge income is related to age and education in a similar way for Internet users and non-users. Willingness to report income increases with education and decreases with age. People who don't use the Internet are more likely to fail to report income for all categories of education and age.

Figure 14-3
Percentage missing income with education and age

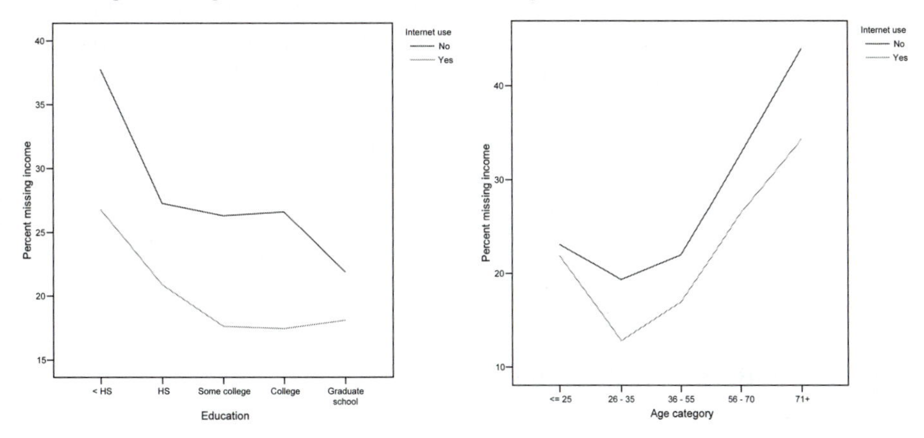

Selecting Missing Value Treatment

From the previous analyses, you can be reasonably confident that the missing values are not missing completely at random. They are associated with education, age, and Internet status. This makes the missing-value problem much more complicated. Simple missing-value treatments that are useful for values missing completely at random aren't appropriate.

You're left with several choices. The easiest is to eliminate income as a predictor variable. This will have a significant impact on your results because it's very likely that income is associated with Internet use, even controlling for age and education. Your model for identifying Internet users will probably not be as good as it would be if you included income in the model. If your goal is to classify future cases for whom income information may also be unavailable, eliminating income as a predictor may be a reasonable strategy, since you won't have the information available for new cases either. You may want to have two sets of equations: one set to use if income is missing and another if income is available.

The best option for the treatment of the missing values is to use sophisticated statistical techniques to impute the missing values for income, provided that they meet the particular missing-value model you've selected. This requires you to have specialized software available and to use it intelligently.

Another strategy, if you have an adequate sample size, is to eliminate cases with missing values for income. You can use only cases with complete information for all of the variables in the analysis. The very serious drawback to this approach is that any conclusions you draw will be restricted to people who report their incomes. That's what you'll do in the following analysis, since sophisticated missing-value procedures are not part of the SPSS Statistics Base system.

SPSS Statistics also offers what is known as the **pairwise** missing-value treatment option. With this treatment, each coefficient in the correlation matrix of the predictor variables is based on all cases that have nonmissing values for the two variables being correlated. This means that cases with missing values only for income are included in the computation of the correlation coefficients that don't involve income. This is also not a good strategy because the relationships between pairs of variables may differ for cases with and without missing values. Because individual correlation coefficients are computed from different subsets of cases, pairwise missing-value treatment can result in a matrix in which the relationships among the coefficients don't make sense.

Examining Descriptive Statistics

Figure 14-4 shows descriptive statistics for each of the groups. All means are based on 1,304 Internet users and 1,012 non-users. You see that Internet users compared to non-users are younger, better educated, have higher incomes, are more likely to be males, have children under the age of 17, live in the suburbs, and are employed full or part time. (Remember that for variables coded 0 and 1, the mean is the proportion of cases with values of 1. For example, 43% of people who don't use the Internet are males and 52% of the people who use the Internet are males.)

Figure 14-4
Group statistics

		Internet use		
		No	Yes	Total
Age	Mean	49.33	40.16	44.17
	Std. Deviation	17.95	13.27	16.14
Gender	Mean	.43	.52	.48
	Std. Deviation	.50	.50	.50
Income	Mean	32.94	58.00	47.05
	Std. Deviation	24.51	29.92	30.35
Children less than 17 years old	Mean	.38	.46	.43
	Std. Deviation	.49	.50	.49
Suburban	Mean	.30	.47	.39
	Std. Deviation	.46	.50	.49
Employed	Mean	.51	.75	.65
	Std. Deviation	.50	.43	.48
Years of education	Mean	12.41	14.30	13.47
	Std. Deviation	2.51	2.40	2.62

Tip: Discriminant analysis considers the predictor variables together. It's possible that even though the group means for a variable are similar, the variable may be useful in distinguishing the groups if the variable is correlated with other predictor variables.

Are the Population Group Means Equal?

Using the column labeled *Sig.* in Figure 14-5, you can test the null hypothesis that the population means for the variables are equal for those who use the Internet and for those who don't. All of the observed significance levels are small, so you can reject the null hypothesis that the population means for the variables are equal. (The F values in Figure 14-5 are the F values you get if you do a one-way analysis of variance for each variable with Internet use as the grouping variable. Because there are two groups, the F value is the square of the equal-variance two-independent-samples t test.)

Warning: For dichotomous variables, the F test is not the statistic of choice for comparing the two proportions. If you do chi-square tests for each of the dichotomous variables, you can reject the null hypothesis that the proportions are equal in the two groups.

Figure 14-5
Test of equality of group means

	Wilks' Lambda	F	df1	df2	Sig.
Age	.921	199.477	1	2314	.000
Gender	.992	17.879	1	2314	.000
Income	.832	466.793	1	2314	.000
Children less than 17 years old	.993	17.446	1	2314	.000
Suburban	.971	68.140	1	2314	.000
Employed	.934	162.420	1	2314	.000
Years of education	.871	342.091	1	2314	.000

The Wilks' lambda statistic in Figure 14-5 is calculated as the ratio of the within-groups sum of squares to the total sum of squares. It is the proportion of the variance not explained by differences between the groups. If all observed group means are equal, lambda is 1. Small values occur when most of the observed variability can be attributed to differences between groups. You see that Wilks' lambda is smallest for income and largest for children less than 17 years old.

Are the Predictor Variables Correlated?

Because interdependencies among the predictors affect most multivariate analyses, you want to examine their correlation matrix. If you see variables with large correlations, you'll know that you won't be able to separate their contributions to the discriminant model. From the correlation matrix in Figure 14-6, you see that the predictor variables are remarkably uncorrelated, with the exception of some expected correlations, such as those between age and having children under 17 and age and working full or part time. It's reassuring to see the positive correlation between education and income.

Figure 14-6
Pooled within-groups correlation matrix

		Age	Gender	Income	Children less than 17 years old	Suburban	Employed	Years of education
Correlation	Age	1.000	-.038	.103	-.354	.047	-.337	.143
	Gender	-.038	1.000	.096	-.122	.008	.167	.041
	Income	.103	.096	1.000	.035	.173	.129	.285
	Children less than 17 years old	-.354	-.122	.035	1.000	-.011	.144	-.113
	Suburban	.047	.008	.173	-.011	1.000	.005	.103
	Employed	-.337	.167	.129	.144	.005	1.000	.085
	Years of education	.143	.041	.285	-.113	.103	.085	1.000

The correlation matrix is called a **pooled within-groups matrix** because it is obtained by averaging the separate covariance matrices for all groups and then computing the correlation matrix from the pooled-covariance matrix. A total correlation matrix is obtained when all cases are treated as if they are from a single sample.

The total and pooled within-groups correlation matrices can be quite different. For example, Figure 14-7 is a plot of two variables for three groups. When each group is considered individually, the correlation coefficient between the two variables is 0; however, the correlation coefficient computed for all cases combined (the total) is 0.97, since groups with larger *x* values also have larger *y* values.

Figure 14-7
No within-groups correlation

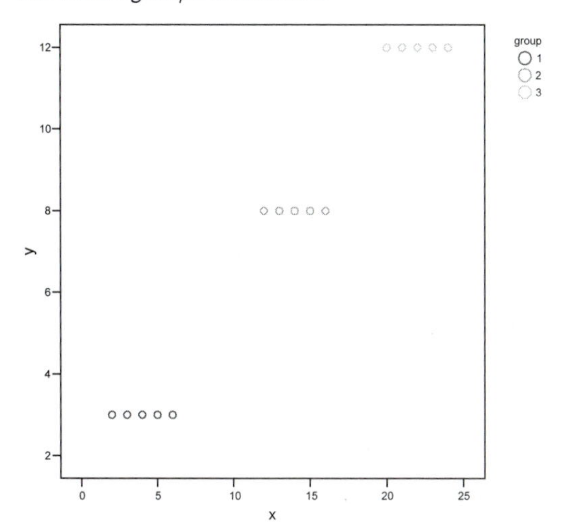

Calculating the Discriminant Function

After all of the preliminaries, you're finally ready to compute the coefficients of the discriminant function. Remember, the goal is to find a linear combination of values of the variables that best separates the Internet users from the non-users. The number of functions that you can calculate to separate the groups is one fewer than the number of groups. Ideally, you would like to assign large scores to people in one group and small scores to people in the other. That way, you can use the scores to predict the group to which a person belongs.

Discriminant Function Coefficients

The unstandardized discriminant function coefficients are shown in Figure 14-8. Because you have two groups, you can have only one set of coefficients.

Figure 14-8
Unstandardized canonical discriminant function coefficients

	Function
	1
Age	-.036
Gender	.030
Income	.021
Children less than 17 years old	-.089
Suburban	.290
Employed	.226
Years of education	.195
(Constant)	-2.231

Using the unstandardized discriminant function coefficients, the score for a 49-year old female—with an income of $30,000 and a child under 17, not living in the suburbs, not employed, and with 12 years of education—is –1.1.

$$\begin{aligned}\text{score} = {} & 49(-0.036) + 0(0.030) + 30(0.021) + 1(-0.089) \\ & + 0(0.290) + 0(0.226) + 12(0.195) - 2.231 = -1.1\end{aligned}$$

You'll use this score to predict whether this person does or does not use the Internet.

Unstandardized Coefficients

As in multiple linear regression, the size of the unstandardized coefficient depends on the units of measurement for each variable. You can't compare the coefficients for age in years and income in thousands of dollars because they are measured in very different units. You also cannot draw any conclusions about the absolute importance of variables, since the value of the coefficient for a variable depends on the other variables in the equation.

Standardized Coefficients

The standardized discriminant function coefficients shown in Figure 14-9 are the coefficients you get when all predictors variables are standardized to a mean of 0 and a within-groups standard deviation of 1.

Figure 14-9
Standardized coefficients

	Function
	1
Age	-.562
Gender	.015
Income	.571
Children less than 17 years old	-.044
Suburban	.140
Employed	.105
Years of education	.477

Standardized coefficients let you compare the magnitude of the coefficients for age and income, although the value of a coefficient for a particular variable still depends on the other variables included in the analysis. For example, the standardized coefficient for years of education goes from 0.48 when income is in the model to 0.73 when income is not in the model.

Tip: If you want to compute your own discriminant function scores based on standardized variables, you must use the pooled within-groups estimate of the standard deviation to standardize the variables. Don't use *z* scores from the Descriptives procedure because these are computed from all cases in all groups. Use the square roots of the diagonal entries from the pooled variance-covariance matrix in Discriminant as the standard deviations.

Signs of the Coefficients

The actual signs of the coefficients are arbitrary. Negative coefficients could just as well be positive if the signs of the positive coefficients were made negative. You have to look at coefficients with the same sign to determine how the variables relate to the groups. You see that income, years of education, employment status, living in the suburbs, and being male all have the same sign, and that age and children under 17 have the opposite sign. That means that young ages, large incomes, many years of education, being male, not having children under 17, and living in the suburbs result in large values for the discriminant score, while older ages, small incomes, limited education, being female, having children under 17, and not living in the suburbs result in small discriminant scores. You'll see that high scores are associated with Internet users and low scores are associated with non-users.

Warning: If two variables are highly correlated with each other, they may have coefficients that are opposite in sign. Their contribution to the discriminant function is shared, and the individual coefficients are not meaningful. Such variables will have the same sign when you compute correlations between the variables and the discriminant function scores.

Relationship to Multiple Regression

When you have two groups of cases, discriminant analysis and multiple regression analysis are closely related. If you compute a multiple linear regression with group as the dependent variable and the same predictor variables, the discriminant function coefficients and the regression coefficients are proportional to each other. Figure 14-10 shows both regression coefficients and discriminant function coefficients for this example, as well as their ratio. The discriminant coefficients can be obtained by multiplying the regression coefficients by 4.4. The ratio between the two sets of coefficients varies from dataset to dataset, but the two sets of coefficients, except for the constant, are always proportional in the two-group case.

Figure 14-10
Regression and discriminant coefficients

Variable	Regression	Discriminant	Ratio
Age	–0.00825	–0.03630	4.4
Gender	0.00686	0.03017	4.4
Income	0.00469	0.02063	4.4
Children less than 17 years old	–0.02025	–0.08909	4.4
Suburban	0.06582	0.28953	4.4
Employed	0.05140	0.22609	4.4
Years of education	0.04432	0.19497	4.4

Warning: The relationship between discriminant analysis and multiple regression is true for only the two-group situation. Never run a multiple regression with group as the dependent variable when there are more than two groups. You won't be able to interpret such results.

Correlations between Scores and Variables

Another way to assess the contribution of a variable to the discriminant score is to compute the Pearson correlations between the values of the function and the values of the variables. This is called a **structure matrix** and is shown in Figure 14-11. The variables are sorted based on the absolute values of the correlation coefficients. You see that income has the highest correlation with the discriminant score and the presence of children under the age of 17 has the smallest correlation coefficient.

Figure 14-11
Structure matrix

	Function
	1
Income	.687
Years of education	.588
Age	-.449
Employed	.405
Suburban	.262
Gender	.134
Children less than 17 years old	.133

Pooled within-groups correlations between discriminating variables and standardized canonical discriminant functions.
Variables ordered by absolute size of correlation within function.

Tip: If you want to compare the relative contribution of the variables and take into account the other variables in the model, use the standardized coefficients. If you want to know the association of the individual variables with the discriminant function ignoring the other variables, use the structure matrix.

Relationship between Scores and Groups

A good discriminant function is one that results in different scores for cases in the different groups. After all, that's the goal of discriminant analysis. In Figure 14-12, you see the analysis-of-variance table for the discriminant scores.

Figure 14-12
Analysis-of-variance table for discriminant scores

Discriminant Scores from Function 1 for Analysis 1

	Sum of Squares	df	Mean Square	F	Sig.
Between Groups	990.215	1	990.215	990.215	.000
Within Groups	2314.000	2314	1.000		
Total	3304.215	2315			

Two statistics that measure the relationship between the discriminant function scores and the groups are based on the analysis-of-variance table.

- The **eigenvalue** is the ratio of the between-groups sum of squares to the within-groups sums of squares. You know that a "good" function has scores that vary a lot between the groups and vary little within a group, resulting in a large eigenvalue. In Figure 14-13, you see that the eigenvalue is

$$\frac{990}{2314} = 0.428$$

  For a two-group situation, there is only one eigenvalue, and it accounts for 100% of the explained variance.

- The **canonical correlation** is another measure of the degree of association between the discriminant scores and the groups. It is the square root of the ratio of the between-groups sum of squares to the total sum of squares. (This is known as the **eta coefficient** in analysis of variance.) In this example, the canonical correlation is

$$\sqrt{\frac{990}{3304}} = 0.55$$

  Values close to 1 indicate that most of the observed variability in the discriminant scores is explained by differences between groups. In the two-group situation, the canonical correlation is simply the Pearson correlation coefficient between the discriminant score and the group variable.

Figure 14-13
Eigenvalues

		Eigenvalue	% of Variance	Cumulative %	Canonical Correlation
Function	1	.428[1]	100.0	100.0	.547

1. First 1 canonical discriminant functions were used in the analysis.

Distribution of Discriminant Scores

The means of the discriminant scores for the two groups are shown in Figure 14-14. The average value for the discriminant function for Internet users is 0.58; for non-users, it is –0.74. This is as large a difference in means as is possible using linear combinations of these predictor variables.

Figure 14-14
Functions at group centroids

		Function
		1
Internet use	No	-.742
	Yes	.576

Unstandardized canonical discriminant functions evaluated at group means

Histograms of the discriminant scores are shown in Figure 14-15. Notice that although Internet users have larger scores than non-users, the scores of the two groups overlap. If the two groups were completely separable, the scores for the two groups wouldn't overlap. The vertical line drawn on each histogram is the line that is used to assign cases to the two groups. Cases to the right of the line are predicted to be Internet users; cases to the left of the line are predicted to be non-Internet users.

Figure 14-15
Histograms of discriminant function scores

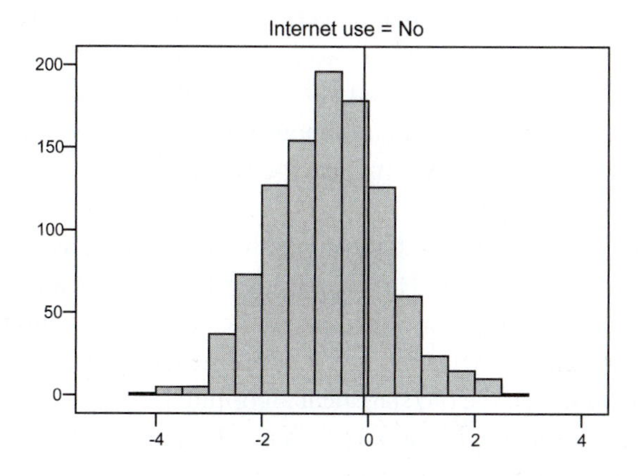

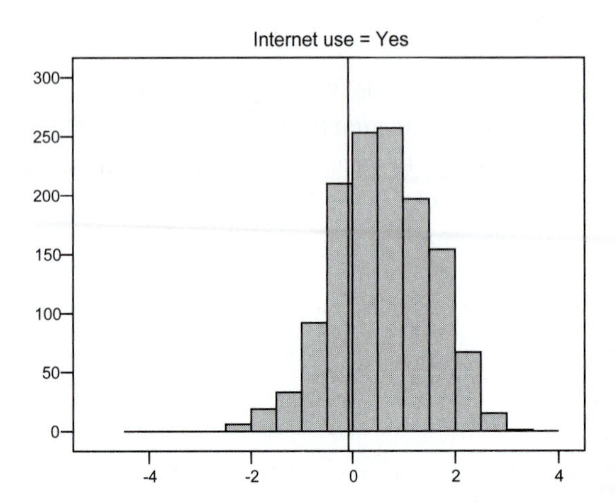

Testing Hypotheses about the Discriminant Function

If your only goal is to classify cases into groups, you don't have to make any assumptions about the distribution of the data values in the population or even assume that the samples come from underlying populations. You can see just how well the discriminant functions classify your data.

If you view your data as samples from the population of Internet users and non-users, you can test various hypotheses about the population values of the discriminant functions. You must assume that each sample is from a population with a multivariate normal distribution and that the variance-covariance matrix of the predictor variables is the same for all populations. Discriminant analysis is robust to violations of the assumption of multivariate normality; dichotomous predictors work reasonably well. Violation of the assumption of equality of variance-covariance matrices can affect both hypothesis testing and classification.

The linear discriminant function is statistically optimal only if the assumptions about the distribution of data values are met. However, linear discriminant analysis (LDA) works well, even when the assumptions that make it the best classification rule are violated. Lim, Loh, and Shih (2000) compared 33 classification algorithms and concluded, "It is interesting that the old statistical algorithm LDA has a mean error rate close to the best. This is surprising because (i) it is not designed for binary valued attributes and (ii) it is not expected to be effective when class densities are multi-modal."

Testing Assumptions

Two assumptions are required for the linear discriminant analysis to be optimal: samples are from multivariate normal populations and from equal variance-covariance matrices in the populations. One test that is used for testing the null hypothesis that the population variance-covariance matrices are equal is Box's *M*. Unfortunately, the test is dependent on the assumption of multivariate normality, and, for large sample sizes, the observed significance level can be small, even for minor departures from equality of variance-covariances matrices.

Tip: You can use binary logistic regression analysis instead of discriminant analysis to model the probability of an event occurring. Logistic regression does not require multivariate normality or equality of variance-covariance matrices. However, if the assumptions required for linear discriminant analysis are met, logistic regression is somewhat inferior to discriminant analysis from a statistical perspective.

The results from Box's *M* are shown in Figure 14-16. Because the sample sizes in the two groups are very large, the observed significance level should be interpreted carefully. Even small differences between the groups that may not affect the analysis can lead you to reject the null hypothesis.

Figure 14-16
Box's M

Box's M		327.409
F	Approx.	11.655
	df1	28
	df2	16442036
	Sig.	.000

Tests null hypothesis of equal population covariance matrices.

The logs of the determinants of the variance-covariance matrices shown in Figure 14-17 are overall measures of the spread of the variance-covariance matrix in each of the groups. You see that, even though Box's *M* led you to reject the null hypothesis, the logs of the determinants are quite similar in value.

Figure 14-17
Log determinant

		Rank	Log Determinant
Internet use	No	7	7.331
	Yes	7	7.534
	Pooled within-groups	7	7.587

The ranks and natural logarithms of determinants printed are those of the group covariance matrices.

Warning: When covariance matrices are not equal, cases will tend to be classified into the group with more variability. Quadratic discriminant analysis can be used when the covariance matrices are unequal. SPSS Statistics does not offer quadratic discriminant analysis, but you can classify cases using separate group covariance matrices.

Testing Equality of Discriminant Function Means

If the data are considered to be a sample from two or more populations, you can test the null hypothesis that the average discriminant scores are equal for all groups in the population. This is done for each of the discriminant functions. The test of the null hypothesis is based on Wilks' lambda, which is the ratio of the within-groups sum of squares to the total sum of squares, in the analysis-of-variance table shown in Figure 14-12 on p. 288.

In this example (from Figure 14-12),

$$\lambda = \frac{2314}{3304} = 0.70$$

This means that about 30% of the variability in discriminant scores is explained by the differences between groups.

Tip: If you calculate a score for each case based on the values of the discriminant function coefficients and then do an analysis of variance on the scores as shown in Figure 14-12, the ratio of between-groups sum of squares to within-groups sum of squares is a maximum. That is, all other sets of coefficients will result in a smaller *F* value.

As shown in Figure 14-18, lambda can be transformed to a variable that has approximately a chi-square distribution with degrees of freedom equal to the number of predictor variables. Based on Wilks' lambda, you can reject the null hypothesis that Internet users and non-users have the same average discriminant function score in the population.

Figure 14-18
Wilks' lambda

		Wilks' Lambda	Chi-square	df	Sig.
Test of Function(s)	1	.700	823.049	7	.000

Tip: If you calculate a multiple regression with group as the dependent variable, multiple $R^2 = 1 - \lambda$.

Classifying Cases into Groups

Even though you reject the null hypothesis that the average discriminant scores in the population are the same in the two groups, this doesn't necessarily mean that the discriminant scores will be useful for predicting to which group a case belongs. The only way to tell how well the discriminant function classifies cases into groups is to use the discriminant function score for each case to assign the case to a group. For the two-group situation, large discriminant scores are assigned to one group and small discriminant scores are assigned to the other.

Tip: If you have cases for whom group membership is not known (or is outside the specified range of group codes) but for whom the independent variables are available, SPSS Statistics will compute discriminant scores and report statistics for ungrouped cases.

Classification Summary

From the footnote in Figure 14-19, you see that 75% of the 2,316 cases that were used in calculating the discriminant analysis are correctly classified, based on the discriminant functions. The rows of the table indicate what category a case really belongs to. That's the value for group in the data file. The columns show the predicted group based on the discriminant functions. Cases that had missing values for the group code or codes that were not included in the discriminant specification but did have valid values for the predictor values are in the last row, labeled as *Ungrouped cases*. From the percentages on the diagonal of the table, you see that 75.3% of the 1,304 Internet users were correctly classified and 74.1% of the non-users were correctly classified.

Tip: You can compute statistics that quantify the strength of the association between the observed and predicted groups or that compare the predictions to chance criteria. Save the predicted group and use the Crosstabs procedure to calculate kappa or similar statistics.

Figure 14-19
Classification results

Classification Results [2]

	Internet use		Predicted Group Membership		Total
			No	Yes	
Original	No	Count	750	262	1012
		%	74.1	25.9	100.0
	Yes	Count	322	982	1304
		%	24.7	75.3	100.0
	Ungrouped cases	Count	1	0	1
		%	100.0	.0	100.0

2. 74.8% of original grouped cases correctly classified.

Leave-One-Out Classification Results

Because models almost always fit the data from which they are estimated better than the population, you know that 74.8% is an overly optimistic guess for the percentage of correctly classified cases. If you applied the model to another sample from the same population, you would expect a smaller correct-classification rate. One way to estimate the true correct-classification rate is to predict the group a case belongs to when the case is excluded from the computation of the discriminant function. You can exclude each case in turn and then compare the observed and predicted groups. This is called **leave-one-out cross-validation**.

Warning: Leave-one-out classification is not the same as jackknifed classification. In jackknifed classification, each case is left out in turn, and all cases are reclassified each time. In leave-one-out classification, only the omitted case is reclassified. Jackknifing provides more accurate estimates of the true misclassification rate.

From Figure 14-20, you see that the cross-validation results for this example are very similar to the original results. When sample sizes in the groups are large, it's unlikely that you will see large differences between the cross-validated and original results because a single case has little effect on the estimation of the coefficients.

Figure 14-20
Cross-validated classification results

Classification Results[3]

				Predicted Group Membership		
				No	Yes	Total
Cross-validated[1]	Internet use	No	Count	747	265	1012
			%	73.8	26.2	100.0
		Yes	Count	324	980	1304
			%	24.8	75.2	100.0

1. Cross validation is done only for those cases in the analysis. In cross validation, each case is classified by the functions derived from all cases other than that case.
3. 74.6% of cross-validated grouped cases correctly classified.

Tip: Bootstrapping, which involves taking repeated samples from the data and calculating misclassification rates, provides a better estimate of the true misclassification rate than the leave-one-out method.

Classifying Rare Events

When one of the groups is much smaller than the other, a large correct-classification rate can occur even when most of the rare group cases are misclassified. The smaller group—diseased individuals, parole violators, honest politicians—is often of primary interest, and you want to classify its members correctly. Instead of minimizing the overall misclassification rate, you want to correctly identify cases in the smaller group. For example, by judging everyone to be disease-free in a cancer-screening program, the error rate will be very small because few people actually have the disease; however, the results are useless if the goal is to identify the diseased individuals.

Tip: The results of different classification rules for identifying "minority" cases can be examined by ranking all cases on the value of their discriminant scores and determining how many minority cases are in the various deciles. If most of the cases are at the extremes of the distribution, a good rule for identifying them can be obtained at the expense of increasing the number of misclassified cases from the larger group. If the intent of the discriminant analysis is to identify persons to receive promotional materials for a new product or to undergo further inexpensive screening procedures, this is a fairly reasonable tactic. Unequal costs for misclassification can also be incorporated into the classification rule by adjusting the prior probabilities to reflect them.

Examining the Probability of Group Membership

Cases are assigned to groups based on numeric scores that are not easily interpretable. You can transform the scores into probabilities of belonging to a particular group, given the score. This is useful information because you can be more confident in your prediction if it is made on the basis of a large probability of belonging to a group rather than if it is based on a small probability. For example, given a particular score, if you know that the probability of being an Internet user is 0.95, you are more confident that a person is an Internet user than if the probability of being an Internet user is 0.52. Two pieces of information are used in transforming the score into a probability: the prior probability and the Mahalanobis distance from the case to each group mean.

Prior Probability

The **prior probability** is the likelihood that a case belongs to a group, apart from any information about the predictor variables. If you know that 70% of U.S. adults are Internet users, that's your prior probability of being an Internet user. Information about the overall likelihood of belonging to a group can be incorporated into the classification rule. The effect of this is to decrease assignment to groups with smaller prior probabilities and to increase assignment to groups with larger prior probabilities. For example, if you're trying to predict whether a person actually drives a Rolls-Royce, you want to incorporate the prior information that the probability of a person driving a Rolls-Royce is very slim. On the other hand, if you are studying the personal characteristics of Rolls-Royce drivers compared to the personal characteristics of drivers of more pedestrian vehicles, you may not be concerned about the widely unequal prior probabilities.

SPSS Statistics gives you two choices for prior probabilities: equal prior probabilities (all groups are equally likely) and prior probability equal to the proportion of cases in each of the groups in your sample. (You must use syntax if you want to specify prior probabilities that are not equal or proportional to size.)

Tip: If you are interested in how well the discriminant function can separate the groups, use equal priors. If you want to estimate what the overall correct classification rate would be if the functions were used in a population where all groups are not equally likely and if your sample is a good estimate of the size of the groups in the population, compute the classification rate with unequal priors, as well.

Statistics for Individual Cases

For each case, you can calculate the probability of belonging to a particular group. In Figure 14-21, you see statistics that can be displayed and saved for each case.

- **Actual group** is the group to which the case really belongs.
- **Predicted group** is the group to which a case is assigned, based on the discriminant score. (If the actual and predicted groups are different, asterisks appear next to the predicted group number.)
- **Discriminant function score** is in the last column labeled *Discriminant Scores, Function 1.*
- **Conditional probability** of the score given the group to which the case is assigned, labeled *P (D >d | G=g)*. This is an estimate of how likely a score that is at least this extreme occurs in the group to which the case is assigned. This is not the probability you are interested in.
- **Posterior probability** is the probability of belonging to the predicted group, given the discriminant score. This is the probability you are interested in. It is labeled *P (G=g | D=d)* and is an estimate of how likely the case is to belong to the group to which it is assigned by the discriminant function. If the probability is large, you are more confident of the assignment than if the probability is small. This calculation involves the prior probabilities of group membership.
- **Mahalanobis distance** is the distance from the case's values for the predictor variables to the average of the values for the group to which a case is assigned. Large values mean the case isn't like the others in the predicted group. Small values mean the case is similar to other cases in the assigned group.

On the right side of Figure 14-21, you see some of the same statistics, calculated for the second-most-likely group a case is predicted to belong to. In this example, because there are only two groups, the second-highest group is always the group that was not predicted by the discriminant function.

Tip: You can save the posterior probabilities for each case and evaluate different classification rules based on the actual probabilities. For example, you can classify cases only if the posterior probability meets a certain threshold; otherwise, you can just put the cases into an "uncertain" category.

Figure 14-21
Casewise results

				Highest Group					Second Highest Group			Discriminant Scores
			Actual Group	Predicted Group	P(D>d | G=g) p	df	P(G=g | D=d)	Squared Mahalanobis Distance to Centroid	Group	P(G=g | D=d)	Squared Mahalanobis Distance to Centroid	Function 1
Original	Case Number	71	0	0	.962	1	.691	.002	1	.309	1.614	-.695
		73	1	1	.475	1	.859	.510	0	.141	4.131	1.290
		79	1	0**	.567	1	.529	.328	1	.471	.556	-.170
		82	1	1	.739	1	.606	.111	0	.394	.970	.243
		83	1	0**	.987	1	.700	.000	1	.300	1.695	-.726
		84	1	1	.738	1	.605	.112	0	.395	.967	.241

** Misclassified case

Automated Model Building

When you have a large number of predictor variables, you may want to identify a smaller subset of variables that can be used to separate the groups. This is particularly important if information is difficult or expensive to obtain (such as certain laboratory results).

Warning: The automated methods that are described in this section do not produce the *best* model in any statistical sense. Many of these methods are criticized for depending on chance associations among variables observed in the sample. They should never substitute for good judgment. Understand your variables before exposing them to an automated model-building algorithm. Treat any model the automated methods produce with a good degree of skepticism. Believe your results when you can validate the model on another dataset. Your final model should make substantive sense and include the smallest number of variables that result in reasonable prediction. The models that result from automated methods are suggestions, not final answers.

SPSS Statistics offers only stepwise variable selection in discriminant analysis.

Setting Criteria

For stepwise variable selection in SPSS Statistics, you have to select both a test statistic criterion for determining what variable enters the model next and significance levels or *F* values that a variable must have to enter a model or to be removed from the model.

Selection Criteria

The following statistics can be used to establish the order in which variables enter into a model:

- **Wilks' lambda**. The change in Wilks' lambda for the model if a variable is entered or removed is evaluated. The F value for the change in Wilks' lambda when a variable is added to a model that contains p independent variables is

$$F_{change} = \left(\frac{n-g-p}{g-1}\right)\left(\frac{1-\frac{\lambda_{p+1}}{\lambda_p}}{\frac{\lambda_{p+1}}{\lambda_p}}\right)$$

- **Rao's *V*.** Also known as the Lawley-Hotelling trace, Rao's V is defined as

$$V = (n_k - g)\sum_{i=1}^{p}\sum_{j=1}^{p} w_{ij}^{*} \sum_{k=1}^{g} (\bar{X}_{ik} - \bar{X}_i)(\bar{X}_{jk} - \bar{X}_j)$$

  where p is the number of variables in the model, g is the number of groups, n_k is the sample size in the kth group, $\bar{X}_{ik}$ is the mean of the ith variable for the kth group, $\bar{X}_i$ is the mean of the ith variable for all groups combined, and w_{ij}^{*} is an element of the inverse of the within-groups covariance matrix. The larger the differences between group means, the larger the Rao's V.

- **Mahalanobis distance.** At each step, the variable that maximizes the Mahalanobis distance between the two closest groups is selected for entry. The squared Mahalanobis distance measures the distance between the centroids of groups.

$$D_{ab}^{2} = (n-g)\sum_{i=1}^{p}\sum_{j=1}^{p} w_{ij}^{*} (\bar{X}_{ia} - \bar{X}_{ib})(\bar{X}_{ja} - \bar{X}_{jb})$$

  where p is the number of variables in the model, $\bar{X}_{ia}$ is the mean for the ith variable in group a, and w_{ij}^{*} is an element from the inverse of the within-groups covariance matrix.

- **Smallest *F* ratio.** At each step, the variable that maximizes the smallest F ratio for pairs of groups enters. The F statistic is

$$F = \frac{(n-1-p)n_1n_2}{p(n-2)(n_1+n_2)}D_{ab}^2$$

  This method weights the Mahalanobis distance by sample size, so the smallest F ratio and Mahalanobis distance may give different results.

- **Sum of unexplained variance.** The sum of the unexplained variation for all pairs of groups can also be used as a criterion for variable selection. The variable chosen for inclusion is the one that minimizes the sum of the unexplained variation. Mahalanobis distance and R^2 are proportional. That is, $R^2 = cD^2$. For each pair of groups, a and b, the unexplained variation is $1 - R_{ab}^2$, where R_{ab}^2 is multiple R^2 when the dependent variable is coded as 0 or 1, depending on whether the case is a member of group a or b.

Warning: The above statistics determine only the order for variable entry. All variables must also meet additional criteria for entry and for removal. These are always based on the change in Wilks' lambda.

Entry Criteria

You have many choices for the statistic whose value determines the order in which variables enter into the model. For a variable to enter the model, it must meet an additional criterion that you control. You can specify the smallest acceptable F value that a variable must have to enter the model (default 3.84) and the smallest F value that a variable must have to remain in the equation (default 2.71). Or you can specify the largest observed significance level that a variable can have to enter the model (default 0.05) and the largest observed significance level that a variable must have to remain in the model (default 0.10). The two criteria are not identical because the significance associated with an F value depends on both the magnitude of the F and its associated degrees of freedom.

Warning: The significance levels for the F statistics cannot be interpreted in the usual fashion because they are based on examining many F values and selecting the largest. The correct significance level is difficult to calculate because it depends on the number of variables, the number of groups, and the correlation among the variables.

Tolerance

Because predictor variables that are strongly related linearly cause computational problems, before a variable is entered into the model, the program checks its tolerance and the tolerance of the variables already in the model to make sure that none of the tolerances become too small if the variable enters next. (Tolerance is $1 - R^2$, where R^2 is the squared multiple correlation coefficient when a variable is predicted from all of the other predictor variables already in the equation.) The smallest tolerance for any variable when the variable in question enters is the *Minimum Tolerance* in Figure 14-23. If the minimum tolerance is less than 0.001, the variable is not allowed to enter.

Following the Steps

The game plan for stepwise variable selection is:

- The first variable included in the analysis has the best value for the selection criterion and meets the entry criterion. (Variable selection stops if no variable meets the criterion.)
- The selection criterion is recomputed for all variables not in the model, and the variable with the best value for the selection criterion enters next if it meets the entry criterion.
- The variables in the model are examined to see if they meet the removal criterion.
- Variables are removed until none remain that meet the removal criterion.
- The process repeats: Variables not in the model are sequentially evaluated for inclusion using the selection criterion, and variables in the model are sequentially evaluated for removal, until no more variables meet the entry or removal criterion.

Stepwise Variable Selection: An Example

Stepwise methods generate a lot of output. At each step, you see separate statistics for variables in the model and for variables not in the model. There is also summary output.

Summary output. Figure 14-22 is a summary of the steps when you use stepwise variable selection for the Internet example with the default values of 3.84 for F-to-enter and 2.71 for F-to-remove. Wilks' lambda is the criterion that establishes the order in which variables are entered. (The column labeled *Removed* is blank because no variables that entered were removed.) Income is the first variable to enter the model; age is the second. Gender and children under 17 are the only predictor variables not entered in the final model.

Figure 14-22
Statistics at each step

Variables Entered/Removed[1,2,3,4]

				Wilks' Lambda							
								Exact F			
		Entered	Removed	Statistic	df1	df2	df3	Statistic	df1	df2	Sig.
Step	1	Income		.832	1	1	2314	467	1	2314	.000
	2	Age		.758	2	1	2314	368	2	2313	.000
	3	Years of education		.707	3	1	2314	320	3	2312	.000
	4	Suburban		.703	4	1	2314	244	4	2311	.000
	5	Employed		.701	5	1	2314	197	5	2310	.000

At each step, the variable that minimizes the overall Wilks' Lambda is entered.

1. Maximum number of steps is 14.
2. Minimum partial F to enter is 3.84.
3. Maximum partial F to remove is 2.71.
4. F level, tolerance, or VIN insufficient for further computation.

Statistics for variables not in the model. Step 1 in Figure 14-23 shows the variables not in the model after income is included. You see that age has the largest F-to-enter value (equivalently the smallest Wilks' lambda), so it enters the equation next because the F-to-enter is larger than 3.84. The column labeled *Wilks' Lambda* in Figure 14-23 is the value of lambda if the variable enters next.

Figure 14-23
Statistics for variables not in the model at step 1

			Tolerance	Min. Tolerance	F to Enter	Wilks' Lambda
Step	1	Age	.989	.989	224.732	.758
		Gender	.991	.991	3.934	.831
		Children less than 17 years old	.999	.999	9.782	.829
		Suburban	.970	.970	17.556	.826
		Employed	.983	.983	83.834	.803
		Years of education	.919	.919	137.865	.785

In step 5 of Figure 14-24, you see that the only variables not in the model are gender and having children less than 17 years old. Neither of these variables meets the entry criterion because the F value for the change in Wilks' lambda when the variable enters

(F-to-enter) is less than 3.84. That's why variable selection stops with five variables in the model.

Figure 14-24
Statistics for variables not in the model at step 5

			Tolerance	Min. Tolerance	F to Enter	Wilks' Lambda
Step	5	Gender	.967	.832	.312	.701
		Children less than 17 years old	.862	.759	1.287	.700

Warning: For large sample sizes, variables that result in very small changes in Wilks' lambda may meet the statistical criterion for entry, since sample size is involved in the computations. For example, after the second step in Figure 14-22, the decreases in Wilks' lambda, although statistically significant, are very small and unlikely to be important. The stepwise algorithm results in a model with five variables, but you should evaluate the model with three variables (*income*, *age*, and *education*) since it's probably just as good.

Statistics for variables in the model. After a variable is entered, the Wilks' lambda for each variable already in the model is checked to make sure that it doesn't meet the criterion for removal. It's possible for a predictor that met the entry criterion to no longer meet the criterion when additional variables are entered. This means the contribution of that variable is no longer important when other variables are present.

Figure 14-25 shows the statistics used to evaluate whether a variable that's already in the equation should be removed for the model at step 5. The F-to-remove column is the F value for the null hypothesis that the change in Wilks' lambda is 0 when the variable is removed. The Wilks' lambda is for the model without the variable. None of the variables meets the removal criterion.

Figure 14-25
Statistics for variables in the model

			Tolerance	F to Remove	Wilks' Lambda
Step	5	Income	.876	214.206	.766
		Age	.846	189.911	.758
		Years of education	.894	152.706	.747
		Suburban	.966	13.097	.705
		Employed	.851	6.623	.703

Classification Results

Figure 14-26 contains classification results, based on only the five variables in the model selected by stepwise variable selection. Notice that the five-variable model classifies the sample data as well as the seven-variable model did.

Figure 14-26
Classification results

Classification Resultsa

				Predicted Group Membership		Total
				No	Yes	
Original	Internet use	No	Count	744	268	1012
			%	73.5	26.5	100.0
		Yes	Count	320	985	1305
			%	24.5	75.5	100.0
		Ungrouped cases	Count	1	0	1
			%	100.0	.0	100.0

a. 74.6% of original grouped cases correctly classified.

Figure 14-27 shows the correct-classification rates for models with smaller numbers of variables. You see that income and age together achieve nearly the same correct-classification rate in the sample as the five-variable model in Figure 14-26.

Figure 14-27
Correct-classification rates (table not from SPSS Statistics)

Variables	**Percentage of cases correctly classified**	**Canonical correlation**
Income	66	0.41
Income and age	72	0.49
Income, age, and education	75	0.55

Discriminant Analysis with Four Groups

When you have more than two groups of cases, the basic idea remains the same. You want to find linear combinations of the predictor variables that separate the groups. Instead of one score for each case, you have several scores. The number of linear combinations you can form is the smaller of one fewer than the number of groups and the number of predictors. The first linear combination has the largest ratio of between-groups to within-groups sums of squares. The second function is uncorrelated with the first and has the next largest ratio, and so on.

As an example, consider Internet users and library patrons. You'll determine whether you can separate four groups: people who use both the Internet and the public library, people who use neither the Internet nor the public library, people who use the Internet but not the library, and people who use the library but not the Internet. The variables are the same as before with the exception of two new dummy variables: *femkids*, which has the value of 1 for women with children under the age of 17, and *student*, which has the value of 1 for students. The new variables are meant to capture the phenomenon that mothers with young children end up in the library more frequently than others, and so do students.

Discriminant Function Coefficients

Standardized discriminant function coefficients are shown in Figure 14-28. Because there are four groups, there are three sets of coefficients. Again, the signs of the coefficients are arbitrary. All positive coefficients could be changed to negative, and all negative coefficients could be changed to positive. Employed young people with many years of education, large incomes, and student status have large values for the first function. The function is reminiscent of the one that you derived to separate Internet users from non-users. Women with small children and people with a lot of education but not large incomes have large values for the second function. Old people with high income or limited education, students, and women with children have large values for the third function.

Figure 14-28
Standardized discriminant function coefficients

	Function		
	1	2	3
Age	-.447	.209	.409
Female with young children	.000	.804	.310
Income	.596	-.312	.672
Employed	.133	-.267	-.100
Years of education	.491	.502	-.693
Student	.236	.258	.557

Function Variable Correlations

If you look at the structure matrix in Figure 14-29, you can see the variables that have the largest correlations with each of the function. The variables are sorted so that variables that have large correlations with the same function are together. This may help in interpreting the functions. You see that income, age, education, and employment are associated with the first function, and women with young children are most strongly associated with the second.

Figure 14-29
Structure matrix

	Function		
	1	2	3
Income	.677*	-.247	.435
Years of education	.595*	.355	-.483
Age	-.417*	.068	.176
Employed	.394*	-.371	-.234
Female with young children	.042	.714*	.227
Student	.277	.226	.372*

Pooled within-groups correlations between discriminating variables and standardized canonical discriminant functions
Variables ordered by absolute size of correlation within function.

*. Largest absolute correlation between each variable and any discriminant function

Function Means

It's also informative to examine the mean values of each of the functions for each group, as shown in Figure 14-30. You use the values of all three of the functions to assign cases to groups. First of all, you see that the four group means for function 3 aren't very different. This is an indication that the function won't contribute very much to the separation of the groups.

Function 1 has means that range from –1.09 for people who don't use either the Internet or the library to 0.60 for people who use both. If you look at the combination of values for functions 1 and 2, you see that the *Neither* group has negative mean values for both functions, the *Both* group has positive mean values for both functions, the *Library only* group has a negative mean value for function 1 and a positive mean value for function 2, and the *Internet only* group has a positive value for function 1 and a negative value for function 2.

Figure 14-30
Functions at group centroids

		Function		
		1	2	3
Internet and library usage	Neither	-1.086	-.145	.077
	Library only	-.613	.208	-.092
	Internet only	.495	-.287	-.084
	Both	.602	.089	.048

Unstandardized canonical discriminant functions evaluated at group means

Eigenvalues and Canonical Correlations

You derived three functions, but that doesn't mean they all contribute to the separation of the groups. You know that the functions go from best to worst in terms of the ratios of the between-groups to within-groups sums of squares. Tests of the functions can be derived from entries shown in the analysis-of-variance table in Figure 14-31.

Figure 14-31
Analysis of variance

		Sum of Squares	df	Mean Square	F	Sig.
Discriminant Scores from Function 1 for Analysis 1	Between Groups	1098.331	3	366.110	366.110	.000
	Within Groups	2185.000	2185	1.000		
	Total	3283.331	2188			
Discriminant Scores from Function 2 for Analysis 1	Between Groups	68.986	3	22.995	22.995	.000
	Within Groups	2185.000	2185	1.000		
	Total	2253.986	2188			
Discriminant Scores from Function 3 for Analysis 1	Between Groups	11.281	3	3.760	3.760	.010
	Within Groups	2185.000	2185	1.000		
	Total	2196.281	2188			

The eigenvalues and canonical correlations for each of the discriminant functions tell you how strongly the functions are related to the groups. As in the single-function case, for each function, the eigenvalue is the ratio of the between-groups to the within-groups sum of squares for the discriminant function scores. For example, from Figure 14-31, for function 1, the eigenvalue is $1098/2185 = 0.50$. The eigenvalues are shown in Figure 14-32.

Figure 14-32
Eigenvalues

		Eigenvalue	% of Variance	Cumulative %	Canonical Correlation
Function	1	.503[1]	93.2	93.2	.578
	2	.032[1]	5.9	99.0	.175
	3	.005[1]	1.0	100.0	.072

1. First 3 canonical discriminant functions were used in the analysis.

You can also look at the canonical correlations for each function, as shown in the last column of Figure 14-32. The canonical correlation is the square root of the between-groups to total sum of squares. The first function always has the largest canonical correlation (in this case, 0.578). There is a large drop in value for the other two functions.

Another criterion on which the functions can be compared is the percentage of between-groups variance that each one explains. Again, the first function is always best and explains the largest proportion of the variability; successive functions explain less. In this example, the first function explains over 90% of the total between-groups sum of squares.

Tip: To calculate the percentage, divide the between-groups sum of squares for each function by the sum of the between-groups sums of squares for all functions.

Testing Equality of Discriminant Function Means

If the populations from which the samples are selected don't differ, the discriminant functions you calculate may represent nothing more than sampling variability. That's why you want to test the null hypotheses that the population means of all of the discriminant functions are the same in all of the groups. For a single function, you used a test based on Wilks' lambda and its associated significance level. Wilks' lambda is used for the multiple-group situation as well, but there are additional considerations.

Overall Test

First, you want to simultaneously test the null hypothesis that the population means for all of the discriminant functions are equal in all of the groups. You can no longer simply compute the ratio of the within-groups to total sum of squares, since there are multiple ratios, one for each function. The Wilks' lambda for multiple functions is the product of the individual Wilks' lambdas for each function. In this example, the Wilks' lambda for all functions considered simultaneously is, using the numbers in Figure 14-31:

$$\lambda = \left(\frac{2185}{3283}\right)\left(\frac{2185}{2254}\right)\left(\frac{2185}{2196}\right) = 0.642$$

This is the lambda shown in Figure 14-33. Based on the small observed significance level, you can reject the null hypothesis that the population mean values for all functions are equal in the four groups.

Figure 14-33
Wilks' lambda

		Wilks' Lambda	Chi-square	df	Sig.
Test of Function(s)	1 through 3	.642	968.110	18	.000
	2 through 3	.964	79.099	10	.000
	3	.995	11.242	4	.024

Testing Successive Functions

When you have more than one function, you can successively test the means of the functions by first testing all means simultaneously, then excluding one function at a time, and then testing the means of the remaining functions. The functions are arranged in descending order, which means that once a set is found not to be significant, all subsequent tests are also not significant. You may find that a subset of discriminant functions account for all of the differences and that additional functions do not represent population differences, only random variation.

The first row of Figure 14-33 is the test for all three function means, the second row is the test for the means of functions 2 and 3, and the last row is the test for just function 3. All function means are found to be statistically different. Even though the last Wilks' lambda is perilously close to 1, it is found to be statistically different in the groups because the sample sizes are so large.

Pairwise Differences between Groups

You can also test whether the population distances between the means of each pair of groups is 0, using the statistics in Figure 14-34. The *F* statistic is proportional to the Mahalanobis distance that measures how far group means are apart in multivariate space. From Figure 14-34, you see that the smallest distance is between the *Internet only* group and the *Both* group. The largest distance is between the *Both* group and the

Neither group. A significance level is printed for each distance, but no adjustment is made for the fact that multiple pairs of groups are compared.

Figure 14-34
Pairwise group comparisons

					Neither	Library only	Internet only	Both
Step	6	Internet and library usage	Neither	F		13.822	87.134	139.415
				Sig.		.000	.000	.000
			Library only	F	13.822		52.374	76.479
				Sig.	.000		.000	.000
			Internet only	F	87.134	52.374		7.872
				Sig.	.000	.000		.000
			Both	F	139.415	76.479	7.872	
				Sig.	.000	.000	.000	

Tip: The *F* for pairwise group comparisons is selected in the Method dialog box. If you want to obtain the matrix for all predictor variables in your list, select Stepwise but set the criterion for variable entry to $p = 0.99$ and the criterion for variable removal to $p = 1$.

Classification

You can see how the first two function values are used for assignment from Figure 14-35. Cases that have values within the area bounded by 1's are assigned to the first group (*Neither*), and cases that have values bounded by 2's are assigned to the second group (*Library only*). Cases are assigned similarly for the other two groups. You can also plot the values of the first two discriminant functions for all cases. If you have a large number of cases without good separation, you'll have a hard time seeing anything.

Figure 14-35
Territorial map

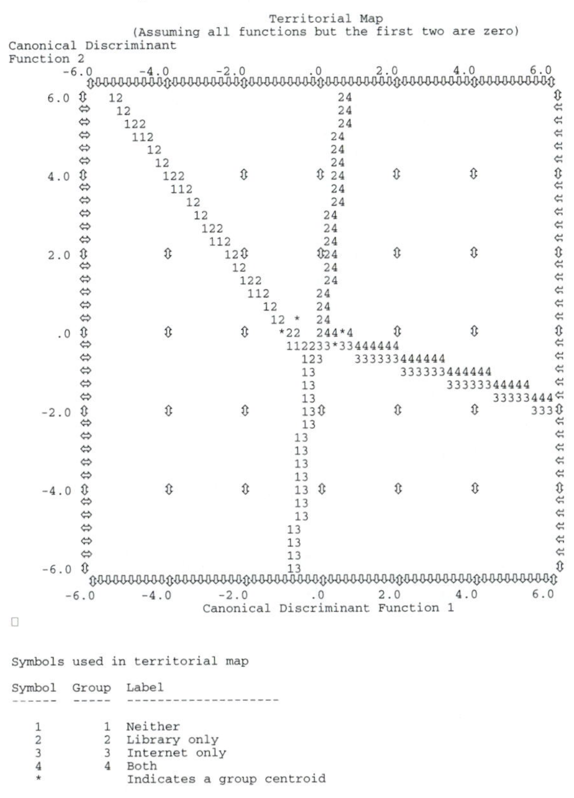

Tip: If you want to plot other pairs of functions, save the discriminant function scores and use the chart facility.

Classification Function Coefficients

When you have more than two groups, the algorithm for classifying cases on the basis of the discriminant function scores is more complicated. It's easier to use the classification function coefficients shown in Figure 14-36. For each case, you calculate its score for each of the classification function coefficients, and then you assign the case to the group for which it has the largest score. These functions are sometimes called **Fisher's linear discriminant functions**. The results are identical to what you get using the discriminant function scores. Fortunately, SPSS Statistics does the assignment of cases to groups for you.

Figure 14-36
Classification function coefficients

	Internet and library usage			
	Neither	Library only	Internet only	Both
Age	.278	.264	.226	.232
Female with young children	4.791	5.328	4.407	5.207
Income	-.025	-.023	.007	.009
Employed	5.055	5.025	5.632	5.415
Years of education	1.812	2.030	2.150	2.212
Student	3.847	4.161	4.561	5.126
(Constant)	-20.468	-22.703	-24.229	-25.528

Fisher's linear discriminant functions

Warning: The proportion of misclassified cases is not a very good summary measure of the effectiveness of the discriminant model because it depends on the kinds of cases that are included in the sample and the proportion of cases in each of the groups. For example, if your sample has many cases that are well-separated, you will be able to classify them well. The same model faced with more cases in the overlapping regions will not classify as well.

Equal Prior Probabilities

Figure 14-37 is the classification table when all four groups are assumed to be equally likely in the population. You see that only 45% of the cases are correctly classified. Of people who don't use either the library or the Internet, 62% are correctly classified. Of people who use only the library, only 33% are correctly classified. Overall, the classification results are not very impressive. (If you use the leave-one-out estimator, the misclassification rate is 44.1%.)

Figure 14-37
Classification results

Classification Results[2,3]

				Predicted Group Membership				Total
				Neither	Library only	Internet only	Both	
Original	Internet and library usage	Neither	Count	263	71	63	27	424
			%	62.0	16.7	14.9	6.4	100.0
		Library only	Count	175	151	74	60	460
			%	38.0	32.8	16.1	13.0	100.0
		Internet only	Count	49	46	204	101	400
			%	12.3	11.5	51.0	25.3	100.0
		Both	Count	80	135	322	368	905
			%	8.8	14.9	35.6	40.7	100.0
		Ungrouped cases	Count	50	30	30	19	129
			%	38.8	23.3	23.3	14.7	100.0

1. Cross validation is done only for those cases in the analysis. In cross validation, each case is classified by the functions derived from all cases other than that case.
2. 45.0% of original grouped cases correctly classified.
3. 44.1% of cross-validated grouped cases correctly classified.

Unequal Prior Probabilities

The classification rate in Figure 14-37 is based on all groups having the same prior probability. Information about how often the various groups occur in the population is ignored. If you set your prior probabilities to be equal to the sample proportions of the groups (424/2189 for *Neither*, 460/2189 for *Library only*, 400/2189 for *Internet only*, and 905/2189 for *Both*) and the proportions in the groups are not approximately equal, the classification results can change considerably. Figure 14-38 shows the prior probabilities.

Figure 14-38
Prior probabilities based on sample size

		Prior	Cases Used in Analysis
			Unweighted
Internet and library usage	Neither	.194	424
	Library only	.210	460
	Internet only	.183	400
	Both	.413	905
	Total	1.000	2189

Tip: If you use syntax and specify prior probabilities that don't add up to 1, SPSS Statistics prints the specified probabilities in an otherwise-mysterious column labeled *Specified Prior* and then divides the specified prior by the sum of the priors so that they would add up to 1. The resulting prior probabilities are displayed in a column labeled *Effective Prior.*

Consider the classification results in Figure 14-39, in which the prior probabilities are equal to the sample proportions. You see that 89% of the cases that used both the Internet and the library are now correctly classified, as compared to 40.7% when equal priors were specified. Correct classification into all of the other groups suffered. Because over 40% of the sample used both the Internet and the public library, that information has overwhelmed the information in the discriminant scores, leading to classification into the largest group. When sample proportions are markedly unequal, their use as prior probabilities always increases the rate of correct classification.

Figure 14-39
Classification results

Classification Results[2,3]

				Predicted Group Membership				
				Neither	Library only	Internet only	Both	Total
Original	Internet and library usage	Neither	Count	238	63	0	123	424
			%	56.1	14.9	.0	29.0	100.0
		Library only	Count	151	114	0	195	460
			%	32.8	24.8	.0	42.4	100.0
		Internet only	Count	34	27	0	339	400
			%	8.5	6.8	.0	84.8	100.0
		Both	Count	58	39	0	808	905
			%	6.4	4.3	.0	89.3	100.0
		Ungrouped cases	Count	42	16	0	71	129
			%	32.6	12.4	.0	55.0	100.0

2. 53.0% of original grouped cases correctly classified.
3. 52.6% of cross-validated grouped cases correctly classified.

Tip: SPSS Statistics offers many additional procedures for analyzing data that have a categorical dependent variable. See the Help system for the modules that you have installed.

Obtaining the Output

To produce the output in this chapter, open the file *library.sav* and follow the instructions below.

Figure 14-2. From the Analyze menu choose Compare Means > Means. Move *age*, *gender*, *income*, *kids*, *suburban*, *work*, and *yearsed* into the Dependent List, and move *internet* into the Independent List. Click OK.

Figures 14-1, 14-4 to 14-6, 14-8, 14-9, 14-11, and 14-13 to 14-21. From the Analyze menu choose Classify > Discriminant. Move *internet* into the Grouping Variable box. Click Define Range and type 0 for Minimum and 1 for Maximum. Click Continue. Move *age*, *gender*, *income*, *kids*, *suburban*, *work*, and *yearsed* into the Independents list. Click Statistics. In the Statistics subdialog box, select Means, Univariate ANOVAs, Box's M, Unstandardized, and Within-groups correlation. Click Continue, and then click Classify. In the Classification subdialog box, select Casewise results and Limit cases to first, and then type the number 20. Also select Summary table, Leave-one-out classification, Combined-groups, Separate-groups, and Territorial map. Click Continue, and then click Save. In the Save subdialog box, select Discriminant scores. Click Continue, and then click OK.

Figure 14-3. From the Graphs menu choose Legacy Dialogs > Line and then click the Multiple icon. Click Define. Move *agecat* into the Category Axis box and *internet* into the Define Lines By box. Select Other statistic and move *missingincome* into the Variable box. Click OK. To generate the other line chart in this figure, repeat the instructions but move *edcat* into the Category Axis box.

Figures 14-7, 14-10, and 14-27 cannot be produced from the data file.

Figure 14-12. From the Analyze menu choose Compare Means > One-Way ANOVA. Move *Dis1_1* into the Dependent List and *internet* into the Factor box. Click OK.

Figures 14-22 to 14-26. From the Analyze menu choose Classify > Discriminant. Move *internet* into the Grouping Variable box. Click Define Range and type 0 for Minimum and 1 for Maximum. Click Continue. Move *age*, *gender*, *income*, *kids*, *suburban*, *work*, and *yearsed* into the Independents list. Click Statistics. In the Statistics subdialog box, select Unstandardized and Separate-groups covariance. Click Continue, and then click Classify. In the Classification subdialog box, select Summary table. Click Continue, and then click Save. In the Save subdialog box, select Discriminant scores. Click Continue, and then click OK.

Figures 14-28 to 14-30, 14-32, 14-33, and 14-35 to 14-37. From the Analyze menu choose Classify > Discriminant. Move *group* into the Grouping Variable box. Click Define Range and type 1 for Minimum and 4 for Maximum. Click Continue. Move *age*, *femkids*, *income*, *work*, *yearsed*, and *student* into the Independents list. Click Statistics. In the Statistics subdialog box, select Unstandardized and Fisher's. Click Continue, and then click Classify. In the Classification subdialog box, select Summary table and Territorial map. Click Continue, and then click Save. In the Save subdialog box, select Discriminant scores. Click Continue, and then click OK.

Figure 14-31. From the Analyze menu choose Compare Means > One-Way ANOVA. Move *Dis1_1*, *Dis1_2,* and *Dis1_3* into the Dependent List and *group* into the Factor box. Click OK.

Figure 14-34. From the Analyze menu choose Classify > Discriminant. Move *group* into the Grouping Variable box. Click Define Range and type 1 for Minimum and 4 for Maximum. Click Continue. Move *age*, *femkids*, *income*, *work*, *yearsed*, and *student* into the Independents list. Select Use stepwise method. Click Classify. In the Classification subdialog box, select Summary table. Click Continue, and then click Method. In the Method subdialog box, select F for pairwise distances. Select Use probability of F. Type 0.9 in the Entry text box and 1 in the Removal text box. Click Continue, and then click OK. The matrix is at step 6.

Chapter

15

Logistic Regression Analysis

Many problems require you to predict the values of a binary dependent variable from a set of independent variables. Multiple regression analysis and discriminant analysis are two related statistical techniques that quickly come to mind; however, these techniques pose difficulties when the dependent variable can have only two values—an event occurring or not occurring.

The **binary logistic regression model** is most frequently used to model the probability that an event occurs. It requires fewer assumptions than discriminant analysis, and even when the assumptions required for discriminant analysis are satisfied, it performs well. (See Hosmer and Lemeshow, 2000, or Kleinbaum and Klein, 2002, for an introduction to logistic regression models.)

Examples

- Can you predict whether a patient is experiencing a myocardial infarction based on laboratory and physical exam findings in the emergency room?
- What variables are related to repeated purchases of products from the same manufacturer?
- What characteristics of high-risk students are useful for predicting graduation from high school?

In a Nutshell

Using the logistic regression model, you can directly estimate the probability that one of two events occurs, based on the values of a set of independent variables that can be either categorical or continuous. Logistic regression always results in estimated probabilities that are between 0 and 1. The only assumptions that are required are that the observations are independent and that the variables are linearly related to the log of the odds that an event occurs. You can use various methods for selecting a subset of variables that are good predictors of the outcome variable for the sample. Once you have estimated a model, you must determine how well it fits the data.

Introduction

When you have an outcome that is a dichotomy (finish or not finish the dissertation, develop or not develop heart disease), you can't use multiple linear regression to study the relationship between the outcome and the independent variables because it is impossible for such data to satisfy the required assumptions. In other words, a binary variable can't be normally distributed with a constant variance. If you ignore the assumption and apply the linear regression model, you will find that the predicted values of the outcome variable can be negative or greater than 1.

Discriminant analysis is often used to predict membership in one of several groups; however, to be statistically optimal, discriminant analysis requires that the independent variables have a multivariate normal distribution. Discriminant analysis doesn't directly estimate probabilities; instead, the discriminant scores must be converted to probabilities.

The logistic regression model is a very flexible tool for studying the relationship between a set of variables that can be continuous or categorical and a categorical outcome. The only assumptions that are needed are that the observations are independent and that the model is correctly specified. That is, all necessary independent variables are included and a linear relationship exists between each of the variables and the log of the ratio of the probability that an event occurs to the probability that it does not occur. Logistic regression is useful for a broad range of experimental designs, including retrospective studies.

Logistic Regression Model

In logistic regression, you directly estimate the probability of an event occurring. For the case of a single independent variable, the logistic regression model is

$$\text{Prob (event)} = \frac{1}{1 + e^{-(B_0 + B_1 X)}}$$

where B_0 and B_1 are coefficients estimated from the data; X is the independent variable; e is the base of the natural logarithms, approximately 2.718; and *Prob (event)* is the predicted probability that an event occurs.

For more than one independent variable, the model is

$$\text{Prob (event)} = \frac{1}{1 + e^{-Z}}$$

where Z is the linear combination

$$Z = B_0 + B_1 X_1 + B_2 X_2 + \ldots + B_p X_p$$

and p is the number of independent variables.

Figure 15-1 is a plot of a logistic regression curve for different values of Z. As you can see, the curve is S-shaped. It closely resembles the curve obtained when the cumulative probability of the normal distribution is plotted. The relationship between the independent variable and the probability is nonlinear. The probability estimate is always between 0 and 1, regardless of the value of Z.

Figure 15-1
Plot of logistic regression curve

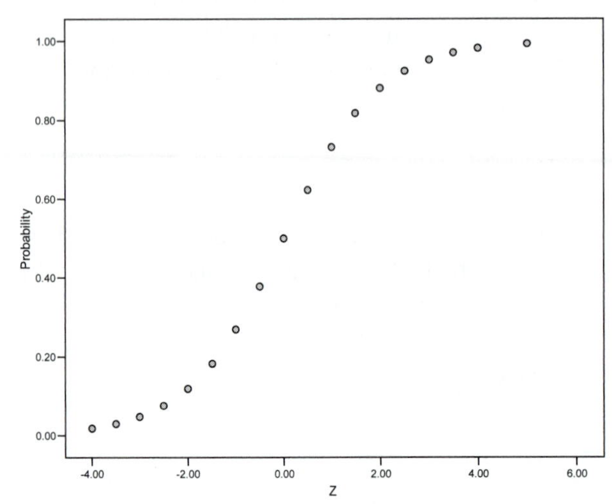

Predicting Node Involvement: The Example

The treatment and prognosis of cancer depends on how much the disease has spread. One of the regions to which a cancer may spread is the lymph nodes. If the lymph nodes are involved, the prognosis is generally poorer than if they are not. For certain cancers, exploratory surgery is done just to determine whether the nodes are cancerous, since this determines what treatment is needed. Predicting whether the nodes are affected or not on the basis of data that are available without surgery avoids considerable discomfort and expense.

In this chapter, you will use data presented by Brown (1980) for 53 men with prostate cancer. For each patient, he reports the age, serum acid phosphatase (a laboratory value that is elevated if the tumor has spread to certain areas), the stage of the disease (an indication of how advanced the disease is), the grade of the tumor (an indication of aggressiveness), and X-ray results. For each of the men in the sample, you know whether the nodes are positive or not, based on surgical evaluation. The goal is to predict whether the nodes are positive for cancer, based on the values of the variables that can be measured without surgery.

Tip: When trying to determine a sample size for logistic regression, you have to worry about both the total sample size and the number of events. The number of events is the smaller of the counts for the values of the binary variable. Peduzzi et al. (1996) suggest that at least 10 events are needed for each parameter that you want to estimate.

Coefficients for the Logistic Model

Figure 15-2 contains the estimated coefficients (under column heading *B*) and related statistics from the logistic regression model that predicts nodal involvement from a constant and the variables *acid*, *age*, *xray*, *grade*, and *stage*. The last three of these variables (*xray*, *grade*, and *stage*) are **indicator variables**, coded 0 or 1. The value of 1 for *xray* indicates positive X-ray findings, the value of 1 for *stage* indicates advanced stage, and the value of 1 for *grade* indicates a more aggressively spreading malignant tumor.

Figure 15-2
Parameter estimates for the logistic regression model

								95.0% C.I.for EXP(B)	
		B	S.E.	Wald	df	Sig.	Exp(B)	Lower	Upper
Step 1[1]	acid	.0243	.013	3.423	1	.064	1.025	.999	1.051
	age	-.0693	.058	1.432	1	.231	.933	.833	1.045
	xray	2.0453	.807	6.421	1	.011	7.732	1.589	37.615
	grade	.7614	.771	.976	1	.323	2.141	.473	9.700
	stage	1.5641	.774	4.084	1	.043	4.778	1.048	21.783
	Constant	.0618	3.460	.000	1	.986	1.064		

1. Variable(s) entered on step 1: acid, age, xray, grade, stage.

Given these coefficients, the logistic regression equation for predicting the probability of nodal involvement is

$$\text{Prob (nodal involvement)} = \frac{1}{1 + e^{-Z}}$$

where

$$Z = 0.0618 + 0.0243(\text{acid}) - 0.0693(\text{age}) + 2.0453(\text{xray}) + 0.7614(\text{grade}) + 1.5641(\text{stage})$$

Calculating a Predicted Probability

You can use the coefficients to predict the probability of positive nodes for a man who is 66 years old, with a serum acid phosphatase level of 48, and values of 0 for the remaining independent variables. You calculate

$$Z = 0.0618 - 0.0693(66) + 0.0243(48) = -3.346$$

and the estimated probability of nodal involvement is then

$$\text{Prob (nodal involvement)} = \frac{1}{1 + e^{-(-3.346)}} = 0.0340$$

Based on this estimate, it appears unlikely that the nodes are malignant.

Warning: Calculation of probabilities applies to the average for all people with that set of values for the independent variables, not for an individual. A particular person either has nodal involvement or does not.

Estimating the Coefficients

In linear regression, you estimate the parameters of the model by using the **least-squares method**. That is, you select regression coefficients that result in the smallest sums of squared distances between the observed and the predicted values of the dependent variable.

In logistic regression, the parameters of the model are estimated by using the **maximum-likelihood method**. The coefficients that make the observed results most likely are selected. Because the logistic regression model is nonlinear, an iterative algorithm is necessary for parameter estimation.

Warning: The predicted probability estimates from logistic regression are based on the proportion of cases in your sample who experience the event. If the mix of cases in your sample does not reflect the mix of cases in the population (for example, you included equal numbers of people who experienced the event and who didn't), the estimated probabilities are not correct for the population, although the estimated coefficients for all of the variables except the constant are correct. If you know the true probability of the event in the population, you can apply a correction to the constant (see Schlesselman, 1982).

Interpreting the Logistic Coefficients

In multiple linear regression, the interpretation of the regression coefficient is straightforward. It's the estimated change in the dependent variable for a one-unit change in the independent variable, assuming that the values of the other independent variables are held constant. The actual value of the coefficient depends on the other independent variables in the model. It is also true in logistic regression that the value of a coefficient depends on the other independent variables in the model; however, the interpretation of the coefficient itself is not as intuitive. It involves knowing the meaning of odds, logits, and odds ratios.

Calculating Odds

The **odds** of an event occurring are defined as the ratio of the probability that an event occurs to the probability that it does not. For example, the odds that nodes are positive are

$$\text{Odds} = \frac{\text{Prob (positive nodes)}}{\text{Prob (negative nodes)}}$$

Warning: Don't confuse this technical meaning of odds with its informal usage to mean simply the probability.

Calculating Logit

The logistic model can be written in terms of the log of the odds, which is called a **logit**:

$$\log\left(\frac{\text{Prob (event)}}{\text{Prob (no event)}}\right) = B_0 + B_1X_1 + \ldots + B_pX_p$$

This results in an equation that looks just like a multiple regression equation on the right side. The dependent variable, however, is the log of the odds ratio. From the equation above, you see that the logistic regression coefficient for a variable is the change in the log odds associated with a one-unit change in the independent variable, when all other independent variables are held constant (assuming there are no other interaction terms in the model involving the variable).

For example, from Figure 15-2, you see that the coefficient for *grade* is 0.76. This tells you that when the grade changes from 0 to 1 and the values of the other independent variables remain the same, the log odds of the nodes being malignant increase by 0.76.

Calculating Change in Odds

As an illustration, calculate the odds of having malignant nodes for a 60-year-old man with a serum acid phosphatase level of 62, a value of 1 for X-ray results, and values of 0 for stage and grade of tumor. The estimated probability that the nodes are malignant is as follows:

$$\text{Prob (malignant nodes)} = \frac{1}{1 + e^{-Z}} = 0.37$$

where

$$Z = 0.0618 + 0.0243(62) - 0.0693(60) + 2.0453(1) = -0.54$$

The probability of not having malignant nodes is 0.63.

The odds of having a malignant node are

$$\text{Odds} = \frac{\text{Prob (event)}}{\text{Prob (no event)}} = \frac{0.37}{1 - 0.37} = 0.59$$

The log odds are –0.53.

If you follow the same procedure and if a case has a value of 1 for grade instead of 0, the estimated probability of malignant nodes is 0.554. Similarly, the estimated odds are 1.24, and the log odds are 0.22.

An increase in the value of grade by one unit results in an increase in the log odds of about 0.75, which is the difference between the two log odds, 0.22 and –0.53. This is the value of the coefficient for grade in Figure 15-2. (Because the precision of the hand calculations is limited, the value of 0.75 isn't exactly equal to the 0.76 value for grade shown in Figure 15-2. With enough precision, you would arrive at exactly the value of the coefficient.)

You can calculate the ratio of the odds when grade changes from 0 to 1. It's just 1.24 divided by 0.59 (or about 2.1). As you'll see, that's also an important number.

Calculating the Odds Ratio

The logistic equation can be written in terms of odds as

$$\frac{\text{Prob (event)}}{\text{Prob (no event)}} = e^{B_0 + B_1X_1 + \ldots + B_pX_p} = e^{B_0}e^{B_1X_1}\ldots e^{B_pX_p}$$

Some simple algebra gives the result that *e* raised to the power B_i is the factor by which the odds change when the *i*th independent variable increases by one unit and all of the other variable values stay the same. This is known as the **odds ratio**. (This is true only if the *i*th variable is not included in any other variables, such as interactions.)

- If B_i is positive, the odds ratio is greater than 1, which means that the odds of the event are increased.
- If B_i is negative, the odds ratio is less than 1, which means that the odds are decreased.
- If B_i is 0, the factor equals 1, and the odds are unchanged.

In the previous section, you calculated the odds ratio for grade. The odds are increased by a factor of 2.1 when grade changes from 0 to 1. This number is in the *Exp(B)* column in Figure 15-2. The 95% confidence interval for the population value of the odds ratio is in the last two columns of Figure 15-2. From the 95% confidence interval, you can see that values anywhere from 0.47 to 9.7 are plausible for the population value of the odds ratio for grade. Because this interval includes the value 1 (no change in odds), you can't conclude based on this sample of data that a unit change in grade in the population is associated with a change in the odds of positive nodes with all other variables held constant. Because the confidence interval for the odds ratio is based on the confidence interval for the corresponding logistic regression coefficient, the confidence interval for the odds ratio will include 1 whenever the confidence interval for the regression coefficient contains 0.

Tip: When an independent variable is continuous (such as age, blood pressure, or years of education), the odds ratio for a unit change in the value of the independent variable may not be useful. For example, you may want to look at the odds ratio associated with a decade change in age or a 5-mm change in blood pressure. The odds ratio for a change of *C* units is e^{CB}, where *B* is the coefficient for a change of one unit.

Testing Hypotheses about the Model

Just as in linear regression, you are interested in the overall test that in the population all of the logistic regression coefficients except the constant are 0, and you are interested in tests for individual coefficients. As in linear regression, the test for a particular coefficient depends on the other variables in the model.

The test that all coefficients in a model are 0 is based on a statistic called the **likelihood-ratio test**. In logistic regression, instead of minimizing the sum of squared residuals as you do in linear regression, you choose parameter estimates that make the observed residuals most likely. The probability of the observed results, given the parameter estimates, is known as the likelihood. It is customary to use -2 times the log of the likelihood ($-2LL$) as a measure of how well the estimated model fits the data. A good model is one that results in a high likelihood of the observed results. This translates to a small value for $-2LL$.

Likelihood-Ratio Test

The change in the likelihood value is used to determine how the fit of a model changes as variables are added or deleted from a model. This is conceptually equivalent to the change in R^2 statistics in linear regression. Instead of looking at the increase (or decrease) in R^2 as terms are added or removed from a model, you're looking at the change in –2 log-likelihood.

The likelihood-ratio statistic, which plays an important role in testing hypotheses in logistic regression (and other statistical procedures), is defined as

$$\text{likelihood ratio} = -2LL(\text{reduced model}) - (-2LL(\text{full model}))$$

For a sufficiently large sample size, the likelihood-ratio statistic has a chi-square distribution with k degrees of freedom, where k is the difference between the number of parameters in the two models. The likelihood ratio tests the null hypothesis that the coefficients of the terms that are excluded from the full model are 0. For example, if the full model contains three independent variables (age, race, and sex) and if the reduced model contains only age, you are testing the null hypothesis that the coefficients for race and sex are 0.

Warning: The likelihood-ratio test can be used only when the reduced model is obtained by dropping terms from the full model; that is, the models are nested.

Are All Coefficients Zero?

It's easy to construct a test of the null hypothesis that all coefficients in a model are 0. The two models that you want to compare are the full model that contains all of the variables and the reduced model that contains only the intercept. To test the null hypothesis that all of the variables in the current example have coefficients of 0, you have to look at $-2LL$ for the constant-only model and $-2LL$ for the full model.

Constant-Only Model

For the logistic regression model that contains only the constant, $-2LL$ is 70.25, as shown in Figure 15-3, the iteration history. From this table, you can also see how the estimates of the coefficients change at each iteration. Because the constant is the only term in the model, you see just the changes in the constant.

Figure 15-3
–2LL for model containing only the constant

Iteration History[1,2,3]

			-2 Log likelihood	Coefficients
				Constant
Iteration	Step 0	1	70.253	-.491
		2	70.252	-.501
		3	70.252	-.501

1. Constant is included in the model.
2. Initial -2 Log Likelihood: 70.252
3. Estimation terminated at iteration number 3 because parameter estimates changed by less than .001.

Tip: The only way to get $-2LL$ for a constant only is to ask for the iteration history.

Full Model

Figure 15-4 shows model summary statistics for the model with all of the independent variables. That's the model you are currently evaluating. The value of $-2LL$ is 48.126.

Figure 15-4
Model summary statistics

		-2 Log likelihood	Cox & Snell R Square	Nagelkerke R Square
Step	1	48.126[1]	.341	.465

1. Estimation terminated at iteration number 5 because parameter estimates changed by less than .001.

Model Change

To calculate the likelihood-ratio test of the null hypothesis that all population coefficients except the constant are 0, you must find the difference between $-2LL$ for the complete model and $-2LL$ for the constant-only model. In this example, $-2LL$ for the model containing only the constant is 70.252 (from Figure 15-3), while for the complete model, it is 48.126 (from Figure 15-4). The difference is 22.126, the model chi-square value in Figure 15-5. The degrees of freedom for the model chi-square is the difference between the six coefficients in the full model and the single coefficient (the constant) in the reduced model. Based on the small observed significance level shown in the last column, you reject the null hypothesis that all of the coefficients in the model are 0.

Figure 15-5
Changes in –2LL

		Chi-square	df	Sig.
Step 1	Step	22.126	5	.000
	Block	22.126	5	.000
	Model	22.126	5	.000

Evaluating Sequential Changes

There are three chi-square entries in Figure 15-5. They are labeled *Step*, *Block*, and *Model*. They test different hypotheses when you are sequentially adding or removing variables from a model.

- **Model chi-square** is the difference between $-2LL$ for the model with only a constant and $-2LL$ for the current model. (If a constant is not included in the model, the likelihood for the model without any variables is used for comparison.) The model chi-square tests the null hypothesis that all coefficients except the constant are 0. This is comparable to the overall F test for regression.
- **Block chi-square** is the change in $-2LL$ between successive entry blocks during model building. If you enter sets of independent variables sequentially into the model using the Enter specification in the Logistic Regression dialog box, this will show you the differences between successive blocks, where a block is defined as all of the terms that enter a model together. In this example, you entered the variables in a single block, so the block chi-square is the same as the model chi-square. If you enter variables in more than one block, these chi-square values will be different.

- **Step chi-square** is the change in $-2LL$ between successive steps of building a model when one of the variable selection algorithms is used. It tests the null hypothesis that the coefficients for the variables added at the last step are 0. The step chi-square test is comparable to the F-change test in stepwise multiple regression.

In this example, you considered only two models: the constant-only model and the model with a constant and five independent variables. That's why the model chi-square, the block chi-square, and the step chi-square values are all the same. If you sequentially consider more than just these two models, using either forward or backward variable selection, the block chi-square and step chi-square will differ.

Summary Measures

In multiple regression, R^2 is an intuitive measure of how well the model predicts the values of the dependent variable. It's the proportion of the variance in the dependent variable explained by the independent variables. Unfortunately, there is no such easily interpretable measure for logistic regression. Two measures that attempt to quantify the proportion of explained variation in the logistic regression model are Cox and Snell R^2 and Nagelkerke R^2.

The Cox and Snell R^2 and the Nagelkerke R^2 statistics are shown in Figure 15-4. They are similar in intent to R^2 in a linear regression model, although the variation in a logistic regression model must be defined differently.

Cox and Snell R^2

The Cox and Snell R^2 is

$$R^2 = 1 - \left[\frac{L(0)}{L(B)}\right]^{2/N}$$

where $L(0)$ is the likelihood for the model with only a constant, $L(B)$ is the likelihood for the model under consideration, and N is the sample size. The problem with this measure for logistic regression is that it cannot achieve a maximum value of 1.

Nagelkerke R^2

Nagelkerke (1991) proposed a modification of the Cox and Snell R^2 so that the value of 1 could be achieved. The Nagelkerke R^2 is

$$R^2 = \frac{R^2}{R^2_{MAX}}$$

where $R^2_{MAX} = 1 - [L(0)]^{2/N}$.

From the Nagelkerke R^2, you can see that about 47% of the variation in the outcome variable is explained by the logistic regression model.

Tip: The values of logistic summary measures are typically much smaller than what you see for a linear regression model. They are difficult to interpret and should be used carefully. If you report them in a paper, they may elicit negative comments from unsophisticated reviewers who expect values that are comparable in magnitude to those obtained in linear regression.

Tests for a Coefficient

In multiple linear regression, you can test the null hypothesis that the population value for a regression coefficient is 0 by using several equivalent statistics (change in R^2, for example, or the ratio of the estimated coefficient to its standard error). All of the statistics necessarily lead you to the same conclusion. That's not the case in logistic regression. SPSS Statistics provides three such tests: the Wald statistic, the likelihood-ratio test, and the score statistic. If the population value of the coefficient is close to 0 and if the sample size is large, the different tests usually agree. If that's not the case, you may get different results; however, the general consensus is that the best test is the likelihood-ratio test. The other two tests are used because they are computationally less intensive, a concern only for large datasets or for very large numbers of variables.

Wald Statistic

For large sample sizes, the test that a particular coefficient is 0 can be based on the Wald statistic, which has a chi-square distribution when squared. When a variable has a single degree of freedom, the Wald statistic is the ratio of the coefficient to its standard error. For categorical variables with more than one degree of freedom, you can calculate an overall Wald statistic with degrees of freedom equal to one less than the number of categories.

For example, the coefficient for *age* is -0.0693, and its standard error is 0.058. (The standard errors for the logistic regression coefficients are shown in the column labeled *S.E.* in Figure 15-2.) The square of the Wald statistic is $(-0.0693/0.058)$ or about 1.43. The significance level for the Wald statistic based on the chi-square distribution is shown in the column labeled *Sig*. In this example, only the coefficients for *xray* and *stage* appear to be significantly different from 0, using a significance level of 0.05.

Warning: The Wald statistic has a very undesirable property. When the absolute value of the regression coefficient becomes large, the estimated standard error is too large. This produces a Wald statistic that is too small, leading you to fail to reject the null hypothesis that the coefficient is 0, when in fact you should. Therefore, whenever you have a large coefficient, you should not rely on the Wald statistic for hypothesis testing. Instead, you should build a model with and without that variable and base your hypothesis test on the change in the log-likelihood (Hauck and Donner, 1977). It is a good idea to examine the change in the log-likelihood when each of the variables is entered into an equation containing the other variables. You can do this by selecting Backward:LR as the method for variable selection and specifying the probability level for removal as 1.

Likelihood-Ratio Test

The best test of the null hypothesis that the coefficient for a variable is 0 is based on the likelihood-ratio test. The full model is the model with all of the variables; the reduced model is the full model without the variable of interest. For example, consider Figure 15-6, which contains, in the *Chi-square* column, the likelihood-ratio test for the null hypothesis that the coefficient for *stage* is 0. In this example, the chi-square value of 4.43 is not too far different from the value of 4.084 shown in Figure 15-2 for the square of the Wald statistic.

Both the step and block statistics are tests of the null hypothesis that the coefficient for *stage* is 0. The model chi-square is a test of the null hypothesis that for the current model, all variables, including *stage*, have coefficients of 0. Based on the small observed significance level, you can easily reject that null hypothesis.

Figure 15-6
Omnibus test of model coefficients

		Chi-square	df	Sig.
Step 1	Step	4.432	1	.035
	Block	4.432	1	.035
	Model	22.126	5	.000

Score Statistic

Rao's efficient score statistic provides a computationally less-intensive test for the null hypothesis that a coefficient is 0 than the previous two statistics. It doesn't require maximum-likelihood estimation of the coefficients, so it is useful in situations where recalculating parameter estimates for many different models would be computationally prohibitive. It's always used by SPSS Statistics for evaluating variables for entry into the model when one of the variable selection algorithms is selected. The score statistic is used for the overall test of the null hypothesis that all coefficients are 0, which is displayed at the beginning of all logistic regression output, as shown in Figure 15-13 on p. 341. The value in the row labeled *Overall Statistics* is not identical to the corresponding log-likelihood statistic. The likelihood-ratio statistic, the Wald statistic, and Rao's efficient score statistic are all equivalent in large samples when the null hypothesis is true (Rao, 1973).

Categorical Variables

In logistic regression, just as in linear regression, the codes for the independent variables must be meaningful. You cannot take a nominal variable, such as religion, assign arbitrary codes from 1 to 5, and then use the resulting variable in the model. You must recode the values of the independent variable by creating a new set of variables that correspond in some way to the original categories.

If you have a two-category variable, such as sex, you can code each case as 0 or 1 to indicate either female or not female (or you could code it as being male or not male). This is called **dummy-variable** or **indicator-variable coding**. *Grade*, *stage*, and *xray* are all examples of two-category variables that have been coded as 0 and 1. The code of 1 indicates that the poorer outcome is present. The interpretation of the resulting coefficients for *grade*, *stage*, and *xray* is straightforward. It tells you the difference between the log odds when a case is a member of the *poor* category and when it is not.

Warning: If you have an indicator variable that is coded 0 and 1, don't use the Define Categorical Variables dialog box to declare it as a categorical variable. By default, the reference (omitted) category is the second category. To preserve your 0 and 1 coding, you must select the alternative reference category, First, in the Define Categorical Variables dialog box. You don't gain anything by declaring an indicator variable as categorical in the Logistic Regression procedure unless you want to change its coding scheme from indicator to something else.

When you have a variable with more than two categories, you must create new variables to represent the categories. The number of new variables required to represent a categorical variable is one less than the number of categories. For example, if instead of the actual values for serum acid phosphatase, you have values of 1, 2, or 3 (depending on whether the value was low, medium, or high), you have to create two new variables to represent the serum phosphatase levels.

Indicator-Variable Coding Scheme

One of the ways you can create two new variables for serum acid phosphatase is to use indicator variables to represent the categories. With this method, one variable would represent the low value, coded 1 if the value is low and 0 otherwise. The second variable would represent the medium value, coded 1 if the value is average and 0 otherwise. The value *high* would be represented by codes of 0 for both of these variables. The choice of the category to be coded as 0 for both variables is arbitrary.

With categorical variables, the only statement that you can make about the effect of a particular category is in comparison to some other category. For example, if you have a variable that represents type of cancer, you can only make statements such as "lung cancer compared to bladder cancer decreases your chance of survival." Or you might say that "lung cancer, compared to all of the cancer types in the study, decreases your chance of survival." You can't make a statement about lung cancer without relating it to the other types of cancer.

If you use indicator variables for coding, the coefficients for the new variables represent the effect of each category compared to a reference category. The coefficient for the reference category is 0. As an example, consider Figure 15-7. The variable *catacid1* is the indicator variable for low serum acid phosphatase, coded 1 for low levels and 0 otherwise. Similarly, the variable *catacid2* is the indicator variable for medium serum acid phosphatase. The reference category is high levels.

Figure 15-7
Indicator variables

		B	S.E.	Wald	df	Sig.	Exp(B)	95.0% C.I.for EXP(B)	
								Lower	Upper
Step 1[1]	age	-.052	.063	.686	1	.407	.949	.839	1.074
	catacid1	-2.008	1.052	3.643	1	.056	.134	.017	1.055
	catacid2	-1.092	.926	1.391	1	.238	.335	.055	2.061
	xray	2.035	.837	5.904	1	.015	7.651	1.482	39.500
	grade	.808	.823	.962	1	.327	2.243	.447	11.262
	stage	1.457	.768	3.597	1	.058	4.294	.952	19.356
	Constant	1.770	3.809	.216	1	.642	5.870		

1. Variable(s) entered on step 1: age, catacid1, catacid2, xray, grade, stage.

With this coding scheme, the coefficient for *catacid1* is the change in log odds when you have a low value compared to a high value. Similarly, *catacid2* is the change in log odds when you have a medium value compared to a high value. The coefficient for the high value is necessarily 0 because it does not differ from itself. In Figure 15-7, you see that the coefficients for both of the indicator variables are negative. This means that compared to high values for serum acid phosphatase, low and medium values are associated with decreased log odds of malignant nodes. The low category decreases the log odds more than the medium category.

The SPSS Statistics Logistic Regression procedure will automatically create new variables for variables declared as categorical. You can choose the coding scheme that you want to use for the new variables.

Warning: You must click the Change button in the Define Categorical Variables dialog box in which you define categorical variables to actually change the type of coding used; otherwise, the change is ignored. A variable that has been declared categorical in the SPSS Statistics procedure has a number in parentheses after it.

Figure 15-8 shows the table that is displayed for each categorical variable. The rows of the table correspond to the categories of the variable. The number of cases with each value is displayed in the column labeled *Frequency*. Subsequent columns correspond to new variables created by the program. The number in parentheses indicates the suffix used to identify the variable in the output. The values of the new variables for each original value are shown under *Parameter coding*.

Figure 15-8
Indicator-variable coding scheme

		Frequency	Parameter coding (1)	(2)
catacid	1	15	1.000	.000
	2	20	.000	1.000
	3	18	.000	.000

From Figure 15-8, you see that there are 20 cases with a value of 2 for *catacid*. Each of these cases will be assigned a code of 0 for the new variable *catacid(1)* and a code of 1 for the new variable *catacid(2)*. Similarly, cases with a value of 3 for *catacid* will be given the code of 0 for both *catacid(1)* and *catacid(2)*.

Tip: Always include all parts of a categorical variable in a model; otherwise, the interpretation of everything changes.

Deviation Coding

The statements you can make based on the logistic regression coefficients depend on how you have created the new variables used to represent the categorical variable. As shown in the previous section, when you use indicator variables for coding, the coefficients for the new variables represent the effect of each category compared to a reference category. If, on the other hand, you want to compare the effect of each category to the average effect of all of the categories, you can select the deviation coding scheme shown in Figure 15-9. This differs from indicator-variable coding only in that the last category is coded as -1 for each of the new variables.

With this coding scheme, the logistic regression coefficients tell you how much better or worse each category is compared to the average effect of all categories, as shown in Figure 15-10. For each new variable, the coefficients now represent the difference from the average effect over all categories. The value of the coefficient for the last category is not displayed, but it is no longer 0. Instead, it is the negative of the sum of the displayed coefficients. From Figure 15-10, the coefficient for the *high* level is calculated as $-(-0.9745 - 0.0589) = 1.0334$.

Figure 15-9
Deviation coding scheme

		Frequency	Parameter coding (1)	(2)
catacid	1	15	1.000	.000
	2	20	.000	1.000
	3	18	-1.000	-1.000

Warning: The parameter coding shown in Figure 15-9 tells you how the data values are transformed to obtain the specified type of parameter estimates. That's how you have to code your data if you want to reproduce the predicted values. The parameter coding table shows you the linear combination of categories corresponding to each parameter estimate only if the contrasts that you specify are orthogonal. Because deviation contrasts are not orthogonal, the parameter coding shown in Figure 15-9 does not tell you what levels of a categorical variable are being compared. It tells you how the categorical variable is being transformed so that deviation contrasts for the parameter estimates are obtained.

Figure 15-10
Coefficients with deviation categorical variables

		B	S.E.	Wald	df	Sig.	Exp(B)
Step 1[1]	age	-.052	.063	.686	1	.407	.949
	xray	2.035	.837	5.904	1	.015	7.651
	grade	.808	.823	.962	1	.327	2.243
	stage	1.457	.768	3.597	1	.058	4.294
	catacid			3.836	2	.147	
	catacid(1)	-.975	.641	2.312	1	.128	.377
	catacid(2)	-.059	.573	.011	1	.918	.943
	Constant	.736	3.735	.039	1	.844	2.088

1. Variable(s) entered on step 1: age, xray, grade, stage, catacid.

Different coding schemes result in different logistic regression coefficients but not in different conclusions. That is, even though the actual values of the coefficients differ between Figure 15-7 and Figure 15-10, they tell you the same thing. Figure 15-6 tells you the effect of category 1 compared to category 3, while Figure 15-9 tells you the effect of category 1 compared to the average effect of all of the categories. You can select the coding scheme to match the type of comparisons that you want to make.

Warning: If you change the coding scheme for a two-valued variable from indicator to one of the other schemes, the odds ratio will change as well. The previous interpretation depended on the variable having codes of 0 and 1. For example, if you code *grade* as deviation instead of indicator, the correct odds ratio for *grade* is e^{2B}, where B is the regression coefficient for the deviation coding.

Creating Interaction Terms

For each variable in the logistic model, you must assume that its effect is the same for all values of other variables in the model. For example, you assume that the effect of a laboratory value is the same over the entire range of ages or the same for men and women. If this is not the case, there is an interaction. Just as in linear regression, you can include terms in the model that represent the interactions. For example, if it made sense, you could include a term for the *acid*-by-*age* interaction in your model by including a product of the two variables.

Interaction terms for categorical variables can also be computed. They are created as products of the values of the new variables. For categorical variables, make sure that the interaction terms created are those of interest. If you are using categorical variables with indicator coding, the interaction terms generated as the product of the variables are generally not those that interest you. Consider, for example, the interaction term between two indicator variables. If you just multiply the variables together, you will obtain a value of 1 only if both of the variables are coded "present." What you would probably like is a code of 1 if both of the variables are present or both are absent.

Tip: If you have a large number of independent variables, examining all pairs of interactions is a laborious procedure and not feasible for a small dataset. Start by identifying variables that are known to be related based on prior research.

Evaluating Linearity

In the logistic regression model, the relationship between the logit and the values of an independent variable must be linear. When you make the assumption of a linear relationship between an independent variable and a dependent variable in linear regression, you can use partial regression plots and residual plots to examine the assumption. The assumption is not as easy to check in logistic regression. You must take additional steps. Because the prostate cancer dataset is small, it's hard to properly check assumptions.

As an example of how to check the assumptions, consider a dataset in which the probability of death for patients diagnosed with sepsis, a blood infection, is predicted using serum albumin levels. Serum albumin measures the presence of protein in the blood. The normal range is between 3.2 and 5.5 mg/dl. Figure 15-11 shows the logistic regression coefficient when serum albumin is used to predict death. Based on the small observed significance level, you reject the null hypothesis that there is no relationship between death and the albumin levels. The question as to whether the relationship between the two variables is really linear remains to be answered.

Figure 15-11
Coefficient for albumin

		B	S.E.	Wald	df	Sig.	Exp(B)
Step 1[1]	albcont	-.602	.117	26.306	1	.000	.548
	Constant	.469	.397	1.398	1	.237	1.598

1. Variable(s) entered on step 1: albcont.

There are many sophisticated ways to approach the problem. You should consult Hosmer and Lemeshow (2000) or (more advanced) Harrell (2001) for discussion of these. As a simple first step, you can create a categorical variable that corresponds to the values of albumin. That's easiest to do in the Visual Bander. Select intervals of equal width so that the expected relationship between the coefficients of the categorical variable and the categories is linear. Then run the logistic regression with the groupings of the continuous variable as a categorical variable with indicator coding. Figure 15-12 shows the results when the highest albumin category is the reference category.

Tip: If you have a variable that has two types of values (0 for people who do not engage in a particular behavior [smoking, exercise] and then the actual number of cigarettes or hours of exercise per month), Robertson et al. (1994) recommend that you include two variables: an indicator variable that has a value of 0 if the amount is 0 and 1 otherwise, and another variable with the actual values.

Figure 15-12
Coefficients for categories of albumin

		B	S.E.	Wald	df	Sig.	Exp(B)
Step 1[1]	albcategories			41.782	4	.000	
	albcategories(1)	1.526	.478	10.208	1	.001	4.600
	albcategories(2)	.567	.481	1.387	1	.239	1.762
	albcategories(3)	-.120	.485	.061	1	.805	.887
	albcategories(4)	.248	.452	.302	1	.582	1.282
	Constant	-1.969	.436	20.423	1	.000	.140

1. Variable(s) entered on step 1: albcategories.

If the relationship between the logit and the values of albumin is linear, the coefficients of the categorical variable should increase or decrease more or less linearly. You see that this is not the case here. Only the first category, the markedly abnormal value for albumin, is significantly different from the last category. If you rerun the analysis using categories 2–5 as the reference category (you have to recode the variable to do that), you find that only the lowest albumin level is significantly different from the rest. This is not an unusual finding. With laboratory values, the relationship between the outcome and the actual value is often not linear. Using three categories of too low, normal, and too high may be useful. The important lesson is that linearity must be examined, and it involves much more than simply plotting residuals.

Tip: You can plot the values of the coefficients by selecting them in the pivot table and then right-clicking your mouse. From the pop-up menu choose Create Graph > Line. This will give you a quick-and-easy plot. Remember that the omitted category has a value of 0 for the coefficient, so you have to mentally place it in the plot.

Automated Model Building

In logistic regression, as in other multivariate statistical techniques, you may want to identify subsets of independent variables that are good predictors of the dependent variable. As always, that entails problems. Model-building algorithms capitalize on the characteristics of the sample selected, so you have no assurances that you would arrive at the same model if you take another sample from the same population. The model also fits the sample better than the population from which it is selected.

Warning: All of the problems associated with variable-selection algorithms in regression and discriminant analysis are found in logistic regression as well. None of the algorithms result in a "best" model in any statistical sense. Different algorithms for variable selection may result in different models. It is a good idea to examine several possible models and choose from among them on the basis of interpretability, parsimony, and ease of variable acquisition.

The SPSS Statistics Logistic Regression procedure has several methods available for model selection. You can enter variables in whatever sequence you want, using multiple enter specifications. (This is sometimes known as sequential logistic regression.) You can also use forward-stepwise selection and backward-stepwise elimination for automated model building. Both forward- and backward-stepwise selection offer you three choices for the criterion to be used for removing variables. For large datasets, maximum-likelihood computations of parameter estimates are intensive, so SPSS Statistics also offers two quicker-to-compute statistics.

Tip: Always use the likelihood ratio as the criterion for variable removal unless it takes too long to compute for your problem. If that's the case, use the conditional method.

Variable Entry and Removal Criteria

SPSS Statistics has a single criterion for variable entry and three criteria for variable removal. The method that is listed in the Method drop-down list specifies how variables are evaluated for removal.

- The score statistic is always used for evaluating variables for entry into the model. You will always see the score statistic for all variables not in the model, regardless of the method you choose for variable removal.
- Variable removal can be based on the Wald statistic, the change in log-likelihood, or a conditional statistic. The conditional statistic is similar to the likelihood-ratio test but uses conditional parameter estimates that don't require reestimation of the model with each variable removed (Lawless and Singhal, 1978).
- All variables that are used to represent the same categorical variable are entered or removed from the model together if they are declared in the Define Categorical Variables dialog box.

Tip: Hosmer and Lemeshow (2000) suggest the following steps for variable selection: A thorough univariate examination of each of the independent variables is conducted; variables with observed significance levels less than 0.25 or those known to be relevant are considered eligible for inclusion into a variable-selection algorithm; at the end of variable selection, each variable is evaluated based on changes in the value of its coefficient when it is in the model alone and when it is included with other variables; all variables not included in the model are entered and changes in coefficients examined. Interaction terms and the appropriateness of linearity of the logits are also evaluated.

Forward-Stepwise Selection

Forward-stepwise variable selection in logistic regression proceeds the same way as in multiple linear regression. You start out with a model that contains only the constant unless the option to omit the constant term from the model is selected. At each step, the variable with the smallest observed significance level for the score statistic, provided that it is less than the chosen cutoff value (by default 0.05), is entered into the model. All variables in the model are then examined to see if they meet the selected removal criterion. For example, if the Wald statistic is used for deleting variables, the Wald statistics for all variables in the model are examined, and the variable with the largest significance level for the Wald statistic, provided that it exceeds the chosen cutoff value (by default 0.1), is removed from the model. If no variables meet removal criteria, the next eligible variable is entered into the model, based on its score statistic.

If a variable is selected for removal and it results in a model that has already been considered, variable selection stops; otherwise, the model is estimated without the deleted variable and the variables are again examined for removal. This continues until no more variables are eligible for removal. Then variables are again examined for entry into the model. The process continues until either a previously considered model is encountered (which means that the algorithm is cycling) or no variables meet entry or removal criteria.

Tip: If you know from experience or prior research findings that certain variables are important for predicting the outcome, you should include them in the model without worrying about their observed significance level.

An Example of Forward Selection

To see what the output looks like for forward selection, consider Figure 15-13, which contains statistics for variables not in the equation when only the constant is included in the model.

Figure 15-13
Variables not in the equation at step 0

			Score	df	Sig.
Step 0	Variables	acid	3.117	1	.077
		age	1.094	1	.296
		xray	11.283	1	.001
		grade	4.075	1	.044
		stage	7.438	1	.006
	Overall Statistics		19.451	5	.002

The last row of the table, labeled *Overall Statistics*, tests the null hypothesis that the coefficients for all variables not in the model are 0. (This statistic is calculated from the score statistics, so it is not exactly the same value as the model chi-square value that you see in Figure 15-5. For large sample sizes, the two statistics are usually similar in value.) If the observed significance level for the overall statistic is large (that is, you don't have reason to reject the hypothesis that all of the coefficients are 0), your success in building a model may be limited. If you continue to build a model, there is a reasonable chance that it will not be useful for other samples from the same population. In this example, you can reject the null hypothesis that all coefficients are 0, so it makes sense to proceed with variable selection.

Tip: Always be on the lookout for coefficients with large standard errors. That's often a sign that there's something wrong.

Following the Steps

As the first example, consider model building when the Wald statistic is used for evaluating variables for removal and when the default criteria for entry ($p = 0.05$) and removal ($p = 0.1$) are used.

Step 1. From Figure 15-13 (the score statistic for each variable if it is entered in the model that contains only the constant), you see that the largest score statistic, or equivalently, the smallest observed significance level is for the *xray* variable. It is less than the default value of 0.05, so *xray* enters the model. This is the model at step 1.

Step 2. All of the variables not in the model at step 1 are evaluated for entry. For each variable not in the model at step 1, the score statistic and its significance level, if the variable were to enter next, is shown in Figure 15-14. You see that the smallest observed significance level is for *stage*. Because it is less than 0.05, *stage* enters the model. At this point, variables already in the model are evaluated for removal, based on the values of the Wald statistic.

Figure 15-14
Variables not in the equation at step 1

			Score	df	Sig.
Step 1	Variables	acid	2.073	1	.150
		age	1.352	1	.245
		grade	2.371	1	.124
		stage	5.639	1	.018
	Overall Statistics		10.360	4	.035

Step 3. All variables not in the model at step 3 are evaluated for entry. From Figure 15-15 (statistics for variables not in the model after step 2), you see that the smallest observed significance level is for *acid*; however, because it is larger than the default value of 0.05, variable selection stops.

Figure 15-15
Variables not in the equation at step 2

			Score	df	Sig.
Step 2	Variables	acid	3.092	1	.079
		age	1.268	1	.260
		grade	.584	1	.445
	Overall Statistics		5.422	3	.143

Model summary statistics at each step are shown in Figure 15-16. (To obtain –2 log-likelihood for the model with only a constant, you must select the iteration history option. For this example, it is 70.252 from Figure 15-3.) In this example, at each step, $-2LL$ decreases.

Figure 15-16
Model summary statistics

		-2 Log likelihood	Cox & Snell R Square	Nagelkerke R Square
Step	1	59.001[1]	.191	.260
	2	53.353[2]	.273	.372

1. Estimation terminated at iteration number 4 because parameter estimates changed by less than .001.
2. Estimation terminated at iteration number 5 because parameter estimates changed by less than .001.

You can use the $-2LL$ values shown in Figure 15-16 to calculate the changes in $-2LL$ values shown in Figure 15-17. The model chi-square tests that all coefficients except the constant are 0. The step chi-square tests that the coefficient for the last variable entered or removed is 0. For example, at step 1, the chi-square value of 11.25 is the difference in $-2LL$ between the constant-only model (70.25) and the model with the constant and *xray* (59.00). At step 2, the model chi-square is the difference in $-2LL$ between the constant-only model and the model that contains the constant, *xray*, and *stage* $(70.252 - 53.353 = 16.899)$. The step chi-square value is the change in $-2LL$ when *stage* is added to a model containing the constant and *xray* $(59.001 - 53.353 = 5.648)$.

Figure 15-17
Omnibus tests of model coefficients

		Chi-square	df	Sig.
Step 1	Step	11.251	1	.001
	Block	11.251	1	.001
	Model	11.251	1	.001
Step 2	Step	5.647	1	.017
	Block	16.899	2	.000
	Model	16.899	2	.000

Tip: If you have interaction terms and want to specify constraints on how terms enter the model, you should use the Multinomial Regression procedure.

Forward Selection with the Likelihood-Ratio Criterion

If you select the likelihood-ratio statistic for deleting variables, the output will look slightly different from that previously described. For variables in the equation at a particular step, output similar to that shown in Figure 15-18 is produced, in addition to the usual coefficients and Wald statistics.

Figure 15-18
Removal statistics

			Model Log Likelihood	Change in -2 Log Likelihood	df	Sig. of the Change
Variable	Step 1	xray	-35.126	11.251	1	.001
	Step 2	xray	-31.276	9.199	1	.002
		stage	-29.500	5.647	1	.017

For each variable in the model, Figure 15-18 contains the log-likelihood for the model, the change in $-2LL$ if the variable is removed, and the observed significance level for the change. If the observed significance level is greater than the cutoff value for remaining in the model, the term is removed from the model and the model statistics are recalculated to see if any other variables are eligible for removal.

Backward Elimination

Forward selection starts without any variables in the model. Backward elimination starts with all of the variables in the model. Then, at each step, variables are evaluated for entry and removal. The score statistic is always used for determining whether variables should be added to the model. Just as in forward selection, the Wald statistic, the likelihood-ratio statistic, or the conditional statistic can be used to select variables for removal.

Warning: Unless the sample size is large, forcing all possible predictors into a model can result in a poorly estimated initial model. The Wald statistic will also be unstable.

Examining the Model

Whenever you build a statistical model, it is important to examine how well the model fits the data. You want to know how well it predicts the dependent variable, whether there are reasons to believe that the requisite assumptions are violated, and whether there are cases that are poorly predicted by the model or are exerting more than their fair share of influence on the estimation of the model. In linear regression, you look at a variety of residuals, measures of influence, and indicators of collinearity.

You also want to know how well the model fits not only the sample of data from which it is derived but also the population from which the sample data were selected. A model almost always fits the sample you used to estimate it better than it will fit the population. For large datasets, it may be feasible to split the data into two parts. You can estimate a model on one part and then apply the model to the other to see how well it fits. There are also other statistical techniques with picturesque names such as *jackknifing* and *bootstrapping* that are useful for assessing how well the model would fit another set of data.

Logistic Regression Diagnostics

In logistic regression, you are not burdened with a large number of assumptions. The main issues you want to address, besides independence of the observations, are:

- Is the relationship between the logit and the variable linear? (See "Evaluating Linearity" on p. 337.)
- How well does the model discriminate between cases that experience the event and cases that don't? (See "Model Discrimination" on p. 346.)
- How well do the predicted probabilities match the observed probabilities over the entire range of values? (See "Model Calibration" on p. 350.)
- Are there unusual cases?

In logistic regression, the evaluation of diagnostics is more complicated than in linear regression because you must compare the predicted probability to the observed probability for all cases with the same values of the independent variables. That's because the model is predicting the proportion of events for a group of similar cases. You can't tell how close the predicted proportion is to the observed proportion from a single case. Diagnostics must be evaluated for each covariate pattern (combination of values of the independent variables). Unless there are adequate numbers of cases with the same covariate patterns, the diagnostic statistics may be difficult to interpret. (See Hosmer and Lemeshow, 2000, for further discussion.) Also, the interpretation of some of the diagnostic statistics depends on the values of the estimated probabilities.

Model Discrimination

Model discrimination evaluates the ability of the model to distinguish between the two groups of cases, based on the estimated probability of the event occurring. That is, you want to know how well the predicted probability of the event separates the cases for whom the event actually occurs and those for whom it does not.

Area Under the ROC Curve: The C Statistic

A perfect model always assigns higher probabilities to cases with the outcome of interest than to cases without the outcome of interest. In other words, the two sets of probabilities do not overlap.

A frequently used measure of the ability of a model to discriminate between the two groups of cases is the *c* statistic. The ***c*** **statistic** can be interpreted as the proportion of pairs of cases with different observed outcomes in which the model results in a higher probability for the cases with the event than for the case without the event. (The *c* statistic is equal to the area under the ROC curve. For further information, see Hanley and McNeil, 1982.) The *c* statistic ranges in value from 0.5 to 1. A value of 0.5 means that the model is no better than flipping a coin for assigning cases to groups. A value of 1 means that the model always assigns higher probabilities to cases with the event than to cases without the event. The *c* statistic can be calculated from Somers' *d* and it is also closely related to the Mann-Whitney *U* statistic.

Tip: To calculate the *c* statistic in SPSS Statistics, you must save the predicted probabilities from the Logistic Regression procedure. Then you can use the ROC facility to calculate the area under the ROC curve.

For the logistic regression model with all of the independent variables, the *c* statistic is 0.845, as shown in Figure 15-19. This means that in almost 85% of all possible pairs of cases in which one case has positive nodes and the other does not, the logistic regression model assigns a higher probability of having positive nodes to the case with positive nodes.

Warning: A model that discriminates well can still be a poor model in terms of how closely the observed and predicted probabilities match. The *c* statistic looks only at whether the probabilities for one group are larger than the predicted probabilities for the other group.

Figure 15-19
Area under the ROC curve

Test Result Variable(s): pre_1 Predicted Value

Area	Std. Error[1]	Asymptotic Sig.[2]	Asymptotic 95% Confidence Interval	
			Lower Bound	Upper Bound
.845	.054	.000	.740	.951

1. Under the nonparametric assumption
2. Null hypothesis: true area = 0.5

Classification Table

One way to assess how well a model fits is to use the predicted probabilities to assign cases to groups. Figure 15-20 shows the observed and predicted group memberships when cases with a predicted probability of 0.5 or greater are classified as having positive nodes. By default, if the estimated probability of the event is less than 0.5, you predict that the event will not occur. If the probability is greater than 0.5, you predict that the event will occur. (In the unlikely event that the probability is exactly 0.5, you can flip a coin for your prediction.)

In practice, the probability cutoff for predicting that nodes are positive depends on the consequences of failing to correctly identify patients with positive nodes as compared to the consequences of falsely identifying patients as having positive nodes.

Figure 15-20
Classification table

Classification Table[1]

				Predicted		
				nodes		Percentage Correct
				Neg	Pos	
Step 1	Observed	nodes	Neg	28	5	84.8
			Pos	7	13	65.0
		Overall Percentage				77.4

1. The cut value is .500

From Figure 15-20, you see that 28 patients without malignant nodes were correctly predicted by the model not to have malignant nodes. Similarly, 13 men with positive nodes were correctly predicted to have positive nodes. The off-diagonal entries of the table tell you how many men were incorrectly classified. A total of 12 men were misclassified in this example—5 men with negative nodes and 7 men with positive nodes. Of the men without diseased nodes, 84.8% were correctly classified. Of the men with diseased nodes, 65% were correctly classified. Overall, 77.4% of the 53 men were correctly classified.

Tip: Sensitivity is the proportion of cases with the event who are correctly predicted to have the event (13/20). **Specificity** is the proportion of cases without the event who are correctly predicted to not have the event (28/33). The **false positive rate** is the proportion of cases without the event who are predicted to have the event (5/33). The **false negative rate** is the proportion of cases with the event who are predicted not to have the event (7/20).

The classification table doesn't reveal the distribution of estimated probabilities for men in the two groups. For each predicted group, the table shows only whether the estimated probability is greater than or less than one-half. For example, you cannot tell from the table whether the seven patients who had false negative results had predicted probabilities near 50% or much lower predicted probabilities. Ideally, you would like the two groups to have very different estimated probabilities. That is, you would like to see small estimated probabilities of positive nodes for all men without malignant nodes and large estimated probabilities for all men with malignant nodes.

Warning: The percentage of cases correctly classified is a poor indicator of model fit, since it does not necessarily depend on how well a model fits. It ignores the actual probability values, replacing them with a cutoff value. It's also possible to add a highly significant variable to the model and have the correct classification rate decrease. You shouldn't use it to compare models from different samples because it depends heavily on the types of cases included in the sample.

Histogram of Estimated Probabilities

Figure 15-21 is a histogram of the estimated probabilities of cancerous nodes. The symbol used for each case designates the group to which the case actually belongs. If you have a model that successfully distinguishes the two groups, the cases for which the event has occurred should be to the right of 0.5, while the cases for which the event has not occurred should be to the left of 0.5. The more the two groups cluster at their respective ends of the plot, the better.

Figure 15-21
Histogram of estimated probabilities

```
         4 ⇳  NN                                                        ⇳
           ⇔  NN                                                        ⇔
           ⇔  NN                                                        ⇔
F          ⇔  NN                                                        ⇔
R        3 ⇳  NN    P          P                                P       ⇳
E          ⇔  NN    P          P                                P       ⇔
Q          ⇔  NN    P          P                                P       ⇔
U          ⇔  NN    P          P                                P       ⇔
E        2 ⇳  NNNN  N   PP     N       P   PP                   P   P   ⇳
N          ⇔  NNNN  N   PP     N       P   PP                   P   P   ⇔
C          ⇔  NNNN  N   PP     N       P   PP                   P   P   ⇔
Y          ⇔  NNNN  N   PP     N       P   PP                   P   P   ⇔
         1 ⇳  NNNNNNNPNNNN NNPNN  N    NN NNN N      P P  P P P NP PP   ⇳
           ⇔  NNNNNNNPNNNN NNPNN  N    NN NNN N      P P  P P P NP PP   ⇔
           ⇔  NNNNNNNPNNNN NNPNN  N    NN NNN N      P P  P P P NP PP   ⇔
           ⇔  NNNNNNNPNNNN NNPNN  N    NN NNN N      P P  P P P NP PP   ⇔
Predicted   ⇩⇩⇩⇩⇩⇩⇩⇩⇩⇩⇩⇩⇩⇩⇩⇳⇩⇩⇩⇩⇩⇩⇩⇩⇩⇩⇩⇩⇩⇩⇳⇩⇩⇩⇩⇩⇩⇩⇩⇩⇩⇩⇩⇩⇩⇳⇩⇩⇩⇩⇩⇩⇩⇩⇩⇩⇩⇩⇩⇩
  Prob:     0             .25            .5             .75            1
  Group:    NNNNNNNNNNNNNNNNNNNNNNNNNNNNNNPPPPPPPPPPPPPPPPPPPPPPPPPPPPPP

              Predicted Probability is of Membership for Pos
              The Cut Value is .50
              Symbols: N - Neg
                       P - Pos
              Each Symbol Represents .25 Cases.
```

From Figure 15-21, you see that there is only one noncancerous case with a high estimated probability of having positive nodes (the case identified with the letter *N* at a probability value of about 0.88). However, there are four diseased cases with estimated probabilities less than 0.25.

By looking at this histogram of predicted probabilities, you can see whether a different rule for assigning cases to groups might be useful. For example, if most of the misclassifications occur in the region around 0.5, you might decide to withhold judgment for cases with values in this region. In this example, it means that you would predict nodal involvement only for cases where you were reasonably sure that the logistic prediction would be correct. You might decide to operate on all questionable cases.

If the consequences of misclassification are not the same in both directions (for example, calling nodes negative when they are really positive is worse than calling nodes positive when they are really negative), the classification rule can be altered to decrease the possibility of making the more severe error. For example, you might decide to call cases "negative" only if their estimated probability is less than 0.3. By looking at the histogram of the estimated probabilities, you can get some idea of how different classification rules might perform. (Of course, when you apply the model to new cases, you can't expect the classification rule to behave exactly the same.)

Model Calibration

Model calibration evaluates how well the observed and predicted probabilities agree over the entire range of probability values. A commonly used test for the goodness of fit of the observed and predicted number of events is the Hosmer-Lemeshow test (Hosmer and Lemeshow, 2000). You divide the cases into 10 approximately equal groups based on the estimated probability of the event occurring (deciles of risk) and see how the observed and expected numbers of events and non-events compare. The chi-square test is used to assess the difference between the observed and expected numbers of events. To use this technique sensibly, you must have a fairly large sample size so that the expected number of events in most groups exceeds five and none of the groups have expected values less than 1. Because the prostate dataset is too small for the test to be useful, we'll consider survival data from 1,085 patients hospitalized for septicemia, a life-threatening infection of the blood stream.*

Figure 15-22 is the table upon which the Hosmer-Lemeshow goodness-of-fit test is based for a logistic regression model that predicts death from septicemia. The cases are divided into 10 approximately equal groups, based on the values for the predicted probability of death. The number of cases in each group is shown in the *Total* column. The groups are not exactly equal in size because cases with the same combination of values for the independent variables are kept in the same group. For each group, the observed and predicted number of deaths and the observed and predicted number of survivors are shown. For example, in the first group of 122 cases, 3 died and 119 survived. Summing the predicted probabilities of death for these 122 cases, the predicted number of deaths is 2.38 and the predicted number of survivors is 119.62.

Warning: You cannot compute the Hosmer-Lemeshow test if you have a small number of distinct predicted probabilities of an event. For example, if you have a small number of categorical variables with a limited number of possible values, you won't be able to create enough groups of cases with different predicted probabilities of the event.

*Thanks to Michael Pine of Michael Pine and Associates for allowing use of these data.

Figure 15-22
Hosmer-Lemeshow table

		Alive		Dead		Total
		Observed	Expected	Observed	Expected	
Step 1	1	119	119.618	3	2.382	122
	2	106	106.314	4	3.686	110
	3	100	103.468	9	5.532	109
	4	101	100.667	8	8.333	109
	5	99	98.215	10	10.785	109
	6	91	89.946	13	14.054	104
	7	92	90.008	17	18.992	109
	8	81	82.663	28	26.337	109
	9	77	70.356	32	38.644	109
	10	32	36.745	63	58.255	95

To calculate the Hosmer-Lemeshow goodness-of-fit chi-square, you compute the difference between the observed and predicted number of cases in each of the cells. You then calculate $(O - E)^2/E$ for each of the cells in the table. The chi-square value is the sum of this quantity over all of the cells. For this example, the chi-square value is 5.81 with 8 degrees of freedom, as shown in Figure 15-23. (The degrees of freedom is calculated as the number of groups minus 2.) The observed significance level for the chi-square value is 0.67, so you do not reject the null hypothesis that there is no difference between the observed and predicted values. The model appears to fit the data reasonably well.

Figure 15-23
Hosmer-Lemeshow chi-square

		Chi-square	df	Sig.
Step	1	5.811	8	.668

The value for the Hosmer-Lemeshow statistic depends on how the cases are grouped. (That's why different software packages may give different values for the same dataset.) If there is a small number of groups, the test will usually indicate that the model fits, even if it does not. If you have a very large number of cases, the value of the Hosmer-Lemeshow statistic can be large, even if the model fits well, since the value of a chi-square statistic is proportional to sample size. In summary, the Hosmer-Lemeshow statistic provides useful information about the calibration of the model, but it must be interpreted with care.

Warning: Hosmer et al. (1997) review additional tests that can be used for comparing the observed and predicted probabilities. They conclude that none of the overall goodness-of-fit tests is especially powerful for samples with fewer than 400 cases.

Diagnostic Statistics for Individual Cases

Just as in linear regression, you can calculate and save a large number of potentially useful diagnostic statistics calculated for each case. Note that SPSS Statistics calculates these statistics for individual cases, not pooled by covariate patterns.

- The **residual** is the difference between the observed probability of the event and the predicted probability of the event, based on the model. For example, if we predict the probability of malignant nodes to be 0.80 for a man who has malignant nodes, the residual is $1 - 0.80 = 0.20$.
- The **standardized residual** is the residual divided by an estimate of its standard deviation. In this case, it is

  $$Z_i = \frac{\text{residual}_i}{\sqrt{P_i(1-P_i)}}$$

  For each case, the standardized residual can also be considered a component of the chi-square goodness-of-fit statistic. If the sample size is large, the standardized residuals should be approximately normally distributed, with a mean of 0 and a standard deviation of 1.
- The **deviance** is computed from

  $$-2 \times \log(\text{predicted probability for the observed group})$$

  The deviance is calculated by taking the square root of the above statistic and attaching a negative sign if the event did not occur for that case. For example, the deviance for a man without malignant nodes and a predicted probability of 0.8 for nonmalignant nodes is

  $$\text{deviance} = -\sqrt{-2\log(0.8)} = -0.668$$

  Large values for deviance indicate that the model does not fit the case well. For large sample sizes, the deviance is approximately normally distributed.
- The **Studentized residual** for a case is approximately the square root of the change in the model deviance if the case is excluded. Discrepancies between the deviance and the Studentized residual may identify unusual cases. Normal probability plots of the Studentized residuals may be useful.

- The **logit residual** is the residual for the model if it is predicted in the logit scale. That is,

$$\text{logit residual}_i = \frac{\text{residual}_i}{P_i(1-P_i)}$$

- The **leverage** in logistic regression is, in many respects, analogous to the leverage in least-squares regression. Leverage values are often used for detecting observations that have a large impact on the predicted values. Unlike linear regression, the leverage values in logistic regression depend on both the dependent variable scores and the design matrix. Leverage values are bounded by 0 and 1. Their average value is p/n, where p is the number of estimated parameters in the model, including the constant, and n is the sample size. For cases with predicted probabilities less than 0.1 or greater than 0.9, the leverage values may be small, even when the cases are influential.
- **Cook's distance** is a measure of the influence of a case. It tells you how much deleting a case affects not only the residual for that case but also the residuals of the remaining cases. Cook's distance (D) depends on the standardized residual for a case, as well as its leverage. It is defined as

$$D_i = \frac{Z_i^2 \times h_i}{(1-h_i)}$$

where Z_i is the standardized residual and h_i is the leverage. In logistic regression, the calculation of Cook's distance is not exact but is an approximation.
- **DfBeta** is the change in the logistic coefficients when a case is deleted from the model. You can compute this change for each coefficient, including the constant. For example, the change in the first coefficient when case i is deleted is

$$\text{DfBeta}(B_1^{(i)}) = B_1 - B_1^{(i)}$$

where B_1 is the value of the coefficient when all cases are included and $B_1^{(i)}$ is the value of the coefficient when the ith case is excluded. Large values for change identify observations that should be examined.

Warning: The DfBeta values calculated by the logistic regression program are an approximation to the true values.

Plotting Diagnostics

All of the diagnostic statistics described in this chapter can be saved for further analysis. In particular, you want to use the Graph procedure to plot the diagnostic statistics against the case numbers and to look for unusual cases.

A plot of the standardized residuals against the case sequence numbers is shown in Figure 15-24. You see cases with large values for the standardized residuals that you can identify in the Data Label Mode in the Chart Editor. Figure 15-25 shows that there is a case that has substantial impact on the estimation of the coefficient for *acid* (case 24). Examination of the data reveals that this case has the largest value for serum acid phosphatase and yet does not have malignant nodes. Because serum acid phosphatase was positively related to malignant nodes, as shown in Figure 15-2, this case is quite unusual. If you remove case 24 from the analysis, the coefficient for serum acid phosphatase changes from 0.0243 to 0.0490. A variable that was, at best, a very marginal predictor becomes much more important.

Figure 15-24
Plot of standardized residual with case ID

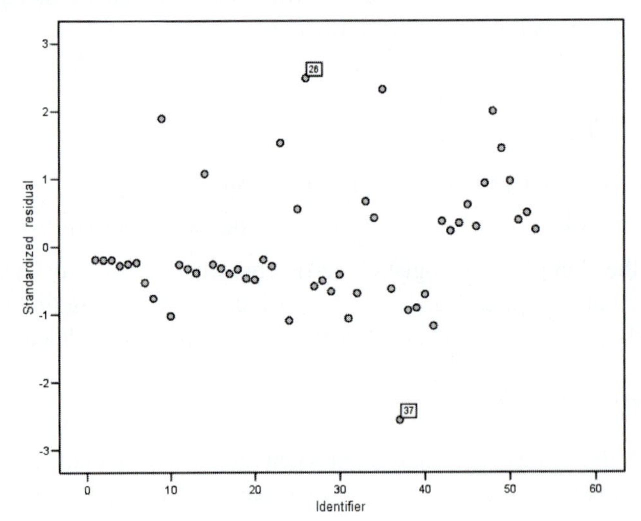

Figure 15-25
Plot of change in acid coefficient with case ID

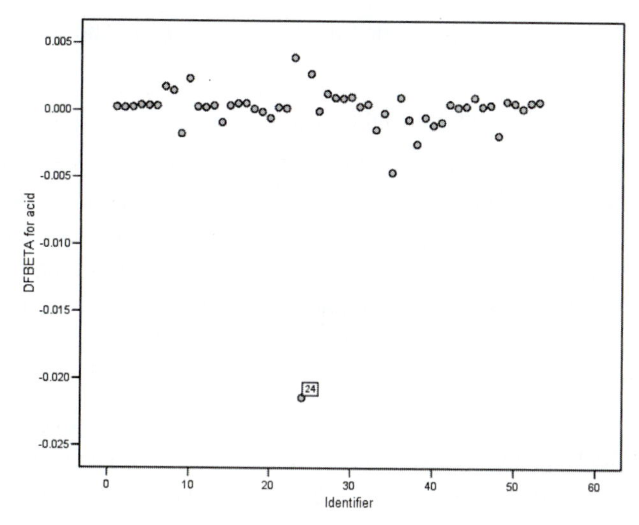

Related Procedures

SPSS Statistics has several procedures for the analysis of a categorical outcome with two or more categories. The Multinomial Logistic Regression procedure in SPSS Statistics can be used to fit a conditional logistic regression model for matched case control pairs, as well as a logistic regression model with more than two outcomes. It also provides for different methods for handling interaction terms in variable selection algorithms. For ordinal dependent variables, the Ordinal Regression procedure in SPSS Statistics can be used to model the dependence of a polytomous ordinal response on a set of predictor variables. See Chapter 20 for further discussion.

Obtaining the Output

To produce the output in this chapter, open the file *prcancer.sav* and follow the instructions below.

Figure 15-1 was not obtained from this data file.

Figures 15-2 to 15-5, 15-20, and 15-21. From the Analyze menu choose Regression > Binary Logistic. Move *nodes* into the Dependent box. Move *acid*, *age*, *xray*, *grade*, and *stage* into the Covariates list. Click Options. In the Options subdialog box, select Classification plots, Iteration history, and CI for exp(B). Click Continue, and then click Save. In the Save subdialog box, select Probabilities in the Predicted Values group, Standardized in the Residuals group, and DfBeta(s) in the Influence group. Click Continue, and then click OK.

Figure 15-6. From the Analyze menu choose Regression > Binary Logistic. Click Reset. Move *nodes* into the Dependent box. Move *acid*, *age*, *xray*, and *grade* into the Covariates list. Click Next. In the Block 2 of 2 group, move *stage* into the Covariates list. Click OK.

Figure 15-7. From the Analyze menu choose Regression > Binary Logistic. Click Reset. Move *nodes* into the Dependent box. Move *age*, *catacid1*, *catacid2*, *xray*, *grade*, and *stage* into the Covariates list. Click OK.

Figure 15-8. From the Analyze menu choose Regression > Binary Logistic. Click Reset. Move *nodes* into the Dependent box. Move *age*, *catacid*, *xray*, *grade*, and *stage* into the Covariates list. Click Categorical. In the Define Categorical Variables subdialog box, move *catacid* into the Categorical Covariates list. Click Continue, and then click OK.

Figures 15-9 and 15-10. Follow the instructions for Figure 15-8, but in the Define Categorical Variables subdialog box, change the Contrast type to Deviation (scroll down the list to see it, if necessary).

Figures 15-11 and 15-12 cannot be produced from these data.

Figure 15-13 to 15-17. From the Analyze menu choose Regression > Binary Logistic. Move *nodes* into the Dependent box. Move *acid*, *age*, *xray*, *grade*, and *stage* into the Covariates list. Set the Method drop-down list to Forward:Wald. Click OK.

Figure 15-18. Follow the instructions for Figures 15-13 to 15-17, but set the Method drop-down list to Forward:LR. Click OK.

Figure 15-19. From the Analyze menu choose ROC Curve. In the ROC Curve dialog box, move *pre_1* to the Test Variable list, and move *nodes* into the State Variable box. Type 1 as the Value of State Variable, and then select Standard error and confidence interval. Click OK.

Figures 15-22 and 15-23 cannot be produced from these data.

Figure 15-24. From the Graphs menu choose Legacy Dialogs > Scatter/Dot, select the Simple Scatter icon, and click Define. In the Simple Scatterplot dialog box, move *ZRE_1* into the Y-Axis box, move *identif* into the X-Axis box, and move *id* into the Label Cases By box. Click OK.

Figure 15-25. From the Graphs menu choose Legacy Dialogs > Scatter/Dot, select the Simple Scatter icon, and click Define. In the Simple Scatterplot dialog box, move *DFB1_1* into the Y-Axis box, move *identif* into the X-Axis box, and move *id* into the Label Cases By box. Click OK.

Chapter

16

Cluster Analysis

Identifying groups of individuals or objects that are similar to each other but different from individuals in other groups can be intellectually satisfying, profitable, or sometimes both. Using your customer base, you may be able to form clusters of customers who have similar buying habits or demographics. You can take advantage of these similarities to target offers to subgroups that are most likely to be receptive to them. Based on scores on psychological inventories, you can cluster patients into subgroups that have similar response patterns. This may help you in targeting appropriate treatment and studying typologies of diseases. By analyzing the mineral contents of excavated materials, you can study their origins and spread.

Tip: Although both cluster analysis and discriminant analysis classify objects (or cases) into categories, discriminant analysis requires you to know group membership for the cases used to derive the classification rule. The goal of cluster analysis is to identify the actual groups. For example, if you are interested in distinguishing between several disease groups using discriminant analysis, cases with known diagnoses must be available. Based on these cases, you derive a rule for classifying undiagnosed patients. In cluster analysis, you don't know who or what belongs in which group. You often don't even know the number of groups.

Examples

- You need to identify people with similar patterns of past purchases so that you can tailor your marketing strategies.
- You've been assigned to group television shows into homogeneous categories based on viewer characteristics. This can be used for market segmentation.

- You want to cluster skulls excavated from archaeological digs into the civilizations from which they originated. Various measurements of the skulls are available.
- You're trying to examine patients with a diagnosis of depression to determine if distinct subgroups can be identified, based on a symptom checklist and results from psychological tests.

In a Nutshell

You start out with a number of cases and want to subdivide them into homogeneous groups. First, you choose the variables on which you want the groups to be similar. Next, you must decide whether to standardize the variables in some way so that they all contribute equally to the distance or similarity between cases. Finally, you have to decide which clustering procedure to use, based on the number of cases and types of variables that you want to use for forming clusters.

For hierarchical clustering, you choose a statistic that quantifies how far apart (or similar) two cases are. Then you select a method for forming the groups. Because you can have as many clusters as you do cases (not a useful solution!), your last step is to determine how many clusters you need to represent your data. You do this by looking at how similar clusters are when you create additional clusters or collapse existing ones.

In *k*-means clustering, you select the number of clusters you want. The algorithm iteratively estimates the cluster means and assigns each case to the cluster for which its distance to the cluster mean is the smallest.

In two-step clustering, to make large problems tractable, in the first step, cases are assigned to "preclusters." In the second step, the preclusters are clustered using the hierarchical clustering algorithm. You can specify the number of clusters you want or let the algorithm decide based on preselected criteria.

Introduction

The term *cluster analysis* does not identify a particular statistical method or model, as do discriminant analysis, factor analysis, and regression. You often don't have to make any assumptions about the underlying distribution of the data. Using cluster analysis, you can also form groups of related variables, similar to what you do in factor analysis. There are numerous ways you can sort cases into groups. The choice of a method depends on, among other things, the size of the data file. Methods commonly used for small datasets are impractical for data files with thousands of cases.

SPSS Statistics has three different procedures that can be used to cluster data: hierarchical cluster analysis, *k*-means cluster, and two-step cluster. They are all described in this chapter. If you have a large data file (even 1,000 cases is large for clustering) or a mixture of continuous and categorical variables, you should use the SPSS Statistics two-step procedure. If you have a small dataset and want to easily examine solutions with increasing numbers of clusters, you may want to use hierarchical clustering. If you know how many clusters you want and you have a moderately sized dataset, you can use *k*-means clustering.

You'll cluster three different sets of data using the three SPSS Statistics procedures. You'll use a hierarchical algorithm to cluster figure-skating judges in the 2002 Olympic Games. You'll use *k*-means clustering to study the metal composition of Roman pottery. Finally, you'll cluster the participants in the 2002 General Social Survey, using a two-stage clustering algorithm. You'll find homogenous clusters based on education, age, income, gender, and region of the country. You'll see how Internet use and television viewing varies across the clusters.

Hierarchical Clustering

There are numerous ways in which clusters can be formed. Hierarchical clustering is one of the most straightforward methods. It can be either agglomerative or divisive. Agglomerative hierarchical clustering begins with every case being a cluster unto itself. At successive steps, similar clusters are merged. The algorithm ends with everybody in one jolly, but useless, cluster. Divisive clustering starts with everybody in one cluster and ends up with everyone in individual clusters. Obviously, neither the first step nor the last step is a worthwhile solution with either method.

In agglomerative clustering, once a cluster is formed, it cannot be split; it can only be combined with other clusters. Agglomerative hierarchical clustering doesn't let cases separate from clusters that they've joined. Once in a cluster, always in that cluster.

To form clusters using a hierarchical cluster analysis, you must select:

- A criterion for determining similarity or distance between cases
- A criterion for determining which clusters are merged at successive steps
- The number of clusters you need to represent your data

Tip: There is no right or wrong answer as to how many clusters you need. It depends on what you're going to do with them. To find a good cluster solution, you must look at the characteristics of the clusters at successive steps and decide when you have an interpretable solution or a solution that has a reasonable number of fairly homogeneous clusters.

Figure Skating Judges: The Example

As an example of agglomerative hierarchical clustering, you'll look at the judging of pairs figure skating in the 2002 Olympics. Each of nine judges gave each of 20 pairs of skaters four scores: technical merit and artistry for both the short program and the long program. You'll see which groups of judges assigned similar scores. To make the example more interesting, only the scores of the top four pairs are included. That's where the Olympic scoring controversies were centered. (For background on the controversy concerning the 2002 Olympic Winter Games pairs figure skating, see *http://en.wikipedia.org/wiki/2002_Olympic_Winter_Games_figure_skating_scandal*. The actual scores are only one part of an incredibly complex, and not entirely objective, procedure for assigning medals to figure skaters and ice dancers.)*

Tip: Consider carefully the variables you will use for establishing clusters. If you don't include variables that are important, your clusters may not be useful. For example, if you are clustering schools and don't include information on the number of students and faculty at each school, size will not be used for establishing clusters.

How Alike (or Different) Are the Cases?

Because the goal of this cluster analysis is to form similar groups of figure-skating judges, you have to decide on the criterion to be used for measuring similarity or distance. **Distance** is a measure of how far apart two objects are, while **similarity** measures how similar two objects are. For cases that are alike, distance measures are small and similarity measures are large. There are many different definitions of distance and similarity. Some, like the Euclidean distance, are suitable for only continuous variables, while others are suitable for only categorical variables. There are also many specialized measures for binary variables. See the Help system for a description of the more than 30 distance and similarity measures available in SPSS Statistics.

Warning: The computation for the selected distance measure is based on all of the variables you select. If you have a mixture of nominal and continuous variables, you must use the two-step cluster procedure because none of the distance measures in hierarchical clustering or *k*-means are suitable for use with both types of variables.

* I wish to thank Professor John Hartigan of Yale University for extracting the data from *www.nbcolympics.com* and making it available as a data file.

To see how a simple distance measure is computed, consider the data in Figure 16-1. The table shows the ratings of the French and Canadian judges for the Russian pairs figure skating team of Berezhnaya and Sikhardulidze.

Figure 16-1
Distances for two judges for one pair

	Long Program		Short Program	
Judge	**Technical Merit**	**Artistry**	**Technical Merit**	**Artistry**
France	5.8	5.9	5.8	5.8
Canada	5.7	5.8	5.8	5.8

You see that, for the long program, there is a 0.1 point difference in technical merit scores and a 0.1 difference in artistry scores between the French judge and the Canadian judge. For the short program, they assigned the same scores to the pair. This information can be combined into a single index or distance measure in many different ways. One frequently used measure is the squared Euclidean distance, which is the sum of the squared differences over all of the variables. In this example, the squared Euclidean distance is 0.02. The squared Euclidean distance suffers from the disadvantage that it depends on the units of measurement for the variables.

Standardizing the Variables

If variables are measured on different scales, variables with large values contribute more to the distance measure than variables with small values. In this example, both variables are measured on the same scale, so that's not much of a problem, assuming the judges use the scales similarly. But if you were looking at the distance between two people based on their IQs and incomes in dollars, you would probably find that the differences in incomes would dominate any distance measures. (A difference of only $100 when squared becomes 10,000, while a difference of 30 IQ points would be only 900. I'd go for the IQ points over the dollars!) Variables that are measured in large numbers will contribute to the distance more than variables recorded in smaller numbers.

Tip: In the hierarchical clustering procedure in SPSS Statistics, you can standardize variables in different ways. You can compute standardized scores or divide by just the standard deviation, range, mean, or maximum. This results in all variables contributing more equally to the distance measurement. That's not necessarily always the best strategy, since variability of a measure can provide useful information.

Proximity Matrix

To get the squared Euclidean distance between each pair of judges, you square the differences in the four scores that they assigned to each of the four top-rated pairs. You have 16 scores for each judge. These distances are shown in Figure 16-2, the proximity matrix. All of the entries on the diagonal are 0, since a judge does not differ from herself or himself. The smallest difference between two judges is 0.02, the distance between the French and Russian judges. (Look for the smallest off-diagonal entry in Figure 16-2.) The largest distance, 0.25, occurs between the Japanese and Canadian judges. The distance matrix is symmetric, since the distance between the Japanese and Russian judges is identical to the distance between the Russian and Japanese judges.

Figure 16-2
Proximity matrix between judges

		Squared Euclidean Distance								
		1:Canada	2:China	3:France	4:Germany	5:Japan	6:Poland	7:Russia	8:Ukraine	9:USA
Case	1:Canada	.000	.210	.200	.070	.250	.140	.240	.220	.220
	2:China	.210	.000	.070	.160	.240	.090	.090	.090	.190
	3:France	.200	.070	.000	.170	.170	.060	.020	.080	.120
	4:Germany	.070	.160	.170	.000	.160	.130	.210	.170	.150
	5:Japan	.250	.240	.170	.160	.000	.150	.190	.150	.070
	6:Poland	.140	.090	.060	.130	.150	.000	.080	.080	.160
	7:Russia	.240	.090	.020	.210	.190	.080	.000	.120	.160
	8:Ukraine	.220	.090	.080	.170	.150	.080	.120	.000	.120
	9:USA	.220	.190	.120	.150	.070	.160	.160	.120	.000

This is a dissimilarity matrix

Tip: In Figure 16-2, the squared Euclidean distance between the French and Canadian judge is computed for all four pairs. That's why the number differs from that computed for just the single Russian pair.

How Should Clusters Be Combined?

Agglomerative hierarchical clustering starts with each case (in this example, each judge) being a cluster. At the next step, the two judges who have the smallest value for the distance measure (or largest value if you are using similarities) are joined into a single cluster. At the second step, either a third case is added to the cluster that already contains two cases or two other cases are merged into a new cluster. At every step, either individual cases are added to existing clusters, two individuals are combined, or two existing clusters are combined.

When you have only one case in a cluster, the smallest distance between cases in two clusters is unambiguous. It's the distance or similarity measure you selected for the proximity matrix. Once you start forming clusters with more than one case, you need to define a distance between pairs of clusters. For example, if cluster A has cases 1 and 4, and cluster B has cases 5, 6, and 7, you need a measure of how different or similar the two clusters are.

There are many ways to define the distance between two clusters with more than one case in a cluster. For example, you can average the distances between all pairs of cases formed by taking one member from each of the two clusters. Or you can take the largest or smallest distance between two cases that are in different clusters. Different methods for computing the distance between clusters are available and may well result in different solutions. The methods available in SPSS Statistics hierarchical clustering are described in "Distance between Cluster Pairs" on p. 370.

Summarizing the Steps: The Icicle Plot

From Figure 16-3, you can see what's happening at each step of the cluster analysis when average linkage between groups is used to link the clusters. The figure is called an **icicle plot** because the columns of *X*'s look (supposedly) like icicles hanging from eaves. Each column represents one of the objects you're clustering. Each row shows a cluster solution with different numbers of clusters. You read the figure from the bottom up. The last row (that isn't shown) is the first step of the analysis. Each of the judges is a cluster unto himself or herself. The number of clusters at that point is 9. The eight-cluster solution arises when the Russian and French judges are joined into a cluster. (Remember they had the smallest distance of all pairs.) The seven-cluster solution results from the merging of the German and Canadian judges into a cluster. The six-cluster solution is the result of combining the Japanese and U.S. judges. For the one-cluster solution, all of the cases are combined into a single cluster.

Warning: When pairs of cases are tied for the smallest distance, an arbitrary selection is made. You might get a different cluster solution if your cases are sorted differently. That doesn't really matter, since there is no right or wrong answer to a cluster analysis. Many groupings are equally plausible.

Figure 16-3
Vertical icicle plot

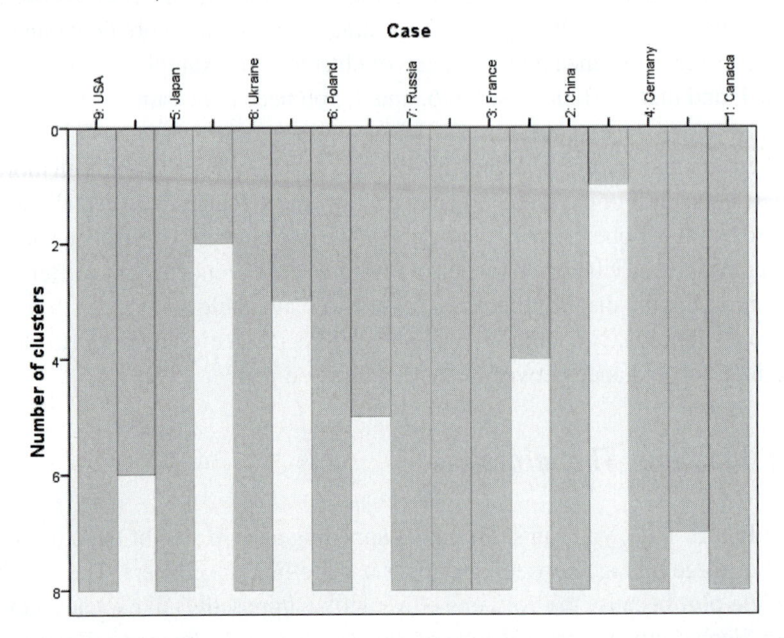

Tip: If you have a large number of cases to cluster, you can make an icicle plot in which the cases are the rows. Specify Horizontal on the Cluster Plots dialog box.

Who's in What Cluster?

You can get a table that shows the cases in each cluster for any number of clusters. Figure 16-4 shows the judges in the three-, four-, and five-cluster solutions.

Figure 16-4
Cluster membership

		5 Clusters	4 Clusters	3 Clusters
Case	1:Canada	1	1	1
	2:China	2	2	2
	3:France	3	2	2
	4:Germany	1	1	1
	5:Japan	4	3	3
	6:Poland	3	2	2
	7:Russia	3	2	2
	8:Ukraine	5	4	2
	9:USA	4	3	3

Tip: To see how clusters differ on the variables used to create them, save the cluster membership number using the Save command and then use the Means procedure, specifying the variables used to form the clusters as the dependent variables and the cluster number as the grouping variable.

Tracking the Combinations: The Agglomeration Schedule

From the icicle plot, you can't tell how small the distance measure is as additional cases are merged into clusters. For that, you have to look at the agglomeration schedule in Figure 16-5. In the column labeled *Coefficients*, you see the value of the distance (or similarity) statistic used to form the cluster. From these numbers, you get an idea of how unlike the clusters being combined are. If you are using dissimilarity measures, small coefficients tell you that fairly homogenous clusters are being attached to each other. Large coefficients tell you that you're combining dissimilar clusters. If you're using similarity measures, the opposite is true: large values are good, while small values are bad.

The actual value shown depends on the clustering method and the distance measure you're using. You can use these coefficients to help you decide how many clusters you need to represent the data. You want to stop cluster formation when the increase (for distance measures) or decrease (for similarity measures) in the *Coefficients* column between two adjacent steps is large. In this example, you may want to stop at the three-cluster solution, after stage 6. Here, as you can confirm from Figure 16-4, the Canadian and German judges are in cluster 1; the Chinese, French, Polish, Russian, and Ukrainian judges are in cluster 2; and the Japanese and U.S. judges are in cluster 3. If you go on to combine two of these three clusters in stage 7, the distance coefficient across the last combination jumps from 0. 093 to 0.165.

Figure 16-5
Agglomeration schedule

		Cluster Combined			Stage Cluster First Appears		
		Cluster 1	Cluster 2	Coefficients	Cluster 1	Cluster 2	Next Stage
Stage	1	3	7	.020	0	0	4
	2	1	4	.070	0	0	8
	3	5	9	.070	0	0	7
	4	3	6	.070	1	0	5
	5	2	3	.083	0	4	6
	6	2	8	.093	5	0	7
	7	2	5	.165	6	3	8
	8	1	2	.188	2	7	0

The agglomeration schedule starts off using the case numbers that are displayed on the icicle plot. Once cases are added to clusters, the cluster number is always the lowest of the case numbers in the cluster. A cluster formed by merging cases 3 and 4 would forever be known as cluster 3, unless it happened to merge with cluster 1 or 2.

The columns labeled *Stage Cluster First Appears* tell you the step at which each of the two clusters that are being joined first appear. For example, at stage 4 when cluster 3 and cluster 6 are combined, you're told that cluster 3 was first formed at stage 1 and cluster 6 is a single case and that the resulting cluster (known as 3) will see action again at stage 5. For a small dataset, you're much better off looking at the icicle plot than trying to follow the step-by-step clustering summarized in the agglomeration schedule.

Tip: In most situations, all you want to look at in the agglomeration schedule is the coefficient at which clusters are combined. Look at the icicle plot to see what's going on.

Plotting Cluster Distances: The Dendrogram

If you want a visual representation of the distance at which clusters are combined, you can look at a display called the dendrogram, shown in Figure 16-6. The dendrogram is read from left to right. Vertical lines show joined clusters. The position of the line on the scale indicates the distance at which clusters are joined. The observed distances are rescaled to fall into the range of 1 to 25, so you don't see the actual distances; however, the ratio of the rescaled distances within the dendrogram is the same as the ratio of the original distances.

The first vertical line, corresponding to the smallest rescaled distance, is for the French and Russian alliance. The next vertical line is at the same distances for three merges. You see from Figure 16-5 that stages 2, 3, and 4 have the same coefficients. What you see in this plot is what you already know from the agglomeration schedule. In the last two steps, fairly dissimilar clusters are combined.

Figure 16-6
The dendrogram

```
Dendrogram using Average Linkage (Between Groups)

                         Rescaled Distance Cluster Combine

  C A S E            0         5        10        15        20        25
Label           Num  +---------+---------+---------+---------+---------+

France            3   -+-------------+
Russia            7   -+             +---+
Poland            6   ---------------+   +-+
China             2   -------------------+ +---------------------+
Ukraine           8   ---------------------+                     +-----+
Japan             5   ---------------+---------------------------+     |
USA               9   ---------------+                                 |
Canada            1   ---------------+---------------------------------+
Germany           4   ---------------+
```

Tip: When you read a dendrogram, you want to determine at what stage the distances between clusters that are combined is large. You look for large distances between sequential vertical lines.

Clustering Variables

In the previous example, you saw how homogeneous groups of cases are formed. The unit of analysis was the case (each judge). You can also use cluster analysis to find homogeneous groups of variables.

Warning: When clustering variables, make sure to select the Variables radio button in the Cluster dialog box; otherwise, SPSS Statistics will attempt to cluster cases, and you will have to stop the processor because it can take a very long time.

Some of the variables used in Chapter 17 are clustered in Figure 16-7, using the Pearson correlation coefficient as the measure of similarity. The icicle plot is read the same way as before, but from right to left if you want to follow the steps in order. At the first step, each variable is a cluster. At each subsequent step, the two closest variables or clusters of variables are combined. You see that at the first step (number of clusters equal to 7), gender and region are combined. At the next step, parental education and wealth are combined. At the four-cluster solution, you may recognize the same grouping of variables as in factor analysis: a cluster for the social variables (region and gender), a cluster for the family variables (parental education and wealth), a cluster for personal attributes (personal education, hard work, and ambition), and ability in a cluster of its own. In factor analysis, ability straddled several factors. It wasn't clearly associated with a single factor. It's always reassuring when several different analyses point to the same conclusions!

Figure 16-7
Horizontal icicle plot

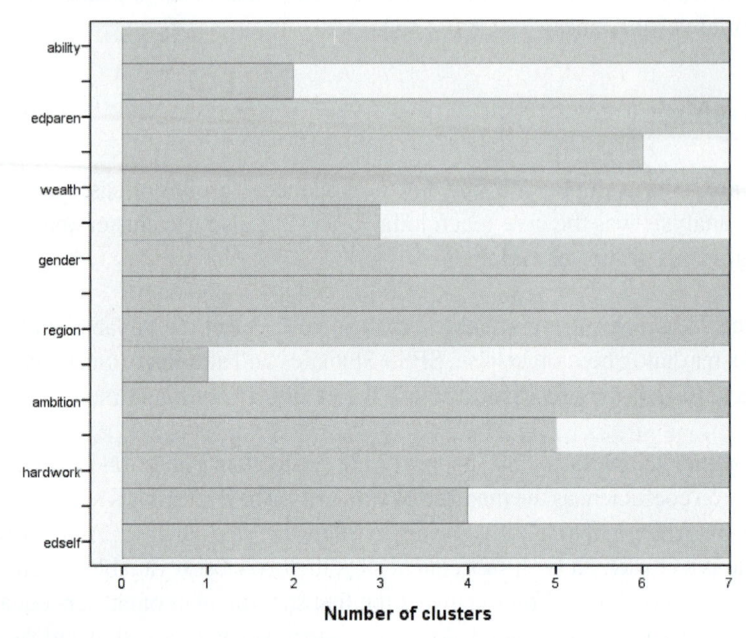

Tip: When you use the correlation coefficient as a measure of similarity, you may want to take the absolute value of it before forming clusters. Variables with large negative correlation coefficients are just as closely related as variables with large positive coefficients.

Distance between Cluster Pairs

The most frequently used methods for combining clusters at each stage are available in SPSS Statistics. These methods define the distance between two clusters at each stage of the procedure. If cluster A has cases 1 and 2 and if cluster B has cases 5, 6, and 7, you need a measure of how different or similar the two clusters are.

Nearest neighbor (single linkage). If you use the nearest neighbor method to form clusters, the distance between two clusters is defined as the smallest distance between two cases in the different clusters. That means the distance between cluster A and cluster B is the smallest of the distances between the following pairs of cases: (1,5), (1,6), (1,7), (2,5), (2,6), and (2,7). At every step, the distance between two clusters is taken to be the distance between their two closest members.

Furthest neighbor (complete linkage). If you use a method called furthest neighbor (also known as complete linkage), the distance between two clusters is defined as the distance between the two furthest points.

UPGMA. The average-linkage-between-groups method, often aptly called UPGMA (unweighted pair-group method using arithmetic averages), defines the distance between two clusters as the average of the distances between all pairs of cases in which one member of the pair is from each of the clusters. For example, if cases 1 and 2 form cluster A and cases 5, 6, and 7 form cluster B, the average-linkage-between-groups distance between clusters A and B is the average of the distances between the same pairs of cases as before: (1,5), (1,6), (1,7), (2,5), (2,6), and (2,7). This differs from the linkage methods in that it uses information about all pairs of distances, not just the nearest or the furthest. For this reason, it is usually preferred to the single and complete linkage methods for cluster analysis.

Average linkage within groups. The UPGMA method considers only distances between pairs of cases in different clusters. A variant of it, the average linkage within groups, combines clusters so that the average distance between all cases in the resulting cluster is as small as possible. Thus, the distance between two clusters is the average of the distances between all possible pairs of cases in the resulting cluster.

The methods discussed above can be used with any kind of similarity or distance measure between cases. The next three methods use squared Euclidean distances.

Ward's method. For each cluster, the means for all variables are calculated. Then, for each case, the squared Euclidean distance to the cluster means is calculated. These distances are summed for all of the cases. At each step, the two clusters that merge are those that result in the smallest increase in the overall sum of the squared within-cluster distances. The coefficient in the agglomeration schedule is the within-cluster sum of squares at that step, not the distance at which clusters are joined.

Centroid method. This method calculates the distance between two clusters as the sum of distances between cluster means for all of the variables. In the centroid method, the centroid of a merged cluster is a weighted combination of the centroids of the two individual clusters, where the weights are proportional to the sizes of the clusters. One disadvantage of the centroid method is that the distance at which clusters are combined can actually decrease from one step to the next. This is an undesirable property because clusters merged at later stages are more dissimilar than those merged at early stages.

Median method. With this method, the two clusters being combined are weighted equally in the computation of the centroid, regardless of the number of cases in each. This allows small groups to have an equal effect on the characterization of larger clusters into which they are merged.

Tip: Different combinations of distance measures and linkage methods are best for clusters of particular shapes. For example, nearest neighbor works well for elongated clusters with unequal variances and unequal sample sizes.

K-Means Clustering

Hierarchical clustering requires a distance or similarity matrix between all pairs of cases. That's a humongous matrix if you have tens of thousands of cases trapped in your data file. Even today's computers will take pause, as will you, waiting for results.

A clustering method that doesn't require computation of all possible distances is *k*-means clustering. It differs from hierarchical clustering in several ways. You have to know in advance the number of clusters you want. You can't get solutions for a range of cluster numbers unless you rerun the analysis for each different number of clusters. The algorithm repeatedly reassigns cases to clusters, so the same case can move from cluster to cluster during the analysis. In agglomerative hierarchical clustering, on the other hand, cases are added only to existing clusters. They're forever captive in their cluster, with a widening circle of neighbors.

The algorithm is called *k*-means, where *k* is the number of clusters you want, since a case is assigned to the cluster for which its distance to the cluster mean is the smallest. The action in the algorithm centers around finding the *k*-means. You start out with an initial set of means and classify cases based on their distances to the centers. Next, you compute the cluster means again, using the cases that are assigned to the cluster; then, you reclassify all cases based on the new set of means. You keep repeating this step until cluster means don't change much between successive steps. Finally, you calculate the means of the clusters once again and assign the cases to their permanent clusters.

Roman Pottery: The Example

When you are studying old objects in hopes of determining their historical origins, you look for similarities and differences between the objects in as many dimensions as possible. If the objects are intact, you can compare styles, colors, and other easily visible characteristics. You can also use sophisticated devices such as high-energy beam lines or spectrophotometers to study the chemical composition of the materials from which they are made. Hand et al. (1994) reports data on the percentage of oxides of five metals for 26 samples of Romano-British pottery described by Tubb et al. (1980). The five metals are aluminum (*Al*), iron (*Fe*), magnesium (*Mg*), calcium (*Ca*), and sodium (*Na*). In this section, you'll study whether the samples form distinct clusters and whether these clusters correspond to areas where the pottery was excavated.

Before You Start

Whenever you use a statistical procedure that calculates distances, you have to worry about the impact of the different units in which variables are measured. Variables that have large values will have a large impact on the distance compared to variables that have smaller values. In this example, the average percentages of the oxides differ quite a bit, so it's a good idea to standardize the variables to a mean of 0 and a standard deviation of 1. (Standardized variables are used in the example.) You also have to specify the number of clusters (*k*) that you want produced.

Tip: If you have a large data file, you can take a random sample of the data and try to determine a good number, or range of numbers, for a cluster solution based on the hierarchical clustering procedure. You can also use hierarchical cluster analysis to estimate starting values for the *k*-means algorithm.

Initial Cluster Centers

The first step in *k*-means clustering is finding the *k* centers. This is done iteratively. You start with an initial set of centers and then modify them until the change between two iterations is small enough. If you have good guesses for the centers, you can use those as initial starting points; otherwise, you can let SPSS Statistics find *k* cases that are well-separated and use these values as initial cluster centers. Figure 16-8 shows the initial centers for the pottery example.

Figure 16-8
Initial cluster centers

	Cluster		
	1	2	3
ZAl	-.03	-.90	2.11
ZCa	.03	1.52	-.76
ZFe	1.05	.40	-1.23
ZMg	.53	.37	-1.11
ZNa	2.60	-.88	-.43

Warning: *K*-means clustering is very sensitive to outliers, since they will usually be selected as initial cluster centers. This will result in outliers forming clusters with small numbers of cases. Before you start a cluster analysis, screen the data for outliers and remove them from the initial analysis. The solution may also depend on the order of the cases in the file.

After the initial cluster centers have been selected, each case is assigned to the closest cluster, based on its distance from the cluster centers. After all of the cases have been assigned to clusters, the cluster centers are recomputed, based on all of the cases in the cluster. Case assignment is done again, using these updated cluster centers. You keep assigning cases and recomputing the cluster centers until no cluster center changes appreciably or the maximum number of iterations (10 by default) is reached.

From Figure 16-9, you see that three iterations were enough for the pottery data.

Figure 16-9
Iteration history

		Change in Cluster Centers		
		1	2	3
Iteration	1	.760	1.370	1.126
	2	.852	.098	.000
	3	.000	.000	.000

Tip: You can update the cluster centers after each case is classified, instead of after all cases are classified, if you select the Use Running Means check box in the Iterate dialog box.

Final Cluster Centers

After iteration stops, all cases are assigned to clusters, based on the last set of cluster centers. After all of the cases are clustered, the cluster centers are computed one last time. Using the final cluster centers, you can describe the clusters. In Figure 16-10, you see that cluster 1 has an average sodium percentage that is much higher than the other clusters. Cluster 2 has higher-than-average values for calcium, iron, and magnesium, an average value for sodium, and a smaller-than-average value for aluminum. Cluster 3 has below-average values for all of the minerals except aluminum.

Figure 16-10
Final cluster centers

	Cluster		
	1	2	3
ZAl	-.36	-.73	1.09
ZCa	.38	.70	-1.06
ZFe	.87	.72	-1.19
ZMg	.85	.70	-1.15
ZNa	2.71	.18	-.79

Tip: You can save the final cluster centers and use them to classify new cases. In the Cluster dialog box, save the cluster centers by selecting Write Final As and then clicking File to assign a filename. To use these cluster centers to classify new cases, select Classify Only, select Read Initial From, and then click File to specify the file of cluster centers that you saved earlier.

Differences between Clusters

You can compute *F* ratios that describe the differences between the clusters. As the footnote in Figure 16-11 warns, the observed significance levels should not be interpreted in the usual fashion because the clusters have been selected to maximize the differences between clusters. The point of Figure 16-11 is to give you a handle on the differences for each of the variables among the clusters. If the observed significance level for a variable is large, you can be pretty sure that the variable doesn't contribute much to the separation of the clusters.

Figure 16-11
Analysis-of-variance table

	Cluster		Error			
	Mean Square	df	Mean Square	df	F	Sig.
ZAl	9.74	2	.240	23	40.61	.000
ZCa	9.26	2	.282	23	32.89	.000
ZFe	11.43	2	.093	23	122.80	.000
ZMg	10.72	2	.155	23	69.32	.000
ZNa	10.72	2	.154	23	69.42	.000

The F tests should be used only for descriptive purposes because the clusters have been chosen to maximize the differences among cases in different clusters. The observed significance levels are not corrected for this and thus cannot be interpreted as tests of the hypothesis that the cluster means are equal.

You see in Figure 16-12 that 2 cases are assigned to the first cluster, 14 to the second, and 10 to the third cluster. You don't like to see clusters with very few cases unless they are really different from the remaining cases. For each case, you can save the cluster to which it is assigned, as well as the distance to its cluster center.

Figure 16-12
Number of cases in each cluster

Cluster	1	2
	2	14
	3	10
Valid		26
Missing		0

If you plot the distances to their cluster centers for all of the cases, as in Figure 16-13, you can see if there are cases that are outliers. Clusters 2 and 3 are unremarkable. Because cluster 1 has only two cases, you see only one point on the plot. (The cluster center is halfway between the two cases, so their distances from it are equal and are plotted at the same location.)

Figure 16-13
Plot of distances to cluster centers

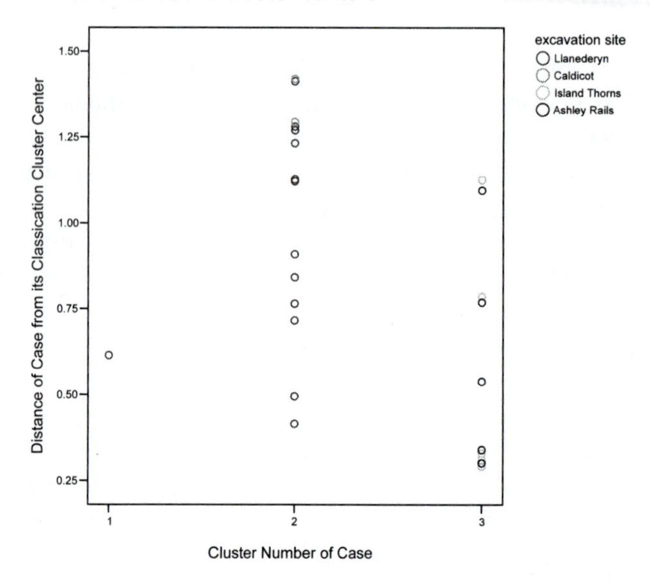

You can decrease the number of clusters to be used to two and that would most likely eliminate the two-case cluster at the expense of making the clusters more heterogeneous. Or you could increase the number of clusters and see how the solution changes. In this example, increasing the number of clusters causes clusters 2 and 3 to split, while cluster 1 remains with the two cases that are characterized by very high levels of sodium.

Locations of the Pottery

The pottery in the dataset was found in one of four locations. To see whether pottery found at the same site is homogeneous with respect to metallic composition, you can crosstabulate the site where a pot was found and the cluster to which it was assigned, as shown in Figure 16-14.

Figure 16-14
Crosstabulation of site and cluster

			excavation site				
			Llanederyn	Caldicot	Island Thorns	Ashley Rails	Total
Cluster Number of Case	1	Count	2	0	0	0	2
		% within excavation site	14.3%	.0%	.0%	.0%	7.7%
	2	Count	12	2	0	0	14
		% within excavation site	85.7%	100.0%	.0%	.0%	53.8%
	3	Count	0	0	5	5	10
		% within excavation site	.0%	.0%	100.0%	100.0%	38.5%
Total		Count	14	2	5	5	26
		% within excavation site	100.0%	100.0%	100.0%	100.0%	100.0%

You see that the anomalous pottery in cluster 1 was found at the first site. Most of the cases in cluster 2 were also from the first site, although two of them were from the second. Pottery from the third and fourth sites were all in cluster 3. It looks like there is a relationship between the metallic composition of a piece of pottery and the site where it was excavated.

Two-Step Cluster

When you have a really large dataset or you need a clustering procedure that can rapidly form clusters on the basis of either categorical or continuous data, neither of the previous two procedures fills the bill. Hierarchical clustering requires a matrix of distances between all pairs of cases, and *k*-means requires shuffling cases in and out of clusters and knowing the number of clusters in advance. The SPSS Statistics TwoStep Cluster Analysis procedure was designed for such applications. It requires only one pass of data (which is important for very large data files), and it can produce solutions based on mixtures of continuous and categorical variables and for varying numbers of clusters.

The clustering algorithm is based on a distance measure that gives the best results if all variables are independent, continuous variables that have a normal distribution, and categorical variables that have a multinomial distribution. This is seldom the case in practice, but the algorithm is thought to behave reasonably well when the assumptions are not met. Because cluster analysis does not involve hypothesis testing and calculation of observed significance levels, other than for descriptive follow-up, it's perfectly acceptable to cluster data that may not meet the assumptions for best performance. Only you can determine whether the solution is satisfactory for your needs.

Warning: The final solution may depend on the order of the cases in the file. To minimize the effect, arrange the cases in random order. Sort them by the last digit of their ID numbers or something similar.

Step 1: Preclustering: Making Little Clusters

The first step of the two-step procedure is formation of preclusters. The goal of preclustering is to reduce the size of the matrix that contains distances between all possible pairs of cases. Preclusters are just clusters of the original cases that are used in place of the raw data in the hierarchical clustering. As a case is read, the algorithm decides, based on a distance measure, if the current case should be merged with a previously formed precluster or start a new precluster. When preclustering is complete, all cases in the same precluster are treated as a single entity. The size of the distance matrix is no longer dependent on the number of cases but on the number of preclusters.

Step 2: Hierarchical Clustering of Preclusters

In the second step, SPSS Statistics uses the standard hierarchical clustering algorithm on the preclusters. Forming clusters hierarchically lets you explore a range of solutions with different numbers of clusters.

Tip: The Options dialog box lets you control the number of preclusters. Large numbers of preclusters give better results because the cases are more similar in a precluster; however, forming many preclusters slows the algorithm.

Clustering Newspaper Readers: The Example

As an example of two-step clustering, you'll consider the General Social Survey data described in Chapter 6. Both categorical and continuous variables are used to form the clusters. The categorical variables are sex, frequency of reading a newspaper, and highest degree. The continuous variable is age.

Some of the options you can specify when using two-step clustering are:

Standardization: The algorithm will automatically standardize all of the variables unless you override this option.

Distance measures: If your data are a mixture of continuous and categorical variables, you can use only the log-likelihood criterion. The distance between two clusters depends on the decrease in the log-likelihood when they are combined into a single cluster. If the data are only continuous variables, you can also use the Euclidean distance between two cluster centers. Depending on the distance measure selected, cases are assigned to the cluster that leads to the largest log-likelihood or to the cluster that has the smallest Euclidean distance.

Number of clusters: You can specify the number of clusters to be formed, or you can let the algorithm select the optimal number based on either the Schwarz Bayesian Criterion or the Akaike information criterion.

Outlier handling: You have the option to create a separate cluster for cases that don't fit well into any other cluster.

Range of solutions: You can specify the range of cluster solutions that you want to see.

Tip: If you're planning to use this procedure, consult the algorithms document in the Help system for additional details.

Examining the Number of Clusters

Once you make some choices or do nothing and go with the defaults, the clusters are formed. At this point, you can consider whether the number of clusters is "good." If you use automated cluster selection, SPSS Statistics prints a table of statistics for different numbers of clusters, an excerpt of which is shown in Figure 16-15. You are interested in finding the number of clusters at which the Schwarz Bayesian Criterion, abbreviated BIC (the I stands for Information), becomes small and the change in BIC between adjacent number of clusters is small. That's not always easy. For this example, the algorithm selected three clusters.

Figure 16-15
Autoclustering statistics

		Schwarz's Bayesian Criterion (BIC)	BIC Change[1]	Ratio of BIC Changes[2]	Ratio of Distance Measures[3]
Number of Clusters	1	6827.387			
	2	5646.855	-1180.532	1.000	1.741
	3	5000.782	-646.073	.547	1.790
	4	4672.859	-327.923	.278	1.047
	5	4362.908	-309.951	.263	1.066
	6	4076.832	-286.076	.242	1.193
	7	3849.057	-227.775	.193	1.130
	8	3656.025	-193.032	.164	1.079
	9	3482.667	-173.358	.147	1.162
	10	3343.916	-138.751	.118	1.240
	11	3246.541	-97.376	.082	1.128
	12	3168.733	-77.808	.066	1.093
	13	3103.950	-64.783	.055	1.022
	14	3042.116	-61.835	.052	1.152
	15	2998.319	-43.796	.037	1.059

1. The changes are from the previous number of clusters in the table.
2. The ratios of changes are relative to the change for the two cluster solution.
3. The ratios of distance measures are based on the current number of clusters against the previous number of clusters.

You can also examine the number of cases in the final cluster solution as shown in Figure 16-16. In this example, you see that the largest cluster has 44% of the clustered cases, and the smallest has 27%. Usually, you don't want many small clusters. (For this example, the excluded number of cases is large because many people were not asked the newspaper question. Only cases with valid values for all variables are included in cluster formation.)

Figure 16-16
Distribution of cases in clusters

		N	% of Combined	% of Total
Cluster	1	396	44.0%	14.3%
	2	244	27.1%	8.8%
	3	259	28.8%	9.4%
	Combined	899	100.0%	32.5%
Excluded Cases		1866		67.5%
Total		2765		100.0%

Warning: When you have cases that are very different from other cases and not necessarily similar to each other, they can have a large impact on cluster formation by increasing the overall number of clusters or making clusters less homogeneous. One solution to the problem is to create an outlier cluster that contains all cases that do not fit well with the rest. SPSS Statistics will do this automatically for you if you select Outlier Treatment in the Options dialog box.

Examining the Composition of the Clusters

Once you've formed clusters, you want to know how they differ. SPSS Statistics offers numerous displays and tables to help you determine the composition of the clusters and the importance of each variable in determining the cluster.

For categorical variables, you get crosstabulations and bar charts of the distribution of the variable within each cluster. For example, Figure 16-17 shows the percentage of males and females in each of the clusters. You see that gender distribution in all of the clusters is fairly similar to the overall distribution. Gender isn't an important variable in forming the clusters.

Figure 16-17
Within-cluster percentage of respondent's gender

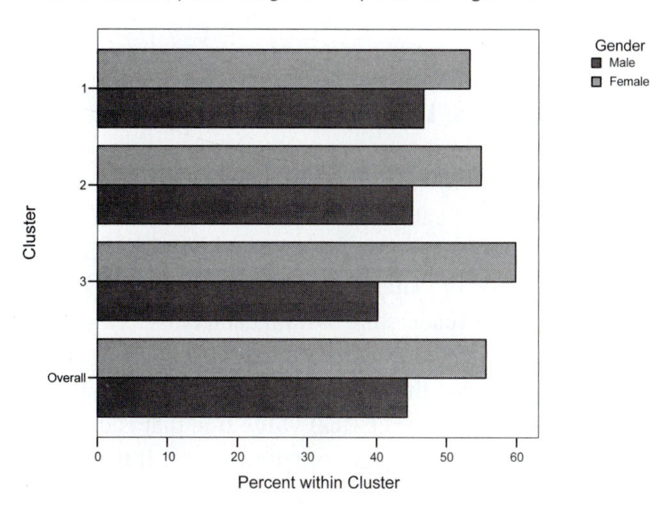

For each continuous variable, you get a plot of the means for each group and simultaneous confidence intervals for the population cluster means. In Figure 16-18, you see that the average age is largest for the second cluster.

Tip: For simultaneous confidence intervals for the means of several groups, the confidence level is for *all* population means being simultaneously included in their respective confidence intervals. Simultaneous confidence intervals for several population means are wider than the individual intervals.

Figure 16-18
Within-cluster percentage of respondent's age

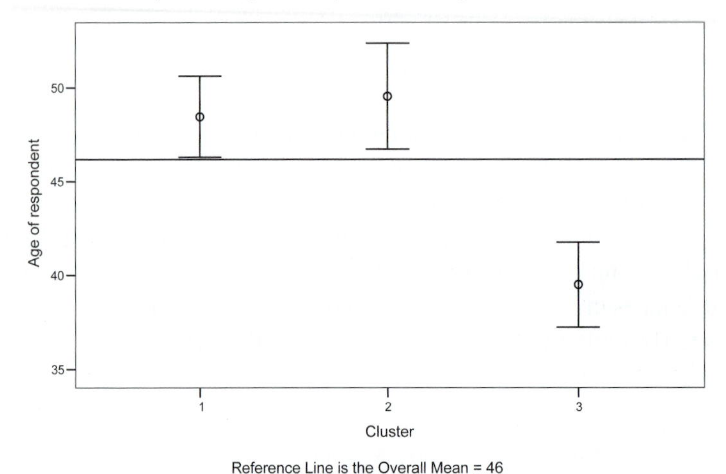

Examining the Importance of Individual Variables

When you cluster cases, you want to know how important the different variables are for the formation of the cluster. For categorical variables, SPSS Statistics calculates a chi-square value that compares the observed distribution of values of a variable within a cluster to the overall distribution of values. For example, Figure 16-19 is a plot of the chi-square statistic for newspaper readership. Within each cluster, the observed distribution is compared to an expected distribution based on all cases. Large values of the statistic for a cluster indicate that the distribution of the variable in the cluster differs from the overall distribution. The **critical value line** that is drawn provides some notion of how dissimilar each cluster is from the average. If the absolute value of the statistic for a cluster is greater than the critical value, the variable is probably important in distinguishing that cluster from the others.

Tip: The values of the statistics are just rough guidelines to help you find out how individual clusters differ from all cases combined. The actual values and the associated significance levels that can be plotted instead of the actual values can't be interpreted in the usual fashion, even when adjustments to the probabilities are made.

Figure 16-19
Importance of newspaper reading to cluster formation

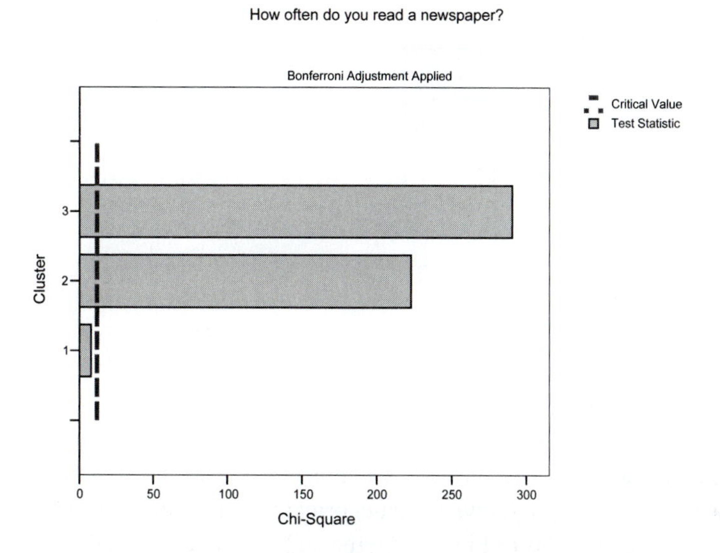

For continuous variables, instead of plots of chi-square values, you can get plots of *t* statistics that compare the mean of the variable in the cluster to the overall mean. Figure 16-20 shows the average age for the three clusters. You see that the average age is statistically different for the three clusters, since the value of the test statistic exceeds the critical value for each of the clusters.

Warning: You must specify a confidence level in the TwoStep Cluster Plots dialog box in order to get the critical value lines plotted.

Figure 16-20
Importance of age

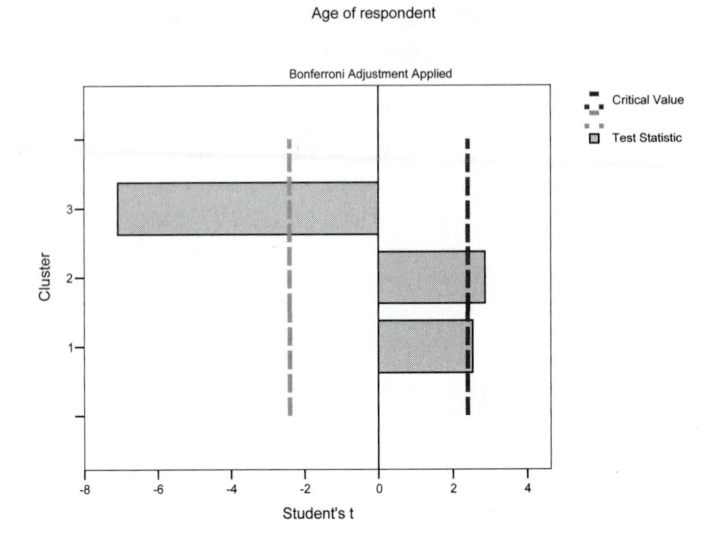

Looking at All Variables within a Cluster

Instead of tracking a single variable across all clusters, you can look at the composition of each cluster. Figure 16-21 shows the categorical variables that make up cluster 3. You see that the distributions of highest degree and frequency of reading a newspaper are different for cluster 3 compared to all clusters. The distribution of gender is not.

Figure 16-21
Importance of variables within a cluster

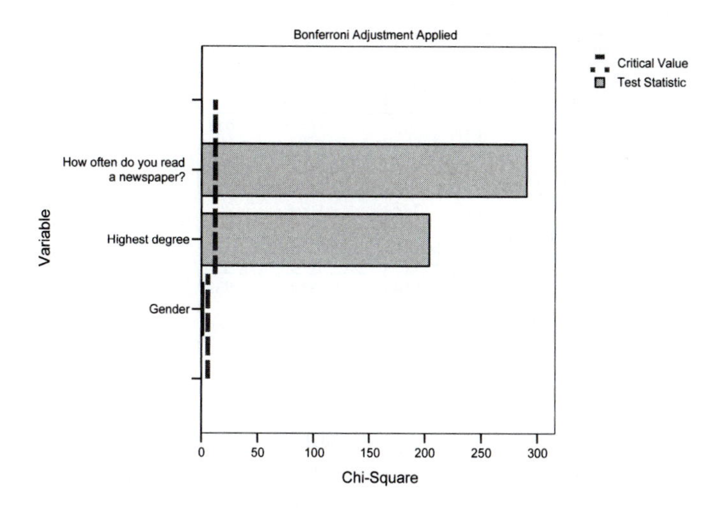

Looking at the Relationship to Other Variables

You can save the cluster membership variable for each case and examine the relationship between the clusters and other variables. For example, Figure 16-22 shows a crosstabulation of cluster membership and use of the Internet. You see that 60% of the members in cluster 1 use the Internet compared to 52% of members in cluster 2. If you want to target people who use the Internet, the cluster solution may help.

Figure 16-22
Crosstabulation of Internet use and cluster membership

			Use WWW other than mail		Total
			Yes	No	
TwoStep Cluster Number	1	Count	238	158	396
		% within TwoStep Cluster Number	60.1%	39.9%	100.0%
	2	Count	127	117	244
		% within TwoStep Cluster Number	52.0%	48.0%	100.0%
	3	Count	147	112	259
		% within TwoStep Cluster Number	56.8%	43.2%	100.0%
Total		Count	512	387	899
		% within TwoStep Cluster Number	57.0%	43.0%	100.0%

From Figure 16-23, you see that you might be able to use the cluster solution to identify people who watch TV for many hours a night. Maybe you can sell them popcorn.

Figure 16-23
Average hours of television viewing by cluster membership

Hours per day watching TV

		Mean	N	Std. Deviation
TwoStep Cluster Number	1	2.77	395	2.468
	2	3.16	244	2.348
	3	3.15	259	2.208
	Total	2.99	898	2.368

Obtaining the Output

To produce the output in this chapter, follow the instructions below.

For Figures 16-2 to 16-6, use the *olympics.sav* data file.

Figures 16-2 to 16-6. From the Analyze menu choose Classify > Hierarchical Cluster. Move all variables except *judgename* into the Variable(s) list. Move *judgename* into the Label Cases By box. In the Cluster group, select Cases. In the Display group, select both Statistics and Plots. Click Statistics. In the Statistics subdialog box, select Agglomeration schedule and Proximity matrix. In the Cluster Membership group, select Range of solutions and specify from 3 through 5 clusters. Click Continue, and then click Plots. In the Plots subdialog box, select Dendrogram, All clusters, and Vertical. Click Continue, and then click OK.

For Figure 16-7, use the *gettingahead.sav* data file.

Figure 16-7. From the Analyze menu choose Classify > Hierarchical Cluster. Move *ability*, *ambition*, *edparen*, *edself*, *hardwork*, *gender*, *region*, and *wealth* into the Variable(s) list. In the Cluster group, select Variables. Select both Statistics and Plots in the Display group. Click Plots. In the Plots subdialog box, select All clusters in the Icicle group and Horizontal in the Orientation group. Click Continue and then click Method. In the Method subdialog box, select Interval in the Measure group, and select Pearson correlation as Interval measure. Click Continue, and then click OK.

For Figures 16-8 to 16-14, use the *pottery.sav* data file.

Figures 16-8 to 16-12. From the Analyze menu choose Classify > K-Means Cluster. Move *ZAl*, *ZCa*, *ZFe*, *ZMg*, and *ZNa* into the Variables list, and specify 3 as Number of Clusters. Click Options and in the K-Means Cluster Analysis Options subdialog box, select both Initial cluster centers and ANOVA table. Click Continue, and then click Save. Select both Cluster membership and Distance from cluster center. Click Continue, and then click OK.

Figure 16-13. From the Graphs menu choose Legacy Dialogs > Scatter/Dot. Select the Simple Scatter icon and click Define. Move *QCL_1* to the X Axis box and *QCL_2* to the Y Axis box. Click OK.

Figure 16-14. From the Analyze menu choose Descriptive Statistics > Crosstabs. Move *QCL_1* into the Row(s) list and *site* into the Column(s) list. Click Cells, and in the Crosstabs Cell Display subdialog box, select Observed in the Counts group and Column in the Percentages group. Click Continue, and then click OK.

For Figures 16-15 to 16-23, use the *gssdata.sav* data file.

Figures 16-15, 16-16, and 16-18 to 16-20. From the Analyze menu choose Classify > TwoStep Cluster. Move *degree*, *news*, and *sex* into the Categorical Variables list, and move *age* to the Continuous Variables list. Click Plots. In the Plots subdialog box, select Within cluster percentage chart, Rank of variable importance, and Confidence level. Click Continue, and then click Output. In the Output subdialog box, select Descriptives by cluster, Cluster frequencies, and Information criterion (AIC or BIC) in the Statistics group, and select Create cluster membership variable in the Working Data File group. Click Continue, and then click OK.

Figure 16-21. Follow the steps above, but in the Plots subdialog box, leave Rank of variable importance selected but now select By variable in the Rank Variables group. Proceed as above.

Figure 16-22. From the Analyze menu choose Descriptive Statistics > Crosstabs. Move the new cluster membership variable, whose name will begin with *TSC_*, into the Row(s) list, and move *usewww* into the Column(s) list. Click Cells, and in the Cells subdialog box, select Row percentages. Click Continue, and then click OK.

Figure 16-23. From the Analyze menu choose Compare Means > Means. Move *tvhours* into the Dependent List, and move the new cluster membership variable, whose name will begin with *TSC_*, into the Independent List. Click OK.

Chapter

17

Factor Analysis

What are creativity, love, and altruism? Unlike variables such as weight, blood pressure, and temperature, you can't measure them on a scale, sphygmomanometer, or thermometer, in units of pounds, millimeters of mercury, or degrees Fahrenheit. Instead, you infer their existence from observed patterns of behavior. For example, an answer of *strongly agree* to items such as he (or she) understands me, reads my manuscripts, laughs at my jokes, and gazes deeply into my soul suggests that the love (or, perhaps, infatuation) factor is present. Love is not a single measurable entity but a construct that is derived from measurement of other, directly observable variables. By identifying such underlying constructs or factors, you can greatly simplify the description and understanding of complex phenomena, such as social interaction. Postulating the existence of something called "love" explains the observed correlations between responses to diverse situations.

Factor analysis is a statistical technique used to identify a relatively small number of factors that explain observed correlations among variables. For example, in a linear factor model, scores on a battery of achievement tests may be expressed as a linear combination of factors that represent verbal skills, mathematical achievement, and perceptual speed. Variables such as consumer ratings of products in a survey can be expressed as a function of factors such as product quality and utility. Factor analysis helps you to identify these not directly observable constructs.

You can also use factor analysis to reduce a large number of correlated variables to a more manageable number of independent factors that you can then use in subsequent analyses. For example, if you've administered a questionnaire with 100 items, your analysis will be greatly simplified if you can replace these 100 items with a smaller number of variables that measure the underlying dimensions.

Tip: The SPSS Statistics Factor Analysis procedure is for identifying factors (exploratory factor analysis). You can't test hypotheses about the fit of the data to a predetermined factor solution (confirmatory factor analysis) with this procedure. However, you can test hypotheses about the number of underlying factors.

Examples

- Determine the dimensions on which consumers rate coffees. These might be heartiness, genuineness, and freshness.
- Determine the characteristics of leaders. These might be efficiency, likability, and respect.
- Describe communities in terms of urbanism, stability, and social welfare.

In a Nutshell

The goal of factor analysis is to reproduce observed correlations among variables by identifying a smaller number of shared factors that account for the observed correlations. You can express each variable as a linear combination of a small number of common factors that are shared by all variables and as a unique factor that is specific to that variable and also represents the error. The correlations between the variables arise from the *sharing* of the common factors. The common factors in turn are estimated as linear combinations of the original variables. To improve the interpretability of the factors, solutions are rotated. A good factor solution is simple and meaningful and accounts for the observed correlations between variables.

Factor analysis involves:

- Computing a correlation matrix
- Extracting factors
- Rotating the factors to make them easier to interpret
- Calculating factor scores

Getting Ahead: The Example

If you ask someone what it takes to get ahead, it's unlikely that the response will be something that you can go out and measure. Even if the answer is money, that's not a simple measurement. What kind of money: family wealth, personal wealth, or income? Many responses will probably be along the lines of good connections, the right kind of family, the proper schooling, and a determination to succeed. These are not variables in the traditional sense because you can't measure them directly. They are constructs. You infer their existence to explain the observed correlations of variables that are related to them. For example, if you ask people to rate the importance of family wealth, personal wealth, and income to "getting ahead," you'll find the responses to these questions are probably correlated because they all involve the money factor. You postulate the existence of the money factor to explain the observed correlations between variables that involve assets. There's a bit of circularity involved here: You assume that sets of variables are correlated because they measure, to a certain extent, the same underlying factor, and you identify the factors by looking at the variables with which they are correlated.

Warning: Don't use factor analysis unless your variables have ordered values for which a correlation matrix is a reasonable summary measure. If you analyze binary variables using Pearson correlation coefficients, you may get too many factors.

As an example of factor analysis, you'll analyze answers that a random sample of 1,335 people gave when the General Social Survey asked them to rate the importance of the variables in Figure 17-1 to "getting ahead in life." The scale was 1 = *essential*, 2 = *very important*, 3 = *fairly important*, 4 = *not very important*, 5 = *not important at all*. Your goal is to summarize the answers by identifying dimensions or factors that people think are related to getting ahead. These factors will explain that the correlations you see among groups of variables. You want to explain the correlations with as few factors as possible, and you want to be able to attach a meaning to the factors. A good factor solution is both simple and interpretable.

In subsequent analyses, if you want to study differences between groups of people about what they think is important, you'll find it much easier to work with a small number of factors than with the original 13 variables. For example, if you identify a "wealth" factor, you can see whether the rich and the poor assign the same importance to the wealth factor for getting ahead.

Figure 17-1
Variable descriptions

Variable name	Question: How important is...
wealth	Coming from a wealthy family?
edparent	Having well-educated parents?
edself	Having a good education yourself?
ambition	Ambition?
ability	Natural ability?
hardwork	Hard work?
knowright	Knowing the right people?
clout	Having political connections?
race	Race?
religion	Religion?
region	Part of the country a person comes from?
gender	Being born a man or a woman?
polbelief	A person's political beliefs?

Tip: Make sure to look at plots of the variables so you can identify points that may distort correlation coefficients. If necessary, consider transforming the data values to achieve linearity. You don't have to assume that your variables are normally distributed except for goodness-of-fit tests, but for small sample sizes, outliers can have a large effect on correlation coefficients.

Step 1: Computing the Observed Correlation Matrix

Figure 17-2 is the observed correlation matrix between answers to all pairs of the 13 questions.

Figure 17-2
Matrix of observed correlation coefficients

	ability	ambition	clout	edparen	edself	gender	hard work	know right	polbelief	race	region	religion	wealth
ability	1.00	.14	.24	.17	.10	.11	.18	.27	.16	.18	.15	.11	.17
ambition	.14	1.00	.01	.05	.28	-.03	.31	.08	-.05	-.01	-.03	-.07	.03
clout	.24	.01	1.00	.28	.10	.33	-.04	.56	.45	.38	.32	.30	.39
edparen	.17	.05	.28	1.00	.31	.17	.02	.29	.22	.20	.18	.13	.41
edself	.10	.28	.10	.31	1.00	.02	.24	.15	.03	.03	.01	-.03	.11
gender	.11	-.03	.33	.17	.02	1.00	-.04	.26	.46	.45	.43	.29	.28
hardwork	.18	.31	-.04	.02	.24	-.04	1.00	.03	.00	-.10	-.03	.02	-.07
knowright	.27	.08	.56	.29	.15	.26	.03	1.00	.29	.32	.24	.24	.40
polbelief	.16	-.05	.45	.22	.03	.46	.00	.29	1.00	.38	.48	.50	.27
race	.18	-.01	.38	.20	.03	.45	-.10	.32	.38	1.00	.45	.40	.34
region	.15	-.03	.32	.18	.01	.43	-.03	.24	.48	.45	1.00	.49	.25
religion	.11	-.07	.30	.13	-.03	.29	.02	.24	.50	.40	.49	1.00	.18
wealth	.17	.03	.39	.41	.11	.28	-.07	.40	.27	.34	.25	.18	1.00

Because the sample size is large, most of the coefficients in Figure 17-2 are statistically different from 0, but that's not what you're looking for. You're looking for groups of variables that are correlated with each other. That's hard to see in a matrix of this size. In Figure 17-2, you don't see very large correlations between variables. The largest correlation coefficient, 0.56, is between the rated importance of clout and the rating of knowing the right people.

Are the Variables Related?

Linear factor analysis depends on sets of variables being linearly related. If that's the case, you expect the observed correlations between pairs of variables to decrease when you control for the linear effects of the other variables. This means that partial correlation coefficients between two variables, controlling for the linear effects of the other variables, should be small compared to the observed correlation coefficient between the two variables. (See Chapter 11.)

Kaiser-Meyer-Olkin Measures

The Kaiser-Meyer-Olkin (KMO) measure of sampling adequacy is an index that compares the sizes of the observed correlation coefficients to the sizes of the partial correlation coefficients. You can compute an overall measure for all of the variables combined and you can compute KMO measures for each of the variables individually.

The overall Kaiser-Meyer-Olkin measure of sampling adequacy is defined as:

$$KMO = \frac{\sum\sum_{i \neq j} r_{ij}^2}{\sum\sum r_{ij}^2 + \sum\sum a_{ij}^2}$$

The numerator is the sum of all of the squared correlation coefficients. The denominator is the sum of all of the squared correlation coefficients plus the sum of all of the squared partial correlation coefficients. If the ratio is close to 1, it means that all of the partial correlation coefficients are small, compared to the ordinary correlation coefficients. That's what you want to see because that indicates that the variables are linearly related. Small values for the KMO measure tell you that a factor analysis of the variables may not be a good idea, since observed correlations between pairs of variables cannot be explained by the other variables. From Figure 17-3, you see that the overall KMO measure is 0.82 for the getting-ahead data, indicating that it's reasonable to go ahead with the factor analysis.

Figure 17-3
KMO measure and Bartlett's test

Kaiser-Meyer-Olkin Measure of Sampling Adequacy.		.820
Bartlett's Test of Sphericity	Approx. Chi-Square	4086.321
	df	78
	Sig.	.000

Tip: Kaiser (1974) declares measures in the 0.90's as marvelous, in the 0.80's as meritorious, in the 0.70's as middling, in the 0.60's as mediocre, in the 0.50's as miserable, and below 0.50 as unacceptable.

The Kaiser-Meyer-Olkin measure for an individual variable is defined as:

$$MSA_i = \frac{\sum_{i \neq j} r_{ij}^2}{\sum r_{ij}^2 + \sum a_{ij}^2}$$

Instead of including all pairs of variables in the sums, only coefficients involving a particular variable are included. Again, reasonably large values are needed for a good factor analysis. The KMO measures for the individual variables are shown in column form in Figure 17-4. You see that all of the measures are reasonably large, so there's no need to consider eliminating any of the variables. You should consider eliminating variables with small values for the measure of sampling adequacy.

Figure 17-4
Kaiser-Meyer-Olkin measures

	Sampling Adequacy
ability	.835
ambition	.637
clout	.835
edparen	.780
edself	.644
gender	.859
hardwork	.573
knowright	.821
polbelief	.840
race	.874
region	.869
religion	.810
wealth	.850

Warning: You can't get SPSS Statistics to put the numbers in a column for you. SPSS Statistics shows KMO measures for individual variables on the diagonal of the anti-image correlation matrix, which is the matrix of the partial correlation coefficients with the signs reversed.

Bartlett's Test of Sphericity

You can test the null hypothesis that the observed data are a sample from a multivariate normal population in which all correlation coefficients are 0 using Bartlett's test of sphericity, also shown in Figure 17-3. If your data come from a population in which all correlation coefficients are 0, there's no point in factor analyzing the data. This test requires the assumption of multivariate normality and is very sensitive to deviations from the assumption. You're better off relying on the KMO measure.

Warning: For large sample sizes, even very small observed correlation coefficients will cause you to reject the null hypothesis that all correlation coefficients are 0.

Step 2: Estimating the Factors

Once you've established that you have variables that are linearly related to each other, you're ready to look for the factors that explain the observed correlations. Remember, the basic assumption of factor analysis is that underlying dimensions, or factors, are responsible for the observed correlations. Observed correlations between variables result from the *sharing* of these factors. For example, the correlation between the getting-ahead variables might be due to factors like personal achievement, family achievement, and basic demographics. Your mission in factor analysis is to identify these factors. Ideally, you'd like to identify a small number of easily interpretable factors.

Warning: You have to have a reasonable sample size to extract factors that are real. Many different recommendations for sample sizes have been suggested, with anywhere from 10 cases per variable to an overall sample size of at least 300 cases. However, an adequate sample size depends on the communalities and number of factors as well (MacCallum et al., 1999).

What's the Model?

The mathematical model for factor analysis looks like a multiple regression equation. Each variable is expressed as a linear combination of factors that are not actually observed. For example, the knowing-the-right-people variable might be expressed as

$$\begin{aligned}\text{knowing the right people} &= a(\text{societal factors}) + b(\text{family achievement})\\ &\quad + c(\text{personal achievement}) + U_{\text{knowing the right people}}\end{aligned}$$

This equation differs from the usual multiple regression because societal factors, family achievement, and personal achievement are not single independent variables. They are labels for groups of variables that characterize these concepts. They are factors. In exploratory factor analysis, factors are not known in advance but are discovered during factor analysis.

Societal factors, family achievement, and personal achievement are called **common factors** because all variables are expressed as functions of them. Variables are correlated because they share these common factors. The *U* is called a **unique factor** because it's that part of knowing-the-right-people that can't be explained by the common factors. The unique factor represents both the variable-specific component and the error. It's unique to the knowing-the-right-people variable.

The general model for the ith standardized variable is

$$X_i = A_{i1}F_1 + A_{i2}F_2 + \ldots + A_{ik}Fk + U_i$$

where the *F*'s are common factors, the *U* is the unique factor, and the *A*'s are the coefficients or loadings used to combine the *k* factors. The unique factors are assumed to be uncorrelated with each other and with the common factors.

The factors are in turn inferred from the observed variables and are estimated as linear combinations of the variables. For example, the personal achievement factor is expressed as

$$\begin{aligned} \text{personal achievement} = {} & W_1(\text{ability}) + W_2(\text{ambition}) \\ & + W_3(\text{clout}) + \ldots + W_{13}(\text{wealth}) \end{aligned}$$

It's possible that all of the variables contribute to personal achievement, but you hope that only a subset of the variables have large coefficients. The *W*'s are known as **factor score coefficients**.

Tip: You can also factor-analyze a covariance matrix. The results from analyzing a covariance matrix may differ from those obtained from analyzing the corresponding correlation matrix.

Principal Components Analysis

There are many different statistical algorithms for extracting factors from a correlation matrix (see Tabachnick and Fidell, 2001). The simplest method is called **principal components analysis**. In principal components analysis, linear combinations of the observed variables are formed. The first principal component is the combination that accounts for the largest amount of variance in the sample. The second principal component accounts for the next largest amount of variance and is uncorrelated with the first. Successive components explain progressively smaller portions of the total sample variance, and all are uncorrelated with each other.

Principal components analysis of a correlation matrix is often the starting point for factor analysis even when other methods are used to extract the final factors. (If you use principal components to extract factors, all output is labeled as *components* instead of *factors*. In this chapter, for simplicity, the terms are used interchangeably.) You can compute as many principal components as you have variables. If you use all of the principal components, you can reproduce each variable exactly, but then you haven't gained anything. You've just replaced the original variables with the same number of principal components.

Tip: The statistical model for principal components analysis is different from that for factor analysis. Factor analysis concerns itself with analyzing shared variance among the variables. Principal components analysis just restructures all of the observed variance by forming linear combinations of the observed variables. Controversy rages about whether principal component models or factor models are best. Wilkinson, Blank, and Gruber (1996) claims, "Principal component and common factor solutions for real data rarely differ enough to matter." See Joreskog (1979) for further discussion.

How Many Factors Do You Need?

Because you can calculate as many principal components as there are variables, you won't gain anything if you replace all of the variables with principal components. You have to determine how many factors you need to adequately represent your data; that is, to reproduce the observed correlations. You have to decide how many factors to keep.

Tip: Regardless of the method you use for factor extraction, the SPSS Statistics default for the number of factors to be used is based on the principal components solution. For extraction methods other than principal components, the maximum number of factors that can be extracted is one fewer than the number of variables.

Looking at Percentage of Variance Explained

The first step in deciding how many factors to keep is to look at the percentage of the total variance in the sample explained by each of the factors. (For simplicity, all variables and factors are expressed in standardized form with a mean of 0 and a variance of 1. Because there are 13 variables in this example, the total variance that can be explained is 13.)

In the *Total* column in Figure 17-5, you see the total variance explained by each factor. The factors are arranged in decreasing order of variance explained. The column labeled *% of Variance* is the percentage of the total variance attributable to each factor. For example, factor 2 has a variance of 1.78, which is 13.66% of the total variance of 13. The *Cumulative %* column is the sum of the percentage variances for that factor and the factors that precede it in the table. You see that about 50% of the total variance is explained by the first three factors. (Remember, the goal is to explain as much variance as possible using as few factors as possible.)

Figure 17-5
Total variance explained

		Initial Eigenvalues		
		Total	% of Variance	Cumulative %
Component	1	3.822	29.402	29.402
	2	1.776	13.660	43.062
	3	1.213	9.332	52.394
	4	.953	7.329	59.723
	5	.793	6.099	65.822
	6	.752	5.788	71.610
	7	.664	5.104	76.714
	8	.654	5.029	81.743
	9	.582	4.479	86.222
	10	.496	3.814	90.036
	11	.488	3.755	93.790
	12	.442	3.401	97.191
	13	.365	2.809	100.000

Extraction Method: Principal Component Analysis.

Warning: There is usually no right answer for the number of factors to use. If in doubt, consider several sensible solutions and evaluate which works best for your problem.

Examining Eigenvalues

Another criterion, aptly named the eigenvalue-greater-than-one criterion, suggests that only factors that account for variances greater than 1 should be included. (Eigenvalues are the variances of the factors.) Factors with a variance of less than 1 are no better than individual variables, since each variable has a variance of 1. This is what SPSS Statistics does if you don't specify other criterion, but it's not always the best strategy. Look at the actual eigenvalues because the difference between an eigenvalue of 0.98 and 1 is very small.

Warning: If you get the error message, "This matrix is not positive definite," it means that some of the eigenvalues of your correlation matrix are not positive numbers. This can occur if one or more of your variables is a linear combination of other variables in the matrix. This will be the case if you have more variables than cases. The SPSS Statistics Factor Analysis procedure uses listwise deletion by default, so make sure you know how many cases are left after cases with missing values are excluded. Pairwise deletion of data can also result in negative eigenvalues. If any eigenvalues are negative, all extraction methods will terminate. Two extraction methods, principal components and unweighted least squares, will extract factors if some eigenvalues are 0.

Looking at the Scree Plot

You can also base your decision on how many factors to keep on the scree plot, as shown in Figure 17-6. This is a plot of the total variance associated with each factor. Typically, the plot shows a distinct break between the steep slope of the large factors and the gradual trailing off of the rest of the factors, the scree that forms at the foot of a mountain. You should use only the factors before the scree begins. In this example, using all of the above criteria, it appears that three factors may be adequate.

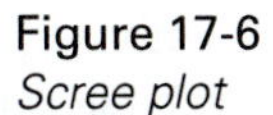

Figure 17-6
Scree plot

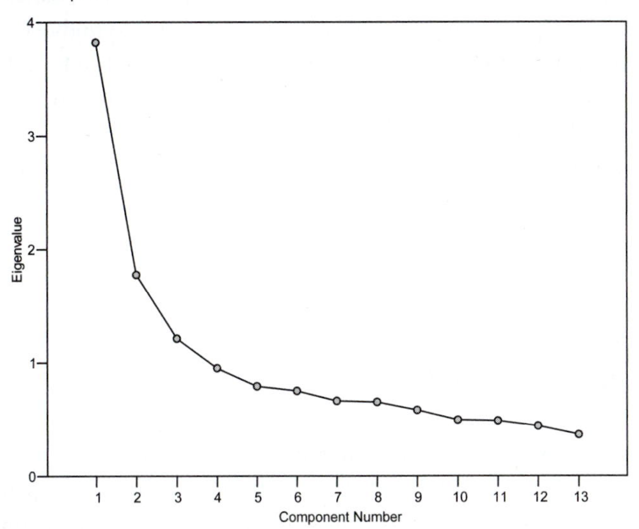

Goodness-of-Fit Test

If you use generalized least squares or maximum likelihood estimation to extract the factors and you are willing to assume that the sample is from a multivariate normal population, you can calculate goodness-of-fit tests for the adequacy of a model with a fixed number of common factors. For large sample sizes, the goodness-of-fit statistic tends to be distributed as a chi-square variate. In most cases, you don't know the number of common factors in advance, so you increase the number of factors until a reasonably good fit is obtained (until the observed significance level is no longer small).

The value of the chi-square statistic is directly proportional to the sample size. The degrees of freedom depends on the number of common factors and the number of variables. For large sample sizes, the goodness-of-fit test may cause rather small discrepancies in fit to be deemed statistically significant, resulting in more factors being extracted than necessary.

The goodness-of-fit chi-square for the getting-ahead example with three common factors and maximum likelihood extraction is shown in Figure 17-7. Because our observed sample size is quite large, the test will reject the null hypothesis even for small deviations from the estimated factor model. You may want to fit a three-factor model anyway if it fits the data reasonably well, since you want a good model that doesn't require many factors.

Figure 17-7
Goodness-of-fit test

Chi-Square	df	Sig.
349.203	42	.000

Selecting a Factor Extraction Method

Once you've decided how many factors you are going to keep, you must estimate the coefficients that relate the factors and variables. You can use the default principal components method to estimate the coefficients or one of the following methods that are available in SPSS Statistics. They are all iterative algorithms, except for image analysis.

- **Principal axis factoring** is very similar to principal components analysis, except that the diagonals of the correlation matrix are replaced by estimates of the communalities, unlike in principal components where the initial communalities are all 1. At the first step, squared multiple correlation coefficients are used as initial estimates of the communalities. Based on these, the requisite number of factors is extracted. The communalities are reestimated from the factor loadings, and factors are again extracted with the new communality estimates replacing the old. This continues until negligible change occurs in the communality estimates.
- **Unweighted least squares** produces, for a fixed number of factors, a factor pattern matrix that minimizes the sum of the squared differences between the observed and reproduced correlation matrices (ignoring the diagonal).
- **Generalized least squares** minimizes the same criterion as unweighted least squares, but correlations are weighted inversely by the uniqueness of the variables. Correlations involving variables with high uniqueness are given less weight than correlations involving variables with low uniqueness.
- **Maximum likelihood factor extraction** produces parameter estimates that are the most likely to have resulted in the observed correlation matrix if the sample is from a multivariate normal distribution. The correlations are weighted by the inverse of the uniqueness of the variables, and an iterative algorithm is used to extract factors.

- **Alpha factor analysis** considers the variables to be a sample from the universe of potential variables. It maximizes the alpha reliability of the factors. This differs from the previously described methods that consider the cases to be a sample from the population and the variables to be fixed.
- **Image factor analysis** analyzes the correlation matrix of the predicted values when each variable is predicted from the others using multiple regression.

Tip: If you get an error message that the solution failed to converge when extracting factors, increase the number of iterations.

How Are the Variables and Factors Related?

Using the coefficients produced by the factor extraction method that you selected, you can express each variable as a linear function of the factors. The coefficients in Figure 17-8 are from the default principal components solution.

Figure 17-8
Component matrix

Component Matrix[a]

	Component		
	1	2	3
ability	.358	.362	.119
ambition	.024	.656	.298
clout	.709	.067	-.204
edparen	.474	.347	-.413
edself	.166	.662	-.032
gender	.633	-.192	.145
hardwork	-.009	.592	.543
knowright	.633	.231	-.268
polbelief	.708	-.187	.237
race	.684	-.172	.059
region	.668	-.234	.315
religion	.606	-.268	.380
wealth	.595	.138	-.454

Extraction Method: Principal Component Analysis.
1. 3 components extracted.

For example,

$$\text{clout} = 0.709(\text{factor1}) + 0.067(\text{factor2}) - 0.204(\text{factor3}) + U_{\text{clout}}$$

$$\text{edself} = 0.166(\text{factor1}) + 0.662(\text{factor2}) - 0.032(\text{factor3}) + U_{\text{edself}}$$

The coefficients are sometimes called **factor loadings** because they tell you how much weight is assigned to each factor for each variable. If the factors are orthogonal (uncorrelated with each other), the factor-loading coefficients are also the correlation coefficients between the factors and the variables.

Tip: For centered factors or variables, the terms *orthogonal* and *uncorrelated* are equivalent. For factors or variables with means other than 0, the terms are not interchangeable.

For example, from Figure 17-8, the variable *clout* has a correlation of 0.709 with factor 1, 0.067 with factor 2, and –0.204 with factor 3. The *edself* variable has a different pattern of correlations with the factors. It is highly correlated with factor 2, somewhat correlated with factor 1, and almost uncorrelated with factor 3. You see that the first factor is most highly correlated with *clout*, *polbelief*, *race*, and *region*. The second factor is most highly correlated with *ambition*, *edself*, and *hardwork*. The matrix of correlations between variables and factors is called the **factor structure matrix**. For orthogonal factors, the factor structure matrix and the factor-loading matrix are one and the same.

Tip: Don't try to attach meaning to the factors yet. It will be easier later.

How Much Variance Is Explained?

Once you've extracted a smaller number of factors, you can evaluate how much of the observed variability in the data is explained by the factor solution. You can look at the percentage of total variance explained and at the percentage of variance explained for each variable.

Percentage of Total Variance Explained

If you use principal components to extract a smaller number of factors (the final factors), the final percentage of variance that is explained is the same as the initial percentage for the same number of factors. You can see this in Figure 17-9 in the columns labeled *Extraction Sums of Squared Loadings.*

Figure 17-9
Extraction statistics for principal components

		Initial Eigenvalues			Extraction Sums of Squared Loadings		
		Total	% of Variance	Cumulative %	Total	% of Variance	Cumulative %
Component	1	3.822	29.402	29.402	3.822	29.402	29.402
	2	1.776	13.660	43.062	1.776	13.660	43.062
	3	1.213	9.332	52.394	1.213	9.332	52.394

Extraction Method: Principal Component Analysis.

If you do not use principal components for extracting the final factors, the percentage of total variance explained will decrease. For example, Figure 17-10 results from using maximum likelihood to extract the three factors. From the *Cumulative %* column under *Extraction Sums of Squared Loadings*, you see that only 38% of the total variance is explained after three factors are extracted.

Figure 17-10
Extraction statistics for maximum likelihood

		Initial Eigenvalues			Extraction Sums of Squared Loadings		
		Total	% of Variance	Cumulative %	Total	% of Variance	Cumulative %
Factor	1	3.822	29.402	29.402	3.255	25.036	25.036
	2	1.776	13.660	43.062	1.094	8.416	33.452
	3	1.213	9.332	52.394	.633	4.871	38.323

Extraction Method: Maximum Likelihood.

Communality: Percentage of Variable's Variance Explained

To judge how well the three-factor model describes the original variables, you can compute the proportion of the variance of each variable that is explained by the three common factors. Because the factors are uncorrelated in this example, the total proportion of variance explained for a variable is just the sum of the variance proportions explained by each factor. (Remember that the square of the correlation coefficient between two variables tells you how much of the variance of one variable is explained by the other.)

For example, consider the wealth variable from Figure 17-8. Factor 1 explains $0.595^2 = 35.4\%$ of its variance, factor 2 explains $0.138^2 = 1.9\%$, and factor 3 explains $0.454^2 = 20.6\%$. The sum of these is 57.9%, the value shown for *wealth* in the column labeled *Extraction* in Figure 17-11.

Figure 17-11
Communalities

	Initial	Extraction
ability	1.000	.274
ambition	1.000	.520
clout	1.000	.548
edparen	1.000	.516
edself	1.000	.467
gender	1.000	.459
hardwork	1.000	.646
knowright	1.000	.526
polbelief	1.000	.592
race	1.000	.501
region	1.000	.600
religion	1.000	.584
wealth	1.000	.579

Extraction Method: Principal Component Analysis.

The proportion of variance explained by the common factors is called the **communality** of the variable. Communalities can range from 0 to 1, with 0 indicating that the common factors don't explain any of the variance, and 1 indicating that all of the variance is explained by the common factors. The variance that is not explained by the common factors is attributed to the unique factor for each variable. That's why it's called the uniqueness. In a successful factor analysis, the communalities for variables should be large. If there are variables with small communalities, that means they can't be predicted by the common factors. If the values are small enough, you should consider removing them from the analysis, since they are not linearly related to the other variables.

Tip: You can use factor analysis to screen questionnaires for items that are not related to the other items.

Initial and Final Communalities

Principal components analysis starts with as many components as there are variables, so the components together explain all of the observed variability in each of the variables. That's why the column labeled *Initial* in Figure 17-11 has values of 1 for the communality of each variable. Reducing the factors to just three substantially reduced the communalities for each of the variables.

If you use a method other than principal components to extract the final factors, the initial communalities will not be 1. As shown in Figure 17-12 for maximum likelihood extraction, the initial communalities are the squared multiple correlation coefficients when each variable is predicted from all of the others.

Figure 17-12
Communalities for maximum likelihood extraction

	Initial	Extraction
ability	.139	.167
ambition	.161	.267
clout	.436	.515
edparen	.263	.258
edself	.197	.247
gender	.333	.355
hardwork	.176	.423
knowright	.382	.485
polbelief	.437	.510
race	.370	.406
region	.390	.505
religion	.370	.461
wealth	.314	.382

Extraction Method: Maximum Likelihood.

Can You Reproduce the Observed Correlations?

In factor analysis, you assume that the observed correlations between variables are due to the sharing of common factors. You can use the estimated correlations between the factors and the variables to reproduce the correlations between pairs of variables. The reproduced correlation between two variables is the sum, for all factors, of the products of the correlation of the two variables with that factor.

$$r_{ij} = \sum_{f=1}^{k} r_{fi} r_{fj} = r_{1i} r_{1j} + r_{2i} r_{2j} + \ldots + r_{ki} r_{kj}$$

For example, the reproduced correlation between wealth and parental education is

$$r_{\text{wealth, edparen}} = (0.595)(0.474) + (0.138)(0.347) + (-0.454)(-0.413) = 0.517$$

You calculate the residual by finding the difference between the observed and reproduced correlations. The observed correlation coefficient between wealth and parental education is 0.409, so the residual is –0.108. Large residuals indicate that the factor model does not fit the data well and should be reevaluated.

You see both the reproduced correlation and the residual in Figure 17-13, which is an excerpt from the reproduced correlation matrix for all variables. The diagonal of the reproduced correlation matrix shows the communalities. Beneath the table, the number and percentage of residuals greater in absolute value than 0.05 is reported. A good solution has a small percentage of large residuals.

Figure 17-13
Reproduced correlations

		edparen	wealth
Reproduced Correlation	edparen	.516[2]	.517
	wealth	.517	.579[2]
Residual[1]	edparen		-.108
	wealth	-.108	

Extraction Method: Principal Component Analysis.

1. Residuals are computed between observed and reproduced correlations. There are 43 (55.0%) nonredundant residuals with absolute values greater than 0.05.
2. Reproduced communalities

Warning: Figure 17-13 is a very small piece of the output that you get when you ask for the reproduced correlation matrix.

Step 3: Making Factors Easier to Interpret: Rotation

If you look at Figure 17-8, you can see that some of the variables are more highly correlated with some factors than others. The correlation patterns between factors and variables are very important because you use them to interpret the factors. Ideally, you would like to see groups of variables with large coefficients for one factor and small coefficients for the others. That makes it easier to assign meanings to the factors.

Warning: You don't want to have factors that are defined by a single variable. A common rule of thumb is that each factor should have at least three variables that load highly on it.

Figure 17-14 is an example of an ideal correlation matrix between factors and variables. (Look for it in your dreams, together with winning lottery numbers.) From Figure 17-14, it's easy to identify the first factor as societal attributes: gender, race, religion, region, and political beliefs. The next set of variables define a family factor: your parents' education, wealth, introductions to the right people, and clout. The last factor might be labeled as personal achievement. It's how hard you work, how much education you have, your abilities, and your ambitions.

Tip: It's the absolute value of the correlation coefficient that matters. Large negative correlations are as desirable as large positive correlations.

Figure 17-14

Hypothetical rotated correlation matrix

	Component		
	1	2	3
region	.900	.000	.000
religion	.800	.000	.000
polbelief	.950	.000	.000
gender	.890	.000	.000
race	.830	.000	.000
wealth	.000	.900	.000
edparen	.000	.850	.000
knowright	.000	.920	.000
clout	.000	.880	.000
hardwork	.000	.000	.940
ambition	.000	.000	.966
edself	.000	.000	.875
ability	.000	.000	.923

Extraction Method: Principal Component Analysis.
Rotation Method: Varimax with Kaiser Normalization.

Your goal in the next step, the rotation phase of factor analysis, is to try to morph Figure 17-8 into Figure 17-14 without resorting to editing the pivot table. The purpose of rotation is to achieve a **simple structure**. In a simple structure, each factor has large loadings in absolute value for only some of the variables, making it easier to identify. You also want each variable to have large loadings for only a few factors, preferably one. That also helps to differentiate the factors from each other. If several factors have high loadings on the same variables, it's difficult to determine how the factors differ.

Rotation does not affect the goodness of fit of a factor solution. The communalities and percentage of total variance accounted for remain the same. The percentage of variance accounted for by each of the factors does, however, change. Rotation redistributes the explained variance for the individual factors. Different rotation methods may result in different factors being identified.

Examining the Basics of Rotation

To see what's involved in rotation, you'll consider two simple examples: one for orthogonal rotation and one for oblique rotation. If you rotate the axes and keep them perpendicular (at right angles) to each other, the rotation is called **orthogonal**. If the axes are not maintained at right angles, the rotation is called **oblique**. You'll rotate the factor axes so that variables have large correlations with a small numbers of factors. You'll try to make large loadings larger and small loadings smaller to make the factors easier to interpret.

Tip: Always start with orthogonal rotation, especially if you are not an experienced factor analyst.

Understanding Orthogonal Rotation

As a simple example of orthogonal rotation, consider Figure 17-15, a plot of factor loadings for four variables (the ratings of the importance of dinner, sleep, vacation, and research) for a random sample of SPSS Statistics users. The factor-loading matrix is shown next to it. The coordinates for each variable on the plot are the correlation coefficients (factor loadings) with the two factors. For example, the coordinates for sleep are (0.5, –0.4). From the plot, it's difficult to interpret either of the two factors, since all variables are related to all factors.

Figure 17-15
Before rotation

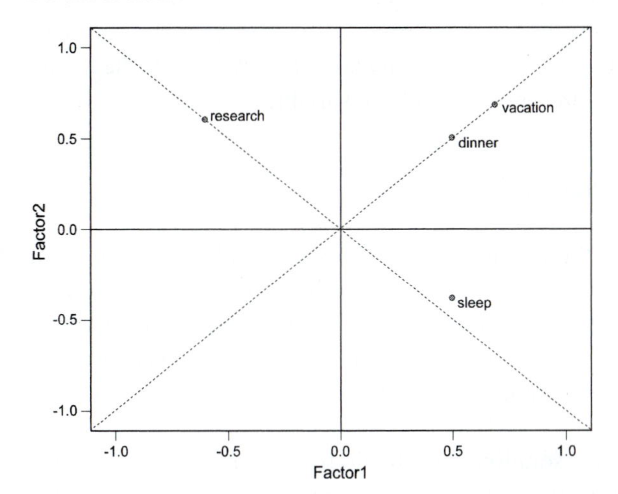

	Factor	
	1	2
Dinner	0.5	0.5
Sleep	0.5	–0.4
Vacation	0.7	0.7
Research	–0.6	0.6

Now consider the plot in Figure 17-16 and the factor matrix next to it. You see that the relationship between factors and variables is much simpler. Dinner and vacations are highly related to factor 1. Research and sleep are highly related to factor 2. By looking at what these pairs of variables have in common, you might identify factor 1 as representing hedonism and factor 2 as scholarship.

Figure 17-16
After rotation

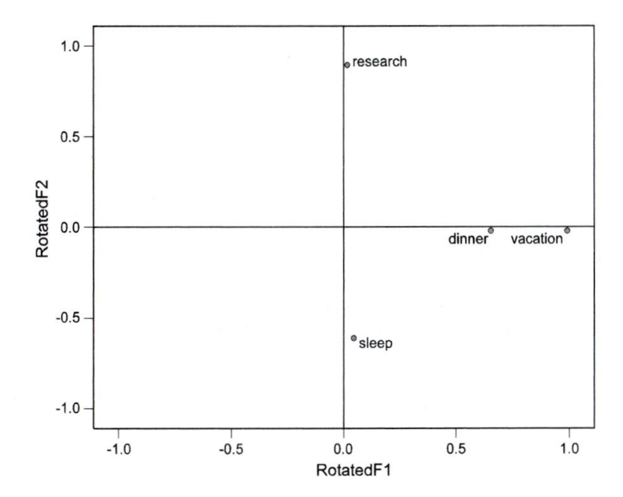

	Factor	
	1	2
Dinner	0.71	–0.02
Sleep	0.05	–0.64
Vacation	0.99	–0.03
Research	0.02	0.85

To get from Figure 17-15 to Figure 17-16, you make the dotted lines in Figure 17-15 the axes. That is, you rotate the axes. Because the axes are perpendicular (90° right angles), the rotation is orthogonal. By rotating the axes, you've achieved a much simpler structure, since variables are highly correlated with only one factor, and factors have large correlations for only some of the variables.

Understanding Oblique Rotation

Figure 17-17 is a plot of the factor loadings for six variables. You can't find orthogonal axes around which the points cluster. You can, however, draw in a new set of axes, shown in the plot. These axes are called **oblique axes** because they are not at right angles to each other. The consequence of an oblique rotation is that the factors are correlated. That makes them more difficult to interpret.

Tip: To get a good factor solution, you should try different methods of extraction and rotation to see which results in the most easily interpretable solution for your problem. You get to determine what the right solution is.

Figure 17-17
Oblique rotation

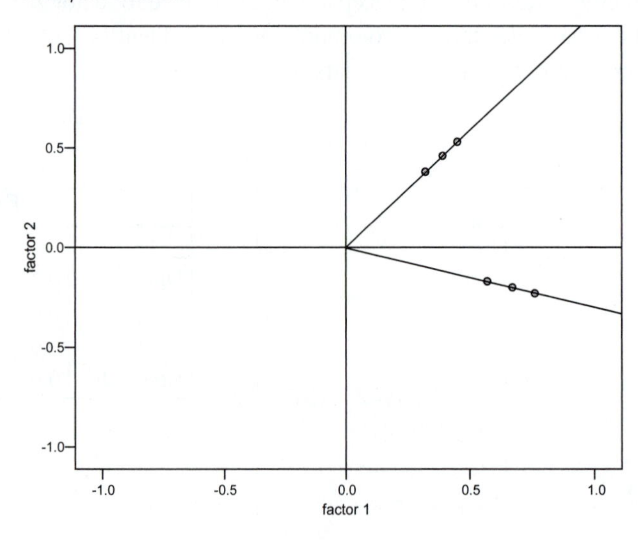

Orthogonal Rotation of Getting-Ahead Data

Now that you understand the basics, you're ready to continue with the example. You'll use varimax orthogonal rotation to rotate the factor loadings in Figure 17-8. Varimax orthogonal rotation is the most frequently used rotation method. It minimizes the number of variables that have high loadings on a factor so that the factors can be interpreted more easily.

Figure 17-18 shows the factor matrix in Figure 17-8 after varimax rotation and after the variables have been sorted by the absolute values of the loadings. It's not the ideal matrix shown in Figure 17-14, but it's much closer to it than the original unrotated matrix. The large coefficients are larger; the small coefficients are smaller.

Figure 17-18
Rotated component matrix

	Component		
	1	2	3
region	.768	.103	.019
religion	.764	.006	.017
polbelief	.743	.199	.024
gender	.641	.216	-.034
race	.631	.316	-.055
wealth	.193	.734	-.055
edparen	.040	.704	.134
knowright	.280	.658	.122
clout	.432	.601	.019
hardwork	.040	-.184	.781
ambition	-.078	.033	.716
edself	-.132	.356	.568
ability	.206	.262	.404

Extraction Method: Principal Component Analysis.
Rotation Method: Varimax with Kaiser Normalization.

To make it easier to identify factors, you can suppress the display of small coefficients. In Figure 17-19, correlations less than 0.3 are not shown. You see three sets of variables. Region, religion, political belief, gender, and race are highly correlated with factor 1. Wealth, parental education, knowing the right people, and clout are associated with factor 2. Hard work, ambition, personal education, and ability are associated with factor 3.

Clout, race, and personal education are not as clearly associated with a single factor as are the other 10 variables. It's not particularly surprising that a person's education is correlated with both the family factor and the personal achievement factor. Clout and race are also related to more than one of these factors.

Figure 17-19
Rotated component matrix with small coefficients suppressed

	Component		
	1	2	3
region	.768		
religion	.764		
polbelief	.743		
gender	.641		
race	.631	.316	
wealth		.734	
edparen		.704	
knowright		.658	
clout	.432	.601	
hardwork			.781
ambition			.716
edself		.356	.568
ability			.404

Extraction Method: Principal Component Analysis.
Rotation Method: Varimax with Kaiser Normalization.

Plotting the Factors

A good way to examine the success of an orthogonal rotation is to plot the variables, using the factor loadings as the coordinates. When the factor solution involves three or more factors, SPSS Statistics produces a three-dimensional plot of the first three factors, as shown in Figure 17-20. The coordinates in Figure 17-20 are the factor loadings for the varimax-rotated solution. You see how the variables cluster. The first two factors have strong clusters of variables associated with them. The variables that define factor 3 don't cluster as tightly as you would like.

Figure 17-20
Component plot in rotated space

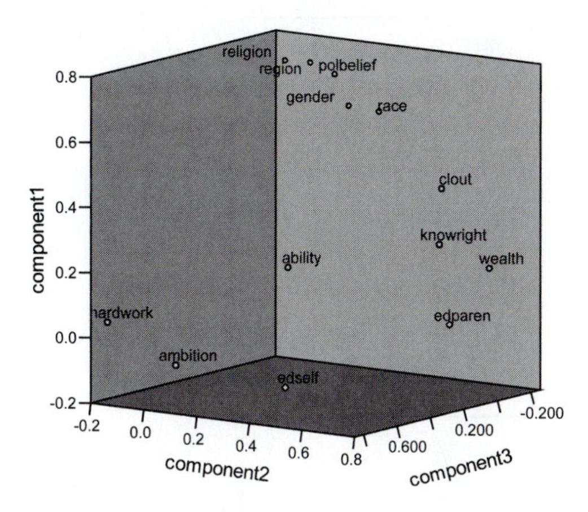

Tip: It's much easier to see the three-dimensional structure of a plot like Figure 17-20 while you rotate it in the Chart Editor.

You can also plot pairs of factors, as shown in Figure 17-21. Ideally, you want to see clusters of variables at +1 and –1, at the tips of the axes, and clusters at the origin. For any pair of factors, you want a variable to have a large correlation with only one of the factors. If you have more than two factors, it's also fine for variables to have small correlations with both of the factors being plotted, provided that they have a large correlation with one of the other factors in the analysis.

Figure 17-21
Pairwise factor plot

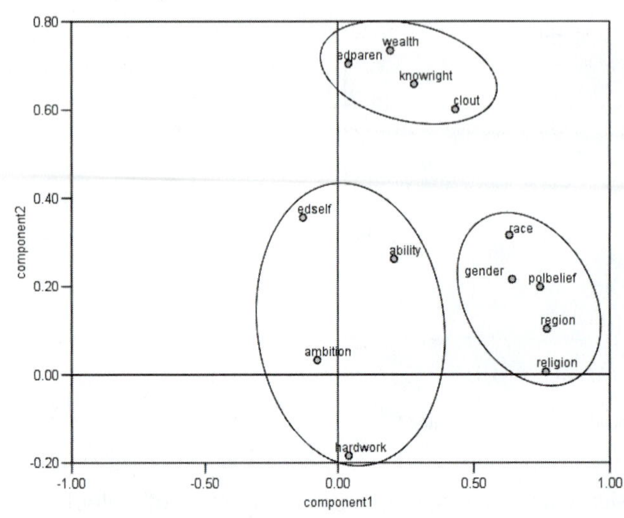

Variance-Explained Statistics

When factors are rotated, the cumulative percentage of explained variance does not change; however, the variance attributable to the individual factor does. The variance is reallocated across the factors, as shown in Figure 17-22.

Figure 17-22
Variance-explained statistics

		Extraction Sums of Squared Loadings			Rotation Sums of Squared Loadings		
		Total	% of Variance	Cumulative %	Total	% of Variance	Cumulative %
Component	1	3.822	29.402	29.402	2.905	22.344	22.344
	2	1.776	13.660	43.062	2.256	17.356	39.699
	3	1.213	9.332	52.394	1.650	12.695	52.394

Extraction Method: Principal Component Analysis.

You see that the second factor accounted for 43% of the variance after extraction and almost 40% of the variance after rotation.

Oblique Rotation of Getting-Ahead Data

Orthogonal rotation results in factors that are uncorrelated. This is an appealing property, but sometimes allowing for correlations among factors does simplify the solution. It is unlikely that influences in nature are uncorrelated, so oblique rotations have often been found to yield meaningful factors. SPSS Statistics provides two methods for oblique rotation: direct oblimin and promax.

Warning: Oblique rotation makes the factors more difficult to interpret because they are correlated with each other.

Oblique rotation preserves the communalities of the variables, as does orthogonal rotation. When oblique rotation is used, however, the factor loading and factor structure matrices are no longer identical. There are now two matrices: a factor pattern matrix with the loadings that relate the factors and the variables, and a factor structure matrix with the correlation coefficients between the factors and the variables.

The factor pattern matrix shown in Figure 17-23 contains the coefficients that relate the factors to the variables after a promax oblique rotation. (That's one of two methods available for oblique rotation in SPSS Statistics.)

Figure 17-23
Pattern matrix

	Component		
	1	2	3
religion	.819	-.159	.039
region	.797	-.054	.031
polbelief	.745	.055	.025
gender	.627	.100	-.038
race	.588	.212	-.070
wealth	.004	.765	-.126
edparen	-.143	.755	.063
knowright	.127	.653	.062
clout	.300	.564	-.030
hardwork	.132	-.256	.807
ambition	-.057	.010	.716
edself	-.211	.385	.532
ability	.170	.218	.386

Extraction Method: Principal Component Analysis.
Rotation Method: Promax with Kaiser Normalization.

The factor structure matrix in Figure 17-24 contains the correlations of the individual variables and the factors. From the structure matrix, you see that oblique rotation resulted in somewhat different factors from orthogonal rotation. You expect that. The first factor is still highly correlated with political beliefs, region, religion, race, gender, and clout. The second factor is most highly correlated with wealth, knowing the right people, parental education, and clout. The third factor has the largest correlation coefficients with hard work, ambition, and personal education.

Figure 17-24
Structure matrix

	Component		
	1	2	3
region	.773	.287	-.003
polbelief	.767	.374	.007
religion	.750	.192	-.010
race	.680	.452	-.062
gender	.671	.361	-.047
wealth	.332	.751	-.027
knowright	.401	.715	.143
edparen	.174	.702	.166
clout	.540	.687	.033
hardwork	-.004	-.096	.770
ambition	-.077	.079	.719
edself	-.066	.365	.589
ability	.248	.340	.409

Extraction Method: Principal Component Analysis.
Rotation Method: Promax with Kaiser Normalization.

The factor correlation matrix in Figure 17-25 tells you how much the factors are correlated. Factors 1 and 2 are the most highly correlated.

Figure 17-25
Factor correlation matrix

		1	2	3
Component	1	1.000	.423	-.034
	2	.423	1.000	.129
	3	-.034	.129	1.000

Extraction Method: Principal Component Analysis.
Rotation Method: Promax with Kaiser Normalization.

Algorithms for Rotation

There are many different algorithms for both orthogonal and oblique rotation. The methods for orthogonal factor rotation available in SPSS Statistics are:

Varimax. Maximizes the variance among the loadings on a factor, which results in loadings that are either large or small and fewer intermediate values.

Quartimax. Minimizes the number of factors needed to explain a variable. A quartimax rotation often results in a general factor with high to moderate loadings on most variables. This is one of its main shortcomings.

Equamax. Is a combination of the varimax method, which simplifies the factors, and the quartimax method, which simplifies variables.

Two algorithms are available for oblique rotation:

Direct oblimin. Is controlled by a parameter (delta) that determines how oblique (far from orthogonal) the axes are. SPSS Statistics accepts numbers less than or equal to 0.8. Usually numbers less than or equal to 0 should be used. The value of 0 corresponds to the most oblique axes within this range. Large negative values of delta will lead to factors that are nearly orthogonal. There is not a single best value because, if there were, you wouldn't be given all these choices!

Promax. Is controlled by a parameter (kappa). A kappa value of 1 gives a varimax orthogonal rotation. Promax starts with an orthogonal rotation and raises the loadings to the specified kappa. The correlation between factors increases with increasing values of kappa. Larger values of kappa result in more loadings that are close to 0 and 1 but at the expense of increasing the correlation between factors. Values less than 4 are generally recommended.

Tip: Try different values of the rotation parameters until you find one that seems to make sense.

Step 4: Computing Factor Scores

Once you've identified the factors and found them sufficiently useful as summary measures, you can calculate factor scores for each case. The factor scores represent how much of each factor a case has. Unfortunately, except for principal components, you can't calculate exact scores. You must estimate the factor scores with imperfect approximations. For example, these approximations can result in estimated factor scores that are correlated even when factors are orthogonal.

Tip: If you want to calculate factor scores for cases that are not included in the analysis, merge them into the data file you used to perform the factor analysis. Create a variable that has different values for cases used in the initial computations for cases that you want scored. Specify this variable in the Selection Variable box in factor and indicate the value it must have for cases to be included in the estimation of the factor solution.

Methods for Calculating Factor Scores

There are several methods for estimating factor score coefficients. The three methods available in SPSS Statistics are regression, Anderson-Rubin, and Bartlett. All three result in scores with a mean of 0. The Anderson-Rubin method always produces uncorrelated scores with a standard deviation of 1, even if the original factors are correlated.

The regression factor scores have a variance equal to the squared multiple correlation between the estimated factor scores and the true factor values (always 1 for principal components). Regression method factor scores can be correlated even when factors are orthogonal. Bartlett factor scores minimize the sum of squares of the unique factors over the variables. They can also be correlated when factors are orthogonal. If principal components extraction is used, all three methods result in the same factor scores (which in this case are no longer estimated but are exact).

Tip: If you want factor scores, consider using a principal components solution and saving exact scores, or you can compute unit weighted factor scores that are just sums of the values of the primary variables that define the factor.

Factor Score Coefficients

Coefficients for regression factor scores are in Figure 17-26. Because you used principal components to extract the factors, all three types of factor scores give the same results.

Figure 17-26
Component score coefficient matrix

	Component		
	1	2	3
ability	.046	.059	.233
ambition	-.011	-.041	.442
clout	.049	.246	-.035
edparen	-.143	.386	.008
edself	-.118	.176	.311
gender	.229	-.021	-.018
hardwork	.094	-.202	.512
knowright	-.027	.304	.016
polbelief	.281	-.063	.025
race	.201	.041	-.042
region	.314	-.123	.034
religion	.335	-.179	.043
wealth	-.091	.388	-.107

Extraction Method: Principal Component Analysis.
Rotation Method: Varimax with Kaiser Normalization.
Component Scores.

The factor score for case *j* for factor *k* is

$$\hat{F}_{jk} = \sum_{i=1}^{p} W_{ji} X_{ik}$$

For the first factor in your example, the score would be

$$\hat{F}_1 = W_1(\text{ability}) + W_2(\text{ambition}) + W_3(\text{clout}) + \ldots + W_{13}(\text{wealth})$$

where the *W*'s are the factor score coefficients for factor 1 in Figure 17-26 and where all variables are standardized.

Tip: You can plot factor scores and look for unusual cases.

Using Factor Scores to Test Hypotheses

You can use the factor scores in place of ordinary variables in subsequent analyses. For example, you can test the null hypothesis that men and women have the same average values for each of the three orthogonal factors. The *t* tests are shown in Figure 17-27.

Figure 17-27
Independent-samples t test

		t-test for Equality of Means				
		t	df	Sig. (2-tailed)	Male-Female Mean Difference	Std. Error Difference
Demographics	Equal variances	.159	1333	.874	.00879	.05523
Family	Equal variances	-4.117	1333	.000	-.22597	.05489
Achievement	Equal variances	1.970	1333	.049	.10866	.05515

You see that there is no significant difference between men and women on the importance of the societal, or demographic, factor (factor 1). However, men and women have significantly different mean scores on the family and personal achievement factors. For each of the original variables, high scores indicate that the variable isn't thought to be important, and low scores indicate that it is considered important for getting ahead. Large factor scores, therefore, indicate that a respondent thought the variables associated with that factor are unimportant for getting ahead. A negative difference between the scores for men and women indicates that women found the factor to be less important than did the men. Women rate personal achievement as more important than men do. Men rank family characteristics as more important than do women. Factor analysis lets you analyze meaningful combinations of the original variables and simplifies the analysis of a large number of variables.

Warning: It's hard to determine *substantive significance* when you're using factor scores because differences in factor scores are difficult to interpret.

Megatip: Entering a Correlation Matrix

If you want to factor-analyze a correlation matrix for which you don't have the raw data, you must figure out the simplest way to feed the matrix to the Factor Analysis procedure. It's easiest to just open a syntax window and paste (or type) your correlation matrix into the Syntax Editor. Then you have to surround it with the appropriate SPSS Statistics commands, as shown in Figure 17-28.

Figure 17-28
SPSS Statistics Syntax Editor

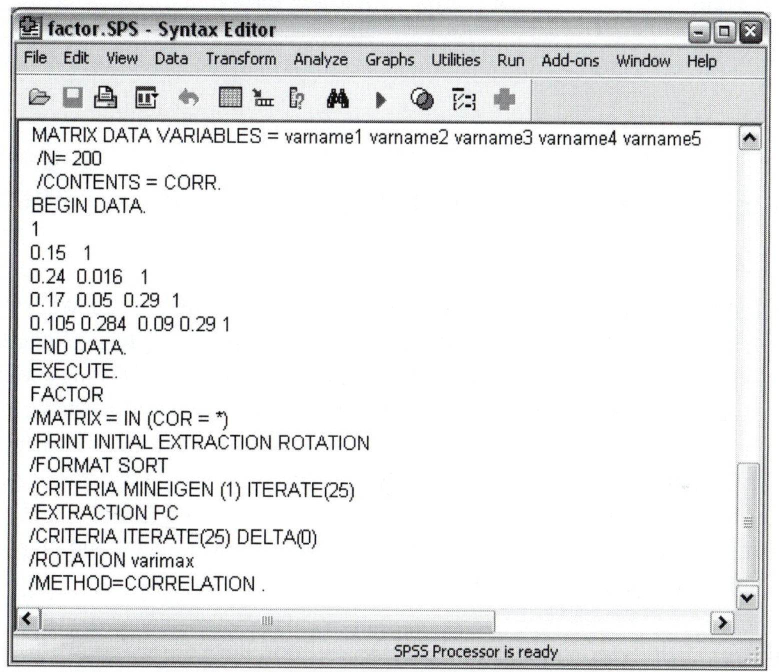

Here's what you do:

- After MATRIX DATA VARIABLES = , list the names of the variables that are in the correlation matrix.
- After N = , specify the sample size from which the matrix was calculated.
- Include CONTENTS = CORR to indicate that a correlation matrix follows.

- Make sure that there is not a period after the variable list or the sample size. Put the periods exactly where the example shows.
- Include BEGIN DATA to signify that the correlation matrix follows.
- Enter the correlation matrix in lower triangular form with 1's on the diagonals. Remember, a correlation matrix is symmetric and has 1's on the diagonals. You want to enter the bottom part and the diagonal. You'll have as many rows as you have variables. (You can use a FORMAT subcommand to indicate other forms of the matrix. In the Help system, search for matrix data to see the syntax.)
- Include END DATA to signify that the correlation matrix has ended.
- Use EXECUTE to tell SPSS Statistics to read the data values.
- Start the Factor specification with FACTOR / MATRIX = IN (COR = *). Don't terminate the command with a period yet.
- Modify the remaining factor specifications to obtain the analysis that you want. The simplest way to do that is to go into the Factor Analysis dialog box, select the analysis that you want, and then click the Paste button. This pastes a FACTOR command into the syntax window. You can then type / MATRIX = IN (COR = *) after the word FACTOR. You can see the FACTOR syntax by clicking Help in the Factor Analysis dialog box, and then clicking the Factor Command Syntax link.
- To run the commands in the Syntax Editor, select the Run menu and make the appropriate choice.

  Warning: The matrix will appear in the Data Editor, but you can't run a factor analysis on it in the usual way. You must use the Syntax Editor each time.

Obtaining the Output

To produce the output in this chapter, open the data file *gettingahead.sav* and follow the instructions below.

Figures 17-2, 17-3, 17-5, 17-6, 17-8, 17-9, 17-11, 17-13, 17-18, 17-20, 17-22, and 17-26. From the Analyze menu choose Dimension Reduction > Factor. Move *ability*, *ambition*, *clout*, *edparen*, *edself*, *gender*, *hardwork*, *knowright*, *polbelief*, *race*, *region*, *religion*, and *wealth* into the Variable(s) list. Click Descriptives. In the Descriptives subdialog box, select Initial solution, Coefficients, Reproduced, Anti-image, and KMO and Bartlett's test of sphericity. Click Continue and then click Extraction.

In the Extraction subdialog box, select Unrotated factor solution and Scree plot, and then click Continue. Click Rotation. In the Rotation subdialog box, select Varimax, Rotated solution, and Loading plot(s). Click Continue, and then click Scores. In the Factor Scores subdialog box, select Save as variables and Display factor score coefficient matrix, and then click Continue. Click Options. In the Options subdialog box, select Sorted by size. Click Continue, and then click OK.

Figure 17-4. Pluck the diagonal entries from the anti-image covariance matrix.

Figures 17-7, 17-10, and 17-12. Repeat the first set of instructions except, in the Extraction subdialog box, select Maximum likelihood.

Figures 17-14 to 17-17 cannot be produced from the data file.

Figure 17-19. Repeat the first set of instructions except, in the Options subdialog box, select Sorted by size and Suppress absolute values less than, then type 0.30 into the box. Click Continue and then click OK.

Figure 17-21. You can rotate the three-dimensional plot to show only two dimensions.

Figures 17-23 to 17-25. Repeat the first set of instructions except, in the Rotation subdialog box, select Promax, Rotated solution, and Loading plot(s).

Figure 17-27. From the Analyze menu choose Compare Means > Independent-Samples T Test. Move *Fac1_1*, *Fac2_1*, and *Fac3_1* into the Test Variable(s) box. Move *sex* into the Grouping Variable box. Click Define Groups. In the Define Groups box, type 1 into the Group 1 box and type 2 into the Group 2 box. Click Continue, and then click OK.

Chapter
18
Reliability Analysis

When you want to evaluate some attribute of a person, such as driving ability, mastery of course materials, or the ability to function independently, you need a measurement device. Often this is a scale or test that consists of multiple individual items. The responses to each of the items can be graded and summed, resulting in a score for each case. A question that frequently arises is: How good is the scale? Two properties of scales are reliability and validity.

A valid test measures what it is supposed to. If you ask students to do mirror drawing and memorize nonsense syllables, that's not a valid indicator of mastery of the concepts of psychology. Similarly, if you ask people about the mechanics of their cars, that's not a valid indicator of driving ability. It's much more complicated to determine whether a particular scale is a valid measure of employee performance, psychological well-being, or parenting skills.

The purpose of a scale is to quantify some underlying dimension. The reliability of a test is a measure of the correlation between scores on the test and the hypothetical "true" value. Three ways of estimating reliability are internal consistency, or scale reliability, which is the degree to which items on the same test measure the same thing; test-retest reliability, which is the degree to which a test yields similar results on several administrations or with parallel tests; and inter-rater reliability, which is the degree to which multiple raters assign the same scores.

Examples

- You have developed a 20-item scale that measures distrust. You want to know how well the overall scale performs.
- You want to determine how much the reliability of a scale changes when you eliminate an item from the scale.
- You want to quantify the agreement among raters in judging the severity of a disease.

In a Nutshell

A scale is composed of many items that are supposed to be tapping into an underlying dimension. A good test results in scores that correlate well with the unknown true score. Among other ways, you can estimate the reliability of a test by examining the internal consistency of the items within a scale or by splitting the scale into parts and looking at the relationship between the two parts.

Using intraclass correlation coefficients, you can summarize the agreement between judges when rating the same items (the cases). The intraclass correlation coefficients can measure either consistency or absolute agreement between judges. The selection of a coefficient to use depends on whether the judges are considered to be fixed or random factors.

Anomie: The Example

Emile Dirkheim, a French sociologist, pioneered the study of **anomie**, the breakdown of social norms. In this example, you'll consider a scale that purports to measure anomie. The scale was administered to 1,334 people by the General Social Survey. It consists of nine statements, all of them pessimistic, and each answered *agree* or *disagree*. The actual statements are given in Figure 18-1. The *agree* responses are coded as 1's; the *disagree* responses are coded as 0's. The value for the scale is the number of questions to which a person answers *agree*. The scale can range in value from 0 to 9.

Tip: Make sure that scale items are coded consistently so that large values for all items indicate the same thing. For example, if your scale has some questions where *strongly agree* points to the presence of some psychiatric disturbance and other questions where *strongly disagree* points to the same disturbance, make sure you recode the values so that large scores have the same meaning on all items.

Figure 18-1 shows means and standard deviations for each of the questions. The highest agreement is for the statement *Don't know whom to trust.* The smallest agreement is for *No right and wrong ways to make money.*

Figure 18-1
Item statistics

	Mean	Std. Deviation	N
Next to health, money is most important	.31	.463	3975
Wonder if anything is worthwhile	.41	.492	3975
No right and wrong ways to make money	.25	.431	3975
Live only for today	.44	.497	3975
Lot of average man getting worse	.58	.493	3975
Not fair to bring child into world	.39	.488	3975
Officials not interested in average man	.63	.482	3975
Don't know whom to trust	.72	.450	3975
Most don't care what happens to others	.56	.497	3975

You can also calculate correlation coefficients and covariances between the items, as well as averages of the pairwise statistics, as shown in Figure 18-2. You see that the average scores for an item range from 25% agreeing to 72% agreeing, a range of 47%. The correlation coefficients between pairs of questions range from 0.11 to 0.44.

Figure 18-2
Summary item statistics

	Mean	Minimum	Maximum	Range	Maximum / Minimum	Variance	N of Items
Item Means	.477	.247	.718	.471	2.905	.024	9
Item Variances	.228	.186	.247	.061	1.327	.000	9
Inter-Item Covariances	.055	.024	.104	.080	4.405	.000	9
Inter-Item Correlations	.241	.105	.438	.332	4.157	.006	9

The covariance matrix is calculated and used in the analysis.

Figure 18-3 shows statistics for the entire scale. The average score is 4.3; the standard deviation is 2.5.

Figure 18-3
Scale statistics

Mean	Variance	Std. Deviation	N of Items
4.29	6.017	2.453	9

Warning: If your scale measures several dimensions of the same construct, you don't want an analysis that is based on pooling all of the items into a single scale. For example, if your total scale is composed of items that measure mathematical ability and items that measure verbal ability, you may want to analyze the two subscales separately because they are tapping into two different dimensions.

How Are the Individual Items Related to the Overall Score?

To see the relationship between the individual items and the composite score, look at Figure 18-4. For each item, the first column shows what the average score for the scale would be if the item were excluded from the scale. You know from Figure 18-3 that the average score for the scale is 4.3. If item 1 is eliminated from the scale, the average score would be the average scale score minus the average item score (4.29 minus 0.31, or 3.98).

Figure 18-4
Item-total statistics

	Scale Mean if Item Deleted	Scale Variance if Item Deleted	Corrected Item-Total Correlation	Squared Multiple Correlation	Cronbach's Alpha if Item Deleted
Next to health, money is most important	3.98	5.151	.309	.111	.735
Wonder if anything is worthwhile	3.88	4.855	.424	.193	.717
No right and wrong ways to make money	4.05	5.089	.381	.160	.724
Live only for today	3.85	5.042	.326	.115	.734
Lot of average man getting worse	3.71	4.761	.471	.254	.708
Not fair to bring child into world	3.91	4.690	.516	.293	.701
Officials not interested in average man	3.66	4.995	.366	.176	.726
Don't know whom to trust	3.58	4.892	.463	.264	.711
Most don't care what happens to others	3.74	4.694	.500	.294	.703

The column labeled *Corrected Item-Total Correlation* is the Pearson correlation coefficient between the score on the individual item and the sum of the scores on the remaining items. For example, the correlation between the score on item 8 and the sum of the scores on all items except 8 (*Don't know whom to trust*) is 0.463. Item 1 has the lowest correlation between the item and the scale. You want individual items to be correlated with the test as a whole. If you have items that do not correlate with the remaining items, you may want to drop them from the scale, since they do not appear to measure the same construct as the other items.

Another way to look at the relationship between an individual item and the rest of the scale is to try to predict a person's score on a particular item, based on scores obtained on the other items. You do this by calculating a multiple regression equation with each item, in turn, as the dependent variable and all the other items as independent variables. The multiple R^2 from this regression equation is displayed in the column labeled *Squared Multiple Correlation*. The largest multiple R^2 for this example is for item 9, a score of 0.29. Ideally, you want the squared multiple correlations to be large because you want the items to measure the same construct.

Reliability Coefficients

From the descriptive statistics, you've learned quite a bit about the scale and the individual items of which it is composed, but you still haven't arrived at an index of how reliable the scale is. Three ways to measure the reliability of a scale are to calculate Cronbach's alpha, to correlate the results from two alternate forms of the same test, or to split the same test into two parts and look at the correlation between the two parts.

Warning: The reliability of a scale depends on the population to which it is administered. Different populations of subjects may result in different scale properties.

Internal Consistency: Cronbach's Alpha

In classical test theory, a subject's response to a particular item is the sum of two components: the true score and the error. The **true score** is the value for the underlying construct that is being measured; the **error** is the part of the response that is due to question-specific factors. For example, answers to the statement, *It's hardly fair to bring a child into the world with the way things look for the future*, are the sum of the measure of anomie and the person's response to this particular question. People who experience anomie but already have children may be more likely to disagree with this question than people who experience anomie but don't have children.

When you calculate the total score, you assume that the errors average out and that you are left with a measure of the true score. A good scale is one that has a lot of true score and little error. If the variance of a scale is much larger than the variance of the individual items, the items are measuring the same construct. If you compare the sum of the item variances to the scale variance, you get an estimate of the lower bound of the true reliability of the scale—the proportion of observed score variability that is due to true score variability.

The index most often used to quantify reliability is Cronbach's alpha, which is

$$\alpha = \frac{k}{k-1}\left(1 - \frac{\sum \text{item variances}}{\text{scale variance}}\right)$$

If the sum of the individual variances is close to the variance of the entire scale, the items in the scale are not correlated, so they are not measuring the same construct. (The variance of the scale is the sum of the individual variances plus the sum of all covariances.) In this case, alpha is close to 0. That's what you would find if you construct a scale from random questions. If the variance of the entire scale is much larger than the sum of the variances of the individual items, this means individual items are correlated; that is, they have a positive covariance and alpha is close to 1. The items are measuring the same thing.

In Figure 18-5, alpha is 0.74 for the anomie scale. That's an acceptable value, but good scales have values larger than 0.8.

Figure 18-5
Cronbach's alpha

Cronbach's Alpha	Cronbach's Alpha Based on Standardized Items	N of Items
.741	.741	9

Warning: Negative values of alpha occur when items are not positively correlated among themselves and the reliability assumption is violated. When analyzing the reliability of a scale, you must have items that are measuring the same thing. Consider also the explanation that the items are not coded in the same direction.

Standardized Alpha

If all of the items are standardized to a variance of 1, Cronbach's alpha is

$$\alpha = \frac{k\bar{r}}{1 + (k-1)\bar{r}}$$

where $\bar{r}$ is the average correlation coefficient between all pairs of items and k is the number of items. For a fixed number of items, the larger the average correlation between items, the larger Cronbach's alpha. In this example, the two alphas are close in value because the variances of the items in the scale are fairly similar. Standardized item alpha is also known as the Spearman-Brown stepped-up reliability coefficient.

Number of Items

The number of items in a scale impacts its reliability. For the same average correlation, alpha increases as you increase the number of items on the scale. If the average correlation is 0.2 on a 10-item scale, alpha is 0.71. If you increase the number of items to 25, alpha is 0.86. If the number of items is large, you can have large reliability coefficients even when the average inter-item correlations are small. That's because the more measurements you take, the closer you should get to the true value, even if the individual items are crude measures.

Warning: You must select Inter-Item statistics in the Statistics dialog box to get standardized alpha.

Interpreting Cronbach's Alpha

Cronbach's alpha can be interpreted in several ways. It's the correlation between this scale and all other possible tests or scales from a hypothetical universe containing the same number of items that measure the characteristic of interest. For the anomie scale, the nine items selected can be viewed as a sample from a universe of many possible items. People could have been asked many other questions that relate to feelings of alienation and rootlessness, other markers of anomie. Cronbach's alpha tells you how much correlation you expect between the present scale and all other possible nine-item scales measuring the same thing.

Another interpretation of Cronbach's alpha is the squared correlation between the score a person obtains on a particular scale (the observed score) and the score he or she would have obtained if questioned on all of the possible items in the universe (the true score). Cronbach's alpha is an estimate of the true alpha, which is a lower bound for the true reliability.

Deleting Items from the Scale

You can determine how much alpha changes when each of the items is deleted from the scale. Items that are not related to the rest cause Cronbach's alpha to increase in value if the item is deleted. If elimination of an item substantially increases Cronbach's alpha, you should consider removing that item from your scale. From Figure 18-4, you see that alpha decreases if any item is deleted.

Tip: One possible explanation for a small alpha is that the scale measures several dimensions. Try factor-analyzing the items to look for several dimensions.

Split-Half Reliability Model

Cronbach's alpha is calculated from the correlation of items on a single scale. It's a measure based on the internal consistency of items. Other measures of reliability are based on randomly splitting the scale into two parts and looking at the correlation between the two parts. Such measures are called **split-half coefficients**. They measure how similar the two parts are. One disadvantage of split-half coefficients is that the results depend on how the scale is split. You can also use split-half methods when the same test is administered twice to the same person or when two equivalent tests are given.

Spearman-Brown Coefficient

Figure 18-6 shows split-half statistics when the first five questions are put in scale 1 and the last four questions are put in scale 2. Separate values of Cronbach's alpha are shown for each of the two parts of the test. They are smaller than the alphas for the entire scale because they are based on fewer items.

Figure 18-6
Split-half statistics

Cronbach's Alpha	Part 1	Value	.580
		N of Items	5[1]
	Part 2	Value	.657
		N of Items	4[2]
	Total N of Items		9
Correlation Between Forms			.534
Spearman-Brown Coefficient	Equal Length		.697
	Unequal Length		.699
Guttman Split-Half Coefficient			.695

1. The items are: Next to health, money is most important, Wonder if anything is worthwhile, No right and wrong ways to make money, Live only for today, Lot of average man getting worse.
2. The items are: Not fair to bring child into world, Officials not interested in average man, Don't know whom to trust, Most don't care what happens to others.

Tip: SPSS Statistics splits the scale into two parts, based on the order in which the variables are listed in the dialog box. If there is an odd number of items, the first part has the larger number of items. You can use syntax to indicate how many items should be in each half. Paste the syntax for a split-half analysis and then insert (*n*) after the word SPLIT, where *n* is the number of items you want in the second part.

The correlation between the two halves is 0.53, as shown in Figure 18-6. The equal-length Spearman-Brown coefficient, which has a value of 0.7 in this example, tells you what the reliability of the nine-item test would be if it were made up of two equal parts, each with a reliability of 0.534. (Remember, the reliability of a test increases as the number of items on the test increases, provided that the average correlation between items does not change.) If the number of items on each of the two parts is not equal, the unequal-length Spearman-Brown coefficient can be used to estimate the reliability of the overall test. In this example, because the two halves are of almost equal length, the two coefficients are similar in value.

The Guttman split-half coefficient is another estimate of the reliability of the overall test. It does not assume that the two parts are equally reliable or have the same variance.

Warning: How you split the scale may affect the reliability coefficient. If the items are administered sequentially, boredom or fatigue may account for differences between two parts formed by including adjacent items.

Guttman's Lower Bounds

Guttman proposed six measures of reliability that all estimate the lower bounds for true reliability. The first is a simple estimate that is the basis for computing some of the other lower bounds. L_3 is a better estimate than L_1 in the sense that it is larger and is equivalent to Cronbach's alpha. L_2 is better than both L_1 and L_3 but is more complex. L_5 is better than L_2 when there is one item that has a high covariance with other items, which, in turn, do not have high covariances with each other. You may see this situation with a test in which each item pertains to one of several different fields of knowledge, plus one question that can be answered with knowledge of any of those fields. L_6 is better than L_2 when the inter-item correlations are low compared to the squared multiple correlation of each item when regressed on the remaining items. L_4 is Guttman's split-half coefficient, as well as a lower bound for the true reliability for any split of the test. Guttman suggests finding the split that maximizes L_4, comparing it to the other lower bounds, and choosing the largest.

Figure 18-7
Guttman's lower bounds

Lambda	1	.659
	2	.745
	3	.741
	4	.695
	5	.725
	6	.730
N of Items		9

Testing Hypotheses about Scales

You can test various hypotheses about the means and variances of the items in your scale. You can fit a strictly parallel model to test that all items in the scale come from populations with the same means and the same variances. You can test the less restrictive hypothesis that all variances are equal by fitting a parallel model.

Parallel and Strictly Parallel Models

Output from the parallel model is shown in Figure 18-8. Based on the observed significance level, you reject the null hypothesis that the parallel model fits. If you reject the parallel model, you'll also reject the strictly parallel model because it adds the additional constraint that means are equal.

Figure 18-8
Test of parallel model

Chi-Square	Value	1018.311
	df	43
	Sig	.000
Log of Determinant of	Unconstrained Matrix	-14.693
	Constrained Matrix	-14.437

Under the parallel model assumption

Cochran's Q: Testing Equality of Related Proportions

If you have a dichotomous scale, as in this example, and you use Cochran's Q test, you can test the null hypothesis that all items have the same mean, as shown in Figure 18-9. If the observed significance level is small, you reject the null hypothesis that all of the proportions are equal.

Figure 18-9
ANOVA with Cochran's Q test

		Sum of Squares	df	Mean Square	Cochran's Q	Sig
Between People		2656.716	3974	.669		
Within People	Between Items	767.377	8	95.922	3892.927	.000
	Residual	5501.067	31792	.173		
	Total	6268.444	31800	.197		
Total		8925.160	35774	.249		

Grand Mean = .48

Tip: Cochran's Q test can be used whenever you have several dichotomous measurements for the same person and want to test whether they are equal. For example, if you record whether a person would buy a product under 10 different incentive scenarios, you can test whether the proportions for all scenarios are equal by using Cochran's Q test.

Testing Equality of Related Means

Cochran's Q tests only the null hypothesis that several related proportions are equal. If your data have more than two values and you want to test the null hypothesis that all means are equal, you can use a mixed-models analysis of variance if your data meet the assumption that the covariance matrix of the items is of a particular type (see Chapter 24). If your data don't satisfy the restrictive assumption about the variance-covariance matrix, you can use Hotelling's T^2 to test the null hypothesis that all items come from a population with equal means. Hotelling's T^2 does require that observations be from a multivariate normal distribution.

Warning: When you have several observations for the same person or item, the observations are not independent, and you need special tests for repeated measurements data (see Chapter 24).

Consider an example from pairs figure skating in the 2002 Olympics. The data are the average scores assigned by each of nine judges to each of 20 pairs of skaters on four variables: technical merit and artistry for both the short program and the long program. Figure 18-10 shows what the data file looks like. Each pair is a case. The average of each of the four scores assigned by the nine judges is a variable. You're interested in examining the relationship between the four sets of scores. The null hypothesis is that the scores are equal for all four aspects; the alternative hypothesis is that they are not.

Figure 18-10

Data arrangement for ANOVA and Tukey's test

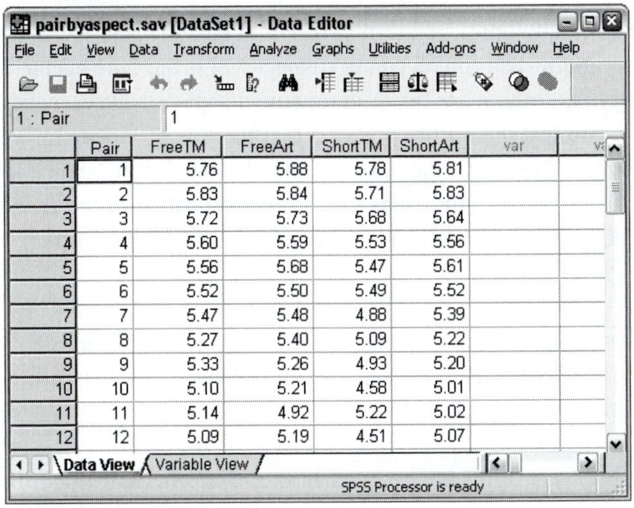

pairbyaspect.sav [DataSet1] - Data Editor

File Edit View Data Transform Analyze Graphs Utilities Add-ons Window Help

1 : Pair 1

	Pair	FreeTM	FreeArt	ShortTM	ShortArt	var
1	1	5.76	5.88	5.78	5.81	
2	2	5.83	5.84	5.71	5.83	
3	3	5.72	5.73	5.68	5.64	
4	4	5.60	5.59	5.53	5.56	
5	5	5.56	5.68	5.47	5.61	
6	6	5.52	5.50	5.49	5.52	
7	7	5.47	5.48	4.88	5.39	
8	8	5.27	5.40	5.09	5.22	
9	9	5.33	5.26	4.93	5.20	
10	10	5.10	5.21	4.58	5.01	
11	11	5.14	4.92	5.22	5.02	
12	12	5.09	5.19	4.51	5.07	

Data View | Variable View

SPSS Processor is ready

Figure 18-11 is the analysis-of-variance table. You are interested in the row labeled *Within People, Between Items*. Based on the small observed significance level, you can reject the null hypothesis that the averages of the four scores are equal.

Figure 18-11

ANOVA and Tukey's test

			Sum of Squares	df	Mean Square	F	Sig
Between People			29.977	19	1.578		
Within People	Between Items		1.074	3	.358	18.470	.000
	Residual	Nonadditivity	.246[1]	1	.246	16.077	.000
		Balance	.858	56	.015		
		Total	1.105	57	.019		
	Total		2.179	60	.036		
Total			32.156	79	.407		

Grand Mean = 4.9817

1. Tukey's estimate of power to which observations must be raised to achieve additivity = 4.898.

Tip: If the data for each case are ranks instead of scores, you can use the Friedman test to test the null hypothesis that the average ranks are the same for all variables. If you want to use the Friedman test and your data are not already ranks, choose K Related Samples from the Nonparametric Tests menu, and then select Friedman. SPSS Statistics will do the ranking for you.

Tukey's Test of Nonadditivity

The question arises of whether the pattern of scores across the four variables differs for the 20 pairs. You might have good reason to suspect that is the case because some pairs are better at artistry and others are better at technical aspects. Using Tukey's test of nonadditivity, you can test the null hypothesis that there is no linear-by-linear interaction between the pairs and the items. If the observed significance level for the test of nonadditivity is small, it indicates that a model for the observed score for an individual must include the main effect for the individual, the main effect for the item, and a subject-by-item interaction.

Warning: Even though the row is labeled *Nonadditivity*, it corresponds to a null hypothesis of additive scores. If the observed significance level for Tukey's test of nonadditivity is small, you conclude that the scores are probably not additive.

From the small observed significance level in the row labeled *Nonadditivity* in Figure 18-11, you can reject the hypothesis that the scores are additive. When you reject the nonadditivity hypothesis, you can calculate a transformation of the data that might eliminate the need for an interaction term. The suggested transformation is shown at the bottom of Figure 18-11. An exponent of 0.5 suggests a square root transformation, an exponent of 2 suggests squaring the data, –0.5 suggests an inverse square root, and so on. In this case, the suggested transformation is raising all scores to the fifth power. The effect of that would be to make all of the scores so large that differences between subjects are masked.

Measuring Agreement

Many endeavors depend on judges (raters) assigning consistent scores to objects or persons. Unless such consistency exists, conclusions about the characteristics of the objects are suspect. For example, if pathologists can't agree on the grading of the aggressiveness of a tumor, studies that involve stratification of people by aggressiveness of their tumors are impossible to design and evaluate. Numerous coefficients that measure agreement among raters or judges are available. They are called **intraclass correlation coefficients**.

Tip: The data structure is the same as for the general reliability analysis. Subjects are the rows of your data file, and judges or raters are columns. For each subject, you have scores for each of the judges. The scores may be a rank or an actual score on some scale.

Intraclass Correlation Coefficients

You can compute intraclass correlation coefficients for either absolute agreement or for consistency. Subjects are always treated as a random factor in the models available in SPSS Statistics. Judges can be either fixed or random factors. If they are fixed, you have a two-way mixed model. If judges are a random factor, you have a two-way random model.

Which One?

The decision as to which coefficient to use to measure agreement of ratings depends on answers to the following questions:

- Do you know which judge assigned which score? If you have only a set of scores for each subject, without information as to which judge assigned which score, you can perform only a one-way analysis of variance, treating person as a random factor. That's also the case if each judge rates only one subject.
- If you know which judge assigned which score, are the judges (raters) a sample of possible judges or are they of interest in their own right? If you have five psychiatrists evaluating patients, it's plausible to consider them as a random sample of psychiatrists that might be evaluating these patients. *Psychiatrist* is then a random factor. If, on the other hand, you are interested in these particular raters because they are the latest recipients of the Nobel Prize in medicine, then *rater* would be a fixed factor. If rater is a fixed factor, you want to estimate a two-way mixed model. If rater is a random factor, then you must estimate a two-way random model.
- Are you interested in absolute agreement among the judges? Absolute agreement means that you are interested in whether the actual scores are identical among the judges. You want to know if they use the same scale in making their ratings. If you are interested in absolute agreement, then you want to select Absolute Agreement as the type for the intraclass correlation coefficient.
- Are you interested in whether the judges rate consistently? That means it doesn't matter whether all judges assign the same actual scores; it just matters that their ratings correlate well with each other. If this is the case, you want to select Consistency as the type for the intraclass correlation coefficient.

Tip: The numerical values of the coefficients are the same for the two-way random and mixed models. What differs is their interpretation. If the raters are a fixed effect, the conclusions you draw are limited to the judges actually included in the study. If the judges are random, the results can be generalized to other raters. When raters are treated as fixed, the ICC (intraclass correlation) estimates for the average measure require the assumption of no rater-by-person interactions (McGaw and Wong, 1996).

Measuring Agreement: Olympic Example

Consider the ratings assigned by judges during the 2002 Olympics to pairs skaters for technical merit during the short program. The data layout is shown in Figure 18-12. The judges are the variables, and the object being rated are the rows. The coefficients calculated by SPSS Statistics always assume that the subjects, in this case the pairs, are a random factor. They are a random sample of pairs skaters. If the judges are considered a random sample of possible judges, the model is a two-way random model. If you are interested in these particular judges, the model is a two-way mixed model.

Figure 18-12
Data for calculating intraclass correlation coefficients

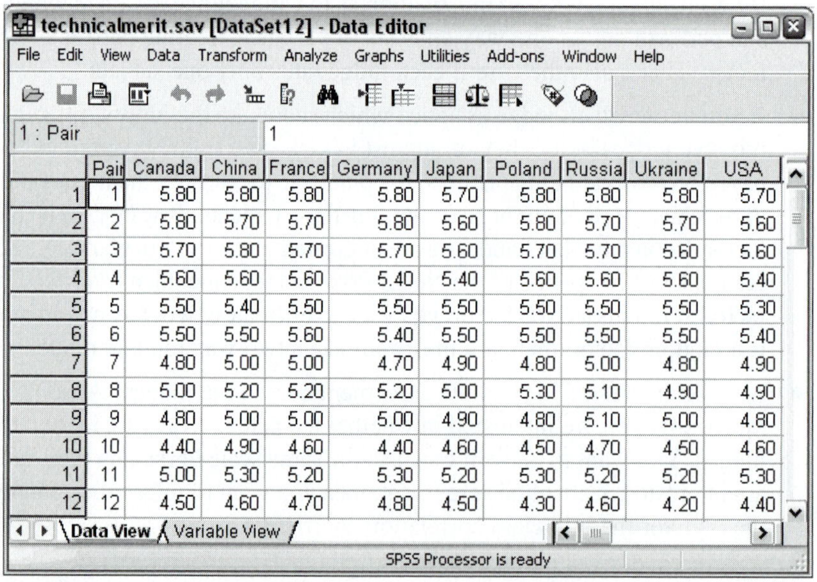

technicalmerit.sav [DataSet12] - Data Editor

File Edit View Data Transform Analyze Graphs Utilities Add-ons Window Help

1 : Pair 1

	Pair	Canada	China	France	Germany	Japan	Poland	Russia	Ukraine	USA
1	1	5.80	5.80	5.80	5.80	5.70	5.80	5.80	5.80	5.70
2	2	5.80	5.70	5.70	5.80	5.60	5.80	5.70	5.70	5.60
3	3	5.70	5.80	5.70	5.70	5.60	5.70	5.70	5.60	5.60
4	4	5.60	5.60	5.60	5.40	5.40	5.60	5.60	5.60	5.40
5	5	5.50	5.40	5.50	5.50	5.50	5.50	5.50	5.50	5.30
6	6	5.50	5.50	5.60	5.40	5.50	5.50	5.50	5.50	5.40
7	7	4.80	5.00	5.00	4.70	4.90	4.80	5.00	4.80	4.90
8	8	5.00	5.20	5.20	5.20	5.00	5.30	5.10	4.90	4.90
9	9	4.80	5.00	5.00	5.00	4.90	4.80	5.10	5.00	4.80
10	10	4.40	4.90	4.60	4.40	4.60	4.50	4.70	4.50	4.60
11	11	5.00	5.30	5.20	5.30	5.20	5.30	5.20	5.20	5.30
12	12	4.50	4.60	4.70	4.80	4.50	4.30	4.60	4.20	4.40

Data View / Variable View

SPSS Processor is ready

The summary statistics for the scores assigned to technical merit in the short program are shown in Figure 18-13. You don't see large differences between the judges for the average scores assigned to the 20 pairs. The Russian judge gave slightly higher scores.

Figure 18-13
Summary item statistics for technical merit scores for short program

	Mean	Std. Deviation	N
Canada	4.76	.73	20
China	4.87	.71	20
France	4.85	.71	20
Germany	4.71	.85	20
Japan	4.71	.82	20
Poland	4.78	.78	20
Russia	4.88	.71	20
Ukraine	4.71	.78	20
USA	4.79	.66	20

From Figure 18-14, you see that the sum of the scores is highly reliable (alpha is close to 1). Most of the differences between the scores can be attributed to differences in skating ability.

Figure 18-14
Cronbach's alpha

Cronbach's Alpha	N of Items
.996	9

Intraclass Correlation Coefficients

The intraclass correlation coefficients for consistency are shown in Figure 18-15. The model is a two-way mixed model in which the skaters are a random effect and the judges are a fixed effect. As you would expect, the coefficients are very large.

Warning: Like the Pearson correlation coefficient, the intraclass correlation coefficient depends on the range of observed values. One of the reasons the coefficients are so large in this example is that the technical skills of the 20 pairs vary greatly. If you restrict your analysis to more homogeneous pairs, such as the top five pairs, the intraclass correlation for a single measure is 0.85; for the top three, the intraclass correlation is 0.59.

Figure 18-15
Intraclass correlation coefficients for consistency

	Intraclass Correlation[1]	95% Confidence Interval		F Test with True Value 0			
		Lower Bound	Upper Bound	Value	df1	df2	Sig
Single Measures	.963[2]	.934	.983	233.383	19.0	152	.000
Average Measures	.996[3]	.992	.998	233.383	19.0	152	.000

Two-way mixed effects model where people effects are random and measures effects are fixed.

1. Type C intraclass correlation coefficients using a consistency definition-the between-measure variance is excluded from the denominator variance.
2. The estimator is the same, whether the interaction effect is present or not.
3. This estimate is computed assuming the interaction effect is absent, because it is not estimable otherwise.

Olympic judges use the same scale for assigning scores, and the actual scores don't differ too dramatically, so you would expect that there would not be too much difference between the intraclass correlation coefficients for consistency and for absolute agreement. The statistics shown in Figure 18-16 are in agreement.

Figure 18-16
Intraclass correlation coefficients for absolute agreement

	Intraclass Correlation[1]	95% Confidence Interval		F Test with True Value 0			
		Lower Bound	Upper Bound	Value	df1	df2	Sig
Single Measures	.957[2]	.923	.980	233.383	19.0	152	.000
Average Measures	.995[3]	.991	.998	233.383	19.0	152	.000

Two-way mixed effects model where people effects are random and measures effects are fixed.

1. Type A intraclass correlation coefficients using an absolute agreement definition.
2. The estimator is the same, whether the interaction effect is present or not.
3. This estimate is computed assuming the interaction effect is absent, because it is not estimable otherwise.

Tip: The average consistency coefficient for the two-way models is identical to Cronbach's alpha.

Single Measures and Average Measures

Each of the five possible combinations of model and coefficient type includes two different intraclass correlation coefficient estimates: one for the reliability of a single rating and one for the reliability for the mean or sum of *k* ratings. The choice as to which one to use depends on whether you plan to rely on a single rater or a combination of *k* ratings. As you would expect, multiple raters produce more reliable measurements.

Obtaining the Output

To produce the output in this chapter, follow the instructions below.

For Figures 18-1 to 18-9, open the file *anomia.sav.*

Figures 18-1 to 18-5 and Figure 18-9. From the Analyze menu choose Scale > Reliability Analysis. Move *anomia1* through *anomia9* into the Items list. Click Statistics to display the Statistics subdialog box. From the Descriptives for group, select Item, Scale, and Scale if item deleted. From the Summaries group, select Means, Variances, Covariances, and Correlations. From the ANOVA Table group, select Cochran chi-square. Click Continue, and then click OK.

Figure 18-6. Repeat the instructions above except select Split-half from the Model drop-down list in the Reliability Analysis dialog box. Click OK.

Figure 18-7. Repeat the instructions for Figure 18-6 except select Guttman from the Model drop-down list.

Figure 18-8. Repeat the instructions for Figure 18-6 except select Parallel from the Model drop-down list.

Figure 18-10. Open the file *pairbyaspect.sav.*

Figure 18-11. From the Analyze menu choose Scale > Reliability Analysis. Move *FreeTM*, *FreeArt*, *ShortTM*, and *ShortArt* into the Items list. Click Statistics to display the Statistics subdialog box. Select Tukey's test of additivity. Click Continue, and then click OK.

For Figures 18-12 to 18-16, open the file *technicalmerit.sav.*

Figures 18-13 to 18-15. From the Analyze menu choose Scale > Reliability Analysis. Move all of the variables except *Pair* into the Items box. Click Statistics to display the Statistics subdialog box. In the Descriptives for group, select Item. At the bottom, select Intraclass correlation coefficient. Then select Two-Way Mixed from the Model drop-down list and select Consistency from the Type drop-down list. Click Continue, and then click OK.

Figure 18-16. Repeat the previous instructions except select Absolute Agreement from the Type drop-down list in the Statistics subdialog box.

Chapter

19

Nonparametric Tests

Many statistical tests require that your data come from a normal population. Some procedures require additional assumptions, such as equality of variances in the groups. The consequences of violating the assumptions depend on the particular test and also on additional factors, such as sample sizes. You know that robust tests will swallow, without choking, data that violate assumptions. However, if your data violate assumptions in ways to which a test is sensitive, especially with small sample sizes or when outliers are present, you need to consider alternative statistical procedures that require less stringent assumptions.

Collectively, these procedures are called distribution-free or **nonparametric tests**. They do not require that the data come from a particular distribution, although some do require assumptions about the shapes of the underlying distributions. The disadvantage of nonparametric tests is that they are less likely to find true differences if the assumptions for parametric procedures are met. Another way of saying this is that nonparametric tests are not as powerful as tests that assume an underlying normal distribution, the so-called **parametric tests**, if the assumptions for parametric procedures are met. The hypotheses tested by nonparametric tests are sometimes somewhat different. For example, the hypothesis may involve medians instead of means.

Tip: When in doubt, do them both! If you reach the same conclusions based on both types of tests, there's nothing to worry about. If the results from the nonparametric test are not significant while those from the parametric test are, try to figure out why. Do you have one or two data values that are much smaller or larger than the rest? If so, they may be affecting the mean and having a large impact on your conclusions. Examine them carefully to make sure they're okay. If the problem is with the non-normal distribution of data values, see if you can transform the data to better conform with the parametric assumptions. If your transformation is successful, you can use one of the more powerful parametric procedures for your analysis.

Examples

- Are births equally distributed over the days of week?
- Do people have the same number of dates before and after signing up for a dating service?
- Is there a relationship between highest degree received and the number of parking tickets received in a year?
- Are parents and grandparents equally likely to be satisfied with the achievements of their children/grandchildren?

In a Nutshell

Most of the tests described in this chapter are alternatives to statistical tests covered in previous chapters. These tests require less restrictive assumptions about the distribution of the data. The tests are not calculated from the actual data values but from the ranks of the values or their differences or counts. These nonparametric tests are less likely to reject a false null hypothesis than the parametric tests if the assumptions for the parametric test are met.

Using the One-Sample Chi-Square Test

You can use the chi-square test of independence to test whether the rows and columns of a crosstabulation are independent. (See Chapter 10.) You compute the statistic from the differences between the observed and expected cell counts. The only assumption you need is that the observations are independent and, of course, randomly selected from the population of interest.

When cases are classified on the basis of a single variable, you can't calculate expected values from row and column totals. You must determine what they should be, based on the null hypothesis that you wish to test. You can test whether all categories are equally likely or you can specify a set of unequal probabilities.

Do Triplicities Recruit Members? The Example

In Chapter 10, you analyzed the relationship between astrological triplicities (fire, air, water, and earth signs) and whether people perceive life as exciting. You could not reject the null hypothesis that there is no relationship between sign and perception of life. It's unlikely that the planets are to blame for your boredom (and certainly not the author of this exciting book).

Specifying Equal Probabilities

Another hypothesis that you can test is whether triplicities are equally likely to capture new entrants into the world; that is, are people equally likely to be born into each of the triplicities? Instead of classifying cases on the basis of two variables, you classify them on only one: their triplicity.

You know how many people in the General Social Survey were born in each triplicity. That takes care of the observed counts. If you want to test the null hypothesis that all triplicities are equally likely to occur in the population, the expected number of cases in all four cells should be equal. Figure 19-1 shows the observed and expected cell counts for each of the four cells. Notice that the expected count is the same for all cells. It's calculated by taking the total number of cases and dividing by four. The differences between the observed and expected counts, the residuals, are in the last column of Figure 19-1.

Figure 19-1
Expected frequencies and residuals

	Observed N	Expected N	Residual
Fire	9246	9281.8	-35.8
Earth	9160	9281.8	-121.8
Air	9484	9281.8	202.3
Water	9237	9281.8	-44.8
Total	37127		

The chi-square test shown in Figure 19-2 is calculated in the usual fashion:

$$\chi^2 = \sum \frac{(\text{observed} - \text{expected})^2}{\text{expected}} = 6.36$$

Based on the asymptotic significance level, you cannot reject the null hypothesis that all four triplicities are equally likely to occur in the population.

Figure 19-2
Chi-square test

	dominant element
Chi-Square[1]	6.358
df	3
Asymp. Sig.	.095

1. 0 cells (.0%) have expected frequencies less than 5. The minimum expected cell frequency is 9281.8.

Specifying Unequal Probabilities

In the previous test, the null hypothesis was that all four triplicities are equally likely in the population. Because the triplicities don't have exactly the same number of days in them, the previous test was not strictly correct. A fairer test can be constructed by setting the predicted value for each triplicity to the number of days that it contains. If the triplicity associated with the sun sign of a date has no effect on the number of births on that date, you would expect 24.9% fire births, 25.5% air births, 24.7% water births, and 24.9% earthlings. Figure 19-3 shows the dialog box into which you must enter the expected values.

Figure 19-3
Dialog box for chi-square test

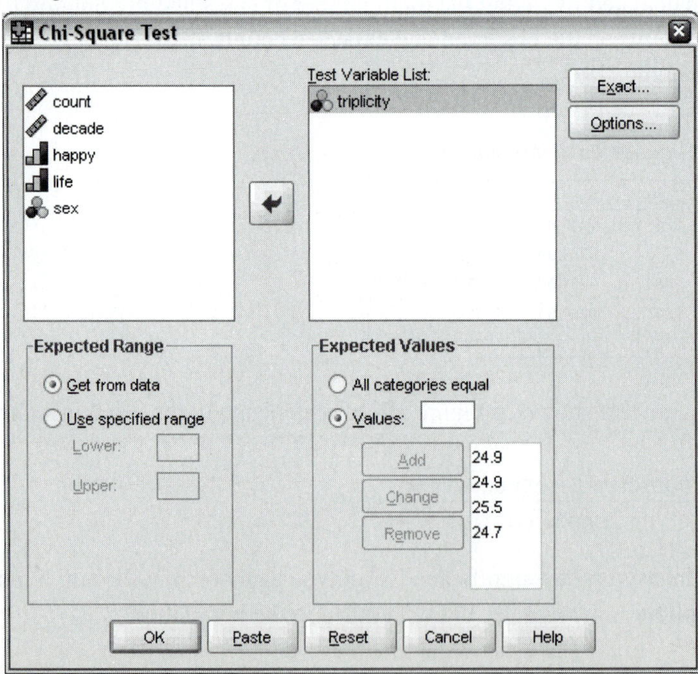

Warning: The expected values, which can be counts, percentages, or proportions, must correspond to the order of the values in your data. For example, the triplicities in the data file are coded 1 = *Fire*, 2 = *Earth*, 3 = *Air*, and 4 = *Water*. The expected values must be in that order.

Figure 19-4 contains the observed counts and the expected counts for the unequal probabilities of each triplicity. The expected count for each triplicity is the product of the total count and the expected percentage. For example, of the total 37,127 births, 24.9% or 9,244.6 are expected to occur in one of the 91 days associated with fire. The same number is expected for earth because it has the same expected proportion. From the chi-square statistic in Figure 19-4, you cannot reject the null hypothesis that birth dates are distributed proportionally among the four triplicities.

Figure 19-4
One-sample chi-square test with unequal expected values

	Observed N	Expected N	Residual
Fire	9246	9244.6	1.4
Earth	9160	9244.6	-84.6
Air	9484	9467.4	16.6
Water	9237	9170.4	66.6
Total	37127		

	dominant element
Chi-Square[1]	1.288
df	3
Asymp. Sig.	.732

1. 0 cells (.0%) have expected frequencies less than 5. The minimum expected cell frequency is 9170.4.

Tip: If the difference between the null hypothesis and the true situation is very small, even for very large sample sizes you may not reject the null hypothesis. You didn't reject the null hypothesis that all triplicities were equally likely even when it was presumably false, since the differences between the equal and true probabilities are very small. However, notice that the observed significance level for the equality null hypothesis is much smaller than the observed significance level when the correct unequal cell probabilities are used.

Using the Binomial Test

Many events can be classified into two mutually exclusive categories: live or die, defective or not defective, pass or fail. Often, you want to draw conclusions about the probability that an event occurs. For example, you want to test the null hypothesis that the probability that an item from your shop is defective is 1% or that a patient survives a difficult surgical procedure is 10%.

The binomial test can be used to test the null hypothesis that the probability of an event is any known number between 0 and 1. To use the binomial test, your observations must be independent. You must also know in advance the null hypothesis value for the probability of the event.

Probability of Using the World Wide Web

The Internet is assuming ever-increasing importance in daily life. The question of what percentage of the population uses the Internet concerns both social scientists who study its impact on factors such as isolation and marketing gurus who want to tap into this addicted audience.

You can use the binomial test to test the null hypothesis that a certain percentage of U.S. adults are Internet users. For example, you want to test the null hypothesis that this percentage is 50%.

Figure 19-5 shows the binomial dialog box for testing the null hypothesis that 50% of U.S. adults use the Internet for functions other than e-mail. (That's the question the General Social Survey asks.)

Figure 19-5
Dialog box for binomial test

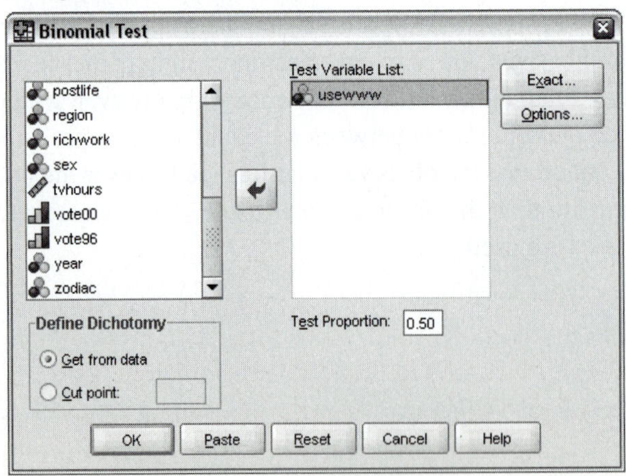

Warning: The test proportion that you specify is for the event that occurs first in the data. For example, if the event coded 1 occurs first in the data, the test proportion is for the event coded 1. If the event coded 0 occurs first in the data, the test proportion is for the event coded 0. Examine the table carefully to make sure that you are testing what you want to test.

The results of the binomial test are shown in Figure 19-6. In the column labeled *Observed Prop.*, you see that 60% of the sample use the Internet. The probability value for the null hypothesis is in the column labeled *Test Prop.* Because the asymptotic two-tailed significance level is less than 0.0005, you can reject the null hypothesis that your sample comes from a population in which 50% of people use the Internet. The percentage is larger.

Figure 19-6
Binomial test with test proportion of 0.5

		Category	N	Observed Prop.	Test Prop.	Asymp. Sig. (2-tailed)
Use WWW other than mail	Group 1	Yes	1643	.60	.50	.000[1]
	Group 2	No	1116	.40		
	Total		2759	1.00		

1. Based on Z Approximation.

Warning: If the test proportion is not 0.50, the significance level displayed is for a one-tailed test. The null hypothesis is that the population value equals the test proportion. The alternative hypothesis depends on whether the observed proportion is greater than or smaller than the test proportion. If the observed proportion is smaller than the test proportion, the alternative hypothesis is that the sample comes from a population with a probability value less than the test proportion. If the observed proportion is greater than the test proportion, the alternative hypothesis is that the sample comes from a population with a probability value greater than the test proportion. You can't double the one-tailed significance level to get a two-tailed test unless the test proportion is a half.

Testing Randomness: The Runs Test

Everyone knows the intuitive meaning of having a "run" of bad luck, good grades, or rainy days. It means that many similar events happen close to each other. You can use the aptly named **runs test** to test the null hypothesis that a sequence of events is random. For example, you can use the runs test to see whether positive and negative residuals occur randomly for observations that have been obtained in sequence. If you see too many runs of positive or negative residuals, you have reason to suspect the independence assumption is violated.

To use the runs test, you must have two events of interest that occur in a known sequence: for example, positive and negative residuals for items being produced; heads or tails in successive coin flips; success or failure at a slot machine. The null hypothesis is that the two values occur randomly. The following (hypothetical) data are for 20 patients who sequentially had more memory chips installed in their skulls. Patients are coded 1 if their ability to memorize nonsense syllables improved from a baseline measurement; 0 if it did not.

000 1 0 11 0000 1 00 111 000

A run is a sequential group of cases with the same value for the variable. You see that in the previous list there are nine runs, which are separated by spaces to make them easier to identify.

The results from the runs test are shown in Figure 19-7. You see that there are nine runs. The probability that you would see a result at least this extreme if the values occur randomly is 0.76, the observed significance level. You can't reject the null hypothesis that the values are independent.

Figure 19-7
Runs test

	Success
Test Value[1]	.50
Total Cases	20
Number of Runs	9
Z	-.305
Asymp. Sig. (2-tailed)	.761

1. User-specified.

Warning: You reject the null hypothesis if there are too many or too few groups of like cases (runs). That's why the test is labeled *2-tailed*. A negative z statistic indicates that you've observed fewer than the expected number of runs. Make sure to examine which is the case for your data. You have too many runs if adjacent values are negatively correlated with each other; for example, a good implant is always followed by a bad implant.

Testing Hypotheses about Two Related Groups

To use the paired-samples t test for small samples, you have to assume that your data come from a normal population and that there aren't any outliers, since the t test is based on means, and means are sensitive to outliers. If the distribution of your data is markedly non-normal or there are outliers, you can use one of two nonparametric alternatives to the paired-samples t test: the sign test or the Wilcoxon test.

You use the sign test to test the null hypothesis that the median difference between the two members of a pair is 0. To use the sign test, you don't have to make any assumptions about the shapes of the distributions from which the data are obtained. The only requirement is that the different pairs of observations be selected independently and that the values can be ordered from smallest to largest. That's because the test is based on seeing which of a pair of values is larger.

Looking at the Sign Test

The sign test is one of the easiest statistical tests to calculate. You find the difference between the values of the variable for the two members of a pair, and then you count the number of positive, negative, and zero differences. You ignore everything except the sign, which is how the name originated. Cases with equal values for the two variables (tied cases) are not used in the computation of the statistic because they don't provide any useful information about differences.

Age at Marriage: The Example

As an example of the sign test, consider the null hypothesis that couples whose marriages are reported in *The New York Times* come from a population of couples whose ages at marriage are equal. For a random sample of 41 couples, you have the age of the bride and the age of groom, as shown in the Data Editor in Figure 19-8.

Figure 19-8
Data Editor with marriage ages

marage.sav [DataSet2] - Data Editor
File Edit View Data Transform Analyze Graphs Utilities Add-ons Window Help

1 : Couple | Carter/Kempf

	Couple	brideage	groomage	difage	olderman
1	Carter/Kempf	32	33	1.00	1.00
2	Chertoff/Tavelinksy	24	26	2.00	1.00
3	Macker/Rideout	29	34	5.00	1.00
4	Krongard/Shreck	29	30	1.00	1.00
5	Bohnen/Allan	27	28	1.00	1.00
6	Schifino/Thornbrough	33	32	-1.00	.00
7	Suzuki/Sasieta	33	33	.00	.
8	Vargas/Sanchez	31	33	2.00	1.00
9	Pitts/Wieland	33	35	2.00	1.00
10	Houlihan/Kelly	40	59	19.00	1.00
11	Martinez/Nohrnberg	34	32	-2.00	.00
12	Rustige/Abbe	29	33	4.00	1.00

Data View | Variable View

SPSS Processor is ready

For each couple, you compute the difference between the age of the groom and the age of the bride. From Figure 19-9, you see that for 7 couples, the age of the groom is less than that of the bride. For 29 couples, the age of the groom is larger than that of the bride. For five couples, the two ages are the same. If the null hypothesis is true, you expect in a sample roughly equal numbers of positive and negative differences.

Figure 19-9
Positive and negative difference counts

		N
groomage - brideage	Negative Differences[1]	7
	Positive Differences[2]	29
	Ties[3]	5
	Total	41

1. groomage < brideage
2. groomage > brideage
3. groomage = brideage

The observed significance level for the test of the null hypothesis that positive and negative differences are equally likely is in Figure 19-10. Because the observed significance level is less than 0.0005, you can easily reject the null hypothesis that the median difference between the two ages is 0. It appears that New York society men tend to marry younger women.

Figure 19-10
Observed significance level for sign test

	groomage - brideage
Z	-3.500
Asymp. Sig. (2-tailed)	.000

Adding Power: The Wilcoxon Test

The only information from the data that the sign test uses is whether the observed difference is positive or negative. Information about the size of the difference is ignored. The **Wilcoxon matched-pairs signed-rank test** is a nonparametric test that uses the information about the size of the difference between the two members of a pair. That's why it's more likely to detect true differences when they exist. However, the Wilcoxon test requires that the differences be a sample from a symmetric distribution. That's a less stringent assumption than requiring normality, since there are many other distributions besides the normal distribution that are symmetric. (It's more restrictive, however, than the assumptions for the sign test.)

Checking Symmetry

Figure 19-11, a histogram of the age differences, suggests that the symmetry assumptions may not be unreasonable for these data. You don't expect a sample from a symmetric distribution to be exactly symmetric. Notice the single large difference: a man who married a woman who is twenty years younger than he is. (Must be a data entry error.) In a parametric test, that would be a signal for real danger, not necessarily for the couple, but certainly for the statistical procedure. The outlying point distorts the mean and standard deviation. In a nonparametric test, the effect of the outlying point is much less because the actual data value are not used, just the ranks.

Figure 19-11
Distribution of differences

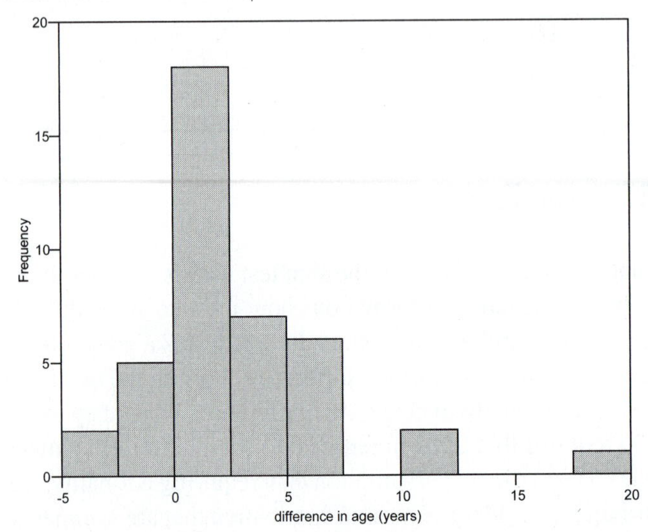

Tip: There's no need to anguish over the symmetry assumption because, based on the sign test, you easily rejected the null hypothesis that the median age difference is 0. There's not much to be gained by doing yet another test unless you want to report a smaller observed significance level.

Calculating the Test

To calculate the Wilcoxon test, first you find the difference between the two values for each pair. Next, for all cases where the difference is not 0, you rank the differences from smallest to largest, ignoring the sign of the differences. That is, the smallest difference in absolute value is assigned a rank of 1, the second smallest difference is assigned a rank of 2, and so on. In the case of ties (equal differences), you assign the average rank to the tied cases.

Once you have the ranks, you calculate the average of the ranks separately for cases with positive differences and for cases with negative differences. If the null hypothesis is true, you expect the mean rank to be similar for the two groups. Because you replace observed differences with ranks, the effect of large differences is less severe than if you used the paired t test.

The mean ranks for the two groups are shown in Figure 19-12. The ranks for the positive differences are quite a bit larger than the ranks for the negative differences. This suggests that the magnitudes of the positive and negative differences are not the same.

Figure 19-12
Wilcoxon signed-rank test summary

		N	Mean Rank	Sum of Ranks
groomage - brideage	Negative Ranks	7[1]	13.86	97.00
	Positive Ranks	29[2]	19.62	569.00
	Ties	5[3]		
	Total	41		

1. groomage < brideage
2. groomage > brideage
3. groomage = brideage

The test of the null hypothesis is based on the statistics in Figure 19-13. Based on the observed significance level, you can reject the null hypothesis that the median difference is 0. New York society women tend to marry older men.

Figure 19-13
Observed significance level for Wilcoxon test

Test Statistics[2]

	groomage - brideage
Z	-3.732[1]
Asymp. Sig. (2-tailed)	.000

1. Based on negative ranks.
2. Wilcoxon Signed Ranks Test

Tip: Because the absolute value of the *z* statistic in Figure 19-13 is somewhat larger than the absolute value of the *z* statistic in Figure 19-10, you know that the observed significance level is smaller for the Wilcoxon test. You can verify this by double-clicking the pivot tables to activate them and then double-clicking the observed significance levels to set them to more decimal places.

Testing Hypotheses about Two Independent Groups

To use the two-independent-samples *t* test, your sample has to come from a normal population or the sample size has to be large enough so that the distribution of sample means is approximately normal. If the sample size is small and markedly non-normal or there are outliers, there are several nonparametric alternatives.

Wilcoxon Test

The Wilcoxon test, also known as the Mann-Whitney *U* test, is the most frequently used alternative to the two-independent-samples *t* test.

Shott (1990) presents data from a study by Williams (1981) that compares the effectiveness of teaching neurological nursing using team nursing and using the standard case assignment method (the control). To compute the Wilcoxon test, you pool all of the data values and rank them from smallest to largest. Then you find the average rank for each of the two groups. If there are tied observations, assign to all of them the average of the ranks that would have been assigned to them if they were not tied.

From Figure 19-14, you see that the average rank for the 22 students in the team nursing group is 16.23, compared to an average rank of 13.5 for the eight students in the control group.

Figure 19-14
Summary statistics for the Wilcoxon test

			N	Mean Rank	Sum of Ranks
Score	Group	Team Nursing	22	16.23	357.00
		Control	8	13.50	108.00
		Total	30		

To see if the observed difference in average ranks is large enough for you to reject the null hypothesis that scores in one population are not larger than scores in the other, look at Figure 19-15. You see that the asymptotic two-tailed significance level is 0.45, so you don't reject the null hypothesis. Because the sample sizes in the two groups are small, you may not be able to detect even larger differences between the two methods.

Figure 19-15
Wilcoxon and Mann-Whitney tests

Test Statistics[2]

	Score
Mann-Whitney U	72.000
Wilcoxon W	108.000
Z	-.758
Asymp. Sig. (2-tailed)	.448
Exact Sig. [2*(1-tailed Sig.)]	.475[1]

1. Not corrected for ties.
2. Grouping Variable: Group

Tip: If you want to test the null hypothesis that the population means are equal, the shape of the distribution must be the same in both groups. This implies that the two population variances are equal. It doesn't matter what the shape of the distribution is, but it has to be the same in the two groups. Strictly speaking, the data should be from a continuous distribution so that there are no ties; however, the test performs reasonably well when there are ties.

Mann-Whitney U Test

The Mann-Whitney U test is closely related to the Wilcoxon test. They are equivalent tests that use different test statistics. The Wilcoxon W is the sum of the ranks in the smaller group. To calculate the U statistic, you count the number of times an observation from the group with the smaller sample size is smaller in value than an observation from the group with the larger sample size. You can calculate the U statistic from the Wilcoxon W. They always have the same observed significance level and are displayed together in Figure 19-15.

Warning: You have to look at the average ranks for the two groups to determine which has the smaller values. The sign of the z statistic won't tell you that.

Wald-Wolfowitz Runs Test

The runs test can also be used to test the null hypothesis that two groups come from the same distribution. The test requires the data to be random samples from continuous distributions. The data values are sorted from largest to smallest, and then the group number takes their place. You determine whether there are too few changes in group number to believe that the samples are from the same population. If one group has larger values than the other, you expect to see too few runs.

For example, if you have the values 10, 20, and 30 in the first group and the values 5, 7, and 15 in the second group, you would order them 5(2), 7(2), 10(1), 15(2), 20(1), and 30(1), where the number in parentheses is the group number. You calculate the number of runs from the group numbers: 22 1 2 11. For this sequence, there are four runs.

Figure 19-16 is the runs test for the nursing data. Notice that there are two observed significance levels, one labeled *Minimum Possible* and one labeled *Maximum Possible*. When cases in the two groups have the same value for the score, different numbers of runs are possible, depending on how the cases are arranged. For example, if group 1 has two cases with the value of 8 and group 2 has two cases with the value of 8, you can arrange the corresponding group numbers as 1,1,2,2 or 1,2,1,2. The number of runs is 2 for the first arrangement and 4 for the second. SPSS Statistics computes observed significance levels for the smallest and largest number of runs that can be computed. Withhold judgement if the two significance levels lead you to different conclusions.

Figure 19-16
Wald-Wolfowitz runs test

Test Statistics[2,3]

		Number of Runs	Z	Exact Sig. (1-tailed)
Score	Minimum Possible	8[1]	-2.031	.024
	Maximum Possible	16[1]	1.807	.965

1. There are 4 inter-group ties involving 15 cases.
2. Wald-Wolfowitz Test
3. Grouping Variable: Group

Tip: For many nonparametric tests with small sample sizes, the observed significance levels are labeled as *Exact*. That's because it is possible to calculate exact probabilities of various configurations of the data if the null hypothesis is true. For larger sample sizes, the test are labeled *Asymptotic* because they are based on the distributions of statistics instead of the data itself.

Testing Hypotheses about Three or More Independent Groups

There are several nonparametric tests for testing hypotheses about three or more independent groups. They require slightly different assumptions about the data, and they test slightly different hypotheses than the corresponding parametric tests.

Kruskal-Wallis Test

Just as one-way analysis of variance is an extension of the two-independent-sample *t* test to more than two groups, the Kruskal-Wallis test is an extension of the Wilcoxon test to more than two groups. You must have random samples from independent groups. You calculate the test the same way as the Wilcoxon. You pool all observations, rank them, and then find the average ranks for each group. In the previous nursing student example, there were three methods of teaching: the team, the usual case management (control), and a primary nursing model. The average ranks for all three groups are shown in Figure 19-17. The team group has the largest average rank (the best score); the control group has the smallest.

Figure 19-17
Average groups for the groups

			N	Mean Rank
Score	Group	Team	22	26.91
		Control	8	22.44
		Primary	20	25.18
		Total	50	

You can test the null hypothesis that all groups have the same distribution of scores, based on the statistics in Figure 19-18. Because the observed significance level, 0.75, is large, you cannot reject the null hypothesis that values from one population are not larger than values from another. If you want to test hypotheses about the population means, you must assume that the shape of the population distributions is the same for all groups. This is less stringent than assuming that the data come from normal populations, although the equality of variance assumption is still needed.

Figure 19-18
Kruskal-Wallis test statistic

Test Statistics[1,2]

	Score
Chi-Square	.578
df	2
Asymp. Sig.	.749

1. Kruskal Wallis Test
2. Grouping Variable: Group

Tip: If your groups are ordered in some way (for example, highest degree or increasing dosage of a drug), the Jonckheere-Terpstra test may be more powerful than the Kruskal-Wallis test.

Median Test

You can test the null hypothesis that two or more groups come from distributions with the same median using the median test. (No surprises here.) To calculate the median test, you pool all of the observations and calculate the combined median. Then you count the number of cases that are above and below the median in each of the groups.

Figure 19-19 is a crosstabulation of the number of cases above and below the median for the nursing data. If the null hypothesis is true, you expect, for each group, roughly equal numbers of cases above and below the median. That seems to be pretty much the case here.

Figure 19-19
Counts for the median test

		Group		
		Team	Control	Primary
Score	> Median	11	3	9
	<= Median	11	5	11

Figure 19-20 contains the median score for the pooled groups (85) and the observed significance level for the test of the null hypothesis. Based on the large observed significance level, you cannot reject the null hypothesis.

Figure 19-20
Test statistic for median test

Test Statistics[2]

	Score
N	50
Median	85.0000
Chi-Square	.382[1]
df	2
Asymp. Sig.	.826

1. 2 cells (33.3%) have expected frequencies less than 5. The minimum expected cell frequency is 3.7.
2. Grouping Variable: Group

Warning: This test is the usual chi-square test for independence of rows and columns, so you have to watch out for cells with very small expected counts.

Three or More Related Groups

If you want to test hypotheses about a set of variables that are measured on the same person or object, you have to use special tests because the values of the variables are not independent. If you have three weight measurements for the same person obtained at three time points (for example, before a weight reduction program, at week six of the program, and one year after the program), you can't analyze the data using the Kruskal-Wallis test because it requires groups to be independent.

Looking for Differences: The Friedman Test

The Friedman test can be used when you have repeated measurements for the same case. For example, Figure 19-21 shows heart rates for nine patients at baseline and after administration of three drugs, including a placebo. (The data are from Shott, 1990, attributed to Platia et al., 1985.)

Figure 19-21
Data arrangement for Friedman test

drugs.sav [DataSet3] - Data Editor

File Edit View Data Transform Analyze Graphs Utilities Add-ons Window Help

1 : Patient 1

	Patient	Baseline	Propranolol	Placebo	Acebutolol	var
1	1	94	67	90	67	
2	2	57	52	69	55	
3	3	81	74	69	73	
4	4	82	59	71	72	
5	5	67	65	74	72	
6	6	78	72	80	72	
7	7	87	75	106	74	
8	8	82	68	76	59	
9	9	90	74	82	80	
10						
11						
12						

Data View / Variable View

SPSS Processor is ready

To calculate the Friedman test, for each case you rank the values of the variables. For example, for patient 1, the rank for baseline is 4, the rank for placebo is 3, and the rank for both propranolol and acebutolol is 1.5. (That's the average of ranks 1 and 2, which are the ranks that propranolol and acebutolol would receive if they were not tied.) The average ranks for each variable are shown in Figure 19-22. Acebutolol and propranolol have smaller observed average ranks than baseline and placebo, meaning that the heart rates are lower when the drugs are administered. That's good for the patient.

Figure 19-22
Average ranks for the four groups

	Mean Rank
Acebutolol	1.89
Baseline	3.44
Placebo	3.11
Propranolol	1.56

The null hypothesis is that the rankings of the variables are equal. The alternative hypothesis is that at least one treatment has larger values than another treatment. Based on the small observed significance level in Figure 19-23, you can reject the null hypothesis. All four treatments don't result in similar heart rates.

Figure 19-23
Friedman test statistic

Test Statistics[1]

N	9
Chi-Square	13.977
df	3
Asymp. Sig.	.003

1. Friedman Test

Warning: You must have repeated measurements of the same variable at different time points or under different conditions. You can't compare heart rates, blood pressures, and serum cholesterol levels.

Measuring Agreement: Kendall's W

Closely related to Friedman's test is Kendall's *W*, also known as Kendall's coefficient of concordance. It measures the agreement of the ranks assigned to each of the related variables across judges. In the heart rate example, each patient can be considered a judge. You measure the agreement between the ranks assigned to the four treatments. A value of 1 for Kendall's *W* means the drugs were ranked in the same order by all of the patients. For example, propranolol had the smallest rank, then acebutolol, then placebo, then propranolol for all of the patients. Kendall's *W* is close to 0 if there is complete disagreement between the rankings of the drugs by the patients. That means different drugs worked differently for different patients. There is no consistent ordering from best to worst across patients.

From Figure 19-24, Kendall's *W* is 0.52, half way between no agreement and perfect agreement. The observed significance level for the test that Kendall's *W* is 0, $p = 0.003$, is always the same as the observed significance for Friedman's test. Kendall's *W* is just another way of summarizing the same results you get using Friedman's test.

Figure 19-24
Kendall's W

N	9
Kendall's W[1]	.518
Chi-Square	13.977
df	3
Asymp. Sig.	.003

1. Kendall's Coefficient of Concordance

Testing Equality of Proportions: Cochran's Q

When the related variables are dichotomous or can be converted to dichotomies, you can calculate Cochran's Q to test the null hypothesis that the distribution of the two values is the same for all the related variables. Consider the anomie scale administered by the General Social Survey. Respondents are asked whether they agree or disagree with the questions listed in Figure 19-25. You want to test the null hypothesis that the proportion of cases who agree with each statement is the same for all of the statements.

Figure 19-25 includes the number of cases who agree and disagree with each of the items. Because the number of cases is the same for all of the variables, you can compare the actual numbers. You see that there's quite a bit of variability in the number of people who agree with each of the statements. Fewest people (25%) agree with the statement that there is no right or wrong way to make money. The largest number of people (72%) agree with the statement that you don't know whom to trust.

Figure 19-25
Counts for Cochran's test

	Value	
	Disagree	Agree
Next to health, money is most important	2736	1239
Wonder if anything is worthwhile	2335	1640
No right and wrong ways to make money	2993	982
Live only for today	2208	1767
Lot of average man getting worse	1655	2320
Not fair to bring child into world	2429	1546
Officials not interested in average man	1458	2517
Don't know whom to trust	1122	2853
Most don't care what happens to others	1767	2208

Figure 19-26 is part of the data file. For each case, you have values for all of the variables.

Figure 19-26
Data arrangement for Cochran's Q

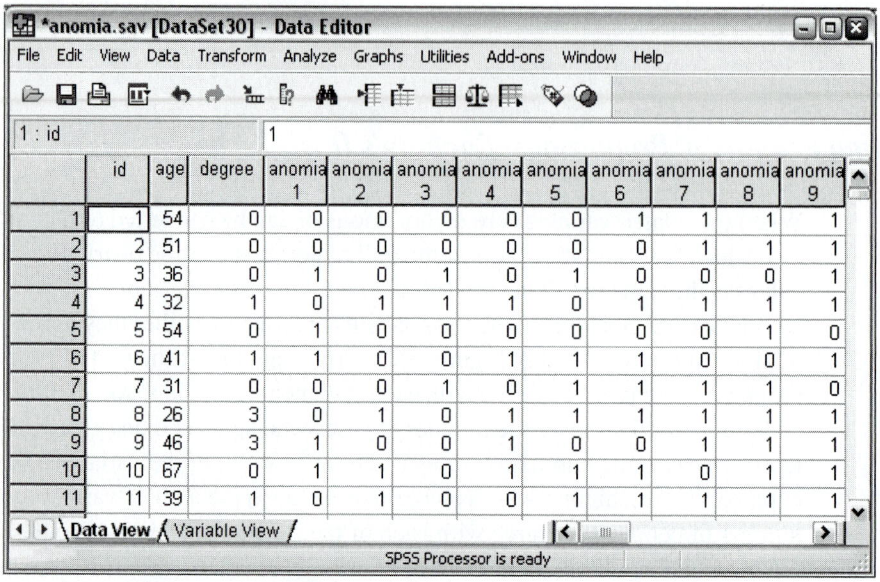

*anomia.sav [DataSet30] - Data Editor

File Edit View Data Transform Analyze Graphs Utilities Add-ons Window Help

1 : id 1

	id	age	degree	anomia 1	anomia 2	anomia 3	anomia 4	anomia 5	anomia 6	anomia 7	anomia 8	anomia 9
1	1	54	0	0	0	0	0	0	1	1	1	1
2	2	51	0	0	0	0	0	0	0	1	1	1
3	3	36	0	1	0	1	0	1	0	0	0	1
4	4	32	1	0	1	1	1	0	1	1	1	1
5	5	54	0	1	0	0	0	0	0	0	1	0
6	6	41	1	1	0	0	1	1	1	0	0	1
7	7	31	0	0	0	1	0	1	1	1	1	0
8	8	26	3	0	1	0	1	1	1	1	1	1
9	9	46	3	1	0	0	1	0	0	1	1	1
10	10	67	0	1	1	0	1	1	1	0	1	1
11	11	39	1	0	1	0	0	1	1	1	1	1

Data View / Variable View

SPSS Processor is ready

Figure 19-27 is a test of the null hypothesis that the percentage of people agreeing (or equivalently disagreeing) with each statement is the same. Based on the small observed significance level, you can reject the null hypothesis.

Figure 19-27
Cochran's Q statistic

N	3975
Cochran's Q	3892.927[1]
df	8
Asymp. Sig.	.000

1. 0 is treated as a success.

Warning: You can't calculate the usual chi-square statistic for these data because the observations are not independent. Each person answered each question, so responses to the questions are not independent.

Obtaining the Output

To produce the output in this chapter, follow the instructions below.

For Figures 19-1 to 19-4, open the file *aggregatedgss.sav.*

Figures 19-1 and 19-2. From the Analyze menu choose Nonparametric Tests > Chi-Square. Move *triplicity* into the Test Variable List. Click OK.

Figure 19-4. From the Analyze menu choose Nonparametric Tests > Chi-Square. Move *triplicity* into the Test Variable List. In the Expected Values group, select Values. Type the four values 24.9, 24.9, 25.5, and 24.7, clicking Add after each value. Click OK.

For Figure 19-6, open the file *gssdata.sav.*

Figure 19-6. From the Analyze menu choose Nonparametric Tests > Binomial. Move *usewww* into the Test Variable List. Type 0.5 into the Test Proportion box. Click OK.

For Figure 19-7, open the file *runstest.sav.*

Figure 19-7. From the Analyze menu choose Nonparametric Tests > Runs. Move *success* into the Test Variable List. In the Cut Point group, deselect Median, select Custom, and type 0.5 into the box. Click OK.

For Figures 19-9 to 19-13, open the file *marage.sav.*

Figures 19-9, 19-10, 19-12, and 19-13. From the Analyze menu choose Nonparametric Tests > 2 Related Samples. Click *brideage* and *groomage* to move them into the Current Selections box, and then click the arrow button to move them into the Test Pair(s) List. In the Test Type group, select both Wilcoxon and Sign. Click OK.

Figure 19-11. From the Graphs menu choose Legacy Dialogs > Histogram. Move *difage* into the Variable box, and then click OK.

For Figures 19-14 to 19-20, open the file *nurses.sav.*

Figures 19-14 and 19-15. From the Analyze menu choose Nonparametric Tests > 2 Independent Samples. Move *Score* into the Test Variable List and *Group* into the Grouping Variable box. Click Define Groups, and in the Define Groups subdialog box, type the values 1 and 2 into the boxes and click Continue. In the Test Type group, select both Mann-Whitney U and Wald-Wolfowitz runs. Click OK.

Figures 19-17 to 19-20. From the Analyze menu choose Nonparametric Tests > K Independent Samples. Move *Score* into the Test Variable List and *Group* into the Grouping Variable box. Click Define Range, and in the Define Range subdialog box, type 1 as the Minimum and 3 as the Maximum. Click Continue. In the Test Type group, select Kruskal-Wallis H and Median. Click OK.

For Figures 19-21 to 19-24, open the file *drugs.sav.*

Figures 19-21 to 19-24. From the Analyze menu choose Nonparametric Tests > K Related Samples. Move *acebutolol*, *baseline*, *placebo*, and *propranolol* into the Test Variables list. In the Test Type group, select Friedman and Kendall's W. Click OK.

For Figures 19-25 and 19-27, open the file *anomia.sav.*

Figures 19-25 and 19-27. From the Analyze menu choose Nonparametric Tests > K Related Samples. Move *anomia1* through *anomia9* into the Test Variables list. In the Test Type group, deselect Friedman and select Cochran's Q. Click OK.

Chapter

20

Ordinal Regression

Many variables of interest are ordinal. That is, you can rank the values, but the real distance between categories is unknown. Diseases are graded on scales from *least severe* to *most severe*. Survey respondents choose answers on scales from *strongly agree* to *strongly disagree*. Students are graded on scales from *A* to *F*.

You can use ordinal categorical variables as predictors, or factors, in many statistical procedures, such as linear regression. However, you have to make difficult decisions. Should you forget the ordering of the values and treat your categorical variables as if they are nominal? Should you substitute some sort of scale (for example, numbers 1 to 5) and pretend the variables are interval? Should you use some other transformation of the values hoping to capture some of that extra information in the ordinal scale?

When your dependent variable is ordinal, you also face a quandary. You can forget about the ordering and fit a multinomial logit model that ignores any ordering of the values of the dependent variable. You fit the same model if your groups are defined by color of car driven or severity of a disease. You estimate coefficients that capture differences between all possible pairs of groups. Or you can apply a model that incorporates the ordinal nature of the dependent variable. However, keep in mind that even when the categories of the dependent variable can be ordered, that doesn't mean that an ordinal model is necessarily the most appropriate model, especially if categories are ordered on more than one dimension, such as strength of opinion and direction, or if categories can be ordered in different ways. (See, for example, Miller and Volker, 1985).

The SPSS Statistics Ordinal Regression procedure, or PLUM (**Pol**ytomous **U**niversal **M**odel), is an extension of the general linear model to ordinal categorical data. You can specify five link functions as well as scaling parameters. The procedure can be used to fit heteroscedastic probit and logit models.

Fitting an Ordinal Logit Model

Before delving into the formulation of ordinal regression models as specialized cases of the general linear model, let's consider a simple example. To fit a binary logistic regression model, you estimate a set of regression coefficients that predict the probability of the outcome of interest. The same logistic model can be written in different ways. The version that shows what function of the probabilities results in a linear combination of parameters is

$$\ln\left(\frac{\text{prob(event)}}{(1 - \text{prob(event)})}\right) = \beta_0 + \beta_1 X_1 + \beta_2 X_2 + \ldots + \beta_k X_k$$

The quantity to the left of the equals sign is called a **logit**. It's the log of the odds that an event occurs. (The odds that an event occurs is the ratio of the number of people who experience the event to the number of people who do not. This is what you get when you divide the probability that the event occurs by the probability that the event does not occur, since both probabilities have the same denominator and it cancels, leaving the number of events divided by the number of non-events.) The coefficients in the logistic regression model tell you how much the logit changes based on the values of the predictor variables.

When you have more than two events, you can extend the binary logistic regression model, as described in the chapter "Multinomial Logistic Regression" in the *SPSS Statistics 17.0 Advanced Statistical Procedures Companion*. For ordinal categorical variables, the drawback of the multinomial regression model is that the ordering of the categories is ignored.

Modeling Cumulative Counts

You can modify the binary logistic regression model to incorporate the ordinal nature of a dependent variable by defining the probabilities differently. Instead of considering the probability of an individual event, you consider the probability of that event and all events that are ordered before it.

Consider the following example: A random sample of Vermont voters was polled. The voters were asked to rate their satisfaction with the criminal justice system in the state (Doble and Green, 1999). They rated judges on the scale: *Poor* (1), *Only fair* (2), *Good* (3), and *Excellent* (4). Each voter also indicated whether he or she or anyone in his or her family was a crime victim in the last three years. You want to model the relationship between their rating and having a crime victim in the household.

Defining the Event

In ordinal logistic regression, the event of interest is observing a particular score *or less*. For the rating of judges, you model the following odds:

θ_1 = prob(score of 1) / prob(score greater than 1)
θ_2 = prob(score of 1 or 2) / prob(score greater than 2)
θ_3 = prob(score of 1, 2, or 3) / prob(score greater than 3)

The last category doesn't have an odds associated with it since the probability of scoring up to and including the last score is 1.

All of the odds are of the form:

$$\theta_j = \text{prob}(\text{score} \leq j) / \text{prob}(\text{score} > j)$$

You can also write the equation as

$$\theta_j = \text{prob}(\text{score} \leq j) / (1 - \text{prob}(\text{score} \leq j)),$$

since the probability of a score greater than j is 1 minus probability of a score less than or equal to j.

Ordinal Model

The ordinal logistic model for a single independent variable is then

$$\ln(\theta_j) = \alpha_j - \beta X$$

where j goes from 1 to the number of categories minus 1.

It is not a typo that there is a minus sign before the coefficients for the predictor variables instead of the customary plus sign. That is done so that larger coefficients indicate an association with larger scores. When you see a positive coefficient in SPSS Statistics output for a dichotomous factor, you know that higher scores are more likely for the first category. A negative coefficient tells you that lower scores are more likely. For a continuous variable, a positive coefficient tells you that as the values of the variable increase, the likelihood of larger scores increases. An association with higher scores means smaller cumulative probabilities for lower scores, since they are less likely to occur.

Each logit has its own α_j term but the same coefficient β. That means that the effect of the independent variable is the same for different logit functions. That's an assumption you have to check. That's also the reason the model is also called the proportional odds model. The α_j terms, called the threshold values, often aren't of much interest. Their values do not depend on the values of the independent variable for a particular case. They are like the intercept in a linear regression, except that each logit has its own intercept. They're used in the calculations of predicted values. From the previous equations, you also see that combining adjacent scores into a single category won't change the results for the groups that aren't involved in the merge. That's a desirable feature.

Examining Observed Cumulative Counts

Before you start building any model, you should examine the data. Figure 20-1 is a cumulative percentage plot of the ratings, with separate curves for those whose households experienced crime and those who didn't. The line for those who experienced crime is above the line for those who didn't. Figure 20-1 also helps you visualize the ordinal regression model. It models a function of those two curves.

Consider the rating *Poor.* A larger percentage of crime victims than nonvictims chose this response. (Because it is the first response, the cumulative percentage is just the observed percentage for the response.) As additional percentages are added (the cumulative percentage for *Only fair* is the sum of *Poor* and *Only fair),* the cumulative percentages for the crime victim households remain larger than for those without crime. It's only at the end, when both groups must reach 100%, that they must join. Because the victims assign lower scores, you expect to see a negative coefficient for the predictor variable, *hhcrime* (household crime experience).

Figure 20-1
Plot of observed cumulative percentages

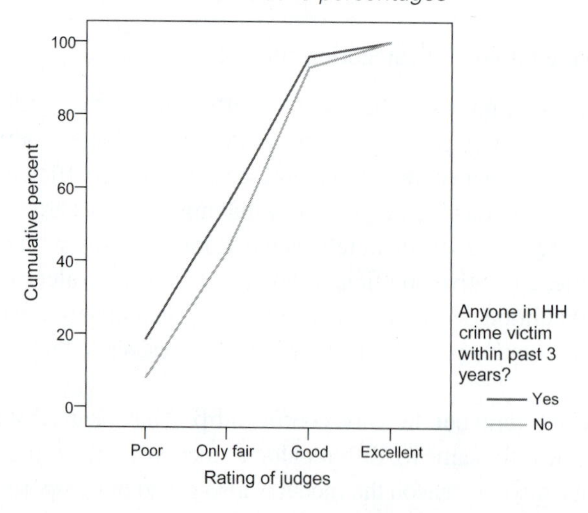

Specifying the Analysis

To fit the cumulative logit model, open the file *vermontcrime.sav* and from the menus choose:

Analyze
  Regression
    Ordinal...

▶ Dependent: rating
▶ Factors: hhcrime

Options...
Link: Logit

Output...
  Display
  ☑ Goodness of fit statistics
  ☑ Summary statistics
  ☑ Parameter estimates
  ☑ Cell information
  ☑ Test of Parallel Lines
  Saved Variables
  ☑ Estimated response probabilities

Parameter Estimates

Figure 20-2 contains the estimated coefficients for the model. The estimates labeled *Threshold* are the α_j s, the intercept equivalent terms. The estimates labeled *Location* are the ones you're interested in. They are the coefficients for the predictor variables. The coefficient for *hhcrime* (coded 1 = *yes*, 2 = *no*), the independent variable in the model, is –0.633. As is always the case with categorical predictors in models with intercepts, the number of coefficients displayed is one less than the number of categories of the variable. In this case, the coefficient is for the value of 1. Category 2 is the reference category and has a coefficient of 0.

The coefficient for those whose household experienced crime in the past three years is negative, as you expected from Figure 20-1. That means it's associated with poorer scores on the rankings of judges. If you calculate $e^{-\beta}$, that's the ratio of the odds for lower to higher scores for those experiencing crime and those not experiencing crime. In this example, exp(0.633) = 1.88. This ratio stays the same over all of the ratings.

The Wald statistic is the square of the ratio of the coefficient to its standard error. Based on the small observed significance level, you can reject the null hypothesis that it is zero. There appears to be a relationship between household crime and ratings of judges. For any rating level, people who experience crime score judges lower than those who don't experience crime.

Figure 20-2
Parameter estimates

		Estimate	Std. Error	Wald	df	Sig.	95% Confidence Interval	
							Lower Bound	Upper Bound
Threshold	[rating = 1]	-2.392	.152	248.443	1	.000	-2.690	-2.095
	[rating = 2]	-.317	.091	12.146	1	.000	-.495	-.139
	[rating = 3]	2.593	.172	228.287	1	.000	2.257	2.930
Location	[hhcrime=1]	-.633	.232	7.445	1	.006	-1.088	-.178
	[hhcrime=2]	0[1]	.	.	0	.	.	.

Link function: Logit.

1. This parameter is set to zero because it is redundant.

Testing Parallel Lines

When you fit an ordinal regression, you assume that the relationships between the independent variables and the logits are the same for all the logits. That means that the results are a set of parallel lines or planes—one for each category of the outcome variable. You can check this assumption by allowing the coefficients to vary, estimating them, and then testing whether they are all equal.

The result of the test of parallelism is in Figure 20-3. The row labeled *Null Hypothesis* contains –2 log-likelihood for the constrained model, the model that assumes the lines are parallel. The row labeled *General* is for the model with separate lines or planes. You want to know whether the general model results in a sizeable improvement in fit from the null hypothesis model.

The entry labeled *Chi-Square* is the difference between the two –2 log-likelihood values. If the lines or planes are parallel, the observed significance level for the change should be large, since the general model doesn't improve the fit very much. The parallel model is adequate. You don't want to reject the null hypothesis that the lines are parallel. From Figure 20-3, you see that the assumption is plausible for this problem. If you do reject the null hypothesis, it is possible that the link function selected is incorrect for the data or that the relationships between the independent variables and logits are not the same for all logits.

Figure 20-3
Test of parallel lines

Test of Parallel Lines[1]

Model	-2 Log Likelihood	Chi-Square	df	Sig.
Null Hypothesis	30.793			
General	28.906	1.887	2	.389

The null hypothesis states that the location parameters (slope coefficients) are the same across response categories.

1. Link function: Logit.

Does the Model Fit?

A standard statistical maneuver for testing whether a model fits is to compare observed and expected values. That is what's done here as well.

Calculating Expected Values

You can use the coefficients in Figure 20-2 to calculate cumulative predicted probabilities from the logistic model for each case:

$$\text{prob(event } j) = 1 / (1 + e^{-(\alpha_j - \beta x)})$$

Remember that the events in an ordinal logistic model are not individual scores but cumulative scores. First, calculate the predicted probabilities for those who didn't experience household crime. That means that β is 0, and all you have to worry about are the intercept terms.

$$\text{prob(score 1)} = 1 / (1 + e^{2.392}) = 0.0838$$
$$\text{prob(score 1 or 2)} = 1 / (1 + e^{0.317}) = 0.4214$$
$$\text{prob(score 1 or 2 or 3)} = 1 / (1 + e^{-2.59}) = 0.9302$$
$$\text{prob(score 1 or 2 or 3 or 4)} = 1$$

From the estimated cumulative probabilities, you can easily calculate the estimated probabilities of the individual scores for those whose households did not experience crime. You calculate the probabilities for the individual scores by subtraction, using the formula:

$$\text{prob(score} = j) = \text{prob(score less than or equal to } j) - \text{prob(score less than } j).$$

The probability for score 1 doesn't require any modifications. For the remaining scores, you calculate the differences between cumulative probabilities:

prob(score = 2) = prob(score = 1 or 2) – prob(score = 1) = 0.3376
prob(score = 3) = prob(score 1, 2, 3) – prob(score 1, 2) = 0.5088
prob(score = 4) = 1 – prob(score 1, 2, 3) = 0.0698

You calculate the probabilities for those whose households experienced crime in the same way. The only difference is that you have to include the value of β in the equation. That is,

$$\text{prob(score} = 1) = 1 / (1 + e^{(2.392 - 0.633)}) = 0.1469$$
$$\text{prob(score} = 1 \text{ or } 2) = 1 / (1 + e^{(0.317 - 0.633)}) = 0.5783$$
$$\text{prob(score} = 1, 2, \text{ or } 3) = 1 / (1 + e^{(-2.593 - 0.633)}) = 0.9618$$
$$\text{prob(score} = 1, 2, 3, \text{ or } 4) = 1$$

Of course, you don't have to do any of the actual calculations, since SPSS Statistics will do them for you. In the Options dialog box, you can ask that the predicted probabilities for each score be saved.

Figure 20-4 gives the predicted probabilities for each cell. The output is from the Means procedure with the saved predicted probabilities (*EST1_1*, *EST2_1*, *EST3_1*, and *EST4_1*) as the dependent variables and *hhcrime* as the factor variable. All cases with the same value of *hhcrime* have the same predicted probabilities for all of the response categories. That's why the standard deviation in each cell is 0. For each rating, the estimated probabilities for everybody combined are the same as the observed marginals for the rating variable.

Figure 20-4
Estimated response probabilities

Anyone in HH crime victim within past 3 years?		Estimated Cell Probability for Response Category: 1	Estimated Cell Probability for Response Category: 2	Estimated Cell Probability for Response Category: 3	Estimated Cell Probability for Response Category: 4
Yes	Mean	.1469	.4316	.3833	.0382
	N	76	76	76	76
	Std. Deviation	.00000	.00000	.00000	.00000
No	Mean	.0838	.3378	.5089	.0696
	N	490	490	490	490
	Std. Deviation	.00000	.00000	.00000	.00000
Total	Mean	.0922	.3504	.4920	.0654
	N	566	566	566	566
	Std. Deviation	.02154	.03202	.04284	.01071

For each rating, the estimated odds of the cumulative ratings for those who experience crime divided by the estimated odds of the cumulative ratings for those who didn't experience crime is $e^{-\beta} = 1.88$. For the first response, the odds ratio is

$$\frac{0.1469/(1-0.1469)}{0.0838/(1-0.0838)} = 1.88$$

For the cumulative probability of the second response, the odds ratio is

$$\frac{(0.1469+0.4316)/(1-0.1469-0.4316)}{(0.0838+0.3378)/(1-0.0838-0.3378)} = 1.88$$

Comparing Observed and Expected Counts

You can use the previously estimated probabilities to calculate the number of cases you expect in each of the cells of a two-way crosstabulation of rating and crime in the household. You multiply the expected probabilities for those without a history by 490, the number of respondents who didn't report a history. The expected probabilities for those with a history are multiplied by 76, the number of people reporting a household history of crime. These are the numbers you see in Figure 20-5 in the row labeled *Expected*. The row labeled *Observed* is the actual count.

The Pearson residual is a standardized difference between the observed and predicted values:

$$\text{Pearson residual} = \frac{O_{ij} - E_{ij}}{\sqrt{n_i \hat{p}_{ij}(1-\hat{p}_{ij})}}$$

Figure 20-5
Cell information

Frequency

Anyone in HH crime victim within past 3 years?		Rating of judges			
		Poor	Only fair	Good	Excellent
Yes	Observed	14	28	31	3
	Expected	11.16	32.800	29.134	2.903
	Pearson Residual	.919	-1.112	.440	.058
No	Observed	38	170	248	34
	Expected	41.05	165.501	249.4	34.094
	Pearson Residual	-.497	.430	-.123	-.017

Link function: Logit.

Goodness-of-Fit Measures

From the observed and expected frequencies, you can compute the usual Pearson and Deviance goodness-of-fit measures. The Pearson goodness-of-fit statistic is

$$\chi^2 = \Sigma\Sigma\frac{(O_{ij} - E_{ij})^2}{E_{ij}}$$

The deviance measure is

$$D = 2\Sigma\Sigma O_{ij}\ln\left(\frac{O_{ij}}{E_{ij}}\right)$$

Both of the goodness-of-fit statistics should be used only for models that have reasonably large expected values in each cell. If you have a continuous independent variable or many categorical predictors or some predictors with many values, you may have many cells with small expected values. SPSS Statistics warns you about the number of empty cells in your design. In this situation, neither statistic provides a dependable goodness-of-fit test.

If your model fits well, the observed and expected cell counts are similar, the value of each statistic is small, and the observed significance level is large. You reject the null hypothesis that the model fits if the observed significance level for the goodness-of-fit statistics is small. Good models have large observed significance levels. In Figure 20-6, you see that the goodness-of-fit measures have large observed significance levels, so it appears that the model fits.

Figure 20-6
Goodness-of-fit statistics

	Chi-Square	df	Sig.
Pearson	1.902	2	.386
Deviance	1.887	2	.389

Link function: Logit.

Including Additional Predictor Variables

A single predictor variable example makes explaining the basics easier, but real problems almost always involve more than one predictor. Consider what happens when additional factor variables—such as *sex*, *age2* (two categories), and *educ5* (five categories)—are included as well.

Recall the Ordinal Regression dialog box and select:

▶ Dependent: rating
▶ Factors: hhcrime, sex, age2, educ5

Options...
Link: Logit

Output...
Display
☑ Goodness of fit statistics
☑ Summary statistics
☑ Parameter estimates
☑ Test of Parallel Lines
Saved Variables
☑ Predicted category

The dimensions of the problem have quickly escalated. You've gone from 8 cells, defined by the four ranks and two crime categories, to 160 cells. The number of cases with valid values for all of the variables is 536, so cells with small observed and predicted frequencies will be a problem for the tests that evaluate the goodness of fit of the model. That's why the warning in Figure 20-7 appears.

Figure 20-7
Warning for empty cells

There are 44 (29.7%) cells (i.e., dependent variable levels by combinations of predictor variable values) with zero frequencies.

Overall Model Test

Before proceeding to examine the individual coefficients, you want to look at an overall test of the null hypothesis that the location coefficients for all of the variables in the model are 0. You can base this on the change in –2 log-likelihood when the variables are added to a model that contains only the intercept. The change in likelihood function has a chi-square distribution even when there are cells with small observed and predicted counts.

From Figure 20-8, you see that the difference between the two log-likelihoods—the chi-square—has an observed significance level of less than 0.0005. This means that you can reject the null hypothesis that the model without predictors is as good as the model with the predictors.

Figure 20-8
Model-fitting information

Model	-2 Log Likelihood	Chi-Square	df	Sig.
Intercept Only	322.784			
Final	288.600	34.183	7	.000

Link function: Logit.

You also want to test the assumption that the regression coefficients are the same for all four categories. If you reject the assumption of parallelism, you should consider using multinomial regression, which estimates separate coefficients for each category. Since the observed significance level in Figure 20-9 is large, you don't have sufficient evidence to reject the parallelism hypothesis.

Figure 20-9
Test of parallelism

Model	-2 Log Likelihood	Chi-Square	df	Sig.
Null Hypothesis	288.600			
General	276.115	12.485	14	.567

The null hypothesis states that the location parameters (slope coefficients) are the same across response categories.

Examining the Coefficients

From the observed significance levels in Figure 20-10, you see that sex, education, and household history of crime are all related to the ratings. They all have negative coefficients. Men (code 1) are less likely to assign higher ratings than women, people with less education are less likely to assign higher ratings than people with graduate education (code 5), and persons whose households have been victims of crime are less likely to assign higher ratings than those in crime-free households. Age doesn't appear to be related to the rating.

Figure 20-10
Parameter estimates for the model

		Estimate	Std. Error	Wald	df	Sig.
Threshold	[rating = 1]	-3.630	.335	117.579	1	.000
	[rating = 2]	-1.486	.302	24.265	1	.000
	[rating = 3]	1.533	.311	24.378	1	.000
Location	[hhcrime=1]	-.643	.238	7.318	1	.007
	[hhcrime=2]	0[1]	.	.	0	.
	[sex=1]	-.424	.163	6.758	1	.009
	[sex=2]	0[1]	.	.	0	.
	[age2=0]	.076	.176	.186	1	.666
	[age2=1]	0[1]	.	.	0	.
	[educ5=1]	-1.518	.389	15.198	1	.000
	[educ5=2]	-1.256	.288	19.004	1	.000
	[educ5=3]	-.941	.310	9.188	1	.002
	[educ5=4]	-.907	.302	9.015	1	.003
	[educ5=5]	0[1]	.	.	0	.

Link function: Logit.

1. This parameter is set to zero because it is redundant.

Measuring Strength of Association

There are several R^2-like statistics that can be used to measure the strength of the association between the dependent variable and the predictor variables. They are not as useful as the R^2 statistic in regression, since their interpretation is not straightforward. Three commonly used statistics are:

- Cox and Snell R^2

$$R^2_{CS} = 1 - \left(\frac{L(\mathbf{B}^{(0)})}{L(\hat{\mathbf{B}})}\right)^{\frac{2}{n}}$$

- Nagelkerke's R^2

$$R^2_N = \frac{R^2_{CS}}{1 - L(\mathbf{B}^{(0)})^{2/n}}$$

- McFadden's R^2

$$R^2_M = 1 - \left(\frac{L(\hat{\mathbf{B}})}{L(\mathbf{B}^{(0)})}\right)$$

where $L(\hat{\mathbf{B}})$ is the log-likelihood function for the model with the estimated parameters and $L(\mathbf{B}^{(0)})$ is the log-likelihood with just the thresholds, and *n* is the number of cases (sum of all weights). For this example, the values of all of the pseudo *R*-square statistics are small.

Figure 20-11
Pseudo R-square

Cox and Snell	.059
Nagelkerke	.066
McFadden	.027

Link function: Logit.

Classifying Cases

You can use the predicted probability of each response category to assign cases to categories. A case is assigned to the response category for which it has the largest predicted probability. Figure 20-12 is the classification table, which is obtained by crosstabulating *rating* by *pre_1*. (This is sometimes called the confusion matrix.)

Figure 20-12
Classification table

Count

Rating of judges	Predicted Response Category		Total
	Only fair	Good	
Poor	15	36	51
Only fair	42	156	198
Good	33	246	279
Excellent	2	35	37
Total	92	473	565

Of the 198 people who selected the response *Only fair,* only 42 are correctly assigned to the category using the predicted probability. Of the 279 who selected *Good*, 246 are correctly assigned. None of the respondents who selected *Poor* or *Excellent* is correctly assigned. If the goal of your analysis is to study the association between the grouping variable and the predictor variables, the poor classification should not concern you. If your goal is to target marketing or collections efforts, the correct classification rate may be more important.

Generalized Linear Models

The ordinal logistic model is one of many models subsumed under the rubric of generalized linear models for ordinal data. The model is based on the assumption that there is a latent continuous outcome variable and that the observed ordinal outcome arises from discretizing the underlying continuum into *j*-ordered groups. The thresholds estimate these cutoff values.

The basic form of the generalized linear model is

$$\text{link}(\gamma_j) = \frac{\theta_j - [\beta_1 x_1 + \beta_2 x_2 + \ldots + \beta_k x_k]}{\exp(\tau_1 z_1 + \tau_2 z_2 + \ldots + \tau_m z_m)}$$

where γ_j is the cumulative probability for the *j*th category, θ_j is the threshold for the *j*th category, $\beta_1 \ldots \beta_k$ are the regression coefficients, $x_1 \ldots x_k$ are the predictor variables, and k is the number of predictors.

The numerator on the right side determines the **location** of the model. The denominator of the equation specifies the scale. The $\tau_1 \ldots \tau_m$ are coefficients for the scale component, and $z_1 \ldots z_m$ are m predictor variables for the scale component (chosen from the same set of variables as the x's).

The **scale component** accounts for differences in variability for different values of the predictor variables. For example, if certain groups have more variability than others in their ratings, using a scale component to account for this may improve your model.

Link Function

The link function is the function of the probabilities that results in a linear model in the parameters. It defines what goes on the left side of the equation. It's the link between the random component on the left side of the equation and the systematic component on the right. In the criminal rating example, the link function is the logit function, since the log of the odds results is equal to the linear combination of the parameters. That is,

$$\ln\left(\frac{\text{prob(event)}}{(1 - \text{prob(event)})}\right) = \beta_0 + \beta_1 x_1 + \beta_2 x_2 + \ldots + \beta_k x_k$$

Five different link functions are available in the Ordinal Regression procedure in SPSS Statistics. They are summarized in the following table. The symbol γ represents the probability that the event occurs. Remember that in ordinal regression, the probability of an event is redefined in terms of cumulative probabilities.

Function	**Form**	**Typical application**
Logit	$\ln\left(\frac{\gamma}{1-\gamma}\right)$	Evenly distributed categories
Complementary log-log	$\ln(-\ln(1-\gamma))$	Higher categories more probable
Negative log-log	$-\ln(-\ln(\gamma))$	Lower categories more probable
Probit	$\Phi^{-1}(\gamma)$	Analyses with explicit normally distributed latent variable
Cauchit (inverse Cauchy)	$\tan(\pi(\gamma - 0.5))$	Outcome with many extreme values

If you select the probit link function, you fit the model described in the chapter "Probit Regression" in the *SPSS Statistics 17.0 Advanced Statistical Procedures Companion*. The observed probabilities are replaced with the value of the standard normal curve below which the observed proportion of the area is found.

Probit and logit models are reasonable choices when the changes in the cumulative probabilities are gradual. If there are abrupt changes, other link functions should be used. The complementary log-log link may be a good model when the cumulative probabilities increase from 0 fairly slowly and then rapidly approach 1. If the opposite is true, namely that the cumulative probability for lower scores is high and the approach to 1 is slow, the negative log-log link may describe the data. If the complementary log-log model describes the probability of an event occurring, the log-log model describes the probability of the event not occurring.

Fitting a Heteroscedastic Probit Model

Probit models are useful for analyzing signal detection data. Signal detection describes the process of detecting an event in the face of uncertainty or "noise." You must decide whether a signal is present or absent. For example, a radiologist has to decide whether a tumor is present or not based on inspecting images. You can model the uncertainty in the decision-making process by asking subjects to report how confident they are in their decision.

You postulate the existence of two normal distributions: one for the probability of detecting a signal when only noise is present and one for detecting the signal when both the signal and the noise are present. The difference between the means of the two distributions is called *d*, a measure of the sensitivity of the person to the signal.

The general probit model is

$$p(Y \le k|X) = \Phi\left(\frac{c_k - d_n X}{\sigma_s^x}\right)$$

where Y is the dependent variable, such as a confidence rating, with values from 1 to K; X is a 0–1 variable that indicates whether the signal was present or absent; c_k are ordered distances from the noise distribution; d_n is the scaled distance parameter; and σ_s is the standard deviation of the signal distribution. The model can be rewritten as

$$\Phi^{-1}[p(Y \le k|X)] = \frac{c_k - d_n X}{e^{ax}}$$

where Φ^{-1} is the inverse of the cumulative normal distribution and a is the natural log of σ_s. The numerator models the location; the denominator models the scale.

If the noise and signal distributions have different variances, you must include this information in the model. Otherwise, the parameter estimates are biased and inconsistent. Even large sample sizes won't set things right.

Modeling Signal Detection

Consider data reported from a light detection study by Swets, et al. (1961) and discussed by DeCarlo (2003). Data are for a single individual who rated his confidence that a signal was present in 591 trials when the signal was absent and 597 trials when the signal was present.

In Figure 20-13, you see the cumulative distribution of the ratings under the two conditions (signal absent and signal present). The noise curve is above the signal curve, indicating that the low confidence ratings were more frequent when a signal was not present.

Figure 20-13
Plot of cumulative confidence ratings

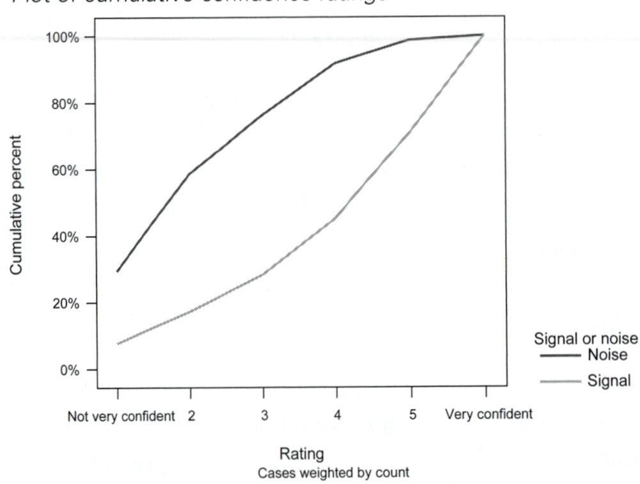

Fitting a Location-Only Model

If you assume that the variances of the noise and signal distributions are equal, you can fit the usual probit model. Open the file *swets.sav.* The data are aggregated. For each possible combination of *signal* and *response*, there is a *count* of the number of times that each response was chosen.

You must weight the data file before proceeding. From the menus choose:

Data
  Weight Cases...

⊙ Weight cases by
▶ count

Analyze
Regression
Ordinal...

▶ Dependent: response
▶ Covariate(s): signal

Options...
Link: Probit

Output...
Display
☑ Parameter Estimates
☑ Goodness of fit statistics

Examining the Goodness of Fit

Since the model has only 12 cells, and no cells have zero frequencies, you can examine the goodness-of-fit statistics without concern that the expected counts are too small for the chi-square approximation to be valid. From Figure 20-14, you see that the model does not fit well. The observed significance level is less than 0.0005.

Figure 20-14
Goodness-of-fit statistics

	Chi-Square	df	Sig.
Pearson	33.728	4	.000
Deviance	32.972	4	.000

Link function: Probit.

One of the reasons the model may fit poorly is because the variance of the two distributions of responses may be different. You need a model that allows the variance of the underlying variable to vary as a function of one or more of the independent variables.

You can select a model for the standard deviation such that

$$\sigma_i = e^{Z_i \gamma}$$

where Z_i is a vector of covariates selected from the predictor variables.

Fitting a Scale Parameter

To fit a model that allows for different variances in the two groups, you must specify a model for the scale parameters. To fit a model with different variances in the two groups, recall the dialog box and select:

Scale...
  Scale model: signal

Because you have only one independent variable, *signal*, separate variances are estimated for each of the two categories of *signal*. If you have several predictor variables, you can specify a separate model for the scale component.

The goodness-of-fit statistics in Figure 20-15 indicate that the model fits much better than the location-only model. The variability of the distributions are an important consideration in this problem.

Figure 20-15
Goodness of fit with scale model

	Chi-Square	df	Sig.
Pearson	1.497	3	.683
Deviance	1.482	3	.687

Link function: Probit.

Parameter Estimates

When you fit a model with scale parameters as well as location parameters, parameter estimates for both are displayed.

Figure 20-16
Parameter estimates for model with location and scale parameters

Parameter Estimates

							95% Confidence Interval	
		Estimate	Std. Error	Wald	df	Sig.	Lower Bound	Upper Bound
Threshold	[response = 1]	-.533	.054	98.809	1	.000	-.638	-.428
	[response = 2]	.204	.050	16.979	1	.000	.107	.301
	[response = 3]	.710	.053	182.311	1	.000	.607	.813
	[response = 4]	1.366	.067	414.418	1	.000	1.235	1.498
	[response = 5]	2.294	.113	409.475	1	.000	2.072	2.516
Location	signal	1.519	.096	250.110	1	.000	1.331	1.707
Scale	signal	.348	.063	30.711	1	.000	.225	.472

Link function: Probit.

The threshold values are distances of the response criteria from the mean of the noise distribution. The location parameter estimate is the estimate of the detection parameter, d_n . To convert the scale parameter to an estimate of the ratio of the noise to signal standard deviations, you must compute $e^{-0.348}$, which is 0.71.

Model-Fitting Information

The overall test of the model is shown in Figure 20-17. When there is a scale parameter, the null hypothesis is that *both* the location parameters and the scale parameters are 0. A scale parameter of 0 means that the variances are equal. Based on the small observed significance level, you can reject this composite null hypothesis. Consult DeCarlo (2003) for further discussion of this example and for other examples of using the Ordinal Regression procedure in signal detection.

Figure 20-17
Model-fitting information

Model	-2 Log Likelihood	Chi-Square	df	Sig.
Intercept Only	461.699			
Final	59.261	402.437	2	.000

Link function: Probit.

Chapter

21

General Loglinear Analysis

When you want to study the relationship between two categorical variables, you usually start by looking at a crosstabulation (see Chapter 10). You can easily test whether the two variables are independent and quantify the strength of the association between them. When you add more variables, the situation quickly grows more complex, since you have a separate two-way table for each combination of values of the additional variables and the relationships within a table depend on which variables are rows, columns, and layers. Detecting and describing interactions of various orders by simply looking at cell counts in two-way tables "controlled" for additional variables is not for the faint of heart.

A statistical technique called **loglinear modeling**, although not as glamorous as the name may suggest, is particularly suited for examining relationships among many categorical variables. You look at interactions of all orders among all possible combinations of variables. You identify large interactions and can then fit a parsimonious model that describes the observed data well. Loglinear models do not require you to stereotype variables as dependent and independent, although a special class of loglinear models, called **logit models**, allow such designations.

SPSS Statistics has a number of procedures that are useful for modeling count data. The three procedures that appear on the Loglinear option from the menus differ in important ways:

- Use the General Loglinear Analysis procedure for estimating parameters of any loglinear model, including those with observed and/or predicted cell counts of 0.
- Use the Model Selection procedure to select a hierarchical loglinear model from all possible hierarchical models. (See the "Model Selection Loglinear Analysis" chapter in the *SPSS Statistics 17.0 Advanced Statistical Procedures Companion* for more information.)

- Use the Logit procedure to estimate the parameters of a model when one of the variables is considered to be the dependent variable. (See the "Logit Loglinear Analysis" chapter in the *SPSS Statistics 17.0 Advanced Statistical Procedures Companion* for more information.)

The general loglinear model facility described in this chapter can be used for both hierarchical and nonhierarchical loglinear models. It has many options that are not available in the Model Selection facility. Different types of parameter estimates are obtained from the General Loglinear and Model Selection facilities, so the discussion of the loglinear model and corresponding parameter estimates is somewhat different in the two chapters.

Examples

- How are self-esteem, religious preference, education, and gender related?
- Is there a relationship among the incidence of blood clots, smoking, and immobility?
- What is the relationship between buying cars, homes, boats, and cereal types?
- Are people equally likely to move from one area of the country to another?

In a Nutshell

In a general loglinear model, the dependent variable is the log of the number of cases in a cell of a multiway crosstabulation. The independent variables are the categorical variables and their interactions. The variability in observed cell counts is explained by main effects and interactions of the independent variables. The cell means of continuous variables (covariates) can also be incorporated into the model. You estimate coefficients for each category of the independent variable and interactions of the categories, and you test hypotheses about the population values of the coefficients. Using a subset of all possible terms, you can identify a model that predicts the dependent variable well.

Many hypotheses about crosstabulated data, such as symmetry of square tables and quasi-independence, can be formulated as tests of parameters in loglinear models. Loglinear models can also be used to standardize crosstabulations to a new set of marginals.

Origins of Data

When you look at a crosstabulation, you usually can't tell much about how the observations crept into it. The same table can result from different experimental designs. For example, if you are studying the relationship between type of treatment and cancer recurrence within one year of treatment, you can gather data in several ways. You can randomly select the records of 1,000 eligible patients from a cancer registry and determine what the initial treatment was and whether they had a recurrence within one year. In this case, your sample size of 1,000 patients is fixed. Or you may decide to look at the records of all eligible cancer patients diagnosed in a particular two-year period. In this case, the sample size is not fixed, since you do not know how many eligible patients you will find in this two-year period. Another way you might obtain data is to select 500 eligible patients with recurrences and 500 eligible patients without recurrences and determine which treatment each patient received. This time the sample size is fixed not only for the total number of patients but also for the number of patients who had a recurrence and the number who did not.

The design of the study determines, in part, which sampling model is appropriate for the data. (A sampling model is a mathematical model that describes the distribution of the data.) The two distributions most often used to describe counts in a crosstabulation are the Poisson distribution and the multinomial distribution. Fortunately, except for the constant, the parameter estimates are usually the same for these two different sampling models.

Poisson Distribution

The **Poisson distribution** is useful for modeling rare events, such as suicides, deaths, or the number of raisins in a tablespoon of cereal. A Poisson sampling model for a crosstabulation arises when:

- The total sample size is not fixed.
- The number of cases in each cell of the table is independent of the others and has a Poisson distribution.

The counts arising from the study in which you obtained records for cases diagnosed in a two-year period may have a Poisson distribution, since the total sample size was not fixed.

Multinomial Distribution

The **multinomial distribution** is a generalization of the binomial distribution to more than two events. Under a multinomial sampling model, each cell of the crosstabulation table has a probability that indicates how likely an observation is to fall into it. The sum of the probabilities across all cells is 1. In a multinomial distribution:

- The total sample size is fixed.
- The cell counts are not independent, since they must sum to the total.

In a multinomial sampling model, you know how many cases will be included in the study before you start. If you fix the row or column totals—such as when you selected 500 cases with a recurrence and 500 cases without a recurrence—the counts for each row or column have a multinomial distribution and the distribution for the entire table is called the product multinomial distribution.

When you specify multinomial sampling in the General Loglinear Analysis procedure, the constant in the model is treated as known. That's why standard errors and confidence intervals are not calculated for it. For further discussion of sampling models, as well as a discussion of loglinear models, see Agresti (2002).

Tip: If you don't know how the data were obtained, treating them as coming from a Poisson distribution is usually a good strategy. That's the default in the General Loglinear Models procedure.

Anatomy of a Loglinear Model

Loglinear modeling follows a plot line similar to most episodes of statistical modeling. You have an equation which relates a set of independent variables to a dependent variable using a set of parameters. You estimate the parameters from your data. You attempt to identify the smallest number of parameters that you need to adequately represent the data.

In loglinear models, the dependent variable is the log of the number of cases in each of the cells of the table. The independent variables are categorical. That is, they have a limited number of distinct values. The equation has many of the same intriguing characters you've encountered before. There are constants, main effects, and various orders of interactions. There are also many different notations that are used for writing loglinear model scripts. The notation used in this chapter may not be mathematically elegant, but it aims to be easy to follow for those who may tune out when confronted with a swarm of subscripts and Greek symbols.

If you have many cells in a multiway crosstabulation, the loglinear model for the table can be quite cumbersome. That's why you'll start with data in a two-way table with two values for each of the variables. Then, as your math teacher told you in high school, you can easily extrapolate it to a zillion variables without loss of generality.

Dissecting a Table

Figure 21-1 is a crosstabulation of whether a person voted in a presidential election by whether or not they have a college degree. The data are from the General Social Survey (Davis and Smith, 2000).

Figure 21-1
Percentage voting by degree

			College Degree		
			No	Yes	Total
Vote	No	Count	369	50	419
		% within College Degree	35.9%	11.8%	28.9%
	Yes	Count	659	372	1031
		% within College Degree	64.1%	88.2%	71.1%
Total		Count	1028	422	1450
		% within College Degree	100.0%	100.0%	100.0%

You can write a loglinear model that predicts the log of the number of cases in each cell from a set of estimated parameters that are collectively denoted as λ. You need a lambda parameter for each value of the row and column variables and for each combination of the row and column variables. The label above the lambda tells you the category, or combination of categories, that the parameters represent.

Warning: Don't believe what you're told. Even for presidential elections, a turnout of over 70% is highly unusual. It's unlikely that everyone who says they voted actually did.

For example, the natural log of the expected number of cases, m_{ij}, in each of the four cells of Figure 21-1 is:

$$\ln(m_{11}) = \mu + \lambda^{\text{nonvoter}} + \lambda^{\text{without degree}} + \lambda^{\text{nonvoter without degree}}$$

$$\ln(m_{12}) = \mu + \lambda^{\text{nonvoter}} + \lambda^{\text{with degree}} + \lambda^{\text{nonvoter with degree}}$$

$$\ln(m_{21}) = \mu + \lambda^{\text{voter}} + \lambda^{\text{without degree}} + \lambda^{\text{voter without degree}}$$

$$\ln(m_{22}) = \mu + \lambda^{\text{voter}} + \lambda^{\text{with degree}} + \lambda^{\text{voter with degree}}$$

Lambda Parameters

In the above equations, the term μ is comparable to the intercept term in a regression model. It tells you the predicted number of cases in a cell if the values of all of the lambda (λ) parameters are 0. The lambda parameters represent the increments or decrements from the base value (intercept) for particular combinations of values of the row and column variables.

Each individual category of the row and column variables has an associated lambda parameter. There is a lambda for voters and a lambda for nonvoters. Similarly, there is a lambda for those with college degrees and for those without college degrees. Since these lambda parameters are for individual variables, they are called **main effect** parameters. There is a main effect for voting and a main effect for education.

There are also lambda parameters for all possible combinations of values of the row and column (main effect) variables. They are called **interactions**. In this example, there are four voter-by-degree interactions:

$\lambda^{\text{voter without degree}}$

$\lambda^{\text{voter with degree}}$

$\lambda^{\text{nonvoter with degree}}$

$\lambda^{\text{nonvoter without degree}}$

For example, $\lambda^{\text{voter with degree}}$ represents how many more cases measured on the log scale (or fewer cases, if λ is negative) are expected in the cell than would be predicted from just the μ, λ^{voter}, and $\lambda^{\text{with degree}}$ parameters.

You are usually interested in the interaction parameters in a loglinear model, since they tell you about the relationships between combinations of variables. For example, the interaction parameters for voting and degree tells you whether education and voting behavior are related.

If you count the number of lambda parameters used to describe the four cells, you'll come up with 12. It's not possible to uniquely estimate 12 parameters from only four observed cell counts. You have to impose constraints on some of the parameters. For example, you can force some parameters to have the value of 0, or you can force certain sets of parameters to sum to 0. This is discussed further in "Reference Categories" on p. 502.

Model Types

A model that contains lambda parameters for all of the categories of the independent variables as well as for all possible combinations of the categories of the independent variables is called a **saturated model**. It can't hold any more terms! A saturated model by definition exactly reproduces the observed cell counts. The model in this example is a saturated model, since it contains terms for the values of voting and education as well as their interaction.

If any of the possible lambda parameters is not present in the model, the model is called an **unsaturated model**. In this example, if the interaction between voting and education is not included in the model, the model is unsaturated. An unsaturated model usually doesn't reproduce the observed counts exactly unless it fits the data perfectly.

Although it is possible to delete any terms in a saturated model, a frequently used type of model, a **hierarchical model**, puts restrictions on what terms can be deleted. In a hierarchical model, if you include a term for the interaction of two or more variables, you must include all lower-order interactions that are a subset of the variables in the interaction, as well as their main effects. If you have three or more variables and include the three-way interaction of the variables *A*, *B*, and *C*, you must also include all main effects (*A*, *B*, and *C*) and all possible two-way interactions (*AB*, *AC*, and *BC*). Hierarchical models are discussed in greater detail in the "Model Selection Loglinear Analysis" chapter in the *SPSS Statistics 17.0 Advanced Statistical Procedures Companion*.

Fitting a Saturated Model

To fit a saturated loglinear model to the data in Figure 21-1, open the data file *vote.sav* and from the menus choose:

Analyze
  Loglinear
    General...

▸ Factor(s): vote, college

Options...
  Display
    ☑ Estimates
  Criteria
    Delta: 0

The estimated lambda parameters for Figure 21-1 are shown in Figure 21-2, in the column labeled *Estimate*. The standard error is in the next column, labeled *Std. Error*. The z statistic is the ratio of the estimate to its standard error. Under the null hypothesis that the population value for the parameter is 0, the z statistic has a normal distribution. The observed significance level for the test of the null hypothesis that the parameter value is 0 is shown in the column labeled *Sig*.

Warning: If you do not set delta to 0 in the Criteria dialog box, all of your observed and expected cell counts will have 0.5 added to them. SPSS Statistics adds a half to all cells in the crosstabulation for computational purposes. For unsaturated models, you don't see the half in the observed and expected frequencies. For the saturated model, you do.

In Figure 21-2, there are several parameter estimates that have footnotes associated with them. All of these parameter estimates have values of 0 and a missing code for all other statistics. These parameters are set to 0 so that a unique solution to the problem can be obtained.

Examining the Lambda Parameters

One of the primary reasons for fitting a saturated model to data is so that you can examine the parameter estimates and see which ones are not significantly different from 0. You may then want to build a model that excludes these parameters. For example, if the voting-with-degree interaction parameter in the previous example was not significant, you could build a model that contained only the main effects for voting and degree, since the independence model holds. Terms should be removed sequentially, since parameter estimates for terms that remain in the model may change when higher-order interaction terms are removed. In this example, all of the parameter estimates are significantly different from 0.

Figure 21-2
Estimated lambda parameters

Parameter Estimates [2,3]

Parameter	Estimate	Std. Error	Z	Sig.	95% Confidence Interval	
					Lower Bound	Upper Bound
Constant	5.919	.052	114.159	.000	5.817	6.021
[vote = 0]	-2.007	.151	-13.324	.000	-2.302	-1.712
[vote = 1]	0[1]	.	.	.	.	.
[college = 0]	.572	.065	8.818	.000	.445	.699
[college = 1]	0[1]	.	.	.	.	.
[vote = 0] * [college = 0]	1.427	.164	8.698	.000	1.105	1.748
[vote = 0] * [college = 1]	0[1]	.	.	.	.	.
[vote = 1] * [college = 0]	0[1]	.	.	.	.	.
[vote = 1] * [college = 1]	.000	.	.	.	.	.

1. This parameter is set to zero because it is redundant.
2. Model: Poisson
3. Design: Constant + vote + college + vote * college

Interpreting the Parameters

When you build a model that involves categorical variables, the coefficients for the categorical variables are trickier to interpret than the coefficients for continuous variables. For example, if you have four regions of the country in a model, you can't make an absolute statement about the effect of a particular region. You must compare it to other regions. You can say that sales in region 1 are higher than the average of sales across all regions. Or that sales in region 1 are higher than sales in region 4. Any comment about region 1 depends on a frame of reference. The same is true for the lambda parameters in a loglinear model. The lambda parameters must have a frame of reference.

Reference Categories

The General Loglinear Analysis procedure establishes a frame of reference when some of the lambda parameters are set to 0 so that unique estimates can be computed. For example, the lambda parameter for voters is 0. This means that the lambda parameter for nonvoters uses voters as a frame of reference. A positive value indicates that there are more nonvoters than voters among people with a college degree. A value of 0 tells you that the number of voters and nonvoters is exactly the same. A negative value tells you that the number of nonvoters is smaller than the number of voters. Similarly, in the region example, if the lambda for region 4 is set to 0, then all other regions are compared to region 4. Usually, effects involving the last category of a variable are set to 0. All other lambda parameters that involve the last category of a variable, are also set to 0. If there are cells without any cases, additional parameters may also be set to 0. Parameters which are set to 0 are called **aliased** or **redundant parameters**.

In this example, from Figure 21-2, you see that the parameters for voting, having a college degree, and all interactions that involve either voting or having a college degree are set to 0.

Warning: Depending on the characteristics of the model and your data, parameters other than those for the last category may be set to 0. Always look at the parameter estimate table to see which parameters are set to 0.

Calculating Expected Values

In many statistical procedures, you can calculate predicted, or expected, values based on the parameters you estimate from your data. For each case, you then compare the expected values to those actually observed to see how close they are. You can use the values of the estimated parameters for a loglinear model to calculate expected cell counts for each of the cells.

If you substitute the values of the parameter estimates in Figure 21-2 into the right side of the previous equation, you will reproduce exactly the natural logs of the observed cell counts, since there are as many estimated parameters as there are cells. (By analogy, think of a linear regression with only two points. You can draw only one line through two points, and the line fits the data perfectly, since you have as many parameters—the slope and the intercept—as you have data points. In this example, you have four cells and four nonzero parameters.) Since your observed and expected cell counts are equal, all residuals (differences between observed and expected cell counts) are 0, as shown in the following examples:

$$\begin{aligned}
\ln(369) &= \mu + \lambda^{\text{non voter}} + \lambda^{\text{without degree}} + \lambda^{\text{non voter without degree}} \\
&= 5.919 - 2.007 + 0.572 + 1.427 \\
\ln(50) &= \mu + \lambda^{\text{non voter}} + \lambda^{\text{with degree}} + \lambda^{\text{non voter with degree}} \\
&= 5.919 - 2.007 + 0 + 0 \\
\ln(659) &= \mu + \lambda^{\text{voter}} + \lambda^{\text{without degree}} + \lambda^{\text{voter without degree}} \\
&= 5.919 + 0 + 0.572 + 0 \\
\ln(372) &= \mu + \lambda^{\text{voter}} + \lambda^{\text{with degree}} + \lambda^{\text{voter with degree}} \\
&= 5.919 + 0 + 0 + 0
\end{aligned}$$

If the model is not saturated, the observed and expected cell counts will not be equal unless the model happens to fit perfectly.

Parameters and Odds

Each of the parameter estimates in a loglinear model can be expressed in terms of expected cell counts. For example, if you set all of the parameters that involve voters and degrees to 0 and then solve for the remaining parameters, you obtain:

$$\hat{\lambda}^{\text{non voter}} = \ln\left(\frac{m_{12}}{m_{22}}\right) = \ln\left(\frac{50}{372}\right) = -2.01$$

$$\hat{\lambda}^{\text{without degree}} = \ln\left(\frac{m_{21}}{m_{22}}\right) = \ln\left(\frac{659}{372}\right) = 0.57$$

$$\hat{\lambda}^{\text{non voter without degree}} = \ln\left(\frac{m_{11}m_{22}}{m_{12}m_{21}}\right) = \ln\left(\frac{369 \times 372}{50 \times 659}\right) = 1.43$$

Tip: You can create linear combinations of expected values by specifying a contrast variable in the main dialog box.

Ratios of cell counts are important for describing the relationships in tables. Many of them have names. Here's a list to help you get your bearings:

- The **odds** for an event is the ratio of the number of people who experience the event to the number of people who do not. For example, from Figure 21-1, the odds of being a voter are 1031/419. The odds of having a college degree are 422/1,028. For a college graduate, the odds of not voting is 50/372.
- The **log odds** is the log of the odds ratio. An odds ratio of 1 corresponds to a log odds of 0 because the logarithm of 1 is 0. The parameter estimate $\hat{\lambda}^{nonvoter}$ is the log odds of not voting for people with a college degree.
- The **odds ratio** is the ratio of two odds. For example, the odds that a person with a degree votes are 372/50=7.44. The odds that a person without a degree votes are 659/369=1.79. The ratio of the two odds is 7.44/1.79=4.15. The odds of a person with a degree voting are more than four times the odds of a person without a degree voting. If there is no association between voting and degree, you expect this ratio to be close to 1.
- The **log odds ratio** is the log of the odds ratio. The log of the odds ratio that a person with a degree votes to the odds that a person without a degree votes is 1.42, the log of 4.15. That's the parameter estimate you see in Figure 21-2 for the non-voter by no-degree interaction. The confidence interval for the log-odds ratio is from 1.10 to 1.75. Notice that the confidence interval does not include the value of 0. So you can reject the null hypothesis that the log odds are 0. This is equivalent to rejecting the null hypothesis that the odds ratio is 1.

Tip: You can convert log odds to odds by finding e^{λ}, where e is the base of the natural logs. For example, a log odds of 0.58 corresponds to an odds of 1.79, since $e^{0.58}$ is 1.79. You can compute a confidence interval for the odds by finding e^{lower} and e^{upper}, where lower and upper are the values for the lower and upper bounds of the confidence interval for the log odds.

Fitting an Unsaturated Model

In the previous example, since the observed significance level for the voting-by-degree parameter estimate is small, indicating that you can reject the null hypothesis that the parameter is zero, a model without the interaction term does not fit the data well. It appears that people with a college degree are more likely to vote than those without.

Obviously, you didn't need a loglinear model to arrive at this conclusion—the chi-square test of independence is a much simpler path—but it was easy to see the basic elements of a loglinear analysis.

Now that you understand the basics of fitting a loglinear model, consider the relationship between three variables: voting, college degree, and gender. The saturated loglinear model has lambda parameters for each of the categories of the variables individually, for each pair of interactions between the variables, and for the three-way interaction. You want to know which of the interaction parameters you need to adequately represent the data.

Start by fitting the default saturated model and examining the parameter estimates and their observed significance levels. From the menus choose:

Analyze
 Loglinear
  General...

▶ Factor(s): vote, college, male

Options...
 Display
  ☑ Estimates
 Criteria:

Delta: 0

Results from the Saturated Model

From Figure 21-3, you see that the coefficients for *male*, for the *male*-by-*vote* interaction, and for the three-way interaction of *male*, *vote,* and *degree* are not significantly different from 0. (All coefficients that are set to 0 have been deleted from the table to make it easier to read.) The test of the coefficient for the two-way interaction between *male* and *vote* tells you whether the two genders are equally likely to vote for those at the highest level of *college*. Since the observed significance level is large, you can't reject the null hypothesis that there is no interaction between gender and voting. Men and women appear to be equally likely to vote.

The three-way interaction between gender, voting, and degree tells you whether the relationship between voting and degree is the same for both genders. Since the observed significance level for the test that it is 0 is large, you don't have enough evidence to believe that, overall, voting depends on gender or that the relationship between voting and degree is different for males and females. You can see that this is the case based on the crosstabulation in Figure 21-4. The percentage voting is similar for males and females, for both college graduates and nongraduates.

Tip: When you have three or more variables in an interaction, you can describe the same interaction (or its absence) in various ways. Instead of saying that the relationship between voting and college degree is the same for males and females, you can say that the relationship between gender and voting is the same for all degree categories or that the relationship between degree and gender is the same for all categories of voting.

Figure 21-3
Parameter estimates for saturated voting, college, and gender model

Parameter Estimates[2,3]

		Estimate	Std. Error	Z	Sig.
Parameter	Constant	5.220	.074	71.005	.000
	[vote = 0]	-2.176	.230	-9.449	.000
	[college = 0]	.352	.096	3.666	.000
	[male = 0]	.011	.104	.104	.917
	[vote = 0] * [college = 0]	1.628	.252	6.464	.000
	[vote = 0] * [male = 0]	.312	.305	1.024	.306
	[college = 0] * [male = 0]	.399	.131	3.049	.002
	[vote = 0] * [college = 0] * [male = 0]	-.365	.332	-1.099	.272

2. Model: Poisson
3. Design: Constant + vote + college + male + vote * college + vote * male + college * male vote * college * male

Based on the tests of the parameters, you can remove the three-way interaction and reestimate the model. If the *vote*-by-*sex* interaction remains nonsignificant when the three-way interaction is removed, you can remove it as well. You want to include the main effects if you have interaction terms that involve them.

Figure 21-4
Percentage voting by gender and education

			Female	Male
			Vote	Vote
			Yes	Yes
	College Degree	No	64.6%	63.4%
		Yes	86.6%	89.8%
	Total		70.3%	72.1%

Warning: Higher-order interactions are difficult to interpret. Highly statistically significant interactions can arise when you have a large table with few cases or even a single cell with a very large or small count.

Specify the Unsaturated Model

If you don't specify a model in the SPSS Statistics General Loglinear Model procedure, you get a saturated model. The footnote below the table of parameter estimates tells you what terms are in the model. To specify your own custom model, without the three-way interaction and without the two-way *male*-by-*vote* interaction, from the menus choose:

Analyze
  Loglinear
    General...

▸ Factor(s): vote, college, male

Model...
  ⊙ Custom
  ▸ Terms in Model: (Main effects): vote college male
    (Interaction): college*male, vote*college

Options...
  Display
    ☑ Frequencies
    ☑ Estimates
    ☑ Residuals
  Plots: adjusted residuals

Parameter estimates for the unsaturated model are shown in Figure 21-5. Except for the coefficient for being a male, all parameter estimates are now significantly different from 0.

Tip: For unsaturated models, an iterative algorithm is used to calculate maximum-likelihood estimates. The convergence information table provides information about the criteria used. If convergence is not achieved, increase the number of iterations.

Figure 21-5
Parameter estimates for unsaturated model

Parameter Estimates[2,3]

		Estimate	Std. Error	Z	Sig.
Parameter	Constant	5.202	.072	72.32	.000
	[college = 0]	.382	.090	4.24	.000
	[male = 0]	.047	.097	.49	.626
	[vote = 0]	-2.007	.151	-13.32	.000
	[college = 0] * [male = 0]	.343	.116	2.95	.003
	[vote = 0] * [college = 0]	1.427	.164	8.70	.000

2. Model: Poisson
3. Design: Constant + college + male + vote + college * male + vote * college

Determine How Well the Model Fits

A saturated model by definition reproduces the data perfectly. The observed and expected cell counts are always equal. That's not the case for an unsaturated model. It may, or may not, fit the data. There are several statistics, based on the observed and expected counts, that you can use to evaluate how well the overall model fits. You should also always look at the fit of the individual cells. (See "Residuals" on p. 510.)

The General Loglinear Analysis procedure starts by counting the number of observed cases in each cell defined by the factor variables, regardless of whether the factors are used in the model specification. The goodness-of-fit statistics are calculated based on the number of cells defined by the factor list. If you include a variable on the factor list but don't use it in your model, you'll get different results than if you didn't include the variable in your factor list. This feature allows you to compare different models for the same table.

Overall Goodness-of-Fit Tests

Two frequently used statistics for measuring goodness of fit are

- The **Pearson chi-square**, defined as

$$\chi^2 = \sum \frac{(O - E)^2}{E}$$

where *O* is the observed count in a cell and *E* is the predicted count based on the model. The summation is taken over all cells in the table that do not have 0 for the expected count.

- The **likelihood ratio chi-square**, defined as

$$G^2 = 2\sum O \ln\left(\frac{O}{E}\right)$$

For large sample sizes, these statistics are equivalent. The advantage of the likelihood-ratio chi-square is that, like the total sum of squares in analysis of variance, it can be subdivided into interpretable parts that add up to the total. The likelihood ratio chi-square is also called the deviance chi-square.

The degrees of freedom for both of the statistics are the same. It depends on the number of cells in the table defined by the factor list, the number of parameters not set to 0, and the number of cells with expected counts of 0. If the model fits the data well, the observed significance levels for the goodness-of-fit statistics should be large. That is, you don't reject the null hypothesis that, in the population, the observed and expected counts are equal.

Warning: If you have small expected values in many cells of the table, both chi-square statistics may not have the correct observed significance level.

From Figure 21-6, you see that the observed significance level is quite a bit larger than 0.05—it is 0.54. This means that these indices do not detect a lack of fit of the model. A model with the main effects for gender, degree, and voting; an interaction term between degree and gender (more males have degrees in the sample than females); and a vote-by-degree interaction (people with degrees are more likely to vote than people without degrees) seems to adequately represent the data, although further diagnostics are needed.

Tip: You want the overall goodness-of-fit statistics to have a large observed significance level, indicating that the model fits. You want individual parameter estimates to have small observed significance levels so that you can reject the null hypothesis that their population values are 0.

Figure 21-6
Goodness-of-fit tests

Goodness-of-Fit Tests[1,2]

	Value	df	Sig.
Likelihood Ratio	1.221	2	.543
Pearson Chi-Square	1.216	2	.544

1. Model: Poisson
2. Design: Constant + college + male + vote + college * male + vote * college

Residuals

The overall goodness-of-fit test tells you if the model appears to fit the data. It doesn't tell you whether there are particular cells that the model fits poorly or whether there is a systematic lack of fit. To see how well the model fits the individual cells, you must examine the residuals for each cell.

Types of Residuals

As in regression analysis, there are different types of residuals that can be used to examine how well the model fits the data on a cell-by-cell basis.

- The **raw residual** is just the difference between the observed and expected cell frequencies. The size of the raw residual depends not only on how well the model fits but also on the number of cases in a particular cell. For example, a raw residual of 5 may indicate poor fit if the observed number of cases in a cell is 4 but excellent fit if the observed number of cases in the cell is 12,000.

Warning: Under the Poisson assumption, raw residuals have variances that are equal to their means, so even small raw residuals don't necessarily indicate a good fit.

- A **standardized residual** is computed by dividing the raw residual by an estimate of the standard deviation of the observed count. If the model is correct, for a Poisson model the standard deviation of the observed count is the square root of the predicted count. For a multinomial model, the standard deviation is $\sqrt{E(1-E/N)}$, where E is the predicted count and N is the total number of cases in the table. Standardized residuals are sometimes called **Pearson residuals** because for a Poisson model, if you square them and sum them over all cells, they equal the Pearson chi-square statistic. For large sample sizes, if the model fits, the distribution of standardized residuals is normal, with a mean of 0 and a standard deviation of less than 1.
- The **adjusted residual** is the standardized residual divided by an estimate of its standard error. Since for a complex model the standard deviation of the standardized residuals can be quite a bit less than 1, the adjusted residual is a better diagnostic aid. For large samples, its distribution is normal, with a mean of 0 and a standard deviation of 1.

- The **deviance residual** can also be used for examining departures from fit in a model. The deviance residual is the cell's contribution to the likelihood-ratio chi-square, maintaining the sign of the raw residual. The sum of the squared deviances for all cells is equal to the likelihood-ratio chi-square. For large sample sizes, the distribution of the deviances is normal, with a mean of 0 and a standard deviation of 1.

Tip: Plot standardized residuals against predicted value to look for poor-fitting cells, which might be the result of errors in data entry.

Examining Residuals

Figure 21-7 is a table that contains observed and expected counts for the voting example. For each cell, you see the three types of residuals. The largest difference between the observed and expected cell counts is about 3.5.

Tip: When you include a term in the loglinear model, the observed and expected marginal counts for the table defined by that term are equal. For example, if you include a gender-by-vote interaction term, the observed and predicted marginal counts for the gender-by-vote table are equal.

Figure 21-7
Observed, expected, and residual values

Cell Counts and Residuals[1,2]

						Observed		Expected					
						Count	%	Count	%	Residual	Standardized Residual	Adjusted Residual	Deviance
Vote	No	College Degree	No	Male	No	217	15.0%	220.036	15.2%	-3.036	-.205	-.402	-.205
					Yes	152	10.5%	148.964	10.3%	3.036	.249	.402	.248
			Yes	Male	No	29	2.0%	25.592	1.8%	3.408	.674	1.027	.659
					Yes	21	1.4%	24.408	1.7%	-3.408	-.690	-1.027	-.707
	Yes	College Degree	No	Male	No	396	27.3%	392.964	27.1%	3.036	.153	.402	.153
					Yes	263	18.1%	266.036	18.3%	-3.036	-.186	-.402	-.186
			Yes	Male	No	187	12.9%	190.408	13.1%	-3.408	-.247	-1.027	-.248
					Yes	185	12.8%	181.592	12.5%	3.408	.253	1.027	.252

1. Model: Poisson
2. Design: Constant + college + male + vote + college * male + vote * college

Plotting Residuals

Figure 21-8 is a scatterplot matrix of the adjusted residuals and observed and expected counts for the voting behavior loglinear model. From the plots, you don't see any obvious pattern between the adjusted residuals and the counts. The observed and expected values fall about a diagonal line.

Figure 21-8
Scatterplot matrix of adjusted residuals

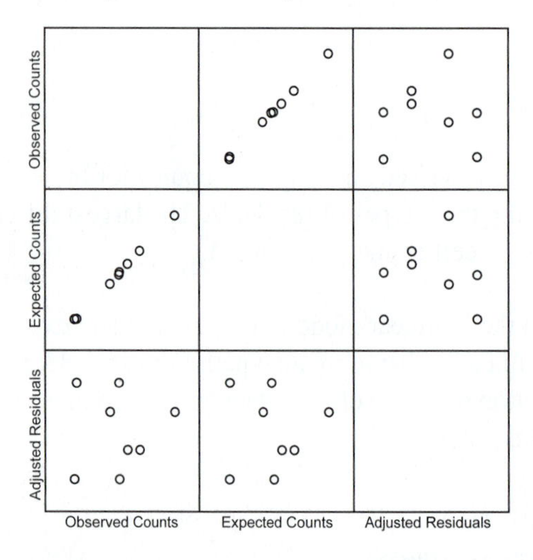

Warning: The saved values are for the aggregated data (cells in the contingency table), even if the data are recorded in individual observations in the Data Editor. If you save residuals or predicted values for unaggregated data, you should aggregate the data to obtain the appropriate cell statistics.

Models for Ordinal Data

In many situations, the categorical variables used in loglinear models are ordinal in nature. Ordinal variables may result from grouping values of interval variables, such as income or education, or they may arise when ordering (but not distance) between categories can be established. Happiness, interest, and opinions on various issues are measured on ordinal scales. Although *a lot* is more than *some*, the actual distance between the two response categories cannot be determined. The additional information contained in the ordering of the categories can be incorporated into loglinear models, which may result in a better representation of the data using fewer parameters.

Job Satisfaction and Income

As an illustration of some common models for ordinal data, you'll analyze data on job satisfaction and income from the General Social Survey. Income is grouped into four categories. Job satisfaction is measured on a four-point scale: very dissatisfied, a little dissatisfied, moderately satisfied, and very satisfied.

Figure 21-9 is a crosstabulation of job satisfaction and income categories. You can see that as salary increases, so does job satisfaction. A total of 20% of people earning less than $15,000 were dissatisfied with their jobs, while only 4.5% of those earning $50,000 or more were dissatisfied with theirs. Similarly, almost 59% of those earning over $50,000 were very satisfied with their jobs, while only 44% of those earning less than $15,000 were very satisfied.

Figure 21-9
Job satisfaction by earnings

			Job satisfaction				
			Very dissatisfied	A little dissatisfied	Moderately satisfied	Very satisfied	Total
Earning category	less than $15,000	Count	9	31	72	88	200
		Row %	4.5%	15.5%	36.0%	44.0%	100.0%
	$15,000 to $29,999	Count	11	26	75	105	217
		Row %	5.1%	12.0%	34.6%	48.4%	100.0%
	$30,000 to $49,999	Count	2	19	87	119	227
		Row %	.9%	8.4%	38.3%	52.4%	100.0%
	$50,000 +	Count	2	7	74	117	200
		Row %	1.0%	3.5%	37.0%	58.5%	100.0%
Total		Count	24	83	308	429	844
		Row %	2.8%	9.8%	36.5%	50.8%	100.0%

Results of the chi-square test of independence from the Crosstabs procedure are shown in Figure 21-10. Based on the very small observed significance level, you can reject the null hypothesis that the variables are independent. Another way of saying this is that the main effects of income and job satisfaction do not adequately represent the data.

Figure 21-10
Chi-square test of independence

	Value	df	Asymp. Sig. (2-sided)
Pearson Chi-Square	32.352[1]	9	.000
Likelihood Ratio	34.964	9	.000
Linear-by-Linear Association	23.185	1	.000
N of Valid Cases	844		

1. 0 cells (.0%) have expected count less than 5. The minimum expected count is 5.69.

Types of Models

You can build a variety of loglinear models that use the ordering of the job satisfaction and income variables. The models depend on the "scores" assigned to each category of response. Often these are arbitrary, since the actual distances between the categories are unknown. In this example, scores from 1 to 4 are assigned to both the income and job satisfaction categories. You can use other scores, such as the midpoints of the salary categories.

There are three common ordinal models that you can fit to the data. The **linear-by-linear association model** incorporates the ordinal nature of both variables. The **row-effects model** uses only the ordering of the column variable. The **column-effects model** uses only the ordering of the row variable.

Linear-by-Linear Association Model

The linear-by-linear association model for two variables is

$$\ln(m_{ij}) = \mu + \lambda_i^{income} + \lambda_j^{job} + BU_iV_j$$

where the scores U_i and V_j are assigned to row *i* and column *j*. In this model, μ and the two lambda parameters are the usual loglinear terms for the constant and the main effects of income and job satisfaction. What differs is the inclusion of the term involving *B*. The coefficient *B* is essentially a regression coefficient that, for a particular cell, is multiplied by the scores assigned to that cell for income and job satisfaction. If the two variables are independent, the coefficient should be close to 0.

(However, a coefficient of 0 does not necessarily imply independence, since the association between the two variables may be nonlinear.) If the coefficient is positive, more cases are expected to fall into cells with large scores or small scores for both variables than would be expected if the two variables were independent. If the coefficient is negative, an excess of cases is expected in cells that have small values for one variable and large values for the other.

Specifying the Model

To fit the linear association model, open the file *gss.sav* and from the menus choose:

Transform
  Compute...
    B=earncat*jobsatisfact

Analyze
  Loglinear
    General...

▶ Factor(s): earncat, jobsatisfact
▶ Cell covariate: B

Model...
  ⊙ Custom
    ▶ Terms in Model (Main effects): earncat, jobsatisfact, B

Options...
  Display
    ☑ Frequencies
    ☑ Estimates

Figure 21-11 contains parameter estimates for the linear-by-linear association model for job satisfaction and income. The parameter labeled *B* is the regression coefficient. The coefficient is positive and large when compared to its standard error, indicating that there is a positive association between income and job satisfaction. As income increases or decreases, so does job satisfaction. The goodness-of-fit statistics displayed in Figure 21-12, as well as the cell counts and residuals in Figure 21-13, indicate that the linear-by-linear interaction model fits the data reasonably well. Inclusion of one additional parameter in the model has changed the observed significance level of less than 0.00005 for the independence model (see Figure 21-10) to 0.18 for the linear-by-linear association model (see Figure 21-12).

Figure 21-11
Parameter estimates for the linear-by-linear association model

Parameter Estimates[2,3]

		Estimate	Std. Error	Z	Sig.
Parameter	Constant	1.555	.656	2.372	.018
	B	.202	.043	4.745	.000
	[earncat = 1]	2.038	.442	4.609	.000
	[earncat = 2]	1.465	.313	4.680	.000
	[earncat = 3]	.829	.181	4.580	.000
	[earncat = 4]	0[1]	.	.	.
	[jobsatisfact = 1]	-1.483	.343	-4.319	.000
	[jobsatisfact = 2]	-.663	.233	-2.841	.004
	[jobsatisfact = 3]	.183	.133	1.375	.169
	[jobsatisfact = 4]	0[1]	.	.	.

1. This parameter is set to zero because it is redundant.
2. Model: Poisson
3. Design: Constant + B + earncat + jobsatisfact

Warning: In the Model dialog box, you cannot create interaction terms between covariates. You must do this with Compute from the Transform menu.

Figure 21-12
Goodness of fit for linear association model

Goodness-of-Fit Tests[1,2]

	Value	df	Sig.
Likelihood Ratio	11.401	8	.180
Pearson Chi-Square	10.991	8	.202

1. Model: Poisson
2. Design: Constant + B + earncat + jobsatisfact

Interpreting the Regression Coefficient

The regression coefficient B in the linear-by-linear association model is a log odds ratio. If two adjacent rows have scores that differ by +1, and two adjacent columns have scores that differ by +1, then

$$e^B = \frac{m_{ij}/m_{i(j+1)}}{m_{(i+1)j}/m_{(i+1)(j+1)}}$$

The numerator of the above odds ratio is for row i, the odds of being in column j instead of column $j+1$. Similarly, the denominator is for row $i+1$ the odds of being in column j instead of column $j+1$.

Consider an example. Figure 21-13 contains observed and expected cell counts for the linear-by-linear association model.

Figure 21-13
Observed and expected counts

Cell Counts and Residuals[1,2]

				Observed		Expected	
				Count	%	Count	%
Earning category	less than $15,000	Job satisfaction	Very dissatisfied	9	1.1%	10.099	1.2%
			A little dissatisfied	31	3.7%	28.096	3.3%
			Moderately satisfied	72	8.5%	80.108	9.5%
			Very satisfied	88	10.4%	81.698	9.7%
	$15,000 to $29,999	Job satisfaction	Very dissatisfied	11	1.3%	6.968	.8%
			A little dissatisfied	26	3.1%	23.733	2.8%
			Moderately satisfied	75	8.9%	82.850	9.8%
			Very satisfied	105	12.4%	103.449	12.3%
	$30,000 to $49,999	Job satisfaction	Very dissatisfied	2	.2%	4.518	.5%
			A little dissatisfied	19	2.3%	18.842	2.2%
			Moderately satisfied	87	10.3%	80.530	9.5%
			Very satisfied	119	14.1%	123.109	14.6%
	$50,000 +	Job satisfaction	Very dissatisfied	2	.2%	2.415	.3%
			A little dissatisfied	7	.8%	12.329	1.5%
			Moderately satisfied	74	8.8%	64.512	7.6%
			Very satisfied	117	13.9%	120.745	14.3%

1. Model: Poisson
2. Design: Constant + B + earncat + jobsatisfact

Since you've assigned sequential scores from 1 to 4 to the rows and columns, adjacent rows and columns differ by a score of 1. Consider two adjacent rows: those for the very satisfied and moderately satisfied categories. Consider also two adjacent incomes: more than $50,000 and income between $30,000 and $50,000. For moderately satisfied people, the predicted odds of having an income between $30,000 and $50,000, compared to over $50,000 is, from the expected counts in Figure 21-13, $80.53/64.51 = 1.24$. Similarly, for people who are very satisfied, the same predicted odds are $123.11/120.74 = 1.02$. That means that as job satisfaction increases, the odds of being in the lower income category decrease. The ratio of the two odds is 1.22. That's the value of $e^{0.202}$, where 0.202 is the parameter estimate for *B* in Figure 21-11. See Agresti (2002) for further discussion of a similar example.

Warning: The General Loglinear Analysis procedure analyzes data on a cell-by-cell basis. If you have covariate values for individual cases, such as age in years or family income, and you specify the variable as a covariate, the average value for a cell will be used. That is, all people will be assigned the average covariate values for the cell. If you want to use the individual values, you must respecify the analysis so that each case is treated as a cell. You do this by creating a subject ID variable, specifying it in the factor list but not in the Model dialog box. All predictors must be treated in the same way. That means that you must code them as dummy or effect coded variables and then specify them as cell covariates in the General Loglinear Analysis procedure.

Fitting Row- and Column-Effects Models

In a row-effects model, you use only the ordinal nature of the column variable. For each row, a separate slope based on the values of the column variables is estimated. The magnitude and sign of the coefficient indicate whether cases are more or less likely to fall into a column with a high or low score, as compared to the independence model. For a column-effects model, only the ordinal nature of the row variable is used. The row- and column-effects models are particularly useful when only one classification variable is ordinal or when both variables are ordinal but a linear trend across categories exists for only one.

Consider a column-effects model when job satisfaction is the column variable and earnings are the row variable. That means you'll use only the ordering of the earnings variable and ignore the ordering of the job satisfaction variable. The model contains the factor variables *satisfact* and *earncat* and a term for the interaction between the covariate, *cov,* which is equal to *earncat* for a case, and *jobsatisfact*.

Transform
  Compute...
    cov=earncat

Analyze
  Loglinear
    General...
    ▶ Factor(s): earncat, jobsatisfact
    ▶ Cell covariate: cov

Model...
  ⊙ Custom
  ▶ Terms in Model: (Main effects) earncat, jobsatisfact
                    (Interactions) jobsatisfact*cov

Options...
  Display
    ☑ Frequencies
    ☑ Estimates
  Plots: adjusted residuals

Tip: You can't specify the same variable as both a factor and a covariate, so it was necessary to compute a new variable, *cov* (equal to the factor *earncat*), for use in this analysis.

The coefficients for each column are displayed in Figure 21-14. The first interaction coefficient is negative, indicating that very dissatisfied people are less likely to be in high-income categories than the independence model would predict. The next two coefficients are also negative but smaller in value. The coefficient for the third job satisfaction category is not statistically different from 0. The coefficient for the fourth job satisfaction category is 0 because it is the reference category. All of the other regression coefficients measure association compared to the last category.

Figure 21-14
Parameter estimates for the column-effects model

Parameter Estimates[2,3]

		Estimate	Std. Error	Z	Sig.
Parameter	Constant	4.743	.087	54.675	.000
	[earncat = 1]	-.285	.138	-2.066	.039
	[earncat = 2]	-.077	.114	-.678	.498
	[earncat = 3]	.059	.101	.588	.556
	[earncat = 4]	0[1]	.	.	.
	[jobsatisfact = 1]	-1.389	.464	-2.995	.003
	[jobsatisfact = 2]	-.441	.277	-1.594	.111
	[jobsatisfact = 3]	-.138	.192	-.720	.471
	[jobsatisfact = 4]	0[1]	.	.	.
	[jobsatisfact = 1] * cov	-.669	.216	-3.098	.002
	[jobsatisfact = 2] * cov	-.519	.117	-4.422	.000
	[jobsatisfact = 3] * cov	-.075	.069	-1.091	.275
	[jobsatisfact = 4] * cov	0[1]	.	.	.

1. This parameter is set to zero because it is redundant.
2. Model: Poisson
3. Design: Constant + earncat + jobsatisfact + jobsatisfact * cov

Overall, the column-effects model fits reasonably well, as shown in the goodness-of-fit statistics in Figure 21-15. The observed significance level, 0.48, is quite large. However, the number of parameters you estimated increased, as compared to the linear-by-linear association model, since you estimated a slope for each job satisfaction category.

Figure 21-15
Goodness of fit for the column effects model

Goodness-of-Fit Tests [1,2]

	Value	df	Sig.
Likelihood Ratio	5.492	6	.482
Pearson Chi-Square	5.543	6	.476

1. Model: Poisson
2. Design: Constant + earncat + jobsatisfact + jobsatisfact * cov

Tip: When a covariate is in the model, SPSS Statistics applies the mean covariate value for cases in a cell to that cell. To analyze an equiprobability model, one in which all cells are equally likely, set the covariate to a constant of 1, specify the factors of interest, and enter the covariate into the model alone.

Incomplete Tables

All models examined so far have been based on complete tables. That is, all cells of the crosstabulations can have nonzero observed frequencies. This is not always the case. For example, if you are studying the association between types of surgery and sex of the patient, the cell corresponding to caesarean sections for males must be 0. This is termed a **fixed-zero** or **structural-zero cell**, since no cases can ever fall into it. Also, certain types of models "ignore" cells by treating them as if they were fixed zeros. Structural-zero cells lead to incomplete tables, and special provisions must be made during analysis. Cells that are structural zeros will always have 0 for the observed and expected counts.

Cells in which the observed frequency is 0 but in which it is *possible* to have cases are sometimes called **random zeros** or **sampling zeros**. For example, in a cross-classification of occupation and ethnic origin, the cell corresponding to Lithuanian sword-swallowers would probably be a random zero. Random zeros are used in the computation of expected values and may have nonzero expected values.

There are many types of models for analyzing incomplete tables. In the next section, we will consider one of the simplest, the quasi-independence model, in which the diagonal entries of a square table are set to be fixed zeros.

Testing Quasi-Independence

As an illustration of testing whether the rows and columns of a table are independent, when the diagonal terms are excluded, consider Figure 21-16, the crosstabulation of the number of children a respondent has and the number of children he or she perceives as ideal.

Figure 21-16
Crosstabulation of the number of real and ideal children

			Real				Total
			0-1	2	3-4	5+	
Ideal	0-1	Count	33	3	4	2	42
		% of Total	4.0%	.4%	.5%	.2%	5.2%
	2	Count	222	135	72	18	447
		% of Total	27.2%	16.6%	8.8%	2.2%	54.8%
	3-4	Count	117	49	119	19	304
		% of Total	14.4%	6.0%	14.6%	2.3%	37.3%
	5+	Count	7	3	2	10	22
		% of Total	.9%	.4%	.2%	1.2%	2.7%
Total		Count	379	190	197	49	815
		% of Total	46.5%	23.3%	24.2%	6.0%	100.0%

You can test the hypothesis that the actual (real) number of children and the ideal number are independent using the chi-square test of independence. From Figure 21-17, the Pearson chi-square value is 143, with 9 degrees of freedom. The observed significance level is very small, leading you to reject the null hypothesis that the two variables are independent.

Figure 21-17
Chi-square test of independence

	Value	df	Asymp. Sig. (2-sided)
Pearson Chi-Square	142.926[1]	9	.000
Likelihood Ratio	109.140	9	.000
Linear-by-Linear Association	53.507	1	.000
N of Valid Cases	815		

1. 2 cells (12.5%) have expected count less than 5. The minimum expected count is 1.32.

Defining Structural Zeros

There are many reasons why the independence model may not fit the data. One possible explanation is that, in fact, the real and ideal sizes are fairly close to one another and most cases fall on the diagonal of the table. In this example, about 36% of the respondents have achieved their ideal family size. A question that remains to be answered is whether there is a relationship between the two variables if you ignore the cases on the diagonal. For example, is there a tendency for people with small families to want large families or people with large families to want small families?

To test this hypothesis, you ignore the diagonal entries of the table and test for independence of the remaining cells using the test of **quasi-independence.** That is, you set the diagonal cells to structural zero by creating a new variable that has a value of 0 for cells that are structural zeros, and 1 otherwise. This is specified as a cell structure variable in the General Loglinear dialog box. Open the file *gss.sav* and from the menus choose:

Transform
  Compute...
    cellweight=1

Transform
  Compute...

If...

  ⊙ Include if case satisfies condition:
  ▸ (real = ideal) cellweight=0

Analyze
  Loglinear
    General...

  ▸ Factor(s): real ideal
  ▸ Cell structure: cellweight

Model...
  ⊙ Custom
  ▸ Terms in Model: (Main effects) real ideal

Options...
  Display
    ☑ Frequencies
    ☑ Estimates

A description of the cells in the multiway crosstabulation is shown in Figure 21-18. There are 16 cells (four categories of *real* by four categories of *ideal*). Four of these have been set to structural zero. (The large number of missing values is a consequence of not everyone in the General Social Survey being asked all of the questions.)

Figure 21-18
Data information table

		N
Cases	Valid	815
	Missing	1950
	Weighted Valid	518
Cells	Defined Cells	16
	Structural Zeros	4
	Sampling Zeros	0
Categories	ideal	4
	real	4

Looking at the diagonal entries in Figure 21-19 (the cells for which *real* = *ideal*), you see that both the observed and expected cell frequencies are set to 0. That's the result of the cell structure variable.

Tip: Always examine the table of observed and expected frequencies to make sure that you have correctly specified the structural zeros.

Figure 21-19
Observed and expected counts for the quasi-independence model

Cell Counts and Residuals[1,2]

				Observed Count	Observed %	Expected Count	Expected %
real	0-1	ideal	0-1	0	.0%	.000	.0%
			2	222	42.9%	215.855	41.7%
			3-4	117	22.6%	123.186	23.8%
			5+	7	1.4%	6.960	1.3%
	2	ideal	0-1	3	.6%	4.203	.8%
			2	0	.0%	.000	.0%
			3-4	49	9.5%	48.080	9.3%
			5+	3	.6%	2.716	.5%
	3-4	ideal	0-1	4	.8%	3.596	.7%
			2	72	13.9%	72.080	13.9%
			3-4	0	.0%	.000	.0%
			5+	2	.4%	2.324	.4%
	5+	ideal	0-1	2	.4%	1.201	.2%
			2	18	3.5%	24.065	4.6%
			3-4	19	3.7%	13.734	2.7%
			5+	0	.0%	.000	.0%

1. Model: Poisson
2. Design: Constant + ideal + real

The goodness-of-fit statistics are shown in Figure 21-20. The model seems to fit reasonably well, since there aren't any cells with very large residuals, and the overall test of fit has a reasonably large observed significance level.

Figure 21-20
Goodness-of-fit statistics

Goodness-of-Fit Tests[1,2]

	Value	df	Sig.
Likelihood Ratio	4.930	5	.424
Pearson Chi-Square	5.048	5	.410

1. Model: Poisson
2. Design: Constant + ideal + real

Tests for Square Tables

Whenever you have a square table in which the same possible responses make up the rows and columns, you can test more complicated hypotheses than just independence or quasi-independence. You can ask questions about the relationships of corresponding cells above and below the diagonal. There are different hypotheses you can test depending on the constraints that you put on row and column margins. For example, you can test for symmetry of the cells above and below the diagonal. Some simple hypotheses can be tested by viewing the square table as being composed of two triangles. You'll first test for symmetry using this data structure since it's easier to understand the underlying concepts. See "Two-Dimensional Representations" on p. 532 for additional information on performing the tests using a two-dimensional representation of the data.

Tip: Make sure each case appears only once in the table.

Testing for Symmetry

Figure 21-21 contains data from a much analyzed study of distance vision in women (Bishop, Feinberg, and Holland, 1975). Distance vision in each eye was graded on a scale from 1 to 4, where 1 is the best vision and 4 is the worst. Using these data, you can answer a variety of questions about the relationship of vision in the right eye and the left eye.

From the large number of cases on the diagonal (both eyes having similar vision), you can tell even without a formal statistical test that the two eyes are not independent ($\chi^2 = 8{,}097$, $df = 9$). You can test the null hypothesis that the two eyes are independent if you ignore the perfect matches by following the previously described procedure of setting all of the diagonal cells to structural zeros. You can reject the null hypothesis of quasi-independence as well ($\chi^2 = 198$, $df = 5$). From the table, you can see that for any row or column, as you move further from the diagonal, the number of cases decreases, so it's not surprising that you reject the hypothesis of quasi-independence.

Figure 21-21
Crosstabulation of eye-testing data

Count

		lefteye				Total
		1	2	3	4	
righteye	1	1520	266	124	66	1976
	2	234	1512	432	78	2256
	3	117	362	1772	205	2456
	4	36	82	179	492	789
Total		1907	2222	2507	841	7477

An interesting question still remains: Is there reason to believe that when the eyes are not identical, the better eye is likely to be the right eye (or the left eye)?

A simple strategy is to count the number of cases above and below the diagonal, but that's a crude measure that doesn't take into account the placement of cases. You really need to look at the off-diagonal cells, pair by pair. For example, you have to compare the count of cases with the values (*right* = 1, *left* = 2) to the number of cases with (*right* = 2, *left* = 1). You have to test whether the table is symmetric. In a symmetric table, corresponding cells above and below the diagonal are equal. That is, the observed count for cell (*i,j*) is the same as the observed count for cell (*j,i*).

To do this in the General Loglinear Analysis procedure, you can rearrange the data into two triangles like those shown in Figure 21-22. Each cell in the table is identified now by three values: whether the right eye or left eye is the better eye (variable *Eye,* the score for the better eye, *Bettereye*, and the score for the worse eye, *Worseeye).* The table is now a three-dimensional representation of the data.

Figure 21-22
Three-dimensional representation

Count

				Bettereye			Total
				1	2	3	
Eye	Left eye better	Worseeye	2	234	0	0	234
			3	117	362	0	479
			4	36	82	179	297
		Total		387	444	179	1010
	Right eye better	Worseeye	2	266	0	0	266
			3	124	432	0	556
			4	66	78	205	349
		Total		456	510	205	1171

To test the symmetry hypothesis, the data can be entered (or transformed) to correspond to the three-dimensional representation, as shown in Figure 21-23.

Figure 21-23
Data Editor for symmetry model

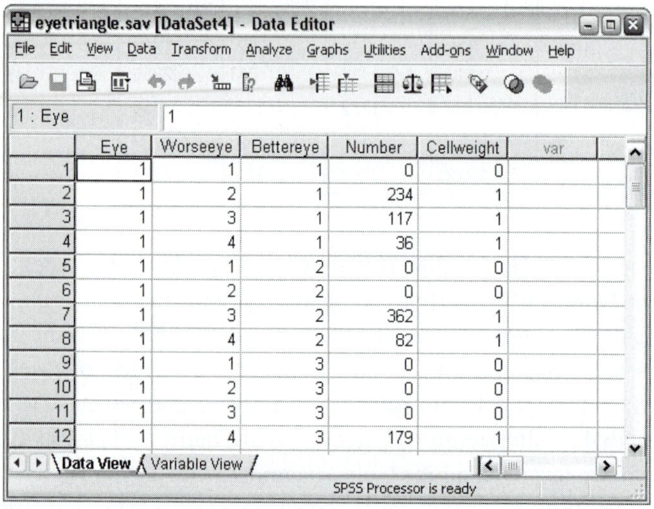

eyetriangle.sav [DataSet4] - Data Editor

File Edit View Data Transform Analyze Graphs Utilities Add-ons Window Help

1 : Eye 1

	Eye	Worseeye	Bettereye	Number	Cellweight	var
1	1	1	1	0	0	
2	1	2	1	234	1	
3	1	3	1	117	1	
4	1	4	1	36	1	
5	1	1	2	0	0	
6	1	2	2	0	0	
7	1	3	2	362	1	
8	1	4	2	82	1	
9	1	1	3	0	0	
10	1	2	3	0	0	
11	1	3	3	0	0	
12	1	4	3	179	1	

Data View / Variable View

SPSS Processor is ready

For each cell in the table, you have the values of the three variables, the number of cases in the cell (*Number*), and whether the cell is a structural 0 (*Cellweight*). All cells in which the score of the worse eye is better or equal to the score of the better eye are set to structural zeros, since that's an impossibility.

Loglinear Model

The loglinear representation of the symmetry model is

$$\ln(m_{ijk}) = \mu + \lambda^{\text{worseeye}} + \lambda^{\text{bettereye}} + \lambda^{\text{worsebybetter}}$$

Note that the variable *eye* does not appear in the model. This implies that the main effects, as well as their interaction, are the same for the two eyes. The probability that a case falls into cell (*i,j*) is the same as the probability that a case falls into cell (*j,i*), when there is symmetry.

Tip: If the assumption of symmetry is met, so is the assumption of marginal homogeneity. That is, the probability of a particular row is equal to the probability of that same column. Except for a two-by-two table, it's not the case that marginal homogeneity implies symmetry.

Obtaining the Analysis

To fit the symmetry model in the General Loglinear Analysis procedure, open the file *eyetriangle.sav* and from the menus choose:

Data
  Weight Cases...
  ⊙ Weight cases by
  ▶ Frequency Variable: Number

Analyze
 Loglinear
  General...

▶ Factor(s): bettereye, worseeye, eye
▶ Cell structure: cellweight

Model...
  ⊙ Custom
  ▶ Terms in Model: (Main effects) bettereye, worseeye
                    (Interaction effects) bettereye, worseeye

Options...
  Display
   ☑ Frequencies

Warning: Although the variable *eye* is not included in the model, it must be specified in the factor list, since it is used to define the cells. SPSS Statistics will warn you that the variable is not included in the model. Click OK and ignore the warning.

Figure 21-24 is an edited table of the observed and expected counts. Rows with zero observed and expected frequencies are deleted to make the table easier to read.

Figure 21-24
Observed and predicted counts for symmetry model

						Observed Count	Expected Count
Bettereye	1	Worseeye	2	Eye	Left eye better	234	250.000
					Right eye better	266	250.000
			3	Eye	Left eye better	117	120.500
					Right eye better	124	120.500
			4	Eye	Left eye better	36	51.000
					Right eye better	66	51.000
	2	Worseeye	3	Eye	Left eye better	362	397.000
					Right eye better	432	397.000
			4	Eye	Left eye better	82	80.000
					Right eye better	78	80.000
	3	Worseeye	4	Eye	Left eye better	179	192.000
					Right eye better	205	192.000

Notice how the expected value is computed for a cell. It is the average of the matching off-diagonal elements in Figure 21-21, the average of the observed counts for when the left eye is better and when the right eye is better. Notice that except for the cell corresponding to *bettereye* = 2 and *worseeye* = 4, the count for the right eye being better is larger than the count for the left eye being better. So it's not surprising, as you see from Figure 21-25, that the symmetry model does not fit well.

Figure 21-25
Goodness-of-fit statistics for symmetry model

Goodness-of-Fit Tests [1,2]

	Value	df	Sig.
Likelihood Ratio	19.249	6	.004
Pearson Chi-Square	19.107	6	.004

1. Model: Poisson
2. Design: Constant + Bettereye + Worseeye + Bettereye * Worseeye

Tip: The Pearson chi-square for testing the fit of the symmetry model is also known as Bowker's test. For two rows and columns, this is McNemar's test.

Adjusted Model of Quasi-Symmetry

In the symmetry model, you are testing whether the probability of falling into cell (*i,j*) is the same as falling into cell (*j,i*) in Figure 21-21. You ignore the fact that there are 1,171 cases above the diagonal and 1,010 cases below the diagonal, so the probabilities of falling into the upper or lower triangle are not equal.

The symmetry test can be modified to test whether the probability of falling into corresponding cells is equal, adjusting for the unequal numbers of cases in the triangles. You test whether the probability of falling into cell (*i,j*) is the same as the probability for falling into cell (*j,i*), if the probability of falling into the upper and lower triangles is equal. This is known as the **adjusted model of quasi-symmetry**. (Bishop, Feinberg, and Holland, 1975) or as the **conditional symmetry model** (McCullagh, 1978).

The model is

$$\ln(m_{ijk}) = \mu + \lambda^{eye} + \lambda^{worseye} + \lambda^{bettereye} + \lambda^{worsebybetter}$$

It's just like the previous model, except that the term λ^{eye} is added. The effect of this term is to preserve the totals in the upper and lower triangles. The expected counts in Figure 21-24 are multiplied by 0.537 (1,171 / 2,181) for the right eye, and by 0.463 for the left eye.

Tip: To specify this model in the General Loglinear Analysis procedure, add the variable *eye* as a main effect in the model.

The goodness-of-fit statistics for this model are shown in Figure 21-26. The model appears to fit better than the simple symmetry model. The Pearson chi-square value has decreased from 19.1 to 7.26.

Figure 21-26
Goodness of fit for adjusted quasi-symmetry

Goodness-of-Fit Tests [1,2]

	Value	df	Sig.
Likelihood Ratio	7.353	5	.196
Pearson Chi-Square	7.261	5	.202

1. Model: Poisson
2. Design: Constant + Bettereye + Eye + Worseeye + Bettereye * Worseeye

Warning: The test of adjusted quasi-symmetry is not the same as the test of quasi-symmetry, which preserves marginal totals in the original two-dimensional table. To obtain the test of quasi-symmetry, you must use the two-dimensional representation outlined below.

Two-Dimensional Representations

The previous tests of symmetry used a three-dimensional representation of a two-way table. It made it easier to see what was being tested and the meaning of the interactions. Now that you understand the basics, you can use General Loglinear Models to perform the same tests using a two-dimensional representation of the data. Here are the steps:

▶ If you enter the data as a weighted crosstabulation, weight the data file using the count variable.

▶ For each cell in the table, compute a series of indicator variables that define the position of the cell in the table. Look at Figure 21-23, which contains the eye-testing example. The variable *s1* has a value of 1 if either the row or the column number is 1. Variables *s2* to *s4* are defined similarly. The next set of indicator variables identifies the positions of the cells. Note that this is done only for the upper triangle (column number greater than row number.)

Figure 21-27
Data Editor for testing symmetry

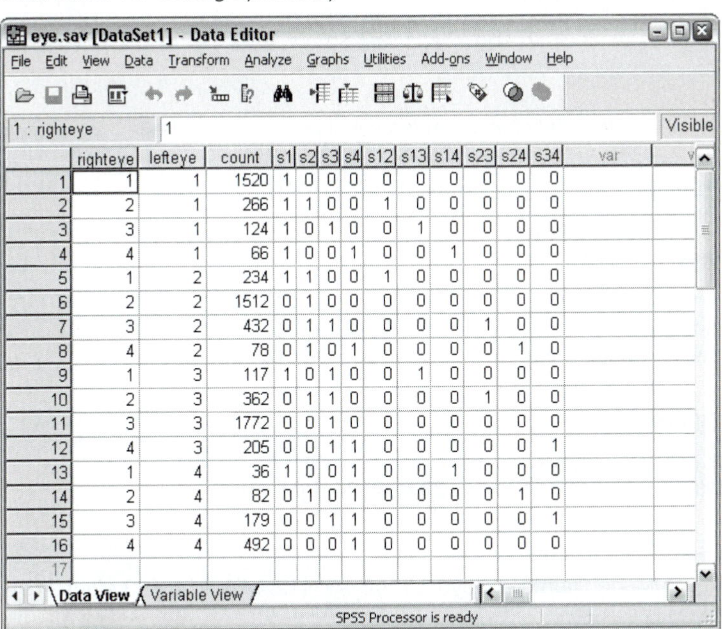

	righteye	lefteye	count	s1	s2	s3	s4	s12	s13	s14	s23	s24	s34
1	1	1	1520	1	0	0	0	0	0	0	0	0	0
2	2	1	266	1	1	0	0	1	0	0	0	0	0
3	3	1	124	1	0	1	0	0	1	0	0	0	0
4	4	1	66	1	0	0	1	0	0	1	0	0	0
5	1	2	234	1	1	0	0	1	0	0	0	0	0
6	2	2	1512	0	1	0	0	0	0	0	0	0	0
7	3	2	432	0	1	1	0	0	0	0	1	0	0
8	4	2	78	0	1	0	1	0	0	0	0	1	0
9	1	3	117	1	0	1	0	0	1	0	0	0	0
10	2	3	362	0	1	1	0	0	0	0	1	0	0
11	3	3	1772	0	0	1	0	0	0	0	0	0	0
12	4	3	205	0	0	1	1	0	0	0	0	0	1
13	1	4	36	1	0	0	1	0	0	1	0	0	0
14	2	4	82	0	1	0	1	0	0	0	0	1	0
15	3	4	179	0	0	1	1	0	0	0	0	0	1
16	4	4	492	0	0	0	1	0	0	0	0	0	0
17													

▶ Specify the model in the General Loglinear Analysis procedure. (The data are in the file *eye.sav*.)

Model Specification

Many common models can be easily estimated with this data representation. Here's how you define them for the eye example: (Variable *s4* is not included in the model because it is completely dependent on the values of *s1, s2,* and *s3*. That is, if you know the values of *s1*, *s2*, and *s3*, you know the value of *s4*.)

Tip: The chi-square value for the test of marginal homogeneity is the difference in chi-square values between the test of symmetry and the test of quasi-symmetry. The degrees of freedom are one less than the number of rows or columns in the square table.

Symmetry:

▶ Factor(s) righteye, lefteye
▶ Covariates: s1 s2 s3 s12 s13 s14 s23 s24 s34

Model...
⊙ Custom
▶ Terms in Model: (only the variables specified as covariates)

Quasi-symmetry:

▶ Factor(s): righteye, lefteye
▶ Covariates: s12 s13 s14 s23 s24 s34

Model...
⊙ Custom
▶ Terms in Model: righteye, lefteye, s12, s13, s14, s23, s24, s34

Adjusted quasi-symmetry:

Factor(s): righteye, lefteye
Covariates: s1 s2 s3 s12 s13 s14 s23 s24 s34 rightbettereye

Model...
⊙ Custom
▶ Terms in Model: rightbettereye, s1, s2, s3, s12, s13, s14, s23 s24 s34

Tip: The easiest way to create the interaction variables is to use the Compute facility with expressions like these:

```
compute s1= any(1,righteye,lefteye).
compute s2= any(2,righteye,lefteye).
compute s3= any(3,righteye,lefteye).
compute s4= any(4,righteye,lefteye).
compute s12= s1*s2.
compute s13= s1*s3.
compute s14= s1*s4.
compute s23= s2*s3.
compute s24= s2*s4.
compute s34= s3*s4.
```

See the Help system for further information.

Tip: The popular Bradley-Terry model for paired comparisons is a logit formulation of the quasi-symmetry model.

Poisson Regression: Modeling Rates

The loglinear models you've fit thus far predict the expected number of cases in a cell based on values of categorical independent variables and covariates. The fitted model is

$$\text{predicted count} = e^{\mu + \Sigma\lambda^i}$$

where the summation is over the lambda parameters for that cell.

You can also use loglinear models to model rates. For example, you can model the incidence of lung cancer based on duration and intensity of smoking. Or you can model the unemployment rate based on sex, education, and age. In both of these situations, the number of events for each combination of values of the independent variables is assumed to follow a Poisson distribution (see "Origins of Data" on p. 495).

In the rate model, you have two values for each cell of the table: the number of events and the number at risk. For example, if you follow 100,000 heavy smokers for one year and observe two cases of lung cancer, the cell "heavy smokers" has two values: 100,000 cases at risk and two lung cancer cases. The model for the rate in a particular cell is:

$$\frac{\text{predicted count}}{\text{number at risk}} = e^{\mu + \Sigma\lambda^i}$$

where the summation is over the lambda parameters for that cell. Taking the logarithm of both sides results in

$$\ln(\text{predicted count}) - \ln(\text{number at risk}) = \mu + \Sigma\lambda^i$$

The term *ln(number at risk)* is called an **offset term**. In the General Loglinear Analysis procedure, the number at risk is specified as a cell structure variable.

Smoking Duration, Intensity, and Death

As an example, consider data presented by Frome (1983) and discussed by Kleinbaum, et al. (1997). The number of lung cancer deaths was recorded in 63 subgroups of cases formed by nine duration-of-smoking groups and seven intensity-of-smoking groups. Figure 21-28 shows the data file.

Figure 21-28
Data file for lung cancer data

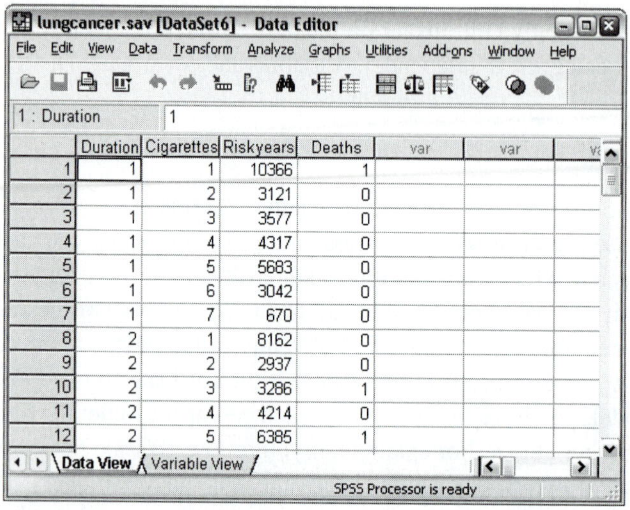

lungcancer.sav [DataSet6] - Data Editor

File Edit View Data Transform Analyze Graphs Utilities Add-ons Window Help

1 : Duration 1

	Duration	Cigarettes	Riskyears	Deaths	var	var
1	1	1	10366	1		
2	1	2	3121	0		
3	1	3	3577	0		
4	1	4	4317	0		
5	1	5	5683	0		
6	1	6	3042	0		
7	1	7	670	0		
8	2	1	8162	0		
9	2	2	2937	0		
10	2	3	3286	1		
11	2	4	4214	0		
12	2	5	6385	1		

Data View / Variable View

SPSS Processor is ready

For each of the subgroups defined by *duration* and *cigarettes*, you have two values: the number of lung cancer deaths and the person's years at risk.

Warning: For rate models, you must have your data in an aggregated form before you perform a general loglinear analysis. You can't have the data at the case level. If you do have data at a case level, use the aggregate facility to count the number of events and the number of cases for each combination of values of the independent variables.

Rate Model

The model you'll fit is

$$\frac{m_{ij}}{N_{ij}} = e^{\mu + \lambda_i^{duration} + \lambda_j^{smoke}}$$

where N_{ij} is the number at risk and m_{ij} is the number of predicted events. You are modeling the rate of the occurrence of an event for cases in cells of a multiway crosstabulation. For example, you can model the automobile accident rate for cases classified by gender and education. The lambda parameters are the usual loglinear model parameters.

You can write this as the loglinear model:

$$\ln(m_{ij}) - \ln(N_{ij}) = \mu + \lambda_i^{duration} + \lambda_j^{smoke}$$

This model assumes that the effects of smoking and duration categories are independent. That means that the duration effect is the same for all categories of smoking. Similarly, the effect of the intensity of the smoking is the same for all categories of duration.

To perform the analysis, open the file *lungcancer.sav* and from the menus choose:

Data
  Weight Cases...
    ⊙ Weight cases by
    ▶ deaths

Analyze
  Loglinear
    General...
    ▶ Factor(s): duration, cigarettes
    ▶ Cell Structure: riskyears

Model...
    ⊙ Custom
    ▶ Terms in Model (Main effects): duration, cigarettes

Options...
    Display
      ☑ Frequencies
      ☑ Estimates

Goodness of Fit

Before examining the individual parameter estimates, you want to see if the model fits. The goodness-of-fit statistics for this model are shown in Figure 21-29. Based on the likelihood-ratio chi-square, it appears that overall the model fits reasonably well, although you must look at the observed and expected values in each of the cells to determine if there are particular combinations of values for which the model fits poorly.

Figure 21-29
Goodness-of-fit statistics for lung cancer deaths

Goodness-of-Fit Tests[1,2]

	Value	df	Sig.
Likelihood Ratio	51.471	48	.340
Pearson Chi-Square	65.661	48	.046

1. Model: Poisson
2. Design: Constant + Age + Cigarettes

Interpreting Parameter Estimates

You see the parameter estimates for the model in Figure 21-30.

Figure 21-30
Parameter estimates for lung cancer deaths

Parameter Estimates[2,3]

		Estimate	Std. Error	Z	Sig.	95% Confidence Interval	
						Lower Bound	Upper Bound
Parameter	Constant	-3.559	.277	-12.840	.000	-4.102	-3.016
	[Cigarettes = 1]	-3.606	.605	-5.962	.000	-4.791	-2.420
	[Cigarettes = 2]	-2.386	.447	-5.336	.000	-3.262	-1.509
	[Cigarettes = 3]	-1.507	.322	-4.680	.000	-2.138	-.876
	[Cigarettes = 4]	-1.297	.313	-4.137	.000	-1.911	-.683
	[Cigarettes = 5]	-.705	.229	-3.085	.002	-1.153	-.257
	[Cigarettes = 6]	-.490	.225	-2.179	.029	-.930	-.049
	[Cigarettes = 7]	0[1]	.	.	.	.	.
	[Duration = 1]	-5.413	1.024	-5.287	.000	-7.420	-3.407
	[Duration = 2]	-4.466	.618	-7.229	.000	-5.677	-3.256
	[Duration = 3]	-3.712	.464	-8.000	.000	-4.621	-2.802
	[Duration = 4]	-2.210	.298	-7.416	.000	-2.795	-1.626
	[Duration = 5]	-2.171	.311	-6.988	.000	-2.780	-1.562
	[Duration = 6]	-1.205	.274	-4.395	.000	-1.742	-.667
	[Duration = 7]	-.966	.286	-3.375	.001	-1.527	-.405
	[Duration = 8]	-.509	.297	-1.715	.086	-1.090	.073
	[Duration = 9]	0[1]	.	.	.	.	.

1. This parameter is set to zero because it is redundant.
2. Model: Poisson
3. Design: Constant + Cigarettes + Duration

Notice that the parameters for the last categories of smoking and duration are set to 0. That means that they serve as the frame of reference for the other parameter estimates. The last categories correspond to people who smoke the most and for the longest period of time.

The parameter estimates for the intensity of smoking categories tells you how much more (or less) likely people of all smoking duration groups are to die from lung cancer for that particular smoking category as compared to heavy smokers (the last category). For example, the parameter estimate for the first smoking category compares non-smokers to heavy smokers. The parameter estimates are all negative, indicating that for all groups, the predicted death rates are smaller than for the heaviest smoking category. The parameter estimates also decrease in value as the number of cigarettes smoked increases, indicating that as the number of cigarettes smoked approaches the "heaviest" category, the predicted death rates from lung cancer, compared to the last category, become more similar.

If you calculate for nonsmokers $e^{\lambda} = e^{-3.61} = 0.027$, the interpretation is even simpler. Over all age groups, the ratio of the predicted death rate from lung cancer for the nonsmokers to the heaviest smokers is 0.027. This means that the predicted death rate from lung cancer in nonsmokers is 2.7% of the death rate in the heaviest smokers. Since duration and smoking intensity are independent in the model, the effect of smoking intensity is assumed to be the same for all duration groups. For the second to the last category (25–34 cigarettes per day), the parameter estimate is –0.49. Since $e^{-0.49}$ is 0.61, the predicted death rate from lung cancer for this group is 61% of the predicted death rate from lung cancer for the last group, smokers who smoke 35 or more cigarettes per day.

The parameter estimates for the duration categories are interpreted in the same way as for the smoking categories. For example, the predicted death rate from lung cancer in the first duration group (15–19 years) is only 0.4% of the predicted death rate in the last category (55–59 years), since $e^{-5.41} = 0.004$. Again, since the effects of duration and smoking intensity are independent in the model, the effect of duration is assumed to be the same for all categories of smoking. If you look at the parameter estimate for duration 8 (50–54 years), you will see that its 95% confidence interval includes 0. This means that the predicted death rate for this group is not significantly different from that of the last group (duration 55–59 years).

Tip: If you want to compare the parameters for any two categories, the simplest approach is to recode the data values so that one of the categories of interest is the reference category. The test that the parameter is 0 for the nonreference category is then equivalent to the test that the two parameters are equal. You can also create contrast variables.

Standardizing Tables

The margins (row and column totals) impact the ease of interpretation of a table. For example, you don't know what a large number of cases in a particular cell means unless you also know the distribution of cases in the margins of a table. Knowing that 1,250 people identify themselves as very happy with life and very happily married tells you little unless you also know how many people in your sample are very happy with life and how many are very happily married.

When you calculate percentages, you standardize the table to have the same number of cases in each row (or column) and can then compare them more sensibly. Calculating percentages is most useful when one of the variables can be identified as an independent variable and the other as a dependent variable. Then you can interpret the percentages as the results you would get when you have 100 cases with each value of the independent variable.

When you don't have a clearly identified dependent and independent variable, the problem becomes more complicated. For example, look at Figure 21-31, a crosstabulation of general happiness and happiness with one's marriage. (General Social Survey data, file *happymarriage.sav*.) The table is difficult to interpret, since the marginal distributions are not uniform. It would be easier to examine the relationship between the two variables if you had equal numbers of cases in each row and column while maintaining the observed relationship between the two variables.

Figure 21-31
Crosstabulation of general happiness and happiness with marriage

Count

		General Happiness			
		Very happy	Pretty happy	Not too happy	Total
Happiness of Marriage	Very happy	220	136	14	370
	Pretty happy	25	173	17	215
	Not too happy	2	9	7	18
Total		247	318	38	603

Adjusting the internal cells of a crosstabulation to fit specified marginals is known as **table standardization**. You may want to fit equal margins so that you can better see the relationships in the table or so that you can compare tables from different samples. Or you may want to fit margins obtained from a different sample or census. This lets you estimate the expected number of cases in each cell of the table if the table comes from the population with the known marginals. You can do both of these with the General Loglinear Analysis procedure.

Fitting Homogeneous Margins

To make Figure 21-31 easier to interpret, let's standardize the table so that it has 100 cases in each row and column.

Warning: You can have the same number of cases in all rows and columns only if the table is square (that is, if the number of rows is equal to the number of columns.)

Setting Up the Data File

The data file is shown in Figure 21-32. The first three variables define each observed cell in the table: the row number, the column number, and the number of cases (see "Megatip: Entering Tables Directly" on p. 193 in Chapter 10 for more information).

Figure 21-32
Data file for homogeneous margins

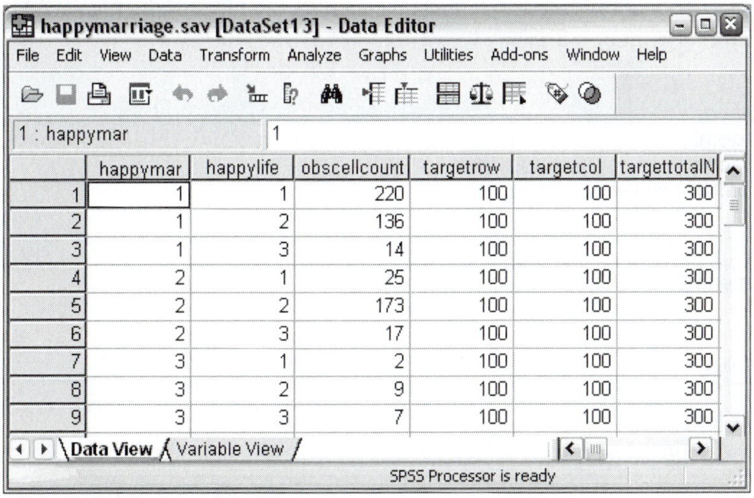

happymarriage.sav [DataSet13] - Data Editor

File Edit View Data Transform Analyze Graphs Utilities Add-ons Window Help

1 : happymar 1

	happymar	happylife	obscellcount	targetrow	targetcol	targettotalN
1	1	1	220	100	100	300
2	1	2	136	100	100	300
3	1	3	14	100	100	300
4	2	1	25	100	100	300
5	2	2	173	100	100	300
6	2	3	17	100	100	300
7	3	1	2	100	100	300
8	3	2	9	100	100	300
9	3	3	7	100	100	300

Data View / Variable View

SPSS Processor is ready

The next two variables are the row and column margins that you want for each cell in the standardized table. In this example, you want each cell to be in a row with 100 cases and in a column with 100 cases. The variable *targettotalN* is the total number of cases that you want in the standardized table. In this example, since you have 100 cases in each of three rows and three columns, the total target sample size is 300.

Tip: Since all cells have the same target values for the standardized table, you can use the Compute facility to easily compute the variables.

Specifying the Analysis

To standardize the table, you have to open the file *happymarriage.sav* and follow these steps:

▶ Calculate starting values for the number of cases in each cell of the target (standardized) table. The independence model is often used as the starting point.

Transform
 Compute...
  start=targetrow*targetcol/targettotalN

▶ Weight the data by the starting values.

Data
 Weight Cases...
 ⊙ Weight cases by
 ▶ start

▶ Run the General Loglinear Analysis procedure, specifying the observed cell count as the cell structure variable.

Analyze
 Loglinear
  General...
  ▶ Factor(s):happylife happymar
  ▶ Cell Structure: obscellcount

Model...
  ⊙ Custom
  ▶ Terms in Model (Main effects):happylife happymar

Options...
  Display
   ☑ Frequencies
   ☑ Estimates
  Plots: *deselect all plots*

Save...
  ☑ Predicted values

You see the standardized table in Figure 21-33. The observed counts are the starting values for each cell. The expected values are the standardized values.

Figure 21-33
Standardized table to equal marginals

				Observed		Expected	
				Count	%	Count	%
General Happiness	Very happy	Happiness of Marriage	Very happy	33.333	11.1%	71.269	23.8%
			Pretty happy	33.333	11.1%	18.341	6.1%
			Not too happy	33.333	11.1%	10.390	3.5%
	Pretty happy	Happiness of Marriage	Very happy	33.333	11.1%	20.234	6.7%
			Pretty happy	33.333	11.1%	58.292	19.4%
			Not too happy	33.333	11.1%	21.474	7.2%
	Not too happy	Happiness of Marriage	Very happy	33.333	11.1%	8.497	2.8%
			Pretty happy	33.333	11.1%	23.367	7.8%
			Not too happy	33.333	11.1%	68.135	22.7%

You can see the results of standardization better if you make a crosstabulation of the predicted values saved in the General Loglinear Analysis procedure. From the menus choose:

Data
  Weight Cases...
  ⊙ Weight cases by: pre_1.
  ▶ pre_1.

Analyze
  Descriptive Statistics
    Crosstabs...
    ▶ Row(s): happymar
    ▶ Columns(s): happylife

Cells...
  Noninteger weights
  ⊙ No adjustment

Figure 21-34
Standardized table from Crosstabs

Count

		General Happiness			Total
		Very happy	Pretty happy	Not too happy	
Happiness of Marriage	Very happy	71.269	20.234	8.497	100.000
	Pretty happy	18.341	58.292	23.367	100.000
	Not too happy	10.390	21.474	68.135	100.000
Total		100.000	100.000	100.000	300.000

You can more easily see that you have a table in which all rows and columns sum to 100. The relationships among the cell counts in the table have stayed the same in the standardized and unstandardized tables. In the standardized table, you can compare the diagonal entries and see that the strongest association between the happiness categories is for very happy people. You couldn't do that in the unstandardized table since the marginal frequencies are so different.

Tip: Standardization is very useful for comparing tables with different marginal distributions. For example, you can compare the relationship between general happiness and happiness of marriage for surveys with different sample sizes by standardizing all of the tables.

Fitting Arbitrary Marginals

You can standardize a table so that the marginals match those from another survey or from census data. For example, you can compare tables from several regions or countries by standardizing them to a common set of marginals. Instead of using the same target value of 100 for all rows and columns, for each cell enter the target numbers for its row and its column. Then follow exactly the same procedure as outlined before, obtaining starting values from the target row and target column values.

Chapter

22

GLM Univariate

Using the General Linear Model Univariate procedure, you can test various hypotheses about the means of a single dependent variable when cases are classified into groups based on one or more factor variables, some of which may be random. You can also include continuous covariates in the model as well as their interactions with the factors and each other. For regression analysis, the independent (predictor) variables are specified as covariates. Many standard analyses are available through the dialog boxes. Additional analyses can be specified using syntax commands.

Both balanced and unbalanced models can be tested. A design is balanced if each cell in the model contains the same number of cases. Four types of sums of squares are available. In addition to testing hypotheses, GLM Univariate produces estimates of parameters and has numerous procedures for testing differences among specific means. You can estimate the predicted mean values for the cells in the model and plot these means to easily visualize relationships.

Residuals, predicted values, Cook's distance, and leverage values can be saved as new variables in your data file for checking assumptions. WLS Weight allows you to specify a variable used to give observations different weights for a weighted least-squares (WLS) analysis.

This chapter includes the following examples:

Example 1: Regression with two independent variables. This study considers the effects of outdoor temperature and insulation thickness on the amount of heating oil consumed in individual homes. Because both predictors (temperature and insulation thickness) are measured on continuous scales, they are entered as covariates. A model with only covariates entered as predictors is a regression model. Residual plots are used to check assumptions about the data.

Example 2: Two-way analysis of variance (ANOVA) with equal sample sizes. In stores, often the shelf location of a product has an effect on sales, as does the size of the store. The data for this example include weekly sales totals classified by four types of shelf location and three store sizes. The results of the analysis indicate that the shelf location and the store size each affect the sales, but the interaction between location and store size is not a significant effect.

Example 3: Univariate ANOVA: A randomized complete block design with two treatments. The effects on the volume of baked bread of all possible combinations of three fats and three surfactants in the dough are investigated. Flour samples from four different sources are used as blocking factors. The *fat*surfactant* interaction effect is shown to be significant using the default method (Type III sums of squares). After the interaction is found to be significant, custom contrast coefficients are discussed to find out which combinations of levels of the factors are different.

Example 4: Univariate ANOVA: A randomized complete block design with empty cells. This study provides another look at bread baking, this time with some combinations of fats and surfactants not available. This creates empty cells in the design and illustrates the difference in data that calls for the Type IV sum-of-squares method.

Example 5: Analysis-of-covariance (ANCOVA) and nesting using the interaction operator. This study investigates the effects of three types of fertilizer on the final height of tomato plants. The initial height of each plant affects its final height, and the model uses this information by including the initial height as a covariate. A nested model is studied by using an interaction effect instead of syntax.

Example 6: A mixed-effects nested design model. A study involving machines and the strains on their glass cathode supports investigates the effects of the four randomly selected heads mounted on each machine. (Heads are nested within machines, since the same head is not used on different machines.) The *head* effect is treated as a random factor. The ANOVA statistics indicate that the machines are not significantly different, but there is an indication that the heads on the same machine may be different. The differences between machines are not significant, but for two of the machines, the differences between head types are significant.

Example 7: Univariate repeated measures analysis using a split-plot design approach. This experiment was intended to determine the effect of a person's anxiety rating on test performance in four successive trials. The data are set up to be studied as subsections of the whole dataset (the "whole plot"). The *anxiety* effect is determined to be not significant, but the *trial* effect is significant.

Example 1
Regression Model with Two Independent Variables

How does the consumption of home heating oil depend on temperature and the thickness of attic insulation? The data for this example are taken from Berenson and Levine (1992). Fifteen similar homes built by one developer in various locations around the United States were evaluated in the study. The builders recorded the amount of oil consumed in January, the average outside temperature (in degrees Fahrenheit), and the number of inches of attic insulation in each home. The data are shown in Figure 22-1.

Figure 22-1
Oil consumption data

Case	Avg. Temperature	Insulation (inches)	Oil Consumed in January (gallons)
1	40	3	275.3
2	27	3	363.8
3	40	10	164.3
4	73	6	40.8
5	64	6	94.3
6	34	6	230.9
7	9	6	366.7
8	8	10	300.6
9	23	10	237.8
10	63	3	121.4
11	65	10	31.4
12	41	6	203.5
13	21	3	441.1
14	38	3	323.0
15	58	10	52.5

First, consider the scatterplot matrix in Figure 22-2 to see how the data are distributed. The chart was created by using the Graphs menu, with regression lines added in the Chart Editor window.

Figure 22-2
Scatterplot matrix of the oil consumption data

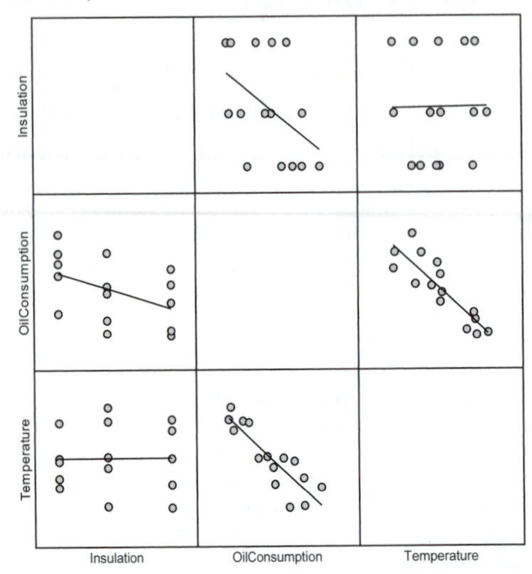

From the scatterplots, you can see a definite negative relationship between temperature and oil consumed and a less well-defined negative relationship between insulation and oil consumed. There is no apparent relation between insulation and temperature. We will use the GLM Univariate procedure to examine the following model:

$$y = \beta_0 + \beta_1 x_1 + \beta_2 x_2 + \varepsilon$$

or

$$\text{oil consumed} = \beta_0 + \beta_1 \text{temperature} + \beta_2 \text{insulation} + \varepsilon$$

This model uses temperature and insulation to predict the amount of heating oil consumed. The two predictor variables in this example are entered as covariates, creating a regression model.

To produce the output for this model, from the menus choose:

Analyze
  General Linear Model
    Univariate...

▶ Dependent Variable: oil
▶ Covariate(s): tempf insu

Model...
  ⊙ Custom
  ▶ Model (Main effects): insu tempf

Options...
  Display
    ☑ Parameter estimates
    ☑ Residual plot

The tests from the output are shown in Figure 22-3.

Figure 22-3
Tests of between-subjects effects

Dependent Variable: Oil consumed (gallons)

Source	Type III Sum of Squares	df	Mean Square	F	Sig.
Corrected Model	228014.626[1]	2	114007.313	168.471	.000
Intercept	480653.404	1	480653.404	710.272	.000
insu	49390.202	1	49390.202	72.985	.000
tempf	176938.161	1	176938.161	261.466	.000
Error	8120.603	12	676.717		
Total	939175.680	15			
Corrected Total	236135.229	14			

1. R Squared = .966 (Adjusted R Squared = .960)

The table contains rows for the components of the model that contribute to the variation in the dependent variable. The row labeled *Corrected Model* contains values that can be attributed to the regression model, aside from the intercept. The rows labeled *insu* and *tempf* are the effects in the model. *Error* displays the component attributable to the residuals, or the unexplained variation. *Total* shows the sums of squares of all of the dependent values. *Corrected Total* (sum of squared deviations from the mean) is the sum of the component due to the model and the component due to the error.

Most of the columns in this table are equivalent to those in the ANOVA table in Linear Regression (see Chapter 13). The sums of squares and degrees of freedom are given for each effect listed in the *Source* column. The mean square is the sums of squares divided by the degrees of freedom, and the F statistic is the mean square divided by the error mean square. Its significance (p value) is given in the *Sig.* column.

In Figure 22-3, the *F* statistic for the corrected model, 168.471, is highly significant ($p < 0.0005$), indicating rejection of the simultaneous test that each coefficient is 0. The footnote shows that R^2 is 0.966 and adjusted R^2 is 0.960. Thus, about 96% of the variation in oil consumption is explained by this model.

Figure 22-4
Parameter estimates

Dependent Variable: Oil consumed (gallons)

					95% Confidence Interval	
Parameter	B	Std. Error	t	Sig.	Lower Bound	Upper Bound
Intercept	562.151	21.093	26.651	.000	516.193	608.109
insu	-20.012	2.343	-8.543	.000	-25.116	-14.908
tempf	-5.437	.336	-16.170	.000	-6.169	-4.704

The parameter estimates are shown in Figure 22-4. From the estimates in the column labeled *B*, the model for estimating the value of oil consumption is

$$\text{oil consumption} = 562.151 - 20.012\text{insulation} - 5.437\text{temperature}$$

You could use this model to predict the January oil consumption of a house in a given temperature zone, for different values of insulation within the studied range of 3 to 10 inches.

Diagnostics

The matrix scatterplot of residuals is shown in Figure 22-5. The plots are useful for checking assumptions about the data.

Figure 22-5
Residuals plots

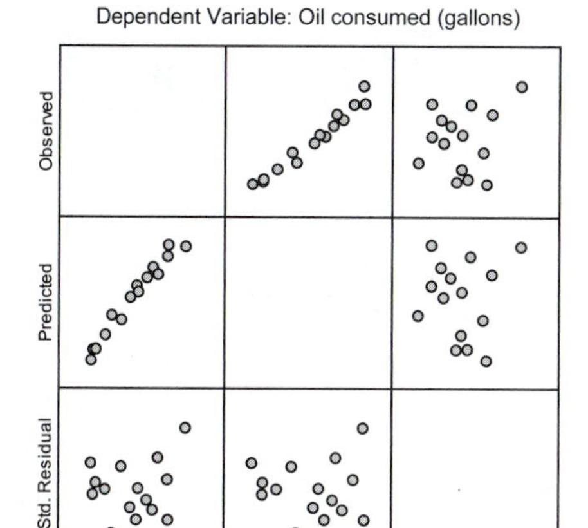

The plot of predicted versus observed values (left middle plot) is close to a straight line, which indicates a good fit. The standardized residuals versus the predicted values (middle lower plot) shows some tendency to increase variability as the dependent variable increases (heteroscedasticity). See Chapter 13 for a more detailed discussion of analyzing various plots in regression analysis.

Example 2
Two-Way Analysis of Variance (ANOVA) with Equal Sample Sizes

When the independent (predictor) variables are categorical, you can enter them in the GLM Univariate procedure as factors. Consider a supermarket chain that is interested in various effects on weekly sales. Does the sales volume of a particular product depend on its shelf location? If it does, is the size of the store a significant factor? And is there any interaction between the shelf location and the size of the store? Berenson and Levine (1992) consider this problem, with sample sizes equal for each combination of shelf location and size. The data are shown in Figure 22-6. You can see that there are two entries for weekly sales for small stores and shelf location A, and there are two entries for each of the other combinations.

Figure 22-6
Weekly sales data

	Shelf Location			
Store Size	**A**	**B**	**C**	**D**
Small	45	56	65	48
	50	63	71	53
Medium	57	69	73	60
	65	78	80	57
Large	70	75	82	71
	78	82	89	75

In the SPSS Statistics data file used for this study, variables include weekly sales (*sales*), shelf location (*location*), and store size (*size*). One case was entered for each value of weekly sales. The two predictor variables in this study, *location* and *size*, are categorical, which means that they should be entered as factors in the GLM Univariate procedure. For this example, assuming that these variables include all of the categories of interest to the supermarket chain makes them **fixed** factors.

To produce the output, from the menus choose:

Analyze
  General Linear Model
    Univariate...

▶ Dependent Variable: sales
▶ Fixed Factor(s): location size

Model...
  ⊙ Custom
  ▶ Model (Main effects): location size
  ▶ Model (Interaction): location*size

Plots...
  ▶ Horizontal Axis: size
  ▶ Separate Lines: location (Click Add)

Options...
  ▶ Display Means for: location*size

The tests displayed in the output are shown in Figure 22-7.

Figure 22-7
Tests of between-subjects effects

Dependent Variable: Weekly sales

Source	Type III Sum of Squares	df	Mean Square	F	Sig.
Corrected Model	3019.333[1]	11	274.485	12.767	.000
Intercept	108272.667	1	108272.667	5035.938	.000
location	1102.333	3	367.444	17.090	.000
size	1828.083	2	914.042	42.514	.000
location * size	88.917	6	14.819	.689	.663
Error	258.000	12	21.500		
Total	111550.000	24			
Corrected Total	3277.333	23			

1. R Squared = .921 (Adjusted R Squared = .849)

The *location*size* interaction term is not significant ($p = 0.663$), so the location and size effects will be assumed to be consistent across levels of the other factor. Since the design is balanced, main-effect estimates are the same here as they would be if the model was refitted without the interaction. Some statisticians would prefer to fit the main-effects-only model anyway, with the interaction sums of squares and degrees of freedom going into the error term. In this case, it would make no difference in the main-effects tests—we see that there is indeed a location effect as well as a size effect.

Estimated Marginal Means and Profile Plots

To further explore the interaction effects, look at the table of estimated marginal means in Figure 22-8 and the profile plot of the same values in Figure 22-9.

Figure 22-8
Estimated marginal means

Shelf location * Store size

Dependent Variable: Weekly sales

Store size	Shelf location	Mean	Std. Error	95% Confidence Interval	
				Lower Bound	Upper Bound
Small	A	47.500	3.279	40.356	54.644
	B	59.500	3.279	52.356	66.644
	C	68.000	3.279	60.856	75.144
	D	50.500	3.279	43.356	57.644
Medium	A	61.000	3.279	53.856	68.144
	B	73.500	3.279	66.356	80.644
	C	76.500	3.279	69.356	83.644
	D	58.500	3.279	51.356	65.644
Large	A	74.000	3.279	66.856	81.144
	B	78.500	3.279	71.356	85.644
	C	85.500	3.279	78.356	92.644
	D	73.000	3.279	65.856	80.144

Figure 22-9
Estimated marginal means of weekly sales

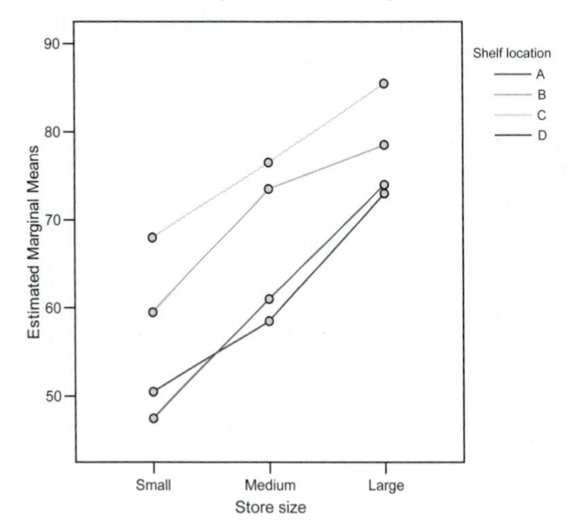

Each line in the chart represents one type of shelf location for the product. You can see that shelf location C produces the highest estimates of weekly sales and that shelf locations A and D produce the lowest. Because the lines are almost parallel, it is apparent that the differences between the shelf locations do not change much with the size of the store. The parallelism in this chart indicates that there is little or no interaction between the two factors. This conclusion reinforces the previous statements about the *F* test.

Example 3
Univariate ANOVA: A Randomized Complete Block Design with Two Treatments

How is the volume of baked bread affected by different combinations of fats and surfactants in the dough? This example uses data from Chapter 12 in Milliken and Johnson (1992). The data are in the SPSS Statistics file *Fat surfactant.sav.* The object of the experiment was to study differences in the specific volume of bread baked with different combinations of ingredients. The doughs were mixed using each of the nine possible combinations of three fats and three surfactants. (Surfactants are often used either to reduce surface tension of a liquid so it absorbs flour and other ingredients better or as containers of emulsifiers to promote a better mix of oil and water.) In statistical terminology, the fats and surfactants are called **treatments**.

Flour is a major ingredient of bread, but differences in flour were not the focus of this experiment. However, the bakers were concerned that differences between fats and surfactants in one batch of flour might not hold for flour from a different source. Therefore, four flours of the same type but from different sources were used in order to **randomize** unsuspected sources of variation in the effects of the two treatments among the flour sources. To create a **complete block design**, all combinations of the treatments were randomly assigned within each block (source) of flour. Because all possible levels of each factor are represented, this is a **fixed-effects model**. Random effects are illustrated in "Mixed-Effects Nested Design Model" on p. 575. Figure 22-10 shows the results of the current experiment.

Figure 22-10
Fats, surfactants, flour, and volume of bread

		Flour			
Fat	**Surfactant**	**1**	**2**	**3**	**4**
	1	6.7	4.3	5.7	-
1	2	7.1	-	5.9	5.6
	3	-	5.5	6.4	5.8
	1	-	5.9	7.4	7.1
2	2	-	5.6	-	6.8
	3	6.4	5.1	6.2	6.3
	1	7.1	5.9	-	-
3	2	7.3	6.6	8.1	6.8
	3	-	7.5	9.1	-

Notice that there are empty cells in the data (indicated by hyphens) due to an ineffective yeast container. Thus, the number of trials for each combination (subclass) across a row varies. However, all nine possible *fat*surfactant* treatment combinations were observed at least once. Therefore, estimation of the *fat*surf* interaction is not affected by the empty cells.

The SPSS Statistics data file *Fat surfactant.sav* uses four variables: *fat* (fat), *surf* (surfactant), *flour* (flour), and *spvol* (specific volume). The dependent variable is *spvol*, and the factors are *fat*, *surf*, and *flour*. Each specific volume is represented by one case in the data file. Empty data cells have no case. The *fat* and *surf* factors each have three levels and *flour* has four.

To produce the output in this example, from the menus choose:

Analyze
  General Linear Model
    Univariate...

▶ Dependent Variable: spvol
▶ Fixed Factor(s): fat surf flour

Model...
  ⊙ Custom
  ▶ Model: flour fat surf fat*surf

In this model, the *fat* factor and the *surf* factor are the main effects of interest, and the *flour* factor serves as a randomized blocking effect to reduce experimental error. Both main effects, *fat* and *surf*, and their interaction are in the model. The blocking effect, *flour*, is also entered. The intercept is included by default. The model does not include interactions between the blocking effect and the treatment effects.

Following Milliken and Johnson's approach, since all *fat*surfactant* treatment combinations are observed at least once, and since some of the same treatment combinations are observed in each block, the test hypotheses for Type III sums of squares are of interest for this design. Because the default method in SPSS Statistics is Type III sums of squares, the method was not changed in the Model dialog box.

Figure 22-11
Tests of between-subjects effects

Dependent Variable: spvol

Source	Type III Sum of Squares	df	Mean Square	F	Sig.
Corrected Model	22.520[1]	11	2.047	12.376	.000
Intercept	1016.981	1	1016.981	6147.938	.000
flour	8.691	3	2.897	17.513	.000
fat	10.118	2	5.059	30.583	.000
surf	.997	2	.499	3.014	.082
fat * surf	5.639	4	1.410	8.522	.001
Error	2.316	14	.165		
Total	1112.960	26			
Corrected Total	24.835	25			

1. R Squared = .907 (Adjusted R Squared = .833)

The analysis-of-variance table is shown in Figure 22-11. The first column of the table, labeled *Source,* lists the effects in the model, and the second column shows the *Type III Sum of Squares* for each effect. The degrees of freedom for each sum of squares is presented in the column labeled *df.* The *Mean Square* column shows the mean square of each effect, which is calculated by dividing the sum of squares by its degrees of freedom. The *F* column shows the *F* statistic for each effect, and its corresponding significance is shown in the next column. In a fixed-effects model, the *F* statistic is calculated by dividing the mean square by the mean square error term at the bottom of the *Mean Square* column.

In this example, the *F* value for the interaction effect *fat*surfactant* is 8.522. The *p* value 0.001 is significant at the conventional level, $\alpha = 0.05$. Thus, in further analysis, the surfactant effect should be compared at each level of fat, and the fat effect should be compared at each level of surfactant.

Estimated Marginal Means and Profile Plots

To study the surfactant effect at each level of fat, the estimated population marginal means of the dependent variable *spvol* at the cell combinations of surfactant and fat are useful, as are plots of the estimated marginal means.

To produce the output in this section, recall the GLM Univariate dialog box and make the following additional selections:

Options...
  Estimated Marginal Means
  ▶ Display Means for: fat*surf

Plots...
  ▶ Horizontal Axis: fat
  ▶ Separate Lines: surf (Click Add)

The estimated population marginal means are the predicted cell means from the model. The population marginal means estimates* for the surfactant and fat combination are the average of the estimated population cell means over the *flour* factor in the model. The cells that are involved in the calculation are those that have at least one observation.

Figure 22-12
*Estimated marginal means of spvol for fat*surfactant*

fat * surfactant

Dependent Variable: spvol

				95% Confidence Interval	
fat	surfactant	Mean	Std. Error	Lower Bound	Upper Bound
1	1	5.536	.240	5.021	6.052
	2	5.891	.239	5.378	6.404
	3	6.123	.241	5.605	6.641
2	1	7.023	.241	6.505	7.541
	2	6.708	.301	6.064	7.353
	3	6.000	.203	5.564	6.436
3	1	6.629	.301	5.984	7.274
	2	7.200	.203	6.764	7.636
	3	8.589	.300	7.945	9.233

*In calculating the estimated marginal means, SPSS Statistics uses the modified definition based on Searle, Speed, and Milliken (1980). Because the estimated marginal means in Milliken and Johnson (1992) use another definition of means, the values shown in SPSS Statistics output may be different from theirs.

The list of the estimated population marginal means and their standard errors is displayed in Figure 22-12. The table shows, for example, that with fat 2 and surfactant 1, the volume of bread predicted from this model is 7.023 units, and with the same fat (fat 2) and surfactant 3, the predicted volume is 6.0 units.

The same information can be shown graphically. Figure 22-13 shows a profile plot of the estimated population marginal means of the dependent variable *spvol* at each level of fat, with a separate line drawn for each type of surfactant.

Figure 22-13
Estimated population marginal means of spvol

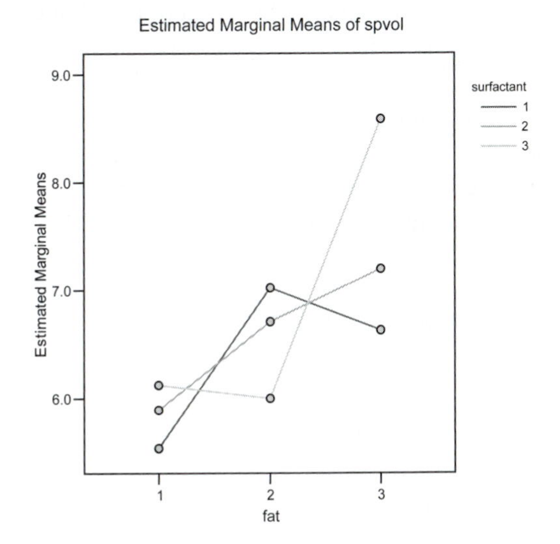

The plot indicates that when surfactant was fixed at level 2, the volume increased with the fat level. At level 1 of surfactant, the volume increased from fat level 1 to level 2 but decreased at fat level 3. On the other hand, the volume at level 3 of surfactant decreased slightly from fat level 1 to level 2 and increased sharply at fat level 3.

Custom Hypothesis Testing

The crossed lines in the profile plot indicate that the surfactant levels do not vary in the same way for all fats in the study. This means that custom hypotheses are needed. This section provides some theory and discusses how to customize your analysis by using syntax commands that are not available in the dialog boxes. You may want to skip this section.

To determine whether there is a significant difference among the levels of surfactant at each level of fat, you can specify three custom hypotheses. Each one tests the surfactant levels for one level of fat. To specify each of these three custom hypotheses, the LMATRIX subcommand is used. The LMATRIX subcommand can be used to customize the hypothesis by specifying the coefficients of the **L** matrix in the general linear hypothesis **LB** = **0**, where **B** is the vector of parameters. In this example, the mathematical model can be written as

$$\text{spvol}_{ijkl} = \mu + \text{flour}_i + \text{fat}_j + \text{surf}_k + \text{fat*surf}_{jk} + \text{error}_{ijkl}$$

where μ denotes the overall mean, flour_i denotes the effect of the *i*th level of flour ($i = 1, 2, 3, 4$), fat_j denotes the *j*th level of fat ($j = 1, 2, 3$), surf_k denotes the *k*th effect of surfactant ($k = 1, 2, 3$), fat*surf_{jk} denotes the *(j,k)*th interaction effect of fat and surfactant, and error_{ijkl} denotes the random effect of the *l*th bread in the *i*th level of flour at the *(j,k)*th fat and surfactant treatment combination. The values of μ, flour_i, fat_j, surf_k, and fat*surf_{jk} are free parameters that are to be estimated from the data. The vector of parameter **B** is the column vector that consists of all of the above parameters.

$$\begin{aligned} B' = (&\mu, \text{flour}_1, \text{flour}_2, \text{flour}_3, \text{flour}_4 \\ &\text{fat}_1, \text{fat}_2, \text{fat}_3, \text{surf}_1, \text{surf}_2, \text{surf}_3 \\ &\text{fat*surf}_{11}, \text{fat*surf}_{12}, \text{fat*surf}_{13} \\ &\text{fat*surf}_{21}, \text{fat*surf}_{22}, \text{fat*surf}_{23} \\ &\text{fat*surf}_{31}, \text{fat*surf}_{32}, \text{fat*surf}_{33}) \end{aligned}$$

To set up the custom hypothesis, the coefficients of the **L** matrix in the expression **LB** must be specified.

$$\mathbf{LB} = l_1\mu +$$
$$l_2\text{flour}_1 + l_3\text{flour}_2 + l_4\text{flour}_3 + l_5\text{flour}_4$$
$$l_6\text{fat}_1 + l_7\text{fat}_2 + l_8\text{fat}_3$$
$$l_9\text{surf}_1 + l_{10}\text{surf}_2 + l_{11}\text{surf}_3$$
$$l_{12}\text{fat*surf}_{11} + l_{13}\text{fat*surf}_{12} + l_{14}\text{fat*surf}_{13}$$
$$l_{15}\text{fat*surf}_{21} + l_{16}\text{fat*surf}_{22} + l_{17}\text{fat*surf}_{23}$$
$$l_{18}\text{fat*surf}_{31} + l_{19}\text{fat*surf}_{32} + l_{20}\text{fat*surf}_{33}$$

where

$$\mathbf{L} = (l_1, l_2, l_3, l_4, l_5, l_6, l_7, l_8, l_9, l_{10}, l_{11}, l_{12}, l_{13}, l_{14}, l_{15}, l_{16}, l_{17}, l_{18}, l_{19}, l_{20})$$

To compare level 1 of surfactant with its level 3 at fat 1, the **L** coefficients (l_9 and l_{12}) associated with the parameters that involve surfactant 1 (surf_1) and the combination of surfactant 1 at fat 1 (fat*surf_{11}) are set to 1, while the **L** coefficients (l_{11} and l_{14}) associated with the parameters that involve surfactant 3 (surf3) and the combination of surfactant 3 at fat 1 (fat*surf_{13}) are set to -1 . All other **L** coefficients that do not associate with the above parameters are irrelevant and are set to 0. The expression **LB** that corresponds to comparing surfactant 1 with 3 at fat 1 is

$$\mathbf{LB} = (\text{surf}_1 - \text{surf}_3) + (\text{fat*surf}_{11} - \text{fat*surf}_{13})$$

Notice that the first part of the above expression compares the main effect of surfactant 1 with surfactant 3 and that the second part specifies the surfactant comparison at fat 1. Similarly, the **LB** expression that corresponds to comparing surfactant 2 with 3 at fat 1 is

$$\mathbf{LB} = (\text{surf}_2 - \text{surf}_3) + (\text{fat*surf}_{12} - \text{fat*surf}_{13})$$

Testing the above two expressions together provides a custom hypothesis test to study the effects of surfactant at fat 1.

To produce the custom hypotheses in SPSS Statistics, from the menus choose:

Analyze
  General Linear Model
    Univariate...

▸ Dependent Variable: spvol
▸ Fixed Factor(s): fat surf flour

Model...
  ⊙ Custom
  ▸ Model: flour fat surf fat*surf

Paste

The pasted syntax is

```
UNIANOVA SPVOL BY FAT SURF FLOUR
  /METHOD = SSTYPE(3)
  /CRITERIA = ALPHA(.05)
  /DESIGN = FLOUR FAT SURF FAT*SURF.
```

The METHOD and CRITERIA subcommands give default specifications. They are not necessary and are not shown in the syntax below.

To specify the custom hypothesis, add an LMATRIX subcommand. The label *Surfactant Difference at Fat 1* is specified within the apostrophes. It is used to identify this custom hypothesis in the output. Next, the effects of SURF and FAT*SURF related to this custom hypothesis are specified, each of them followed by the values of the coefficients of **L** associated with it. For example, the values following the SURF effect are 1, 0, and –1. They are the **L** coefficients associated with the parameters $surf_1$, $surf_2$, and $surf_3$ in the **LB** expression, respectively. The values for all *fat*surf* coefficients that would come after these are set to 0.

```
UNIANOVA SPVOL BY FAT SURF FLOUR
 /LMATRIX = 'Surfactant Difference at Fat 1'
  SURF 1 0 -1 FAT*SURF 1 0 -1 0 0 0 0 0 0;
  SURF 0 1 -1 FAT*SURF 0 1 -1 0 0 0 0 0 0
/DESIGN = FLOUR FAT SURF FAT*SURF.
```

Similarly, the LMATRIX subcommands for studying the effects of surfactants for fat 2 and the effects of surfactants for fat 3 can be specified as follows:

```
UNIANOVA SPVOL BY FAT SURF FLOUR
/LMATRIX = 'Surfactant Difference at Fat 2'
  SURF 1 0 -1 FAT*SURF 0 0 0 1 0 -1 0 0 0;
  SURF 0 1 -1 FAT*SURF 0 0 0 0 1 -1 0 0 0
/LMATRIX = 'Surfactant Difference at Fat 3'
  SURF 1 0 -1 FAT*SURF 0 0 0 0 0 0 1 0 -1;
  SURF 0 1 -1 FAT*SURF 0 0 0 0 0 0 0 1 -1
/DESIGN = FLOUR FAT SURF FAT*SURF.
```

The custom hypothesis labeled *Surfactant Difference at Fat 2* is the custom hypothesis for studying the effects of surfactants at fat 2; the one labeled *Surfactant Difference at Fat 3* is the custom hypothesis for studying the effects of surfactants at fat 3. Notice that the values of the coefficients of **L** following the FAT*SURF effect are 0, 0, 0, 1, 0, and –1 (for the one with coefficients 1, 0, and –1 for SURF). The first three zeros represent the **L** coefficients associated with the parameters fat*surf_{11}, fat*surf_{12}, and fat*surf_{13}. Since the effects of surfactants at fat 2 are being studied, the **L** coefficients associated with parameters related to fat 1 are set to 0. Similarly, in the last LMATRIX subcommand, there are two sets of three zeros for fat 1 and fat 2. The results of the three custom hypotheses are displayed in Figure 22-14, with captions added to identify them.

Figure 22-14
Custom hypothesis test results

Dependent Variable: spvol

Source	Sum of Squares	df	Mean Square	F	Sig.
Contrast	.488	2	.244	1.475	.262
Error	2.316	14	.165		

Surfactant Difference at Fat 1

Dependent Variable: spvol

Source	Sum of Squares	df	Mean Square	F	Sig.
Contrast	1.831	2	.915	5.534	.017
Error	2.316	14	.165		

Surfactant Difference at Fat 2

Dependent Variable: spvol

Source	Sum of Squares	df	Mean Square	F	Sig.
Contrast	3.817	2	1.909	11.539	.001
Error	2.316	14	.165		

Surfactant Difference at Fat 3

The resulting *F* statistics show that there are statistically significant effects of surfactant on the dependent variable *spvol* at fat 2 and fat 3. The effect of surfactant on *spvol* is not significant at fat 1.

Example 4
Univariate ANOVA: A Randomized Complete Block Design with Empty Cells

How is the volume of baked bread affected by different combinations of fats and surfactants in the dough when data for some of the *fat*surfactant* combinations are not available? This example uses the data from Chapter 15 in Milliken and Johnson (1992). In "Univariate ANOVA: A Randomized Complete Block Design with Two Treatments" on p. 555, we studied the effect of combining three different fats with each of three surfactants on the specific volume of bread. All combinations of fats and surfactants were included at least once. Here, we have deleted some of the data. The Type III sum-of-squares method is no longer appropriate because there are empty cells in the design. The data, with some treatment combinations removed, are shown in Figure 22-15.

Figure 22-15
Fat and surfactant combinations missing

		Flour			
Fat	Surfactant	1	2	3	4
1	1	6.7	4.3	5.7	-
	2	7.1	-	5.9	5.6
	3	-	-	-	-
2	1	-	5.9	7.4	7.1
	2	-	-	-	-
	3	6.4	5.1	6.2	6.3
3	1	7.1	5.9	-	-
	2	7.3	6.6	8.1	6.8
	3	-	7.5	9.1	-

Data missing across two rows

Notice that all data in the treatment combinations (fat 1 and surfactant 3) and (fat 2 and surfactant 2) were removed. They are treated as missing data in this example. Figure 22-16 illustrates the treatment combinations available for hypothesis testing.

Figure 22-16
Treatment combinations available

	Surfactant		
Fat	**1**	**2**	**3**
1	Data available	Data available	
2	Data available		Data available
3	Data available	Data available	Data available

Since some of the treatment combinations are empty, the test hypotheses for Type III sums of squares are not suitable for this set of data; they may not have meaningful interpretations. Instead, the test hypotheses for Type IV sums of squares are more suitable here. The Type IV sum-of-squares method is designed to construct test hypotheses that are meaningful in the presence of empty cells in the model. Type IV hypotheses compare the levels of one treatment by averaging over one or more common levels of the other treatments. Such a hypothesis is constructed by using contrasts of available treatment combinations. The next section considers all of the Type IV hypotheses for fat.

Constructing Hypotheses

The mathematical model for the design considered in this example can be written as

$$\text{spvol}_{ijkl} = \mu + \text{flour}_i + \mu_{jk} + \text{error}_{ijkl}$$

where μ denotes the overall mean, flour_i denotes the effect of the ith level of flour ($i = 1, 2, 3, 4$), μ_{jk} denotes the effect of the treatment combination of the jth level of fat and kth level of surfactant, and error_{ijkl} denotes the random effect of the lth bread in the ith level of flour at the (j,k)th fat and surfactant treatment combination. The treatment combination effect can then be written as

$$\mu_{jk} = \text{fat}_j + \text{surf}_k + \text{fat*surf}_{jk}$$

where fat_j denotes the jth level of the effect of fat ($j = 1, 2, 3$), surf_k denotes the kth level of the effect of surfactant ($k = 1, 2, 3$), and fat*surf_{jk} denotes the (j,k)th interaction effect of fat and surfactant. The means μ_{13} and μ_{22} do not exist because there is no data available. Thus, any hypothesis that involves these two treatment combinations cannot be tested. Using Figure 22-16, all possible meaningful Type IV hypotheses are illustrated in Figure 22-17.

Figure 22-17
All possible Type IV hypotheses for fat

Hypothesis		
$H_{10}:$	$\mu_{11} = \mu_{21}$	$fat_1 - fat_2 + fat*surf_{11} - fat*surf_{21} = 0$
$H_{20}:$	$\mu_{21} = \mu_{31}$	$fat_2 - fat_3 + fat*surf_{21} - fat*surf_{31} = 0$
$H_{30}:$	$\mu_{12} = \mu_{32}$	$fat_1 - fat_3 + fat*surf_{12} - fat*surf_{32} = 0$
$H_{40}:$	$\mu_{23} = \mu_{33}$	$fat_2 - fat_3 + fat*surf_{23} - fat*surf_{33} = 0$
$H_{50}:$	$\mu_{11} + \mu_{12} = \mu_{31} + \mu_{32}$	$(fat_1 - fat_3) + 0.5(fat*surf_{11} + fat*surf_{12} - fat*surf_{31} - fat*surf_{32}) = 0$
$H_{60}:$	$\mu_{21} + \mu_{23} = \mu_{31} + \mu_{32}$	$(fat_2 - fat_3) + 0.5(fat*surf_{21} + fat*surf_{23} - fat*surf_{31} - fat*surf_{33}) = 0$

Analysis

To produce the output in this section, first modify the data in the *Fat surfactant.sav* file. You can save it under a new name. Then, from the menus choose:

Analyze
General Linear Model
Univariate...

▶ Dependent Variable: spvol
▶ Fixed Factor(s): flour fat surf

Model...

⊙ Custom
▶ Model: flour fat surf fat*surf
Sum of squares: Type IV

The results of the analysis of variance are shown in Figure 22-18.

Figure 22-18
Analysis-of-variance table using Type IV sums of squares

Dependent Variable: spvol

Source	Type IV Sum of Squares	df	Mean Square	F	Sig.
Corrected Model	20.244[1]	9	2.249	11.815	.000
Intercept	830.754	1	830.754	4363.845	.000
flour	7.773	3	2.591	13.609	.001
fat	5.046	2	2.523	13.252	.001
surf	2.500	2	1.250	6.566	.013
fat * surf	5.400	2	2.700	14.183	.001
Error	2.094	11	.190		
Total	930.510	21			
Corrected Total	22.338	20			

1. R Squared = .906 (Adjusted R Squared = .830)

The analysis-of-variance table shows that the effects are significant with $p = 0.001$. The effects *surf* and *fat*surf* are also significant with $p = 0.013$ and $p = 0.001$, respectively.

Contrast Coefficients

To understand the Type IV hypothesis testing for fat in this example, you can look at the contrast coefficients shown in the contrast coefficients table for fat. As in "Univariate ANOVA: A Randomized Complete Block Design with Two Treatments" on p. 555, customized contrast coefficients are available using syntax commands but not in the dialog boxes. To obtain a contrast coefficients table for each effect in the model, recall the GLM Univariate dialog box, paste the syntax, and add a PRINT=TEST(LMATRIX) subcommand.

```
unianova spvol by flour fat surf
 /method = sstype(4)
 /print = test(lmatrix)
 /design = flour fat surf fat*surf.
```

The output contains tables of contrast coefficients for each effect in the design. The table for fat is shown in Figure 22-19. The first column indicates the parameters in the model. For example, *[fat=1]* represents the parameter for fat_1, *[fat=1]*[surf=1]* represents the parameter for $fat*surf_{11}$, and so on. The next two columns, labeled *L6* and *L7*, indicate the two hypotheses on the fat effect that are being tested together in the analysis-of-variance table shown in Figure 22-18.

Figure 22-19
Contrast coefficients

fat[1]

Parameter	Contrast L6	Contrast L7
Intercept	.000	.000
[flour=1]	.000	.000
[flour=2]	.000	.000
[flour=3]	.000	.000
[flour=4]	.000	.000
[fat=1]	1.000	.000
[fat=2]	.000	1.000
[fat=3]	-1.000	-1.000
[surf=1]	.000	.000
[surf=2]	.000	.000
[surf=3]	.000	.000
[fat=1] * [surf=1]	.500	.000
[fat=1] * [surf=2]	.500	.000
[fat=2] * [surf=1]	.000	.500
[fat=2] * [surf=3]	.000	.500
[fat=3] * [surf=1]	-.500	-.500
[fat=3] * [surf=2]	-.500	.000
[fat=3] * [surf=3]	.000	-.500

The default display of this matrix is the transpose of the corresponding L matrix.
Based on Type IV Sums of Squares.

1. The estimable function is not unique

Hypotheses L6 and L7 correspond to hypotheses H_{50} and H_{60} in Figure 22-17. Hypotheses H_{10} to H_{40} can be tested using the LMATRIX subcommand to set up custom hypotheses. For example, to test hypotheses H_{10} and H_{20} together, the following syntax can be used:

```
unianova spvol by flour fat surf
/method = sstype(4)
/LMATRIX = 'H10 and H20'
     fat 1 -1  0  fat*surf  1  0 -1  0   0   0   0;
     fat 0  1 -1  fat*surf  0  0  1  0  -1   0   0
/design = flour fat surf fat*surf.
```

The univariate results of the above custom hypothesis for H_{10} and H_{20} are shown in Figure 22-20.

Figure 22-20
Univariate test results for custom hypotheses

Dependent Variable: spvol

Source	Sum of Squares	df	Mean Square	F	Sig.
Contrast	3.307	2	1.654	8.687	.005
Error	2.094	11	.190		

SPSS Statistics automatically chooses meaningful Type IV hypotheses for you. However, other sets of hypotheses, such as the ones for H_{10} and H_{20}, may be more appropriate for your study. When there are situations where other sets of Type IV hypotheses are possible, SPSS Statistics reminds you by displaying a message saying that the Type IV testable hypothesis is not unique in the tests of between-subjects effects table (see Figure 22-18). You can set up your own Type IV hypotheses by using the LMATRIX subcommand.

Example 5
Analysis of Covariance (ANCOVA) and Nesting Using the Interaction Operator

Do different types of fertilizers have different effects on the height of tomato plants? The data in this example are taken from Searle (1987). They are stored in SPSS Statistics file *Tomato.sav* and shown in Figure 22-21.

Figure 22-21
Tomato plant growth

Fertilizer	Final height	Initial height
1	74	3
1	68	4
1	77	5
2	76	2
2	80	4
3	87	3
3	91	7

The dependent variable is the height of tomato plants at 10 weeks (*height*). The factor of interest is the type of fertilizer used after transplanting (*fert*). Three varieties of fertilizer were applied. Since there was variation in the initial heights of the plants at the time of transplantation, the initial value (*initial*) is entered as a covariate in order to determine whether the type of fertilizer makes a difference in the final height for plants of equal initial height.

Testing Assumptions: Homogeneity of Regression Slopes

To be able to compare final height means for the three types of fertilizer without having to condition on a particular value of the covariate, you must be able to assume that the regression of final height on initial height is the same for all three types of fertilizer. This assumption of equality (homogeneity) of regression slopes can be tested by fitting a model containing main effects of *fert* and *initial*, as well as the *fert*initial* interaction. The interaction term provides the test of the null hypothesis of equal slopes.

To produce the output in this section, from the menus choose:

Analyze
  General Linear Model
    Univariate...

▶ Dependent Variable: height
▶ Fixed Factors(s): fert
▶ Covariate(s): initial

Model...
  ⊙ Custom
  ▶ Model: fert initial fert*initial

The results are shown in Figure 22-22. The interaction term, *fert*initial*, shows no evidence of violation of the equal slopes assumption: the *F* value is 0.023, with a significance level of 0.977. The homogeneity of regressions assumption is not rejected, and you can proceed to estimate the effects of fertilizer type on final height given initial height. (Of course, such a small sample provides little power to detect much; these data are being used simply to illustrate the steps involved in this type of analysis.)

Figure 22-22
Tests of between-subjects effects

Dependent Variable: Final Height

Source	Type III Sum of Squares	df	Mean Square	F	Sig.
Corrected Model	330.500[1]	5	66.100	1.763	.515
Intercept	2932.423	1	2932.423	78.198	.072
fert	30.396	2	15.198	.405	.743
initial	18.000	1	18.000	.480	.614
fert * initial	1.750	2	.875	.023	.977
Error	37.500	1	37.500		
Total	44055.000	7			
Corrected Total	368.000	6			

1. R Squared = .898 (Adjusted R Squared = .389)

Analysis-of-Covariance Model

To produce the analysis-of-covariance results, recall the GLM Univariate dialog box, or from the menus choose:

Analyze
  General Linear Model
    Univariate...

▶ Dependent Variable: height
▶ Fixed Factors(s): fert
▶ Covariate(s): initial

Model...
  ⊙ Full factorial

Options...
  Estimated Marginal Means
  ▶ Display Means for: fert

  Display
    ☑ Parameter estimates

The ANCOVA results are shown in Figure 22-23.

Figure 22-23
Tests of between-subjects effects

Dependent Variable: Final Height

Source	Type III Sum of Squares	df	Mean Square	F	Sig.
Corrected Model	328.750[1]	3	109.583	8.376	.057
Intercept	3796.875	1	3796.875	290.207	.000
initial	18.750	1	18.750	1.433	.317
fert	243.187	2	121.594	9.294	.052
Error	39.250	3	13.083		
Total	44055.000	7			
Corrected Total	368.000	6			

1. R Squared = .893 (Adjusted R Squared = .787)

In the *F* and *Sig.* columns, there is some evidence of a *fert* effect: the *F* value is 9.294, with a significance of 0.052. Using the default Type III sums of squares, the test for the covariate is a test of the common or pooled within-cells regression of final height on initial height. This regression coefficient estimate appears in the parameter estimates table as the *B* coefficient for *initial* (see Figure 22-24).

Figure 22-24
Parameter estimates

Dependent Variable: Final Height

Parameter	B	Std. Error	t	Sig.	95% Confidence Interval	
					Lower Bound	Upper Bound
Intercept	82.750	5.814	14.234	.001	64.248	101.252
initial	1.250	1.044	1.197	.317	-2.073	4.573
[fert=1]	-14.750	3.463	-4.259	.024	-25.771	-3.729
[fert=2]	-8.500	4.177	-2.035	.135	-21.792	4.792
[fert=3]	0[1]	.	.	.	.	.

1. This parameter is set to zero because it is redundant.

In the GLM parameterization, the intercept parameter estimate gives the estimated value of the last category of *fert* (*fert=3*) when the covariate is equal to 0. The *fert=1* and *fert=2* coefficients subtract the level 3 predicted value from the levels 1 and 2 predicted values, respectively. Adding one of these coefficients to the intercept estimate gives the estimated value for that level of *fert*, again when the covariate is equal to 0.

Estimated Marginal Means

The estimated marginal means table (see Figure 22-25) displays the estimated means and standard errors for each level of *fert* when the covariate is at its mean value (which, for this dataset, is 4).

Figure 22-25
Estimated marginal means

Dependent Variable: Final Height

Fertilizer	Mean	Std. Error	95% Confidence Interval	
			Lower Bound	Upper Bound
1	73.000	2.088	66.354	79.646
2	79.250	2.763	70.458	88.042
3	87.750	2.763	78.958	96.542

These means can be calculated by using values from the *B* column in Figure 22-24. The level 1 estimated marginal mean is

$$82.75 - 14.75 + 4 \times 1.25 = 73$$

For level 2,

$$82.75 - 8.5 + 4 \times 1.25 = 79.25$$

and for level 3,

$$82.75 + 0 + 4 \times 1.25 = 87.75$$

The 0 in the level 3 equation is from the aliased coefficient for *fert=3*. This version of adjusted means is based on the definitions given in Searle, Speed, and Milliken (1980). In unbalanced designs, these estimated means usually differ from the adjusted means given by the PMEANS subcommand in the MANOVA procedure by a constant amount. Although both give the same results for all contrasts among covariate adjusted means, the GLM estimated marginal means (*EMMEANS*) definition is more consistent with the statistical literature.

A Nested Model Using the Interaction Operator

Finally, had there been evidence of **heterogeneity** or inequality of regressions, you could estimate a model incorporating separate slopes. The separate slopes estimates could be reconstructed from parameter estimates from the interaction model originally fitted, since this is the same overall model as the separate slopes model, but there are easier ways to obtain the individual slope estimates. If you were to specify the nesting of *initial* within *fert* explicitly, you would have to use syntax. However, due to the parameterization used in GLM, many nested models can be specified using the crossing or interaction operator in the Model dialog box. Specifying the main effect of *fert* and the interaction of *fert*initial* without the main effect *initial* fits the same nested model as one with *fert* and *initial-within-fert* effects, and the *fert*initial* parameter estimates will give the simple slope estimates within each level of *fert*.

To obtain these separate slope estimates, recall the GLM Univariate dialog box, or from the menus choose:

Analyze
  General Linear Model
    Univariate...

▶ Dependent Variable: height
▶ Fixed Factors(s): fert
▶ Covariate(s): initial

Model...
  ⊙ Custom
  ▶ Model: fert fert*initial (initial is removed)
  ☐ Include intercept in model (deselect)

Options...
  ☑ Parameter estimates

The tests of between-subjects effects (ANOVA table) are not of much interest at this point (see Figure 22-26).

Figure 22-26
Tests of between-subjects effects

Dependent Variable: Final Height

Source	Type III Sum of Squares	df	Mean Square	F	Sig.
Model	44017.500[1]	6	7336.250	195.633	.055
fert	3521.963	3	1173.988	31.306	.130
fert * initial	20.500	3	6.833	.182	.899
Error	37.500	1	37.500		
Total	44055.000	7			

1. R Squared = .999 (Adjusted R Squared = .994)

The *fert*initial* effect tests the null hypothesis that all slopes are equal to 0 (a hypothesis already rejected if an interaction between the factor and the covariate is assumed). The *fert* effect tests the null hypothesis that all three estimated means in the separate slopes model are equal to 0 when the covariate is set to 0.

Figure 22-27
Parameter estimates

Dependent Variable: Final Height

Parameter	B	Std. Error	t	Sig.	95% Confidence Interval	
					Lower Bound	Upper Bound
[fert=1]	67.000	17.678	3.790	.164	-157.616	291.616
[fert=2]	72.000	13.693	5.258	.120	-101.987	245.987
[fert=3]	84.000	11.659	7.205	.088	-64.144	232.144
[fert=1] * initial	1.500	4.330	.346	.788	-53.519	56.519
[fert=2] * initial	2.000	4.330	.462	.725	-53.019	57.019
[fert=3] * initial	1.000	2.165	.462	.725	-26.510	28.510

The parameter estimates table (see Figure 22-27) displays the estimated simple slopes (the three *fert*initial* coefficients in the *B* column). Suppressing the intercept term produces *fert* coefficients that give estimated values for the *fert* means when *initial* is 0. Therefore, each *fert* coefficient and the corresponding *fert*initial* coefficient can be combined into a separate prediction equation for that level of the fertilizer factor. For example, for *fert=1*, using values in the *B* column, the equation would be

$$\hat{Y}_1 = 67 + 1.5(\text{initial})$$

Example 6
Mixed-Effects Nested Design Model

Do strain readings of glass cathode supports from five different machines depend on the various heads installed in the machine? An engineer studied the strain readings of glass cathode supports from five different machines, as described by Hicks (1982). Each machine has its own four heads mounted. Since four heads are mounted on each machine, and the types of heads are not interacting with different machines, heads are said to be **nested within** machines. The four types of heads that are mounted on a machine can be chosen from many possible heads. Therefore, the four heads are considered as a random sample that might be used on a given machine. Four strain readings were taken from each head in each machine. The results of this experiment are shown in Figure 22-28.

Figure 22-28
Strain readings of glass cathode supports

	Machine				
Head	**A**	**B**	**C**	**D**	**E**
1	6, 2, 0, 8	10, 9, 7, 12	0, 0, 5, 5	11, 0, 6, 4	1, 4, 7, 9
2	13, 3, 9, 8	2, 1, 1, 10	10, 11, 6, 7	5, 10, 8, 3	6, 7, 0, 3
3	1, 10, 0, 6	4, 1, 7, 9	8, 5, 0, 7	1, 8, 9, 4	3, 0, 2, 2
4	7, 4, 7, 9	0, 3, 4, 1	7, 2, 5, 4	0, 8, 6, 5	3, 7, 4, 0

The SPSS Statistics data file *Glass strain.sav* contains three variables: *strain*, *machine*, and *head*. One case is entered for each strain reading. The variable *strain* is the dependent variable, *machine* is a factor with five levels, and *head* is a random factor with four levels within each level of *machine*.

To indicate a factor that is random, move it to the Random Factor(s) list. To specify a nested design, you must use syntax.

To produce the output in this example, from the menus choose:

Analyze
  General Linear Model
    Univariate...

▶ Dependent Variable: strain
▶ Fixed Factor(s): machine
▶ Random Factor(s): head

Model...
  ⊙ Custom
  ▶ Model: machine

Paste

In the syntax window, at the end of the DESIGN subcommand, type in the nested factor for *head-within-machine*, HEAD(MACHINE). The modified syntax is

```
UNIANOVA STRAIN BY MACHINE HEAD
  /RANDOM = HEAD
  /METHOD = SSTYPE(3)
  /INTERCEPT = INCLUDE
  /CRITERIA = ALPHA(.05)
  /DESIGN MACHINE HEAD(MACHINE).
```

Run the command by clicking the Run button on the Syntax Editor toolbar.

Note that this nested model has no interaction effect because the head factor is not crossed with the five machine levels. It is a **mixed-effects model** in which the machine factor is a fixed effect and the head factor is a random effect. The *head-within-machine* effect is also treated as random because it contains a random factor. If you specify a random factor, all effects that contain the random factor are automatically random.

Figure 22-29
Tests of between-subjects effects

Dependent Variable: Strain Reading

Source		Type III Sum of Squares	df	Mean Square	F	Sig.
Intercept	Hypothesis	2020.050	1	2020.050	107.117	.000
	Error	282.875	15	18.858[1]		
machine	Hypothesis	45.075	4	11.269	.598	.670
	Error	282.875	15	18.858[1]		
head(machine)	Hypothesis	282.875	15	18.858	1.762	.063
	Error	642.000	60	10.700[2]		

1. MS(head(machine))
2. MS(Error)

Univariate tests are shown in Figure 22-29. The table displays the appropriate error terms for a mixed-effects model. These error terms are used in the *F* tests to determine whether the effects are significant. In fact, the appropriate error term for the *F* test is usually a linear combination of the mean squares in the model. When a random factor is specified, each of these combinations of mean squares is calculated and used automatically.

The linear combination of the mean squares that is used as the error term in testing each effect in the model is displayed in footnotes at the bottom of the analysis-of-variance table. For example, to test the *machine* effect, the mean square *MS(head(machine))* is used as the error term, and the *F* statistic for the *machine* effect is

$$\frac{11.269}{18.858} = 0.598$$

You can see that the error term for *machine* is the same as the mean square for *head(machine)*.

Footnote *2* indicates that to test the *head-within-machine* effect, the error mean square *MS(Error)* is used as the error term; the *F* statistic for the *head-within-machine* effect is

$$\frac{18.858}{10.7} = 1.762$$

From the table in Figure 22-29, the *machine* effect is not significant, but the *head-within-machine* effect is significant at the 0.10 alpha level ($p = 0.063$). This result suggests that there may be differences in glass strain between heads on the same machine.

Composition of Error Terms in a Mixed-Effects Model

To see how SPSS Statistics determines the appropriate error term in a mixed-effects model, look at the expected mean square components of each effect in the model. These components are shown in Figure 22-30.

Figure 22-30
Expected mean square components

Expected Mean Squares [1,2]

Source	Variance Component		
	Var(head(machine))	Var(Error)	Quadratic Term
Intercept	4.000	1.000	Intercept, machine
machine	4.000	1.000	machine
head(machine)	4.000	1.000	
Error	.000	1.000	

1. For each source, the expected mean square equals the sum of the coefficients in the cells times the variance components, plus a quadratic term involving effects in the Quadratic Term cell.
2. Expected Mean Squares are based on the Type III Sums of Squares.

The **expected mean square** of an effect in the model is the expectation of the mean squares of that effect in the model. It is usually a linear combination of variance components of the random effects and quadratic terms of the fixed effects in the model. In Figure 22-30, each row represents the linear combination of variance components involved in the expected mean square of the effect in the model.

The expected mean square for the *head-within-machine* effect involves only two variance terms: the variance of the *head-within-machine* effect itself and the variance of error. The coefficient for the variance of *head(machine)* is 4, and the coefficient for the variance of error is 1. This means that the expected mean square for *head(machine)* is

$$4\,\mathrm{Var}(\mathrm{head}(\mathrm{machine})) + 1\,\mathrm{Var}(\mathrm{Error})$$

For the *machine* effect, in addition to the linear combination of the variance components—$4\,\mathrm{Var}(\mathrm{head}(\mathrm{machine})) + 1\,\mathrm{Var}(\mathrm{Error})$—a quadratic term that involves the fixed effect *machine* is also included. Therefore, the expected mean square of the *machine* effect is

$$(4\,\mathrm{Var}(\mathrm{head}(\mathrm{machine})) + 1\,\mathrm{Var}(\mathrm{Error}) + Q(\mathrm{machine}))$$

where *Q(machine)* represents the quadratic term involving the *machine* effect.

Testing the *machine* effect is equivalent to testing whether the quadratic term of *machine* is 0. If Q(machine) = 0, the expected mean square of the *machine* effect is the same as the expected mean square of *head(machine)*. Therefore, the mean square for *head(machine)* is an appropriate error term for testing the *machine* effect. In the univariate tests table (see Figure 22-29), the error term for *machine* has a footnote to indicate the relationship. You can see that the values across the row for the *machine* error term are the same as the values in the *head(machine)* hypothesis row.

Similarly, testing the *head-within-machine* effect is equivalent to testing whether the variance of *head(machine)* is 0. Since the expected mean square of *head(machine)* when Var(head(machine) = 0 is the same as the variance of error, the mean square of error is an appropriate error term for testing the *head-within-machine* effect.

Contrast Coefficients

To detect which machine contains heads that are significantly different, five sets of custom hypotheses (one for each machine) are needed. In each of the five sets of custom hypotheses, the heads within the same machine will be compared. Comparing four heads implies three contrasts. Notice that the degrees of freedom for *head(machine)* is 15, which means that 15 contrasts are involved in this hypothesis. You can partition the hypothesis for *head(machine)* into five parts; each part represents one comparison of heads within a machine with three contrasts. To display all 15 contrasts for testing *head(machine)*, you must use syntax. Click the Dialog Recall toolbar button and select the GLM Univariate dialog box. Click the Paste button and add the subcommand PRINT=TEST(LMATRIX) as well as the HEAD(MACHINE) effect.

```
UNIANOVA strain BY machine head
 /PRINT = TEST(LMATRIX)
 /RANDOM = head
 /DESIGN = machine HEAD(MACHINE).
```

Run the command by clicking the Run button on the Syntax Editor toolbar. SPSS Statistics displays matrices showing all of the contrast coefficients. The part of the matrix with nonzero coefficients for *head(machine A)* has the following entries:

[head=1]([machine=A])	1.000	.000	.000
[head=2]([machine=A])	.000	1.000	.000
[head=3]([machine=A])	.000	.000	1.000
[head=4]([machine=A])	–1.000	–1.000	–1.000

To specify the custom hypothesis for comparing the heads in machine A, the coefficients for the first three contrasts (columns) in the contrast coefficients matrix for *head(machine)* are used. Similarly, the coefficients for the second three contrasts in the contrast matrix are used for machine B, and so on. The remaining coefficients in the table are all equal to 0. The following syntax shows the five custom hypotheses using the LMATRIX subcommand:

```
UNIANOVA STRAIN BY MACHINE HEAD
  /RANDOM = HEAD
  /DESIGN = MACHINE HEAD(MACHINE)
  /LMATRIX = "head(machine A)"
    HEAD(MACHINE) 1 0 0 -1 0 0 0 0 0 0 0 0 0 0 0 0 0 0 0 0;
    HEAD(MACHINE) 0 1 0 -1 0 0 0 0 0 0 0 0 0 0 0 0 0 0 0 0;
    HEAD(MACHINE) 0 0 1 -1 0 0 0 0 0 0 0 0 0 0 0 0 0 0 0 0
  /LMATRIX = "head(machine B)"
    HEAD(MACHINE) 0 0 0 0 1 0 0 -1 0 0 0 0 0 0 0 0 0 0 0 0;
    HEAD(MACHINE) 0 0 0 0 0 1 0 -1 0 0 0 0 0 0 0 0 0 0 0 0;
    HEAD(MACHINE) 0 0 0 0 0 0 1 -1 0 0 0 0 0 0 0 0 0 0 0 0
  /LMATRIX = "head(machine C)"
    HEAD(MACHINE) 0 0 0 0 0 0 0 0 1 0 0 -1 0 0 0 0 0 0 0 0;
    HEAD(MACHINE) 0 0 0 0 0 0 0 0 0 1 0 -1 0 0 0 0 0 0 0 0;
    HEAD(MACHINE) 0 0 0 0 0 0 0 0 0 0 1 -1 0 0 0 0 0 0 0 0
  /LMATRIX = "head(machine D)"
    HEAD(MACHINE) 0 0 0 0 0 0 0 0 0 0 0 0 1 0 0 -1 0 0 0 0;
    HEAD(MACHINE) 0 0 0 0 0 0 0 0 0 0 0 0 0 1 0 -1 0 0 0 0;
    HEAD(MACHINE) 0 0 0 0 0 0 0 0 0 0 0 0 0 0 1 -1 0 0 0 0
  /LMATRIX = "head(machine E)"
    HEAD(MACHINE) 0 0 0 0 0 0 0 0 0 0 0 0 0 0 0 0 1 0 0 -1;
    HEAD(MACHINE) 0 0 0 0 0 0 0 0 0 0 0 0 0 0 0 0 0 1 0 -1;
    HEAD(MACHINE) 0 0 0 0 0 0 0 0 0 0 0 0 0 0 0 0 0 0 1 -1.
```

The results of the custom hypotheses are shown in Figure 22-31. A caption was inserted at the bottom of each table to identify the contrast.

Figure 22-31
Contrasts for head-within-machine

Dependent Variable: Strain Reading

Source	Sum of Squares	df	Mean Square	F	Sig.
Contrast	50.188	3	16.729	1.563	.208
Error	642.000	60	10.700		

head(machine A)

Dependent Variable: Strain Reading

Source	Sum of Squares	df	Mean Square	F	Sig.
Contrast	126.187	3	42.062	3.931	.013
Error	642.000	60	10.700		

head(machine B)

Dependent Variable: Strain Reading

Source	Sum of Squares	df	Mean Square	F	Sig.
Contrast	74.750	3	24.917	2.329	.083
Error	642.000	60	10.700		

head(machine C)

Dependent Variable: Strain Reading

Source	Sum of Squares	df	Mean Square	F	Sig.
Contrast	6.500	3	2.167	.202	.894
Error	642.000	60	10.700		

head(machine D)

Dependent Variable: Strain Reading

Source	Sum of Squares	df	Mean Square	F	Sig.
Contrast	25.250	3	8.417	.787	.506
Error	642.000	60	10.700		

head(machine E)

Using $\alpha = 0.10$, you can see that there are significant differences between heads in machine B and machine C.

Example 7
Univariate Repeated Measures Analysis Using a Split-Plot Design Approach

Does the anxiety rating of a person affect performance on a learning task? An experiment was conducted to study the effect of anxiety on a learning task. Twelve subjects were assigned to one of two anxiety groups according to their anxiety measurement scores. The subjects were given four sets of trials, called blocks of trials. The criterion (score) is the number of errors in each block of trials. The data are taken from Winer, Brown, and Michels (1991) and are shown in Figure 22-32.

Figure 22-32
Error scores by students in two anxiety groups

		Score			
Subject	**Anxiety Group**	**Trial 1**	**Trial 2**	**Trial 3**	**Trial 4**
1	1	18	14	12	6
2	1	19	12	8	4
3	1	14	10	6	2
4	1	16	12	10	4
5	1	12	8	6	2
6	1	18	10	5	1
7	2	16	10	8	4
8	2	18	8	4	1
9	2	16	12	6	2
10	2	19	16	10	8
11	2	16	14	10	9
12	2	16	12	8	8

The SPSS Statistics file *anxiety.sav* contains the four variables *subject, anxiety, score,* and *trial*. Each value of *score* is represented by one case in the data file, as shown in Figure 22-33.

Figure 22-33
Data structure

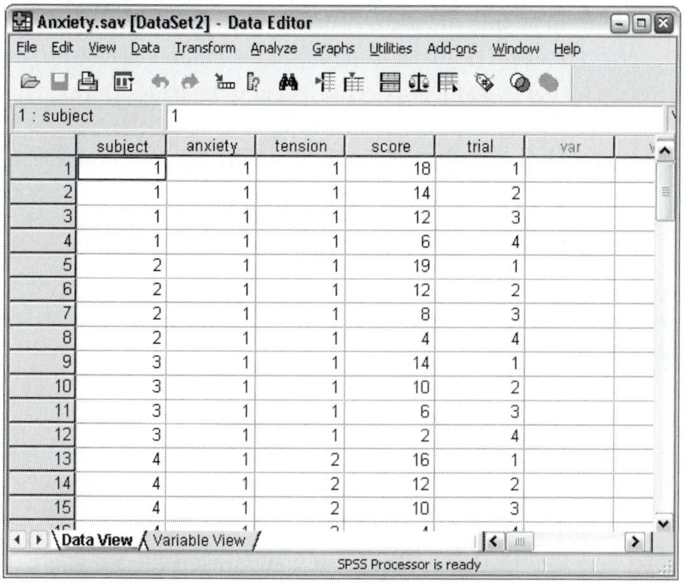

Anxiety.sav [DataSet2] - Data Editor

File Edit View Data Transform Analyze Graphs Utilities Add-ons Window Help

1 : subject 1

	subject	anxiety	tension	score	trial	var
1	1	1	1	18	1	
2	1	1	1	14	2	
3	1	1	1	12	3	
4	1	1	1	6	4	
5	2	1	1	19	1	
6	2	1	1	12	2	
7	2	1	1	8	3	
8	2	1	1	4	4	
9	3	1	1	14	1	
10	3	1	1	10	2	
11	3	1	1	6	3	
12	3	1	1	2	4	
13	4	1	2	16	1	
14	4	1	2	12	2	
15	4	1	2	10	3	

Data View Variable View

SPSS Processor is ready

There are two sizes of experimental units in the data: the subjects, which are the larger experimental units, and the blocks of trials within each subject, which are the smaller experimental units. The *anxiety* effect is considered as the whole-plot treatment, since it is applied to the subjects, the larger experimental unit. Historically, statistical studies divided an agricultural plot into subplots, and the names **whole plot** and **subplot** have carried over into fields of study other than agriculture. The *trial* effect is considered as the subplot treatment, since it is applied to the blocks of trial within each subject, the smaller experimental units.

When people perform the same type of task more than once, there is often a change from one trial to the next, so that performance on one trial is correlated with performance on previous trials. These tendencies can be expressed by saying that there are nonzero correlations among the blocks of trials within the same subject, with higher correlations for blocks that are closer in time. Usually, a split-plot design assumes that these correlations are 0 or at least remain constant over all the blocks within the same subject. In this example, the covariances among the blocks of trials within the same subject are temporarily assumed to be 0.

In this example, the model uses a split-plot design. The *anxiety* effect is the whole plot treatment, and the *subject-within-anxiety* effect, designated as *subject(anxiety)*, serves as the replication term in testing the *anxiety* effect. It should be entered as a random factor. The *trial* effect has four levels. It is the subplot treatment. The *trial*anxiety* effect is the interaction effect between the *anxiety* effect and the *trial* effect.

To produce the output shown in this example, from the menus choose:

Analyze
 General Linear Model
  Univariate...

▸ Dependent Variable: score
▸ Fixed Factor(s): trial anxiety
▸ Random Factor(s): subject

Model...
 ⊙ Custom
 ▸ Model: anxiety trial trial*anxiety

Paste

In the syntax window, on the DESIGN subcommand, insert the *subject-within-anxiety* effect—SUBJECT(ANXIETY)—after ANXIETY. The resulting syntax is

```
UNIANOVA score BY trial anxiety subject
  /RANDOM subject
  /METHOD = SSTYPE(3)
  /CRITERIA = ALPHA(.05)
  /DESIGN anxiety SUBJECT(ANXIETY) trial trial*anxiety.
```

To run the command, click the Run button on the Syntax Editor toolbar. The analysis-of-variance table is shown in Figure 22-34.

Figure 22-34
Tests of between-subjects effects

Dependent Variable: Score

Source		Type III Sum of Squares	df	Mean Square	F	Sig.
Intercept	Hypothesis	4800.000	1	4800.000	280.839	.000
	Error	170.917	10	17.092[1]		
anxiety	Hypothesis	10.083	1	10.083	.590	.460
	Error	170.917	10	17.092[1]		
subject(anxiety)	Hypothesis	170.917	10	17.092	6.652	.000
	Error	77.083	30	2.569[2]		
trial	Hypothesis	991.500	3	330.500	128.627	.000
	Error	77.083	30	2.569[2]		
trial * anxiety	Hypothesis	8.417	3	2.806	1.092	.368
	Error	77.083	30	2.569[2]		

1. MS(subject(anxiety))
2. MS(Error)

Notice from the footnote for the error term of the *anxiety* effect (footnote 2) that the mean square of the *subject-within-anxiety* effect is used as the error term in testing the *anxiety* effect. The *F* statistic is 0.590, which is not significant. Both the *trial* effect and the *trial*anxiety* effect use the error mean square as the error term. The *F* statistic for the *trial* effect is 128.627, which is significant with $p = 0.0001$. The *F* statistic for the *trial*anxiety* effect is 1.092, which is not significant.

Expected Mean Squares

The expected mean squares table is shown in Figure 22-35.

Figure 22-35
Expected mean squares components

Expected Mean Squares [1,2]

Source	Variance Component		
	Var(subject(anxiety))	Var(Error)	Quadratic Term
Intercept	4.000	1.000	Intercept, anxiety, trial, trial * anxiety
anxiety	4.000	1.000	anxiety, trial * anxiety
subject(anxiety)	4.000	1.000	
trial	.000	1.000	trial, trial * anxiety
trial * anxiety	.000	1.000	trial * anxiety
Error	.000	1.000	

1. For each source, the expected mean square equals the sum of the coefficients in the cells times the variance components, plus a quadratic term involving effects in the Quadratic Term cell.
2. Expected Mean Squares are based on the Type III Sums of Squares.

In addition to the variance components, there is a quadratic term associated with each of the fixed effects and other fixed effects containing the effect. For the *anxiety* effect, the other fixed effect that contains it and involves the quadratic term is the higher-order interaction effect *trial*anxiety.* This higher interaction effect is assumed to be 0 when testing the *anxiety* effect. For the *trial* effect, the higher-order interaction effect *trial*anxiety* is the effect that contains it and involves the quadratic term. This effect is assumed to be 0 when testing the *trial* effect. Since there are no other effects in the model that contain the *trial*anxiety* effect, the quadratic term for *trial*anxiety* consists only of itself.

Chapter

23

GLM Multivariate

Multivariate analysis of variance considers the effects of factors on several dependent variables at once, using a general linear model. The factors divide the cases (or subjects) into groups. The hypotheses tested are similar to those in univariate analysis, except that in multivariate analysis, a vector of means replaces the individual means.

In addition to the output in univariate analysis, multivariate *F* tests are available. You can also display the hypothesis and error sums-of-squares and cross-product (SSCP) matrices for each effect in the design, the transformation coefficient table (**M** matrix), Box's *M* test for equality of covariance matrices, and Bartlett's test of sphericity.

This chapter includes the following examples:

Example 1: Multivariate ANOVA: Multivariate two-way fixed-effects model with interaction. The effects of two measurements, the amount of an additive and the rate of extrusion of plastic film, are studied simultaneously on three properties of the manufactured plastic film. The SSCP (sums-of-squares and cross-products) matrices are displayed for the main effects and the interaction effect. These values are used by SPSS Statistics to calculate the *F* values for hypothesis testing on the three dependent variables. The conclusion is that the main effects of amount of additive and rate of extrusion are significant, but their interaction is not.

Example 2: Profile analysis: Setting up custom linear hypotheses. In a survey, 30 couples were given questions designed to rate companionate and passionate love on a five-point scale. A plot of the mean answers of husbands and wives is used to set up hypotheses that will investigate the differences between groups. The **L** and **M** matrices are used together in the model to define contrasts, where **LBM = 0**. The conclusion is that the parallelism and the coincidence of the profiles of husbands and wives is not rejected. However, the equality of means of the four answers is rejected.

Example 1
Multivariate ANOVA: Multivariate Two-Way Fixed-Effects Model with Interaction

How can the optimum conditions for extruding plastic film be evaluated by using a statistical technique called evolutionary operation? A study was conducted to examine how the amount of an additive and the rate of extrusion affected the conditions for extruding plastic film. The data are from Johnson and Wichern (1988). The amount of additive and rate of extrusion are two fixed factors that may interact so that the model is a two-way fixed-effects model with interaction. Three properties of the extruded film were measured: tear resistance, gloss, and opacity. These are the three dependent variables in this multivariate model. Evolutionary operation, the statistical method for process improvement used here, was described by Box and Draper (1969).

Figure 23-1
Effects of rate of extrusion and amount of additive on plastic film

Independent Factors		Dependent Variables		
Change in rate of extrusion	Amount of additive	Tear resistance	Gloss	Opacity
1	1	6.5	9.5	4.4
1	1	6.2	9.9	6.4
1	1	5.8	9.6	3.0
1	1	6.5	9.6	4.1
1	1	6.5	9.2	0.8
1	2	6.9	9.1	5.7
1	2	7.2	10.0	2.0
1	2	6.9	9.9	3.9
1	2	6.1	9.5	1.9
1	2	6.3	9.4	5.7
2	1	6.7	9.1	2.8
2	1	6.6	9.3	4.1
2	1	7.2	8.3	3.8
2	1	7.1	8.4	1.6
2	1	6.8	8.5	3.4
2	2	7.1	9.2	8.4
2	2	7.0	8.8	5.2
2	2	7.2	9.7	6.9
2	2	7.5	10.1	2.7
2	2	7.6	9.2	1.9

The data are recorded in the SPSS Statistics file *Plastic.sav*, in the order shown in Figure 23-1. There are two independent factors in the model: the change in rate of extrusion (*extrusn*) and the amount of additive (*additive*). The *extrusn* factor has two levels: low and high (indicated by 1 and 2, respectively, in the data). The *additive* factor has two levels: 1.0% and 1.5% (indicated by 1 and 2, respectively, in the data). In addition to these two main factors, their interaction *extrusn*additive* is included in the model. Since three dependent variables are studied at the same time, multivariate analysis of variance is used.

SSCP Matrices

In univariate fixed-effects analysis of variance, the total sum of squares of the model is partitioned into sums of squares due to the effects in the model and the error sum of squares. Each effect in the model is then evaluated by using an *F* statistic, which is the ratio of the sum of squares due to the effect and the error sum of squares. In a multivariate model, there is more than one dependent variable; hence, the sums of squares due to the effects in the model and the error sums of squares are no longer scalars. Instead, they are replaced by square matrices. For each of these square matrices, the dimension is equal to the number of dependent variables. In univariate analysis, there is only one dependent variable, and the matrix reduces to a scalar, which is the same as the sum of squares of the corresponding effect.

In a multivariate model, these square matrices are called the sums-of-squares and cross-products (**SSCP**) matrices. Analogous to the univariate test of an effect, the "ratio" of the SSCP matrix due to the effect being tested (**H**) and the SSCP matrix of the appropriate error (**E**) is used to evaluate the effect of interest. Since **H** and **E** are matrices, this "ratio" in a multivariate model is evaluated by the determinant of $\mathbf{HE}^{-1}$, where $\mathbf{E}^{-1}$ is the inverse of **E**. The **H** matrix, which is the SSCP matrix of the effect being tested, is called the hypothesis SSCP matrix, and the **E** matrix, which is the SSCP matrix of the appropriate error, is called the error SSCP matrix. In a fixed-effects multivariate analysis of variance, the hypothesis SSCP matrix for testing any effect in the model is its own SSCP matrix due to that effect, and the error SSCP matrix for all tests in the model is always the SSCP matrix due to error in the model.

Running the Tests

To produce the output shown in this example, from the menus choose:

Analyze
  General Linear Model
    Multivariate...

▶ Dependent Variables: tear_res, gloss, opacity
▶ Fixed Factor(s): extrusn, additive

Options...
  ☑ SSCP matrices

The hypothesis SSCP matrices for testing the effects in the model are shown in Figure 23-2. These are the matrices that are used in calculating the F values.

Figure 23-2
Between-subjects SSCP matrices

			Tear Resistance	Gloss	Opacity
Hypothesis	Intercept	Tear Resistance	920.724	1264.045	533.979
		Gloss	1264.045	1735.384	733.090
		Opacity	533.979	733.090	309.684
	extrusn	Tear Resistance	1.740	-1.505	.855
		Gloss	-1.505	1.301	-.740
		Opacity	.855	-.740	.421
	additive	Tear Resistance	.760	.682	1.930
		Gloss	.682	.612	1.732
		Opacity	1.930	1.732	4.900
	extrusn * additive	Tear Resistance	.000	.016	.044
		Gloss	.016	.544	1.468
		Opacity	.044	1.468	3.960
Error		Tear Resistance	1.764	.020	-3.070
		Gloss	.020	2.628	-.552
		Opacity	-3.070	-.552	64.924

Based on Type III Sum of Squares

The results of the multivariate analysis of variance are shown in Figure 23-3. Each effect is shown as a layer.

Figure 23-3
Multivariate tests

Effect		Value	F	Hypothesis df	Error df	Sig.
Intercept	Pillai's Trace	.999	5950.906[1]	3.000	14.000	.000
	Wilks' Lambda	.001	5950.906[1]	3.000	14.000	.000
	Hotelling's Trace	1275.194	5950.906[1]	3.000	14.000	.000
	Roy's Largest Root	1275.194	5950.906[1]	3.000	14.000	.000
extrusn	Pillai's Trace	.618	7.554[1]	3.000	14.000	.003
	Wilks' Lambda	.382	7.554[1]	3.000	14.000	.003
	Hotelling's Trace	1.619	7.554[1]	3.000	14.000	.003
	Roy's Largest Root	1.619	7.554[1]	3.000	14.000	.003
additive	Pillai's Trace	.477	4.256[1]	3.000	14.000	.025
	Wilks' Lambda	.523	4.256[1]	3.000	14.000	.025
	Hotelling's Trace	.912	4.256[1]	3.000	14.000	.025
	Roy's Largest Root	.912	4.256[1]	3.000	14.000	.025
extrusn * additive	Pillai's Trace	.223	1.339[1]	3.000	14.000	.302
	Wilks' Lambda	.777	1.339[1]	3.000	14.000	.302
	Hotelling's Trace	.287	1.339[1]	3.000	14.000	.302
	Roy's Largest Root	.287	1.339[1]	3.000	14.000	.302

1. Exact statistic

The first column in the table is the effect in the model being tested, and the second column lists the names of the test statistics. Four commonly used test statistics for multivariate analysis are displayed in the output: Pillai's trace, Wilks' lambda, Hotelling's trace, and Roy's largest root. The next column displays the value of the test statistics, followed by the *F* statistic, which is a transformed value of the corresponding test statistic and has an approximate *F* distribution. The hypothesis and error degrees of freedom of the *F* distribution are shown.

Of the four test statistics, Wilks' lambda has the virtue of being convenient and related to the likelihood-ratio criterion. However, for some practical situations (see Olsen, 1976), Pillai's trace may be the most robust and powerful criterion among the others. Choice of these multivariate statistics may depend on the situation. Their *F* statistics may sometimes be the same. Whenever the *F* statistic has an exact *F* distribution, a footnote is displayed as a reminder.

In the test of the interaction effect *extrusn*additive*, the four *F* statistics are all the same, with a value of 1.339. Since the *F* value is not significant at the 0.05 alpha level, we conclude that there is no interaction effect. The *F* statistics for testing *extrusn* and *additive* are 7.554 and 4.256, respectively. Both are significant at $\alpha = 0.05$, indicating that changes in both the rate of extrusion and the amount of additive affect the three dependent variables.

Example 2
Profile Analysis: Setting Up Custom Linear Hypotheses

Do husbands and wives share the same point of view on love and marriage? An investigation was performed to study the points of view of married couples on love and marriage. A set of four questions was answered by a sample of 30 couples. The first two questions concerned the feeling of passionate love, while the second two questions were about the feeling of companionate love. All four questions were rated on a five-point scale, ranging from "None at all" to "A tremendous amount." The data are taken from Johnson and Wichern (1988).

Figure 23-4 shows a plot of the means of the answers to the four questions. The mean rating by husbands and the mean rating by wives for each question are plotted on separate lines.

Figure 23-4
Plot of husbands' and wives' profiles

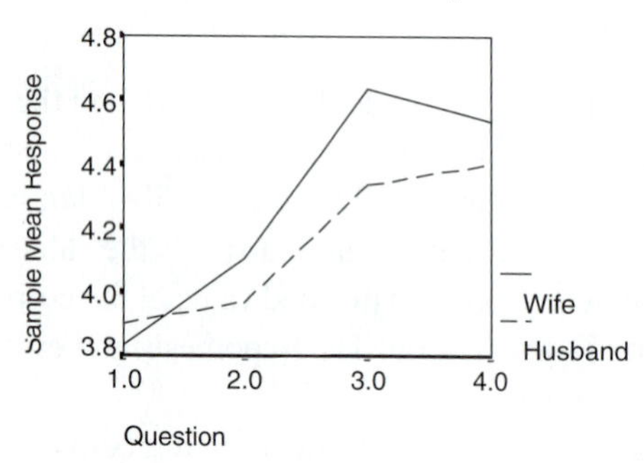

The plot suggests three queries:

- Parallelism of profiles: Are the population mean profiles of husbands and wives parallel to each other?
- Equality of husbands' and wives' profiles: Assuming parallelism, are the profiles of husbands and wives equal?
- Equality of answer means: Assuming parallelism, are the answer means equal?

In order to satisfy parallelism of profiles, the two profile plots should be parallel to each other for each line segment between adjacent questions. This query can be viewed as testing the hypothesis of no response by group interaction. Testing the equality of the two profiles is equivalent to testing whether there is any difference between the husbands' and wives' answers. Testing the equality of answer means is equivalent to testing for any differences between husbands and wives in each of the four survey questions. In the next section, we will study each of the three queries related to the plot by specifying the **L** matrix and the **M** matrix, using a multivariate linear model.*

Setting Up Contrast Matrices

In the SPSS Statistics data file *love.sav*, *Q1, Q2, Q3,* and *Q4* are variables that represent the answers to the four love and marriage questions, on a five-point scale. The variable *spouse* divides the cases into two groups, husbands and wives. You can form a multivariate linear model with *Q1, Q2, Q3,* and *Q4* as the dependent variables and *spouse* as the factor.

To answer the three queries listed in the previous section, you can use the LMATRIX and the MMATRIX syntax subcommands, which correspond to the **L** matrix and **M** matrix of the general multivariate linear hypothesis **LBM** = **0**. The **B** matrix is the parameter matrix in which each column corresponds to the parameter vector for the linear model on each of the dependent variables. In this example, the four dependent variables correspond to the four columns of **B**. Since there is only one factor, *spouse,* in the model, and it has only two levels, *husband* and *wife*, each column of **B** has three parameters to be estimated: the overall intercept and the two parameters for each level of *spouse*. The **L** matrix sets up contrasts for the factor in the model, and the **M** matrix sets up contrasts for the dependent variables.

Testing Parallelism

To test for parallelism, you must set up both the **L** matrix and the **M** matrix. Since the first query asks whether the line segments between adjacent answers are parallel, the **M** matrix should be set for contrasts, which can compare adjacent answers. In other words, comparisons should be made of the mean of *Q1* against *Q2*, *Q2* against *Q3*, and *Q3* against *Q4*. The **M** matrix will be a 4×3 matrix of the following form:

$$\begin{bmatrix} 1 & 0 & 0 \\ -1 & 1 & 0 \\ 0 & -1 & 1 \\ 0 & 0 & -1 \end{bmatrix}$$

*This problem would normally be handled by using a repeated measures effect. However, the example here is used to illustrate the use of LMATRIX and MMATRIX subcommands.

Because the **M** matrix is used to test whether the line segments of the husband group and the wife group are parallel, the **L** matrix compares the husband group with the wife group. The **L** matrix will be a 1×3 matrix of the following form:

$$\begin{bmatrix} 0 & 1 & -1 \end{bmatrix}$$

where the first value, 0, corresponds to the overall intercept parameter in the linear model, and the pair 1 and -1 compares the husband group with the wife group.

To produce the output in this section, from the menus choose:

Analyze
  General Linear Model
    Multivariate...

▶ Dependent Variables: q1, q2, q3, q4
▶ Fixed Factor(s): spouse
Model...
  ⊙ Custom
  ▶ Model: spouse

Paste

To the pasted syntax, add the LMATRIX and MMATRIX subcommands as shown:

```
GLM q1 q2 q3 q4 BY spouse
 /METHOD = SSTYPE(3)
 /INTERCEPT = INCLUDE
 /CRITERIA = ALPHA(.05)
 /DESIGN spouse
 /LMATRIX = spouse 1 -1
 /MMATRIX = q1 1 q2 -1;
            q2 1 q3 -1;
            q3 1 q4 -1.
```

To run the GLM command, click the Run button on the Syntax Editor toolbar. The syntax indicates that there are four dependent variables, *Q1*, *Q2*, *Q3,* and *Q4*, in the model, and one factor, *spouse*. The DESIGN subcommand indicates the model. The subcommands METHOD, INTERCEPT, and CRITERIA are default specifications (with which we are not concerned here).

The **L** matrix is specified by using the LMATRIX subcommand, followed by the factor name *spouse* and the contrast for the factor. The **M** matrix is specified by the MMATRIX subcommand, followed by an equals sign, the name of one dependent variable, its coefficient, the name of the dependent variable being compared with the first variable, and its coefficient. Each of the three comparisons is listed in a similar manner. If a dependent variable and its coefficient are not specified, the coefficient of that dependent variable is assumed to be 0. A semicolon (;) is used to indicate the end of one column in the **M** matrix.

Parallelism Test Results

The results of the custom hypothesis test of parallelism are shown in Figure 23-5.

Figure 23-5
Multivariate test results

	Value	F	Hypothesis df	Error df	Sig.
Pillai's trace	.121	2.580[1]	3.000	56.000	.063
Wilks' lambda	.879	2.580[1]	3.000	56.000	.063
Hotelling's trace	.138	2.580[1]	3.000	56.000	.063
Roy's largest root	.138	2.580[1]	3.000	56.000	.063

1. Exact statistic

With $p = 0.063$, the test is not significant at $\alpha = 0.05$. Parallelism of the mean profiles of husbands and wives will be assumed.

Testing for Equality of Profiles

To test the equality of husbands' and wives' profiles, the means of the four answers for the husband group and the wife group are compared. In this case, the **M** matrix corresponds to the average of the four answers, and the **L** matrix corresponds to the contrast used to compare husbands and wives.

To produce the output in this section, recall the GLM Multivariate dialog box and paste, or from the menus choose:

Analyze
  General Linear Model
    Multivariate...

▶ Dependent Variables: q1, q2, q3, q4
▶ Fixed Factor(s): spouse
Model...
  ⊙ Custom
  ▶ Model: spouse

Paste

To the pasted syntax, add the LMATRIX and MMATRIX subcommands as shown:

```
GLM q1 q2 q3 q4 BY spouse
 /METHOD = SSTYPE(3)
 /INTERCEPT = INCLUDE
 /CRITERIA = ALPHA(.05)
 /DESIGN spouse
 /LMATRIX = spouse 1 -1
 /MMATRIX = q1 .25 q2 .25 q3 .25 q4 .25.
```

To run the GLM command, click the Run button on the Syntax Editor toolbar. The results are shown in Figure 23-6.

Figure 23-6
Univariate test results

Transformed Variable: T1

Source	Sum of Squares	df	Mean Square	F	Sig.
Contrast	.234	1	.234	1.533	.221
Error	8.869	58	.153		

Since the **M** matrix has only one column, the univariate result is used. The *p* value for the test is 0.221, which is not significant at $\alpha = 0.05$. This means that equality of the husbands' and wives' profiles can be assumed. That is, the answers from the husbands and the wives to the four questions can be assumed to be the same.

Testing for Equality of Answer Means

To test the equality of answer means, the means of the husbands' group and wives' group for the four answers are compared with each other. In this case, the **M** matrix is the one used in testing parallelism, while the **L** matrix corresponds to taking the average of each group. It is [1 0.5 0.5]. The value 1 corresponds to the intercept, and the two values of 0.5 correspond to the husbands' group and the wives' group, respectively.

To produce the output in this section, recall the GLM Multivariate dialog box and paste, or from the menus choose:

Analyze
  General Linear Model
    Multivariate...

▶ Dependent Variables: q1, q2, q3, q4
▶ Fixed Factor(s): spouse
Model...
  ⊙ Custom
  ▶ Model: spouse

Paste

To the pasted syntax, add the LMATRIX and MMATRIX subcommands as shown:

```
GLM Q1 Q2 Q3 Q4 BY spouse
 /DESIGN spouse
 /METHOD = SSTYPE(3)
 /INTERCEPT = INCLUDE
 /CRITERIA = ALPHA(.05)
 /LMATRIX = INTERCEPT 1 spouse .5 .5
 /MMATRIX = Q1 1 Q2 -1;
            Q2 1 Q3 -1;
            Q3 1 Q4 -1.
```

The contrast coefficients are specified in the **L** matrix, and the transformation coefficients are specified in the **M** matrix. The results are displayed in Figure 23-7.

Figure 23-7
Multivariate test results

	Value	F	Hypothesis df	Error df	Sig.
Pillai's trace	.305	8.188[1]	3.000	56.000	.000
Wilks' lambda	.695	8.188[1]	3.000	56.000	.000
Hotelling's trace	.439	8.188[1]	3.000	56.000	.000
Roy's largest root	.439	8.188[1]	3.000	56.000	.000

1. Exact statistic

For each test, $p < 0.0005$, which is significant. It means that the equality of answers cannot be assumed.

Figure 23-8
Univariate test results for each individual contrast on the dependent variables

Source	Transformed Variable	Sum of Squares	df	Mean Square	F	Sig.
Contrast	T1	1.667	1	1.667	2.433	.124
	T2	12.150	1	12.150	13.973	.000
	T3	.017	1	.017	.212	.647
Error	T1	39.733	58	.685		
	T2	50.433	58	.870		
	T3	4.567	58	.079		

Figure 23-8 displays the univariate results for this custom hypothesis on each difference of adjacent answers. Each of the three columns in the **M** matrix corresponds to one contrast (or transformation) on the dependent variables. The right side of the first column in Figure 23-8 displays the name of the transformed variable. The first transformed variable, *T1*, corresponds to the first column of the **M** matrix, which compares *q1* with *q2*. The second transformed variable, *T2,* corresponds to the second column of the **M** matrix, which compares *q2* with *q3*. The last transformed variable, *T3,* corresponds to the last column of the **M** matrix, which compares *q3* with *q4*. The three contrast results, taken as a set, imply that the first two population means are the same, but differ from the last two, which do not differ from each other.

Chapter

24

GLM Repeated Measures

Repeated measures analysis applies to situations in which the same measurement is made multiple times on each subject or case. A within-subjects factor that encompasses each set of repeated measurements is defined. Between-subjects factors that divide the cases into groups can be specified as well as covariates.

This chapter includes the following examples:

Example 1: Repeated measures analysis of variance. Twelve students were each tested four times on a learning task. They were also rated for high anxiety or low anxiety. The data are arranged with four trial variables so that the four scores are recorded in one case for each student. A within-subjects factor, *trial,* is defined, and the *anxiety* factor is specified as a between-subjects factor because it divides the group of subjects into two groups. The *anxiety* factor is shown to be not significant, while the *trial* effect is significant. In checking assumptions, Box's *M* and Mauchly's test of sphericity are used.

Example 2: Doubly multivariate repeated measures analysis of variance. A new drug is tested against a placebo in its effects on respiration and pulse. Half of the patients receive the new drug, and half receive the placebo. The variable *drug* divides the subjects into the new-drug group and the placebo group and is used as a between-subjects factor. Both effects, respiration and pulse, are measured together at three different times, making six variables in the data file. The newly defined within-subjects variable is called *time*, with three levels for the three times.

Example 1
Repeated Measures Analysis of Variance

Does the anxiety of a person affect performance on a learning task? Twelve subjects were assigned to one of two anxiety groups on the basis of an anxiety test, and the number of errors made in four blocks of trials on a learning task was measured. In this example, we use a repeated measures analysis-of-variance technique to study the data.

In the data file *Anxiety 2.sav*, there is one case for each subject (*subject*). Four trial variables (*trial1*, *trial2*, *trial3*, and *trial4*) contain the scores for all of the subjects (see Figure 24-1).

Figure 24-1
Data arrangement

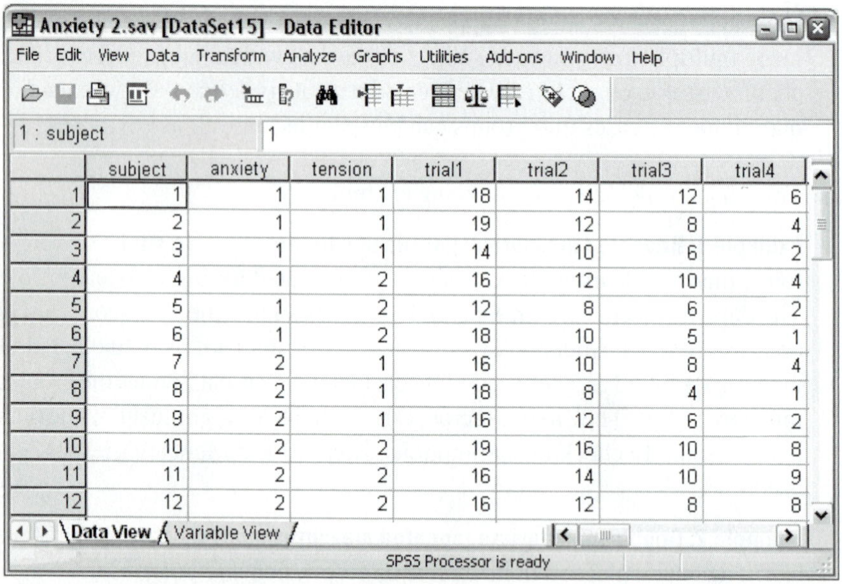

Anxiety 2.sav [DataSet15] - Data Editor

File Edit View Data Transform Analyze Graphs Utilities Add-ons Window Help

1 : subject 1

	subject	anxiety	tension	trial1	trial2	trial3	trial4
1	1	1	1	18	14	12	6
2	2	1	1	19	12	8	4
3	3	1	1	14	10	6	2
4	4	1	2	16	12	10	4
5	5	1	2	12	8	6	2
6	6	1	2	18	10	5	1
7	7	2	1	16	10	8	4
8	8	2	1	18	8	4	1
9	9	2	1	16	12	6	2
10	10	2	2	19	16	10	8
11	11	2	2	16	14	10	9
12	12	2	2	16	12	8	8

Data View Variable View

SPSS Processor is ready

To use the repeated measures analysis-of-variance technique, we distinguish two types of factors in the model: between-subjects factors and within-subjects factors. A **between-subjects factor** is any factor that divides the sample of subjects or cases into discrete subgroups. For example, the factor *anxiety* is a between-subjects factor because it divides the 12 subjects into two groups: one with high anxiety measurement scores and one with low anxiety measurement scores. A **within-subjects factor** is any factor that distinguishes measurements made on the same subject or case rather than distinguishing different subjects or cases. For example, the factor *trial* is a within-subjects factor in this analysis because it distinguishes the four measurements of error taken for each of the subjects.

A repeated measures analysis consists of the analysis of a within-subjects model, which describes the model for the within-subjects factors, plus the analysis of a between-subjects model, which describes the model for the between-subject factors. For example, in the current repeated measures analysis, the within-subjects part of the model consists of the within-subjects factor *trial* and the interaction, and the between-subjects part of the model consists of the between-subjects factor *anxiety*.

In the data file *Anxiety 2.sav*, the variables *trial1*, *trial2*, *trial3,* and *trial4* contain the number of errors made at each of the four trials by each subject.

To produce the output in this example, from the menus choose:

Analyze
 General Linear Model
  Repeated Measures...

Within-Subject Factor Name: trial
Number of Levels: 4

Click Add and click Define.

▶ Within-Subjects Variables (trial): trial1, trial2, trial3, trial4
▶ Between-Subjects Factor(s): anxiety

Options...
 ☑ Homogeneity tests

In the above choices, for the within-subjects factor, a new name, *trial*, is typed, which designates the measurements made on the same subject. The number of levels is the number of measurements made on each subject (4). Each of the four within-subjects variables, *trial1*, *trial2*, *trial3,* and *trial4,* corresponds to one measurement for each subject.

Between-Subjects Tests

For repeated measures analysis, the results for testing the effects in the model from a multivariate approach and a univariate approach are both provided. For testing the between-subjects effects in the model, both approaches yield the same results. This test is constructed by summing all of the within-subjects variables in the model and dividing by the square root of the number of within-subjects variables. Then an analysis of variance is performed on the result. In this example, since there is only one between-subjects factor, *anxiety*, the test for the between-subjects effects is performed by a one-way analysis of variance using

$$\frac{\text{trial}_1 + \text{trial}_2 + \text{trial}_3 + \text{trial}_4}{\sqrt{4}}$$

as the dependent variable. The results from testing the between-subjects effect, *anxiety*, are shown in Figure 24-2. The significance $p = 0.460$ indicates that this effect is not significant at $\alpha = 0.05$.

Figure 24-2
Test of between-subjects effects

Measure: MEASURE_1
Transformed Variable: Average

Source	Type III Sum of Squares	df	Mean Square	F	Sig.
Intercept	4800.000	1	4800.000	280.839	.000
anxiety	10.083	1	10.083	.590	.460
Error	170.917	10	17.092		

Multivariate Tests

The multivariate table contains tests of the within-subjects factor, *trial*, and the interaction of the within-subjects factor and the between-subjects factor, *trial*anxiety*. The results are shown in Figure 24-3.

Figure 24-3
Multivariate tests

Effect		Value	F	Hypothesis df	Error df	Sig.
trial	Pillai's Trace	.961	64.854[1]	3.000	8.000	.000
	Wilks' Lambda	.039	64.854[1]	3.000	8.000	.000
	Hotelling's Trace	24.320	64.854[1]	3.000	8.000	.000
	Roy's Largest Root	24.320	64.854[1]	3.000	8.000	.000
trial * anxiety	Pillai's Trace	.479	2.451[1]	3.000	8.000	.138
	Wilks' Lambda	.521	2.451[1]	3.000	8.000	.138
	Hotelling's Trace	.919	2.451[1]	3.000	8.000	.138
	Roy's Largest Root	.919	2.451[1]	3.000	8.000	.138

1. Exact statistic

Four test statistics—Pillai's trace, Wilks' lambda, Hotelling's trace and Roy's largest root—are provided. In this example, the *F* statistics for all four tests are the same. The *p* value for testing the *trial* effect is less than 0.001, which is significant at $\alpha = 0.05$, while the *p* value for testing the *trial*anxiety* effect is 0.138, which is not significant at $\alpha = 0.05$.

Checking Assumptions

The assumption for the multivariate approach is that the vector of the dependent variables follows a multivariate normal distribution, and the variance-covariance matrices are equal across the cells formed by the between-subject effects. Box's *M* test for this assumption is shown in Figure 24-4. The significance is 0.315, indicating that the null hypothesis (that the observed variance-covariance matrices are equal across the two levels of the *anxiety* effect) is not rejected.

Figure 24-4
Test of equality of covariance matrices

Box's Test of Equality of Covariance Matrices[1]

Box's M	21.146
F	1.161
df1	10
df2	478.088
Sig.	.315

Tests the null hypothesis that the observed covariance matrices of the dependent variables are equal across groups.

1. Design: Intercept+anxiety
Within Subjects Design: trial

Testing for Sphericity

To use the univariate results for testing the within-subjects factor, *trial,* and the interaction of the within-subjects factor and the between-subjects factor, *trial*anxiety*, some assumptions about the variance-covariance matrices of the dependent variables should be checked. The validity of the *F* statistic used in the univariate approach can be assured when the variance-covariance matrix of the dependent variables is *circular* in form (Huynh and Mandeville, 1979). Mauchly (1940) derived a test that verifies this variance-covariance matrix structure by performing a test of sphericity on the orthonormalized transformed dependent variable. This test, which is automatically performed when a repeated measures analysis is used, is shown in Figure 24-5. The significance of this test is 0.053, which is slightly larger the 0.05 alpha level, indicating that the variance-covariance matrix assumption is just barely satisfied.

Figure 24-5
Mauchly's test of sphericity

Mauchly's Test of Sphericity[2]

Measure: MEASURE_1

Within Subjects Effect	Mauchly's W	Approx. Chi-Square	df	Sig.	Epsilon[1]		
					Greenhouse-Geisser	Huynh-Feldt	Lower-bound
trial	.283	11.011	5	.053	.544	.701	.333

Tests the null hypothesis that the error covariance matrix of the orthonormalized transformed dependent variables is proportional to an identity matrix.

1. May be used to adjust the degrees of freedom for the averaged tests of significance. Corrected tests are displayed in the Tests of Within-Subjects Effects table.
2. Design: Intercept+anxiety
Within Subjects Design: trial

What if the significance indicated rejection? If the above sphericity test were to be rejected, sphericity could not be assumed. Then an adjustment value, called epsilon, would be needed for multiplying the numerator and denominator degrees of freedom in the F test. The significance of the F test would then be evaluated with the new degrees of freedom. The last three columns in Figure 24-5 display three possible values of epsilon, based on three different criteria: Greenhouse-Geisser, Huynh-Feldt, and lower-bound. The Greenhouse-Geisser epsilon is conservative, especially for a small sample size. The Huynh-Feldt epsilon is an alternative that is not as conservative as the Greenhouse-Geisser epsilon; however, it may be a value greater than 1. When its calculated value is greater than 1, the Huynh-Feldt epsilon is displayed as 1.000, and this value is used in calculating the new degrees of freedom and significance. The lower-bound epsilon takes the reciprocal of the degrees of freedom for the within-subjects factor. This represents the most conservative approach possible, since it indicates the most extreme possible departure from sphericity. SPSS Statistics displays results for each of the values of epsilon.

Testing Hypotheses

The univariate tests for the within-subjects factor, *trial,* and the interaction term, *trial*anxiety,* are shown in Figure 24-6, with all layers displayed.

Figure 24-6
Tests of within-subjects effects

Measure: MEASURE_1

Source		Type III Sum of Squares	df	Mean Square	F	Sig.
trial	Sphericity Assumed	991.500	3	330.500	128.627	.000
	Greenhouse-Geisser	991.500	1.632	607.468	128.627	.000
	Huynh-Feldt	991.500	2.102	471.773	128.627	.000
	Lower-bound	991.500	1.000	991.500	128.627	.000
trial * anxiety	Sphericity Assumed	8.417	3	2.806	1.092	.368
	Greenhouse-Geisser	8.417	1.632	5.157	1.092	.346
	Huynh-Feldt	8.417	2.102	4.005	1.092	.357
	Lower-bound	8.417	1.000	8.417	1.092	.321
Error(trial)	Sphericity Assumed	77.083	30	2.569		
	Greenhouse-Geisser	77.083	16.322	4.723		
	Huynh-Feldt	77.083	21.016	3.668		
	Lower-bound	77.083	10.000	7.708		

The significance for each test when sphericity is assumed, or when any of the three epsilons is used, is displayed. The third column indicates what type of epsilon is used in evaluating the significance of the test. For example, the row labeled *Sphericity Assumed* indicates that the assumption about the variance-covariance matrix is assumed and that the significance is evaluated using the original degrees of freedom. The row labeled *Greenhouse-Geisser* indicates that the Greenhouse-Geisser epsilon is used and that the significance in this row is evaluated using the Greenhouse-Geisser epsilon adjustment. In this example, because the variance-covariance matrix assumption is assumed to be satisfied (by the result of Mauchly's test), the conclusion can be drawn based on the row labeled *Sphericity Assumed*. The *trial* effect is significant, with a p value of less than 0.001, and the *trial**anxiety* interaction effect is not significant, with a p value of 0.368. Notice that the same conclusion can be drawn using any three of the epsilons.

Contrasts

It is sometimes useful to introduce contrasts among the within-subjects variables in order to study the levels of the within-subjects factors. To specify contrasts for the within-subjects variables, recall the dialog box for Repeated Measures and choose:

Contrasts...
  Factors: trial
  Contrast: Repeated (Click Change)

This set of contrasts does not affect the results of the univariate and multivariate analyses. It is useful only for comparing the levels of the within-subjects factor. The tests of within-subjects contrasts are shown in Figure 24-7.

Figure 24-7
Within-subjects contrasts

Measure: MEASURE_1

Source	trial	Type III Sum of Squares	df	Mean Square	F	Sig.
trial	Level 1 vs. Level 2	300.000	1	300.000	52.023	.000
	Level 2 vs. Level 3	168.750	1	168.750	83.678	.000
	Level 3 vs. Level 4	147.000	1	147.000	78.750	.000
trial * anxiety	Level 1 vs. Level 2	.333	1	.333	.058	.815
	Level 2 vs. Level 3	4.083	1	4.083	2.025	.185
	Level 3 vs. Level 4	16.333	1	16.333	8.750	.014
Error(trial)	Level 1 vs. Level 2	57.667	10	5.767		
	Level 2 vs. Level 3	20.167	10	2.017		
	Level 3 vs. Level 4	18.667	10	1.867		

A repeated contrast compares one level of *trial* with the subsequent level. The first column indicates the effect being tested. For example, the label *trial* tests the hypothesis that, averaged over the two anxiety groups, the mean of the specified contrast is 0.

The contrasts are in the second column. The first contrast, *Level 1 vs. Level 2,* compares the first level of *trial* with the second level of *trial.* Notice that the first dependent variable, *trial1,* corresponds to the observation of the first level of *trial* and that the second dependent variable, *trial2,* corresponds to the observation of the second level of *trial*, and so on. In the table above, all three specified contrasts—*Level 1 vs. Level 2*, *Level 2 vs. Level 3,* and *Level 3 vs. Level 4*—are significant at $\alpha = 0.01$.

The label *trial*anxiety* tests the hypothesis that the mean of the specified contrast is the same for the two anxiety groups. Except for the last specified contrast, *Level 3 vs. Level 4,* which may be significant at $\alpha = 0.05$, all others are not significant at $\alpha = 0.01$.

Example 2
Doubly Multivariate Repeated Measures Analysis of Variance

Can a new drug improve respiratory and pulse scores? A study was performed to determine whether a new drug has any effect on the respiratory score and pulse score of a patient. Twelve randomly selected patients having similar medical conditions were randomly divided into two groups. Patients in one group were treated with the new drug, and patients in another group were treated with a placebo. The respiratory score and the pulse were measured at three different times. The data, created for this example, are shown in Figure 24-8.

Figure 24-8
Respiratory and pulse measurements

	Respiratory Score			Pulse		
Drug	**Time 1**	**Time 2**	**Time 3**	**Time 1**	**Time 2**	**Time 3**
New Drug	3.4	3.3	3.3	77	77	73.5
New Drug	3.4	3.4	3.3	77	77	77
New Drug	3.3	3.4	3.4	80.5	80.5	80.5
New Drug	3.4	3.4	3.4	80.5	80.5	80.5
New Drug	3.3	3.4	3.3	77	77	84
New Drug	3.3	3.3	3.3	70	70	84
Placebo	3.3	3.3	3.3	98	98	94.5
Placebo	3.2	3.3	3.4	91	91	94.5
Placebo	3.2	3.2	3.2	94.5	94.5	94.5
Placebo	3.2	3.2	3.2	91	91	101.5
Placebo	3.2	3.3	3.3	94.5	94.5	101.5
Placebo	3.3	3.2	3.1	91	91	98

The data are stored in the file *New drug.sav*, with one case for each subject. The respiratory scores are stored in three variables (*resp1, resp2,* and *resp3*), one for each time. The pulses are stored in three variables (*pulse1*, *pulse2*, and *pulse3*)—again, one for each time. The variable *drug* has two values, 1 for the group that received the new drug and 2 for the group that received the placebo.

The type of drug divides the patients into two groups; therefore, it is a between-subjects factor. The three time points distinguish measurements taken on the same patient, making *time* a within-subjects factor. Notice that two measurements, respiratory score and pulse score, are taken for all patients at each factor combination—this is a doubly multivariate repeated measures design.

To produce the output shown in this example, from the menus choose:

Analyze
  General Linear Model
    Repeated Measures...

Within-Subject Factor Name: time
Number of Levels: 3 (Click Add)
Measure Name: resp pulse (Click Add for each measure)

Click Define.

▶ Within-Subjects Variables (time): resp1, resp2, resp3, pulse1, pulse2, pulse3
▶ Between-Subjects Factor(s): drug

The within-subjects factor name is a new name that you specify to designate groups of measurements made on the same subject—in this example, *time* for the three times. The number of levels indicates the number of groups for the within-subjects factor. A measure name is a new name that you specify to designate the group of like scores measured at different times. In this example, the measure names indicate the respiratory score and the pulse. Notice the order of the within-subjects variables. All three respiratory score names appear first, and then the three pulse names, matching the order on Measure Name.

The multivariate results of the analysis are shown in Figure 24-9 and Figure 24-10. In this example, the *time* effect and the *time*drug* effect are not significant. However, the *drug* effect is significant at $p < 0.0005$, indicating that the new drug has an effect on the respiratory and pulse scores.

Figure 24-9
Multivariate tests

Multivariate Tests[2]

Effect			Value	F	Hypothesis df	Error df	Sig.
Between Subjects	Intercept	Pillai's Trace	1.000	32480.421[1]	2.000	9.000	.000
		Wilks' Lambda	.000	32480.421[1]	2.000	9.000	.000
		Hotelling's Trace	7217.871	32480.421[1]	2.000	9.000	.000
		Roy's Largest Root	7217.871	32480.421[1]	2.000	9.000	.000
	drug	Pillai's Trace	.956	98.177[1]	2.000	9.000	.000
		Wilks' Lambda	.044	98.177[1]	2.000	9.000	.000
		Hotelling's Trace	21.817	98.177[1]	2.000	9.000	.000
		Roy's Largest Root	21.817	98.177[1]	2.000	9.000	.000
Within Subjects	time	Pillai's Trace	.612	2.761[1]	4.000	7.000	.114
		Wilks' Lambda	.388	2.761[1]	4.000	7.000	.114
		Hotelling's Trace	1.578	2.761[1]	4.000	7.000	.114
		Roy's Largest Root	1.578	2.761[1]	4.000	7.000	.114
	time * drug	Pillai's Trace	.496	1.722[1]	4.000	7.000	.249
		Wilks' Lambda	.504	1.722[1]	4.000	7.000	.249
		Hotelling's Trace	.984	1.722[1]	4.000	7.000	.249
		Roy's Largest Root	.984	1.722[1]	4.000	7.000	.249

1. Exact statistic

2. Design: Intercept+drug
Within Subjects Design: time

Figure 24-10
Between-subjects effects

Transformed Variable: Average

Source	Measure	Type III Sum of Squares	df	Mean Square	F	Sig.
Intercept	resp	391.380	1	391.380	69750.941	.000
	pulse	224.001	1	224.001	14197.254	.000
drug	resp	.100	1	.100	17.871	.002
	pulse	2.454	1	2.454	155.563	.000
Error	resp	.056	10	.006		
	pulse	.158	10	.016		

Bibliography

Agresti, A. 2002. *Categorical data analysis*. 2nd ed. New York: John Wiley and Sons.

Anscombe, F. J. 1973. Graphs in statistical analysis. *The American Statistician*, 27:17–21.

Belsley, D. A., E. Kuh, and R. E. Welsch. 1980. *Regression diagnostics: Identifying influential data and sources of collinearity*. New York: John Wiley and Sons.

Berenson, M. L., and D. M. Levine. 1992. *Basic business statistics: Concepts and applications*. Upper Saddle River, N. J.: Prentice Hall.

Bishop, Y. M. M., S. E. Fienberg, and P.W. Holland. 1975. *Discrete multivariate analysis: Theory and practice*. Cambridge, Mass.: MIT Press.

Bland, J. M., and D. G. Altman. 1995. Comparing methods of measurement: Why plotting difference against standard method is misleading. *The Lancet*, 346:1085–1087.

Box, G. E. P. 1979. Robustness in the strategy of scientific model building. In *Robustness in Statistics*, R. L. Launer and G. N. Wilkinson, eds. New York: Academic Press. 201–236.

Box, G. E. P., and N. R. Draper. 1969. *Evolutionary operation: A statistical method for process improvement*. New York: John Wiley and Sons.

Brown, B. W., Jr. 1980. Prediction analyses for binary data. In *Biostatistics Casebook*, R. G. Miller, B. Efron, B. W. Brown, and L. E. Moses, eds. New York: John Wiley and Sons.

Brozek, J., F. Grande, J. Anderson, and A. Keys. 1963. Densitometric analysis of body composition: Revision of some quantitative assumptions. *Annals of the New York Academy of Sciences*, 110:113–140.

Cedercreutz, C. 1978. Hypnotic treatment of 100 cases of migraine. In *Hypnosis at its bicentennial*, F. H. Frankel and H. S. Zamansky, eds. New York: Plenum.

Cohen, J. 1960. A coefficient of agreement for nominal scales. *Educational and Psychological Measurement*, 20:37–46.

Cook, R. D. 1977. Detection of influential observations in linear regression. *Technometrics*, 19:15–18.

Davis, J. A., and T. W. Smith. 2002. *General social survey*. Chicago: National Opinion Research Center.

DeCarlo, L.T. 2003. Using the PLUM procedure of SPSS to fit unequal variance and generalized signal detection models. *Behavior Research Methods, Instruments & Computers*, 35 (1): 49–56.

Doble, J., and J. Greene. 1999. Attitudes toward crime and punishment in Vermont: Public opinion about an experiment with restorative justice (computer file). Englewood Cliffs, N.J.: Doble Research Associates, Inc. (producer), 2000. Ann Arbor, Mich.: Inter-university Consortium for Political and Social Research (distributor), 2001.

Fleiss, J. L. 1981. *Statistical methods for rates and proportions*. 2nd ed. New York: John Wiley and Sons.

Frome, E. L. 1983. The analysis of rates using Poisson regression models. *Biometrics,* 39:665–674.

Hand, D. J., et al. 1994. *A handbook of small data sets*. D. J. Hand, F. Daly, A. D. Lunn, K. J. McConway, and E. Ostrowski, eds. London: Chapman and Hall.

Hanley, J. A., and B. J. McNeil. 1982. The meaning and use of the area under a receiver operating characteristic (ROC) curve. *Radiology*, 143:29–36.

Harrell, F. E. 2001. *Regression modeling strategies: With applications to linear models, logistic regression, and survival analysis*. New York and Berlin: Springer.

Hauck, W. W., and A. Donner. 1977. Wald's test as applied to hypotheses in logit analysis. *Journal of the American Statistical Association*, 72:851–853.

Hays, W. L. 1994. *Statistics.* 5th ed. Fort Worth, Tex.: Harcourt Brace College Publishers.

Hedeker, D., R. D. Gibbons, and B. R. Flay. 1994. Random-effects regression models for clustered data with an example from smoking prevention research. *Journal of Consulting and Clinical Psychology*, 62:4:757–765.

Hicks, C. R. 1982. *Fundamental concepts in the design of experiments*. 3rd ed. New York: Holt, Rinehart and Winston.

Hosmer, D. W., T. Hosmer, S. Le Cessie, and S. Lemeshow. 1997. A comparison of goodness-of-fit tests for the logistic regression model. *Statistics in Medicine*, 16:965–980.

Hosmer, D. W., and S. Lemeshow. 2000. *Applied logistic regression*. New York: John Wiley and Sons.

Huynh, H., and Feldt, L. S. 1970. Conditions under which mean square ratios in repeated measurements designs have exact *F*-distributions. *Journal of the American Statistical Association*, 65:1582–1589.

Huynh, H., and G. K. Mandeville. 1979. Validity conditions in repeated measures design. *Psychological Bulletin*, 86:984–973.

Johnson, R. A., and D. W. Wichern. 1988. *Applied multivariate statistical analysis*. London: Prentice Hall International, Inc.

Johnson, R. W. 1996. Fitting percentage of body fat to simple body measurements. *Journal of Statistics Education*, 4 (1).

Joreskog, K. G. 1979. Basic ideas of factor and component analysis. In: *Advances in factor analysis and structural equation models*, K. G. Joreskog and D. Sorbom, eds. Cambridge, Mass.: Abt Books.

Kaiser, H. F. 1974. An index of factorial simplicity. *Psychometrika*, 39:31–36.

Kleinbaum, D. G., and M. Klein. 2002. *Logistic regression: A self-learning text.* New York and Berlin: Springer.

Kleinbaum, D. G., L. L. Kupper, and K. E. Muller. 1997. *Applied regression analysis and other multivariable methods.* 3rd ed. Boston: Duxbury Press.

Lawless, J. F., and K. Singhal. 1978. Efficient screening of nonnormal regression models. *Biometrics*, 34:318–327.

Lim, T., W. Loh, and Y. Shih. 2000. A comparison of prediction accuracy, complexity and training time of thirty-three old and new classification algorithms. *Machine Learning*, 40:203–229. Available online at: *www.stat.wisc.edu/~loh/treeprogs/quest1.7/mach1317.pdf.*

Little, R. J. A., and D. B. Rubin. 2002. *Statistical analysis with missing data.* New York: John Wiley and Sons.

MacCallum, R. C., K. F. Widaman, S. Zhang, and S. Hong. 1999. Sample size in factor analysis. *Psychological Methods*, 4 (1): 84–99.

Mauchly, J. W. 1940. Significance test for sphericity of a normal *n*-variate distribution. *Annuals of Mathematical Statistics*, 11:204–209.

McCullagh, P. 1978. A class of parametric models for the analysis of square contingency tables with ordered categories. *Biometrika*, 69:413–418.

McGaw, K. O., and S. P. Wong. 1996. Forming inferences about some intraclass correlation coefficients. *Psychological Methods*, 1 (4): 390.

Miller, P. W., and P. A. Volker. 1985. On the determination of occupational attainment and mobility. *Journal of Human Resources*, 20:197–213.

Milliken, G. A., and D. E. Johnson. 1992. *Analysis of messy data.* Vol. 1, *Designed experiments.* New York: Chapman and Hall.

Moore, D. S. 1995. *The basic practice of statistics.* New York: W. H. Freeman and Company.

Nagelkerke, N. J. D. 1991. A note on the general definition of the coefficient of determination. *Biometrika*, 78 (3): 691–692.

Nicely, G. F., J. Platt, N. A. Hepler, and J. Wells. 2002. Evaluation of a truancy reduction program in Nashville, Tennessee, 1998–2000 [computer file]. ICPSR version. Nashville, Tenn.: Metropolitan Development and Housing Agency and Metropolitan Nashville/Davidson County Juvenile Court [producers], 2001. Ann Arbor, Mich.: Inter-university Consortium for Political and Social Research [distributor].

Norušis, M. J. 2008. *SPSS Statistics 17.0 Advanced Statistical Procedures Companion.* Upper Saddle River, N. J.: Prentice Hall.

_____. 2008. *SPSS Statistics 17.0 Guide to Data Analysis.* Upper Saddle River, N. J.: Prentice Hall.

Olsen, C. L. 1976. On choosing a test statistic in multivariate analysis of variance. *Psychological Bulletin*, 83:579–193.

Paul, O., et al. 1963. A longitudinal study of coronary heart disease. *Circulation*, 28:20–31.

Peduzzi, P. N., J. Concato, E. Kemper, T. R. Holford, and A. Feinstein. 1996. A simulation study of the number of events per variable in logistic regression analysis. *Journal of Clinical Epidemiology*, 99:1373–1379.

Platia, E. V., R. Berdoff, G. Stone, et al. 1985. Comparison of acebutolol and propranolol therapy for ventricular arrhythmias. *Journal of Clinical Pharmacology*, 25:130–137.

Pothoff, R. F., and S. N. Roy. 1964. A generalized multivariate analysis of variance model useful especially for growth curve problems. *Biometrika*, 51:313–326.

Rao, C. R. 1973. *Linear statistical inference and its application*. 2nd ed. New York: John Wiley and Sons.

Raudenbush, S. W., and A. S. Bryk. 2002. Hierarchical linear models: Applications and data analysis methods. Thousand Oaks, Calif.: Sage Publications.

Robertson, C., C. Hseih, P. Maisonneuve, P. Boyle, and G. MacFarlane. 1994. The analysis of case-control studies with continuous exposure variables. *Epidemiology*, 5 (2).

Rodger, E. J., G. D'Elia, C. Jorgensen, and J. Woelfel. 2000. *The impact of the Internet on public library use: An analysis of the current consumer market for library and Internet services*. Final report submitted to the Institute of Museum and Library Services.

Schlesselman, J. J. 1982. *Case-control studies: Design, conduct, analysis*. New York and London: Oxford University Press.

Searle, S. R. 1987. *Linear models for unbalanced data*. New York: John Wiley and Sons.

Searle, S. R., F. M. Speed, and G. A. Milliken. 1980. Population marginal means in the linear model: An alternative to least squares means. *The American Statistician*, 34 (4): 216–221.

Shott, S. 1990. *Statistics for health professionals*. Philadelphia: Saunders.

Singer, J. D. 1998. Using SAS PROC MIXED to fit multilevel models, hierarchical models, and individual growth models. *Journal of Educational and Behavioral Statistics*, 23 (4): 323–355.

Siri, W. E. 1956. Gross composition of the body. In: *Advances in biological and medical physics,* Vol. IV, J. H. Lawrence and C. A. Tobias, eds. New York: Academic Press.

Swets, J. A., W. P. Tanner, and T. G. Birdsall. 1961. Decision processes in perception. *Psychological Review*, 68.

Tabachnick, B. G., and L. S. Fidell. 2001. *Using multivariate statistics*. Boston: Allyn and Bacon.

Tubb, A., A. J. Parker, and G. Nickless. 1980. The analysis of Romano-British pottery by atomic absorption spectrophotometry. *Archaeometry*, 22:153–171.

Tufte, E.R. 1983. *The visual display of quantitative information*. Cheshire, Conn.: Graphics Press.

van Belle, G. 2002. *Statistical rules of thumb*. New York: John Wiley and Sons.

Velleman, P., and L. Wilkinson. 1993. Nominal, ordinal, interval, and ratio typologies are misleading. *The American Statistician*, 47 (1): 65–72.

Wilkinson, L., G. Blank, and C. Gruber. 1996. *Desktop data analysis with SYSTAT*. Upper Saddle River, N. J.: Prentice Hall.

Wilkinson, L., and Task Force on Statistical Inference. 1999. Statistical methods in psychology journals: Guidelines and explanations. *American Psychologist*, 54 (8): 594–604.

Willett, J. B. 1988. Questions and answers in the measurement of change. In: *Review of research in education (1988–1989)*, E. Rothkopf, ed. Washington, D. C.: American Educational Research Association.

Willett, J. F., and J. D. Singer. 2003. *Analyzing longitudinal data*. Oxford: Oxford University Press.

Williams, A. 1981. Student learning: Team vs. primary nursing. *Nursing Management*, 12:48–51.

Winer, B. J., D. R. Brown, and K. M. Michels. 1991. *Statistical principles in experimental design*. New York: McGraw-Hill.

Index